A
VOYAGE
To the Islands
Madera, Barbados, Nieves, S. Christophers
AND
JAMAICA,
WITH THE
Natural History
OF THE
Herbs and Trees, Four-footed Beasts, Fishes,
Birds, Insects, Reptiles, &c.
Of the last of those ISLANDS;

To which is prefix'd An
INTRODUCTION,
Wherein is an Account of the
Inhabitants, Air, Waters, Diseases, Trade, &c
of that Place, with some Relations concerning the Neighbouring Continent, and Islands of *America*.

ILLUSTRATED WITH
The FIGURES of the Things describ'd,
which have not been heretofore engraved;
In large Copper-Plates as big as the Life.

By *HANS SLOANE*, M.D.
Fellow of the *College* of *Physicians* and Secretary
of the *Royal-Society*.

In Two Volumes. Vol. I.

Many shall run to and fro, and Knowledge shall be increased. Dan. xii. 4.

LONDON:
Printed by *B. M.* for the Author, 1707.

A VOYAGE

To the Islands

Madera, Barbados, Nieves, S. Christophers

AND

JAMAICA,

WITH THE

Natural History

OF THE

Herbs and Trees, Four-footed Beasts, Fishes,
Birds, Insects, Reptiles, &c.

Of the last of those ISLANDS;

INTRODUCTION

TO THE

Inhabitants, Air, Waters, Diseases, Trade, &c.
of that Place, with some Relations concerning the Neighbouring Continent, and Islands of *America*.

ILLUSTRATED WITH

The Figures of the Things describ'd,
which have not been heretofore engraved;

In large Copper-plates as big as the Life.

By HANS SLOANE, M. D.

Fellow of the College of Physicians and Secretary
of the Royal-Society.

In Two Volumes. Vol. I.

LONDON:

Printed by B. M. for the Author, 1707.

To Her Most Excellent Majesty,

The QUEEN:

THIS

Natural History of Jamaica,

ONE OF

The Largest and most Considerable

OF

Her Majesty's PLANTATIONS

IN

AMERICA.

Is with all Humility Dedicated,

BY

Her Majesty's most dutiful and
most obedient Subject,

Hans Sloane.

To Her Most Excellent Majesty,

The QUEEN:

THIS
Natural History of Jamaica,
ONE OF
The Largest and most Considerable
OF
Her Majesty's PLANTATIONS
IN
AMERICA,

Is with all Humility Dedicated,

BY

Her Majesty's most Dutiful and
most obedient Subject,

Hans Sloane.

THE
PREFACE.

IT is now Eighteen Years since I went Physician to the Duke of Albemarle to Jamaica. I had from my Youth been very much pleas'd with the Study of Plants, and other Parts of Nature, and had seen most of those Kinds of Curiosities, which were to be found either in the Fields, or in the Gardens or Cabinets of the Curious in these Parts. The Accounts of these strange Things, which I met with in Collections, and, was inform'd, were common in the West-Indies, were not so satisfactory as I desired. I was Young, and could not be so easy, if I had not the pleasure to see what I had heard so much of, especially since it had been a great contentment to me, to see many things cultivated in English Gardens which I had seen grow wild in other Countries, whereof I conceived my self afterwards to be better appris'd, than I was of such as I had not seen common in the Fields, and in plenty. I thought by that means the Ideas of them would be better imprinted in my Mind, and that, upon occasion, both the knowledge of them and their Uses might be afterwards more familiar to me. These Inclinations remain'd with me some time after I had settled my self to practise Physic in London, and had had the Honour to be admitted a Fellow of the College of Physicians, as well as of the Royal Society. These unmerited Favours did not at all alter my mind, but rather incited me to do what I could to be no useless Member, but to cast in my Mite towards the Advancement of Natural Knowlege, and the Faculty of Physic, and by that means endeavour to deserve a Place amongst so many Great and Worthy Persons: This Voyage seem'd likewise to promise to be useful to me, as a Physician; many of the Antient and best Physicians having travell'd to the Places whence their Drugs were brought, to inform themselves concerning them.

His Grace the Duke of Albemarle having obtain'd the supreme Command of the Island of Jamaica, and other Parts of English-
America

The PREFACE.

America, where he should arrive, employ'd Dr. Barwick, *who was his Physician, to look out for one who could take care of Him and his Family in case of Sickness; Dr.* Barwick *spake to me in this matter, enquiring if any Physician of my Acquaintance would undertake it: This seem'd to me to be such an Opportunity as I my self wanted, to view the Places and Things I design'd, and at the same time to prosecute the Practice of Physic; wherefore I desir'd he would give me time to think of it, and after due consideration I resolv'd to go, provided some Preliminaries and Conditions were agreed to, which were all granted.*

I intended so soon as on board to have made several Experiments and Observations in the Voyage, but was prevented by a very long and tedious Sea-Sickness, unless in some particulars, of which I have given an Account in the ensuing Voyage. His Grace the Duke of Albemarl's *Commission and Instructions requiring he should muster all the Forces of the* English-Caribe-*Islands, in his way to* Jamaica, *made him stop some days at most of them, which gave me an opportunity of spending some time in looking after the Curiosities of those Places, which are taken Notice of in the Voyage thither. I have left out most of whatever is related by any Author I had perused, unless what they mention of the Uses of Plants, or such particulars wherein I thought they were mistaken.*

Upon my Arrival in Jamaica, *I took what pains I could at leisure-Hours from the Business of my Profession, to search the several Places I could think afforded Natural Productions, and immediately described them in a Journal, measuring their several Parts by my Thumb, which, with a little allowance, I reckoned an Inch. I thought it needless to be more exact, because the Leaves of Vegetables of the same sorts, Wings of Birds, &c. do vary more from one another, than that does from the exact measure of an Inch; As to Colours, 'tis certain they are very hard to describe: There are so many Varieties of them, that they require new Names to express them. I observed in describing of them, that the Leaves of most, if not all, Plants are Greenest on their uppermost sides, or that exposed to the Sun-Beams; and Lighter and more Whitish underneath. This is not only so in* Jamaica, *but in all places where I have been; when the Leaves or Tops of Plants have not been exposed to the Light and Sun, they are not only whiter, but tenderer, and often digestible by our Stomachs. This appears by the Tops of the Palm [in* Jamaica *and the* West-Indies*] call'd*

The PREFACE.

call'd Cabbage-Trees the Germen, Sprout, Top, Bud or unexpanded Leaves of which, are boil'd, and eat like our Cabbage, or pickl'd. The tops of the Chamæriphe or Dwarf-Palm (of the Leaves of which are made Brooms) are likewise eaten in Spain and Italy. Thus Hops, White-Briony, Smilax aspera, Sparagus, just sprouting, common Cabbage naturally, Seleri, Lettuce, Fennel, Chardons and several Herbs, are render'd White and tender by only being cover'd with Earth from the Sun and Light. I have seen a common Bramble whose Twigs accidentally grew through the chink of a Window into a dark Room, which, because not exposed to the Light or Sun, were white and tender. 'Tis not only so, as I believe, in Greens, but also in other Colours which are heightned by the Sun: for the great variety of Colours observ'd in Tulips, at first sprouting out are all whitish, then greenish, and when the Sun and Light has farther acted upon them, they arrive at that variety we observe in them with pleasure. This I take notice of, because the same Plants I describe, may have in European Gardens different Colours, from what they are in their native Soil, and a warmer Sun.

After I had gather'd and describ'd the Plants, I dried as fair Samples of them as I could, to bring over with me. When I met with Fruits that could not be dried or kept, I employ'd the Reverend Mr. Moore, one of the best Designers I could meet with there, to take the Figures of them, as also of the Fishes, Birds, Insects, &c. in Crayons, and carried him with me into several places of the Country, that he might take them on the place. When I return'd into England, I brought with me about 800 Plants, most whereof were New, with the Designs before-mention'd, &c. And shew'd them very freely to all lovers of such Curiosities: I gave my very particular and intimate Friend Mr. Courten whatever I brought with me, that he wanted in his extraordinary Museum. Amongst other Persons who saw them, was Sir Arthur Rawdon, who observing the great variety of Plants I had brought with me, sent over Mr. James Harlow, a Gardener, (who had formerly gone to Virginia for Mr. Watts) to bring the Plants themselves alive to him, for his Garden at Moyra in Ireland. This Mr. Harlow perform'd, and there they grew and came many of them to great perfection. He not only brought over with him a Ship almost laden with Cases of Trees, and Herbs, planted and growing in Earth, but also a great number of Samples of them very well preserv'd in Paper; and knowing that he that went over on purpose, might meet with some

things

The PREFACE.

things I had not obſerv'd, and ſeen others in different Seaſons from me, I wrote to Sir Arthur Rawdon *and my very learned Friend Dr.* William Sherard, *who then was with him, to deſire the favour of them, that in order to the perfecting my Account of* Jamaica, *I might have a ſight of what of that kind he brought over with him. This Sir* Arthur Rawdon *and Dr.* Sherard *not only granted, but alſo made me a Preſent of ſuch Samples as had been brought over which I wanted. The Reader will find theſe taken notice of, in ſeveral Places of this Hiſtory. After I had peruſed them, they were, together with abundance of other rare Plants, by Dr.* Sherard's *Direction ſent to* Oxford, *where Mr.* Jacob Bobart *has made very good uſe of them in the Hiſtory of Plants he lately publiſh'd there; and leſt there might from dry Samples, come any Confuſion in Natural Hiſtory, Dr.* Sherard *afterwards at my Requeſt gave me the View of ſuch Plants as Mr.* Bobart *had deſcrib'd, which has enabled me to put the Synonymous Names of the aforeſaid Hiſtory of Plants, publiſh'd by Mr.* Bobart, *which the Reader will likewiſe find taken notice of in their due places. Theſe were not the only Favours I had of Dr.* Sherard; *for he furniſhed me with many ſcarce Books he bought in his Travels beyond Sea.*

Many Perſons were deſirous I ſhould publiſh an Account of what I met with towards the forwarding Natural Hiſtory, I conſented to this, provided the Obſervations I had made, ſhould be thought worthy of Publication. I thought the greateſt Judge I could adviſe with, in theſe Circumſtances, was Mr. Ray, *who for his Probity, Learning, Language, &c. ſeem'd to me the propereſt to adviſe with: I therefore waited on him, ſhew'd him ſome of the Plants, and tranſmitted to him my Obſervations on them. I deſired him to correct them and add his Emendations. He was pleaſed to approve of them, and think much better of them than I my ſelf did, or do; ſo that the Emendations I expected, are wanting. I am afraid his Kindneſs, and favourable Opinion of me, may be the cauſe; for I am ſenſible there are herein a great many faults, not only in Hypotheſes or Opinions, which I propoſe only as Conjectures, and ſhall eaſily part with; but there are likewiſe many imperfections in the Language, and in the Obſervations themſelves, which were generally written in haſte, and by me, who know too well how unduly qualified I am for ſuch an Undertaking. However, ſuch as they were, when I was reſolv'd to publiſh them, I thought it neceſſary to look into the Books in ſeveral Languages, which treated of thoſe Subjects either deſignedly or accidentally. Some Men*

ſeem

The PREFACE.

seem to have a great desire to be the first Authors of discovering such or such Plants, and to have them carry their Names in the first Place, but I endeavour'd rather to find if any thing I had observ'd was taken Notice of by other Persons; I therefore look'd into most Books of this Nature, and the greatest part of what I found, is publish'd in the Catalogue of Jamaica Plants I printed about ten Years since, wherein I endeavour'd to do right to the first Authors and the Publick; It is a Catalogue of the Plants I met with at Jamaica, &c. Which I think, for Synonymous Names of the Plants therein mentioned, is somewhat more Copious and exact than any other before it: And which may be of some Use to inquisitive Persons, especially when they shall have this History of the things therein contain'd. I have not reprinted in this Book those Names or Titles, because 'twas great Labour, and is done there already, but have only added the Synonymous Names I found in such Books as are since Publish'd or come to my Hands. The looking after the Observations of Others, to make this more Exact and Useful, has given an Opportunity to several People to anticipate me, by either Publishing such Things as I have shewn them, told them, or communicated to Mr. Courten, from whom they had them; wherein they have committed some Mistakes, for want of full Information or exact Memory.

I have been sometimes short, in giving the Uses of these Plants, but I hope I may be understood, and the Author whose Name is set down (Abbreviated,) may on any Ambiguous or Obscure Passage be consulted, for those Notes were written to bring to my Mind what they said, to be perused at leisure, and modell'd after, which I have never yet had time to do. The principal Design of them was, that the Inhabitants of those Places, might understand what Uses the Plants they have growing Sponte or in Gardens with them, are put to in any of the Countries wherever they grow, that so they may have recourse to them in Cases that require them. It is very hard to carry thither such European Simples as are proper for the Cure of all sorts of Diseases, and People are put to it to find such as are effectual in some of them, and yet will keep their Vertues. This puzzl'd me several times, as may be seen in the following Medical Observations.

The first Volume contains an Introduction, giving an Account of the Situation, Temperature, Diseases, &c. of the Island, which seem'd necessary to be premis'd to the History itself. After this, follows the Voyage thither,

B   and

The PREFACE.

and then the History of the Plants that grow there, in which I have followed mostly the Method of Mr. Ray *in his History of Plants, joining his Genera or Tribes together by the Method of* Rivinus, *or Number of the* Petala *or Leaves of the Flowers; As those which are* Monopetalous *first, those* Dipetalous *next, then the* Tripetalous, Tetrapetalous, Pentapetalous, Hexapetalous, *and such as have no exact Numbers of Coloured Leaves in their Flowers. When this History was begun, and near finished, I wanted many considerable helps, the Method of the great Botanist Dr.* Tournefort *was not printed, much less the Book of new Kinds of* West-India *Plants, lately publish'd by Monf.* Plumier, *who, since my Return from the* West-Indies, *went into the* Caribe-Islands, *and there observ'd and design'd himself many of the Plants very accurately. He has engrav'd them himself, and printed a Book, which consists chiefly of* Ferns; *And has sav'd me a great deal of Trouble, finding his Figures so Good, that I did not judge it necessary the same Plants should be engraven again, but be only referr'd to, in my History.*

In the Second are contained the Trees, *the* Quadrupeds, Birds, Fishes, Testaceous *and* Crustaceous *Animals, and* Insects, *&c.*

It may be ask'd me to what Purposes serve such Accounts, I answer, that the Knowlege of Natural-History, *being Observation of Matters of Fact, is more certain than most Others, and in my slender Opinion, less subject to Mistakes than* Reasonings, Hypotheses, *and* Deductions *are; And on this Occasion I have heard it reported of* Gabriel Naude, *That he used to say he Acquiesc'd in the Ecclesiastical History, Doubted the Civil, and Believ'd the Natural. These are things we are sure of, so far as our Senses are not fallible; and which, in probability, have been ever since the Creation, and will remain to the End of the World, in the same Condition we now find them: They afford great Matter of Admiring the Power, Wisdom and Providence of Almighty God, in Creating, and Preserving the things he has created. There appears so much Contrivance, in the variety of Beings, preserv'd from the beginning of the World, that the more any Man searches, the more he will admire; And conclude them, very ignorant in the History of Nature, who say, they were the Productions of Chance.*

Another

The PREFACE.

Another Use of this History may be, to teach the Inhabitants of the Parts where these Plants grow, their several Uses, which I have endeavour'd to do, by the best Informations I could get from Books, and the Inhabitants, either Europeans, Indians or Blacks. Jamaica had been, before it was taken by the English, in the possession of the Spaniards, almost from the time the West-Indies were discover'd: They had brought many Fruit-Trees from the Main-Continent, where they are Masters, and suffer no other Europeans to come; which throve wonderfully, and now grow as it were Sponte: These they made use of for Food, Physic, &c. And were forc'd to leave with their Habitations, to the English, and the Skill of Using them remain'd with the Blacks and Indians, many of whom came, upon a Proclamation that they should be Free, submitted peaceably, and liv'd with the English after the Spaniards had deserted it. There were among these, several which made small Plantations of their own, wherein they took care to preserve and propagate such Vegetables as grew in their own Countries, to use them as they saw occasion: I made search after these, and what I found, is related in this History. Besides these Helps, some of the Dutch who had liv'd in Brasil, came hither, and others of the Dutch and English from Surinam, where they had observ'd the Effects of some Plants they met with in Jamaica, and used them for the same Purposes they had done in Brasil and Surinam, towards the Relief of the Inhabitants. For this Reason the Reader will find herein, many of the Vertues of Plants to agree with the Observations of Authors, writing of other Parts of the West-Indies.

There is another Use to be made of this Book, which is this; In reading Voyages, and talking with Travellers to the West-Indies, &c. one shall meet with Words, and Names of Things, one has no Notion or Conception of: by looking for such Names in the Index of the Catalogue of Jamaica Plants, you are referr'd to the Page where you find a List of such as have treated of it; And in this History under the first Title of it in the Catalogue, is the History of it. If on the other hand, any Person desires to know who has written of such or such a Plant in Jamaica, let him look into the Catalogue, and under the first Title of the Plant, he will find Citations to direct him to the Pages of most of the Books wherein it is spoken of.

Another

The PREFACE.

Another Advantage, the Knowlege of what is produced there naturally brings, is a Consideration of the Causes of some very strange, but certain, Matters of Fact. It has puzzl'd the Philosophers of all Ages, to give an Account how Parts of Vegetables and Animals, Real Sea-Shells and Substances should be found remote from the Seas, wherein they seem to have been produced and bred. This Phænomenon will appear stranger, when it is made out, that many of the Substances, as for instance, Corals, Echini marini, the Palats and Tongues of Fishes hereafter described, and which now live and breed in the Seas, adjoining to Jamaica, and no nearer than some few Degrees on this side, are found in as great plenty in the Inland Parts of England imbedded in the Earth, Clay, Sand, Chalk, or Stone, as if it had once been the natural Place of their Production and Increase. This I was very much surpris'd to find.

These matters of Fact being certainly laid down, may perhaps afford some Hints for the more clear Reasonings and Deductions of better Heads ; And I know not but that the several Species of new Ferns, and some Plants by me found there, and here describ'd and figur'd, may be discover'd upon some of the Stones or Slates called Werk, *which lie in plenty in the* Strata *over the* Cole-pits *in many Places of* England. *These Plants and their Impressions are of several Kinds, and many of them are not to be reduced to the Ferns or Plants, found in* England *or the adjoining Countries. His Grace the Duke of* Beaufort *has done me the honour to give me some of these Slates he had in* Glocestershire; Dr. Richardson *from* Yorkshire *has communicated several of them to me ; as has also Mr.* Beaumont *from* Somersetshire.

Upon the first Discovery of the West-Indies, *several People went thither on the Account of observing its Natural Productions. The first that I read of is one* Codrus *an* Italian, *who went from* Spain *for that purpose. The Discoveries he made were but few, or not communicated to the World : The only Account we have of him, is in* Peter Martyr's Decads, *wherein that Elegant Writer acquaints the* Pope, *what News they then had in* Spain, *from the* West-Indies. Hernandez *was sent by the* King of Spain, *to search after Natural Productions about* Mexico ; *He design'd and describ'd many of the things he met with, at the Expence of* 60000 Ducats ; *his Papers were put into the hands of* Nardus
Antonius

The PREFACE.

Antonius Recchus, *from whose Manuscripts they were by the* Lyncei *Publish'd at* Rome, *Anno* 1649. *They were chang'd from their first order, as appears by the* Spanish-*Copy, Printed at* Mexico; *and 'tis pity that they were alter'd, and are so short and obscure: Meeting with many of the Plants he describes in* Jamaica, *I had a great mind to be satisfied about them, and being told that the Original Draughts were in the King of* Spain's *Library, in the* Escurial *near* Madrid, *I wrote to Mr.* Aglionby *when he was Envoy from the late King* William *to the Court of* Spain, *to procure a Sight of that Work, and give me an Account of it. He was so obliging as to take the Pains to go thither, and was told that the Book was there, and that he should some time or other see it; which, tho' he endeavour'd several Times, yet he could never effect: Neither had other curious Travellers, better Fortune; for when they had heard of this Book, and (knowing of what importance it would be to see these Originals) did endeavour to procure a Sight of them, the Library-Keepers were so ignorant, to produce to them, some other Book, no ways to the Purpose. Upon the whole matter, I am apt to think the Originals were carried to* Rome, *where the History was Publish'd, and that they remain'd there with* Recchus *his Nephew; where, If my Memory fail me not,* Fabius Columna *says he saw them, and that they are either to be found there, or at* Naples, *where* Columna *liv'd, that wrote Notes on them, or that they are lost. However, it went with the Manuscript, from which that at* Rome *was publish'd; there was a Copy printed at* Mexico *in* Spanish, *in the Year* 1615. Francisco Ximenes, *one who attended the Sick of the Hospital in that City, publish'd it then, with Emendations, Notes, and the Additional Observations he had made of several Simples he had found in* Espaniola, *or* Sto. Domingo, *and other Islands of the* West-Indies. John de Laet *mentions this Book in* Latin, *and from him* Vander Linden, *in his Book* de Scriptis Medicis *has, I suppose, taken its Title, but I verily believe it was never printed in any other Language than* Spanish. John de Laet *takes many things out of him, and puts them very often in a wrong Place, as additional to the Observations of* Marcgrave *in* Brasile, *in the Edition of that Author, publish'd by him in* 1648. *But that Fault may be easily pardon'd, in one who was no more than a Collector and Editor of Books wherein he did not pretend to any great Knowlege.*

The

The PREFACE.

The first I have seen among the French, who made any Natural Observations in the West-Indies to Purpose, was John de Lery; He went to Brasile, which Voyage he wrote, and gives a good and pleasant Account of many Things he met with. The next was Andre Thevet, who publish'd an Account of Brasile, under the Name of France Antartique; he went the same Voyage, I think, the Year after de Lery, and writes of the same things in such a manner, that one would be apt to suspect he had seen Lery's Papers. The next upon this Argument, was Jaques Bouton, who wrote of the Island Martinico, which was the first settl'd of any of the French Islands. His Accounts are but very short; however, from him the other French Writers, I think, took many of their Names of Natural things. J. B. du Tertre wrote of the Caribe-Islands. His Work was pretty Large and Useful, and was contain'd at first in one Volume, printed in in 1654, in 4to and afterwards came out in 3. Vol. in 1667 -71. There are therein many Remarks and Observations upon the French-Islands, and for Natural History, many things are figured, which, tho' not very accurately, yet are truer than those of any before him. Monf. Rochefort printed a Book of the Caribes, after du Tertre; he seems to me to have taken many things from the first Edition of du Tertre, and to agree with him in most things: The Figures he gives, I suppose were not drawn upon the Place, but by Memory, and are, for that Reason, not to be regarded. This Book is printed in English, in Folio, under the Title of the History of the Caribe-Islands.

According to the Division of the New-discover'd World, between the Spaniards and Portuguese, the last possess'd themselves of Brasile; Amongst others of that Nation who went thither, was one, suppos'd to be a Jesuit of Elvas, whose Name is ghess'd to be Tr. Manoel, who had the care of the Sick of the College of Baya, made Observations of that Country, and wrote them in a Book in the Portuguese Language: This Book was taken by Mr. Cook of Dartmouth, translated and publish'd by Purchas, p. 1289. Pilgr. part. 4. Tho' this was in itself a very short Relation, and little more than the Names of things, yet falling into the Hands of the Dutch, by being taken Notice of by de Laet in his America, (who had it out of Purchas's Collections,) it gave them occasion to enquire after many things therein mention'd; when they had

conquer'd

The PREFACE.

conquer'd the greatest Part of Brasile: *Piso who was a Physician, and* Marcgrave *an Industrious Scholar, going out of* Holland *thither, were very curious, and took great care to observe what they thought worth Notice, and had* Marcgrave *liv'd to have publish'd his own Observations, that Work would have been much more considerable; but his Papers falling into the Hands, first of* John de Laet, *who, tho' a Learned Man, yet was ignorant of Natural History; and then of* Piso, *who, tho' he was a Practical Physician, yet had no great Knowlege of Natural things; are not so much to be depended on as one could wish: The Figures and Descriptions are transpos'd, as I have found by examination, and the first Edition is preferable to the Second, wherein* Piso *hath taken too much Liberty of ascribing the Vertues of European Herbs to those of the same Tribe in* Brasile. *'Tis very evident the Vertues of all Plants of the same Tribe, are not always found to correspond; we need go no farther than this History, wherein the* Spanish Patata *eaten commonly in* Jamaica, *is a true* Convolvulus, *the most part of which Family purge; the* Cassada, *of the Root of which Bread is made, is a true* Ricinus, *the Tribe of which, generally speaking, vomit and purge with great Violence,* &c.

It may be objected, that 'tis to no purpose to any in these Parts of the World, *to look after such Herbs,* &c. *because we never see them; I answer, that many of them and their several Parts have been brought over, and are used in Medicines every day, and more may, to the great Advantage of Physicians and Patients, were People inquisitive enough to look after them. The Plants themselves have been likewise brought over, planted, and throve very well at* Moyra, *in* Ireland, *by the Direction of Sir* Arthur Rawdon; *as also by the Order of the Right Reverend Dr.* Henry Compton, *Bishop of* London, *at* Fulham; *at* Chelsea *by Mr.* Doudy; *and* Enfield *by the Reverend Dr.* Robert Uvedale; *and in the Botanic Gardens of* Amsterdam, Leyden, Leipsick, Upsal, &c. *but especially at* Badminton *in* Glocester-shire, *where they are not only rais'd some few handfuls high, but come to Perfection, flower and produce their ripe Fruits, even to my Admiration; and that, by the Direction of her Grace the Duchess of* Beaufort, *who at her leisure Hours, from her more serious Affairs, has taken pleasure to command the raising of Plants in her Garden, where, by means of Stoves and Infirmaries, many of them have come to greater Perfection, than in any Part of* Europe.

It

The PREFACE.

It may be objected, that there is no end of such Discoveries, that every Country, and distant Climate, has different Plants, not to be found in England. I answer it is not so, for tho' England be very remote, both in Longitude and Latitude, from Jamaica, yet I found there many Plants, which grow Sponte, *which also I found to grow wild in England, and more I observ'd to be common to that Place, and the South Parts of* France. *By Collections sent to the very Industrious and Skilful Botanist Mr.* Petiver, *others, and my self, from* Spain, Portugal, Barbary, Guinea, *and the* East-Indies, *as well as by the* Hortus Malabaricus, *I find a great many Plants common to* Spain, Portugal *and* Jamaica, *more common to* Jamaica *and the* East-Indies, *and most of all common to* Jamaica *and* Guinea; *so that the following History may reasonably be suppos'd, not only to describe most of the Natural Productions of* Espaniola, Barbados *and the other hot* American *Isles, but also many of those of* Guinea *and the* East-Indies, *and therefore may be thought to contribute to the more distinct Knowlege of all those Parts.*

I have been so conversant in Matters of this nature for several Years, that I know 'tis impossible to escape the Censure of several sorts of Men, as the Envious and Malitious, who will, I am sure, spare no Pains to find Faults; those who strive to make ridiculous any thing of this kind, and think themselves great Wits, but are very Ignorant, and understand nothing of the Argument. These, if one were afraid of them, and consulted his own Ease, might possibly hinder the Publication of any such Work, the Efforts to be expected from them, making possibly some impression upon Persons of equal Dispositions; but considering that I have the Approbation of Others, whose Judgement, Knowlege, &c. I have great reason to value; And considering that these sorts of Men, have been in all Ages ready to do the like, not only to ordinary Persons, and their Equals, but even to abuse their Princes, and blaspheme their Maker; I shall, as I have, ever since I seriously consider'd this Matter, think of, and treat them with the greatest Contempt.

THE

THE
INTRODUCTION.

THE first Discovery of the *West-Indies*, to me seems to have been accidental, as has happen'd in most other great Discoveries. *Christopher Columbus*, first solicited the King of *Portugal* to set him out; but that Prince being wearied with the Discoveries, and little Advantage of *Guinea*, would not meddle in it. *Christopher* then sent his Brother *Bartholomew* into *England*, on the same Errand, in the Year 1488. I have made as good a search as I could, after a Map which was made at *London*, by this *Bartholomew Columbus*, whereby he would have induc'd King *Henry* VII. to have been the fitter of him out, but I could not hear of it, neither at the Royal Library at St. *James's*, nor any where else. This Map, and this Proposal were not regarded, and 'tis a common Tradition, that instead of discovering the *West-Indies*, there was bought at *Antwerp*, a Suit of fine Tapistry Hangings, with Money that had been set apart, and thought sufficient for that purpose. These Hangings are now said to remain at *Hampton* Court. This Map, which seems to me to have been made from the Letters of *Paolo Fisico* of *Florence*, in 1474. 'tis likely was a Sea-Chart of the Parts of the World then known; wherein those to the West were *Ireland*, the *Azores*, Cape *Verd*, and the Coast of *Guinea*. It had to the East, the Parts of *China*, then lately discover'd, which they then thought were fifteen Hours East of *Portugal*, and they were still of Opinion, that they had not reach'd the extreme Parts of the *East-Indies* or *Cipango*, call'd *Japan*, where, Report had told them, were great Riches. Therefore, there being, as they thought, only about one third of the way to *Cataio* and *China*, if they went by the West, they concluded the best and nearest way to take that Course. On this mistaken Foundation, (*China*, by later Discoveries, being only about eight Hours East) and some probabilities from some Artificial and Natural things, driven on Shore by the Winds and Currents on the *Azores* and *Porto Santo*, which came to the Knowledge of *Columbus*, (who had been wreck'd at *Lisbon*, and was married to the Widow of one

Perestrillo,

Col. cap 8.

The Introduction.

Perestrello, who was the Discoverer of *Madera* and *Porto Santo*) he projected going to the farther Part of the *East-Indies*, not yet discover'd; where common Fame, and the assurance of People come from thence, told them, were Houses covered with Gold, (in some parts some of them at this Day are gilded) Spices and other very rich Commodities in great plenty. He was oppos'd in this Project by learned Men, and with Difficulty was at last believ'd by *Isabella*, then Queen to *Ferdinand* King of *Spain*, who being influenced by her Confessor, *Luigi di S. Angelo*, in favour of this * *Marianus,* Project, pawn'd her Jewels to equip *Columbus*, who * by this means *l. 20. c. 3* got seventeen thousand Ducats. He let out *August* 1492. and went through many Disasters, endured much Pain, Watching and perpetual Labour. He by these means, kept his Men from Mutinies, and at length discover'd some Birds, afterward some Land-herbs and Fruits the Sea, and at last Saint *Salvador* or *Guanahani*, one of the *Lucaie* or *Bahama* Islands, on the 12th of *October*, and on the 15th he came to the North-side of *Hispaniola*. He left there some Men, and took thence, to shew in *Spain*, some *Indians*, Gold, Parrats, Maiz, or *Indian* Corn, and other valuable or strange things. On the 4th of *January*, 1493. he set Sail from *Hispaniola* for *Spain*, and arrived at *Lisbon* the 4th of *March* in the same year, and at *Palos* in *Spain* the the 13th of the same Month, that is, in seven Months and eleven Days from going out.

Columbus, likewise brought into *Europe* in his Ship, and first Voyage, from these places, the Pox, which spread so quickly all over *Europe*, that *Antonius Benivenius*, who was at that time a great and famous Practiser in Physick at *Florence*, in the first Chapter of his Book *P. 318.* *de Abditis nonnullis ac mirandis morborum, & sanationum causis*, tells us, that the *Lues Venerea* then beginning in *Spain*, had spread itself through *Italy*, and *France*, and that in the Year 1496, it had possess'd many People in all the Provinces of *Europe*. *Dodonaeus*, likewise tells us, that this Disease very much raged in the War that *Charles* VIII. King of *France* had with *Alphonsus* King of *Naples* in the Year 1494. † P. 318. and yet thinks *Gulielmus de Saliceto*, † who liv'd in 1270. *Valescus de* § Lib.6.cap.5 *Tarenta*, ‖ who liv'd in 1418. and *Bernardus de Gordonio*, who died in 1305. give us an account of some Symptoms of it.

I am of Opinion notwithstanding what these have said, and some other less material Passages in antient Writers and Historians, and what *Joannes ab Arderne* has written about *An.* 1360. and likewise what * P. 445. *Stow* * says of the Laws of the publick Stews in *Southwark*, that this

The Introduction.

this was a Distemper altogether new in *Europe*, *Africa* and *Asia*, before it was brought from the *West-Indies*. The Diseases mention'd by the before-cited Authors being different from that Distemper, both in Symptoms and Cure, only perhaps communicated somewhat after the same manner; I have seen some such singular Cases, attended with considerable Inconveniences and Fears, and yet not at all pocky.

The 25th of *September*, 1493. *Christopher Columbus* set Sail a second time for *Hispaniola*, and discover'd the *Caribes*. After he came to the Fort he had left, † he found all the *Spaniards* dead, and this account of them from the *Indians*, that so soon as he had sail'd for *Spain*, mortal Discords had arisen about Gold and Women, each of the *Spaniards* pretending to barter for Gold for himself; and to take as many, and what Women he pleas'd, without being satisfied with what was thought reasonable, and allotted them by the *Cacique*, *Indian* Captain, or King; that some of them had gone on these Errands towards the Mines, where one *Caunapo*, a *Cacique*, had killed most of them, and come and burnt their Fort, whereby the remainder flying had been drown'd, and were perished.

† *Col. cap. 19.*

‖ After *Columbus*'s Return to *Hispaniola*, he went to discover the South side of *Cuba*, thinking that to be the Continent, and not an Island.

‖ *Col. c. 54.*

He was inform'd in the other Isles, that in *Jamaica* was Gold, wherefore he went towards it, discover'd it on *Sunday* the 3d of *May*, 1494. and on *Monday* the next day, he came thither. He found none of that Metal, but great Number of Canoes and armed Inhabitants, who had better Understandings than those of the other Islands, and who oppos'd his Landing. Some of them were hurt by Guns, and the rest yielded, and were peaceable. *Columbus*, as he coasted the North side, was extremely pleas'd with this Island, thinking it surpassed any he had yet seen, for Verdure, Fertility, Victuals, *&c.* which he judged to come from its being water'd with Showers drawn thither by the Woods, which he had observ'd to produce the like in the * *Canaries* and *Madera* before their being clear'd of Trees.

* *Oviedo, Cor. c. 48. Linschot. descr. Amer.*

When *Columbus*, in † his third Voyage, had been to discover the Continent, he met with very contrary Winds and Currents (which ran always here Westwardly) so that he was forc'd to come to this Island, where his Ships being worm-eaten, could carry him no farther.

† *Col. c. 104 Martyr.*

He

The Introduction.

He was here in great distress, and his Men had mutiny'd. Some of them he sent in a Canoe to *Hispaniola*, the others remained with him at *Porto Buono*, in the North-side of *Jamaica*, being an extraordinary good Port, and the place which was afterwards called by the *Spaniards Sevilla*, and at this day St. *Anns*. The *Indians* murmured, thinking one Christian eat as much as twenty of them, and would not support them, till *Columbus* ‖ took the advantage of an Eclipse was to be the next night, *viz*. the 29th of *February*, 1504. He told them the God of the Christians was angry with them, and would send them Pestilence and Famine, if they did not relieve his Men. As a sign of the truth of it, next night they should see the Moon eclipsed. The *Indians* brought him Victuals, when they saw the Prediction fulfill'd, desiring he would intercede for them, and promise to fight their Enemies. This he engag'd to do, and came out of his Closet at the time when he knew the Eclipse was to end, saying his Prayers were granted. He liv'd by the effects of this Eclipse till Boats came from *Hispaniola*, and carried him and his Men thither. This Story is the more Authentic, because the Author * was there present with his Father. † This Island was conquered under *Don Diego Colon* by *Juan de Esquivel*, and other Captains. In some small time the *Indian* Inhabitants, to the number of sixty thousand ‖ were all destroyed by the Severities of the *Spaniards*, sending to Mines, *&c*. I have seen in the Woods, many of their Bones in Caves, which some people thought were of such as had voluntarily inclos'd or immured themselves, in order to be starved to death, to avoid the Severities of their Masters.

‖ *Col. c.* 103.

* *Fernan Col. Galvanes*, 1079.
† *Oviedo, Car. cap.* 49.
‖ *Bart. de las Casas*.

However the *Spaniards* planted here as at *Hispaniola*, and from hence * *Garay* Governour of it in 1523. went in three Ships and discover'd *Florida* from the Cape of that name to *Vera-cruz*, and would have planted it, had he not been hindered by *Cortes*. This Island had in it, in the North side † *Sevilla* now called St. *Anns*, the Ruins of which are now to be seen. In this Town were large Houses, and a Cathedral Church, of which the famous *Peter Martyr* was Abbot. ‖ The Abbot of this place was Suffragan to the Archbishop of *Santo Domingo* in *Hispaniola*. On the same side of the Island, about eleven Leagues to the East of *Sevilla* was *Melilla*, and fourteen Leagues from it on the South side was *Oristan*. It was very meanly inhabited by the *Spaniards*, * had no Money † and only fifty Persons in one Town, but they had ‖ *Crawles* and *Stancias*, where was planted *Cassada*. It had Beeves wild, and so consequently Hides and dry'd Beef, *&c*. and was reckoned the most pleasant and fruitful Isle in the *West-Indies*, and a Store-house for the Main. Notwithstanding this

* *Galvanos*, p 1084.
† *Lact. p.*20.
‖ *Ib. p.* 9.
* *Chilton*.
† *Alex. Ursine.* 1419.
‖ *Earl of Cumberland*, p. 1147.
Sir Antony Sherley, p. 596. and 601.

it

The Introduction. v

it was taken in 1596. by one *Shirley*, to whom its people submitted.

The Island *Jamaica*, had this name at the time of its first Discovery by *Columbus* *. The *Spaniards* write it either *Jamaica*, *Ja-*   * Col c. 54.
mayca or *Xamaica*, J Consonant and X, amongst them, being pronounc'd as a *Greek* X. It afterwards was called St. *Jago*; † but soon † *Martyr.*
obtained again its first name, which it retain'd when it was taken *dec.*
by the *English* Army, sent into these parts under General *Venables*
in the year 1655. The Descendents and Posterity of *Columbus*
were, and are still called Dukes of *Veragua* and *Vega*, and Marquesses of *Jamaica*. *Columbus* had this Island given him and his
Heirs by the Crown of *Spain*, in place of several Privileges and
Duties he was by agreement to have had, as first Discoverer and
Admiral of these Seas, which were, after coming to the knowledge
of them, thought too great for a Subject to enjoy. It is called
la Jamaique by the *French*, and *Gjamaica*, by the *Italians*. The Island
of *Antego*, one of the *Antisles* or *Caribes*, had the same name with
this Island given it by the *Indians*, but it was soon changed to that
of *Santa Maria del Antigua* from whence the present name is by corruption ‖ derived.   ‖ *Col. c. 47.*

Jamaica lies in that part of the North Sea, which washes the East
side of the Continent of *America*. This Sea is called the *Mare Boreale,
Septentrionale*, or *Mar del Nort*, to distinguish it from the *Pacific* or
South Sea, called *Mar del Zur*, which lies West of the main Land of
America. It lies nearer the Continent or Main, than most of the
other considerable *American* Isles; which Islands, as it were, guard
it from the violence of the Winds, and great *Atlantic* Ocean, and
render it fitter for the produce of the Manufacture and Trade of
those parts, than any of them. It has many *Cayos*, commonly
called Keys, Shoals and Rocks round it, whereby ignorant Sailers
are incommoded. It lies to the South West of *England* at about
fifteen hundred Leagues, or four thousand five hundred Miles distance from it. It has to the East of it *Hispaniola*, or *Santo Domingo*, about thirty five Leagues distant. To the North *Cuba*
distant about twenty Leagues, to the South *Porto Belo*, and to
the South East *Santa Martha*, both about one hundred and sixty
Leagues off, and it has also *Cartagena* one hundred and forty Leagues
distant. These three last places are on the Continent of *America* and very great places for Trade, *Cartagena* for Gold and Silver,
Portobelo for the same, *Cascarilla*, the Bark of *Peru*, or *Jesuits*
Powder, and *Sarsaparilla*, and *Santa Martha* for Pearls, all which
are brought to *Jamaica*, in exchange for Blacks and *European* Commodities. Besides, it lies near *Campeche* and *Vera Cruz*, the first a
very considerable place for Logwood, and the other being the Port

(b)   Town

The Introduction.

Town to *Mexico*, for its trading in Gold and Silver, *Cochenille*, and *Sarsaparilla*. It has a situation very happy, likewise in this respect, that it is near the *Caymanes*, the *Cayos* or desert Rocks or Isles, of *Cuba*, and the Isle *de Vacas, des Vaches*, or of *Ash*, where the Turtlers seldom fail of getting plenty of Turtle or Tortoises, to furnish the inferior sort of people with good Food, at an easie and moderate price.

The Latitude of St. *Jago de la Vega*, or *Spanish* Town in *Jamaica* is 17°. 30′. North of the Equinoctial, between it and the Tropic of *Cancer*, so that it is placed in the Torrid Zone. Its Inhabitants are *Amphiscii*, that is, have their Shadows thrown South of them, that part of the Year when the Sun is to the North of them, and North the greatest part of the Year, viz. when the Sun is to the Southwards of them. When the Sun is vertical, or directly over their Heads, they are *Ascii*, that is, their Bodies at Noon have no Shadow at all, and this happens twice a year, that is, when the Sun is going to the Tropic of *Cancer*, and returning from the same.

It is easie to find from its Latitude, that 'tis in the *Arabic*, or second Climate, according to that division of the Earth, whereby 10°. of Latitude is allowed each Climate, and which has its Name from *Arabia*, which is comprehended in it. It is also plain, that the days and nights must be very near equal the year round; so that there will be only an hours difference between the length of the longest day in Summer, and the shortest in Winter. This Latitude, likewise demonstrates that the Twilight here, between the Sun's setting, and no perception of any of his Rays at night (which is when it is about 18°. under the Horizon) or some of his Rays being seen, and his Body visible in the morning, will be very short, or not near so long, as the same continues in places that are situated in an oblique Sphere.

The Longitude of the middle of this Island is about 76°. West of *London*, this has been ascertain'd by Observations of several Eclipses by Mr. *Charles Boucher*, who sent these Observations to Mr. *Halley*, so that I cannot but wonder how *Christopher Columbus* could mistake so much as * to say that by an Observation of an Eclipse of the Moon, the East end of *Hispaniola* was five hours twenty three minutes West of *Cadiz*, whereas by Mr. *Boucher's*, and later Observations, it is certain *Jamaica* is not much over five hours West of *London*. That is to say, when 'tis about twelve a Clock at Noon at *London*, 'tis about seven in the morning at *Jamaica*, and when 'tis five a Clock in the evening in *London*, 'tis about twelve at Noon at *Jamaica*. But the value of the *Philippine* Islands, which were to belong to the Crowns of *Spain* or *Portugal*, by the Popes Bull, according

Cap. 5.

The Introduction. vii

cording to their distance from the Line of Division of the World, was the occasion of great Mistakes in the Relations and Maps of Seamen, which have some of them but very lately been rectified, and I doubt some may yet be left wrong placed.

This Island lies *East* and *West*, and is about one hundred forty Miles long, and about sixty broad in the broadest place, which is near to its middle, it decreasing towards both extreams, in many places, being not half so broad. But it is not very easie to be very exact in this, because of the many turnings of the ways, or courses of the Gullies or Torrents, by which those who cross the Country, must pass.

I find by an account given to Sir *Thomas Lynch*, when he was Governor of *Jamaica*, that from Point *Morant* to Point *Negril*, the Course N. W. 81. deg. 140 Miles dist. The breadth from *Portland*-Bay, between *Rio-Nuevo* and *White* River, the Course North o d. 51 ¼ dist. and from Great Point *Pedro*, to *Dunkin's* Cliffs, the Course Northward 48 ½ dist. From Salt-Pond-Point, to the Mouth of *Annotto* River, falling into Port *Antonio* Harbour, over-against *Lynch*-Island, the Course North 26 distance. From *Cabarita*-Point to *Cove*-Harbour, the Course North, 22 ½ dist. From the Southward of Point-*Negril*, to *Negril-Coi*, Course North 10 ½ dist.

The quantity of Acres are 7450000, whereof are *Savanna* 350000 Acres. Manurable 6100000. Unmanurable 100000. Rivers of *Jamaica*, running into the Sea, are eighty.

Rivers in Jamaica, *beginning at Point* Morant, *and so along the South-side, to the Westward.*

SAwl's River, *Cane-Garden* River, *Crawle* River, *Lynch* River, *Seymar's* River, *White* River, *Nut's* River, *Middle* River, *Morant* River, *Lin's* River, *Negro* River, *Clark's* River, *Spring* River, *White* River, *Yallah Little* River, *Black* River, and two small River, more, *Yallah Great* River, *Barbicon* River, *Cane* River, *Hope* River, *Mamee* River, *Rock* River, *Rio Cobre* River, *Rio Pedro* River, *Dry* River, *Rio de Coco Forked* River, *Rio Mana* Gully, *Nelson's* River, *Salt* or *Black* River, *Boure's* River, *Cock Pit* River, *Mino* River, *Rock* River, *Pindar's* River, *Tick* River, St. *Thomas's* River, *Crooked* River, *Green* River, *Milk* River, *Vere* River, *Ryme's* River, *Swift* River, *Black* River, *Caju* River, *Wiet's* River, *Grass* River, *One Eye* River, *Cave Little* River, *Dean's* River, *Cabarita* River, *Three Mile* River, *Roaring* River, *Alegator* River, *Magotty* River, *Flint* River, *Great* River, *Little* River, *Great* River, *Morosse* River, *Mantica*

The Introduction.

tica River, *Negro* River, *Little* River, *Rio Para Matartiboron* River, *Cameron*'s River, *Rio Bueno Still* River, *Bear Garden* River, *Langland*'s River.

Rivers on the North-side are,

ST. *Anne*'s Great River, *Ocharios* or Rivers *White* River, *Stone-Bridge* River, *Rio Nuevo* River, *Rio Sambre Tiabo* River, *Oro Cabeça* River, *Six Mile* River, *Plantain Walk* River, *Negro* River, *Puerto Maria* River, *Crawle* River, *Water*'s River, *Stony* River, *Aqua alta*, *Anotto* River, *Flinty* River *Trunnels* River, *Orange* River, *Ugly* River, *Ginge*'s River, *Little Tom*'s River, *Fox*'s River, *Sandy* River, *Plantain Walk* River, Church River or *Pencarne* River, *Iterboreale* River, *Dry* River, *Haycock* River, *White* River, *Buffe Bay* River, *Spanish* River, *Devil*'s River, *Swift* River, *Great Devil*'s River, *Back* River, *Louise* River, *Daniel*'s River, *Rio Grande White* River, *Fox*'s River, *Sandy* River, St. *Antonio* River, *Cameron* River, *Back* River, *Annotto* River, *Guava* River, *Savanna* River, *Snaky* River, *Negro* River, *Stony* River, *Annotto* River, *Porto Antonio* River, *Turtle Crawle* Rivers, *Priest Man*'s River, *Mulato* River, *Manchinil* River, *Crawle* River, *Driver*'s River, *White* River, *Hector*'s River, *Horse-Savanna* River, *Savage* River, *Plantain-Garden* River, *Muddy* River, *Sulphur* River, *Clark*'s River, *Coquar-Tree* River, *Cove* River.

The whole Island has one continu'd ridge of Hills running East and West through its middle, which are called generally the blue Mountains, from their appearing of that colour, which comes from the Eyes going through a vast quantity of *Æther*, or Air, as looking to the Heavens in a clear day. The tops of some are higher than others, one of the highest is called Mount *Diablo*. Other Hills there are on each side of this Ridge of Mountains, which are lower.

Although this place be situated in the *Torrid Zone*, yet the Air of it may very well be affirm'd temperate, in that the heat of the days is qualified by the length of the nights, which here is about twelve hours long all the year round; so that the Sun has not that length of time to heat the Atmosphere, as where the days are longer, and the nights shorter, or none at all. 'Tis on the same account that the evenings are much hotter than the mornings, although the Sun be equally distant, and the Rays fall the same way, the heat becoming then extraordinary, because it has been warm'd from morning to that time, by the beams of the Sun, a great many of which continue to act with those coming fresh from the Sun. The Breezes blowing all the year between from North-East to South-East, and

rising

The Introduction.

rising gradually as the Sun rises, is another cause of this Air's being the more temperate. There is before Rain, very often here, a great heat (although the Sun be over-clouded) as well as uneasiness on Men's Bodies, especially those that are sickly, which so soon as the Rain falls is quite removed. This is common to other parts of the world as well as this place, and seems to proceed from moisture, or rather from the spring of the humours of the Body distending the Vessels, the Atmosphere being then lighter, and more moist. The Breeze from Land at night is very cooling, though thought very unhealthy by the *Spaniards*, on what account I know not. 'Tis usually argued from Irons rusting very soon here, that the Air is corroding; but this I believe comes from the Heat, whereby most People sweat, which being salt and very penetrating corrodes the Iron, and rusts it when they touch it, or keep it in their Pockets. On the Mountains and high Land 'tis much cooler than in the Valleys; in these last the Sun Beams are reflected, from the sides of the neighbouring rising Grounds so on one another, that they seem to make in several places a kind of *Focus* as in a Burning-Glass. I never found more heat here than as in some Valleys near *Montpelier* where the situation of the Hills in their neighbourhood occasioned excessive heat. The *Savannas* are here likewise the more Temperate; because they are places where few Sun Beams are reflected on the Body, having few small rising Hills to interrupt the course of the Breezes, or reflect the Sun Beams

The Air here, notwithstanding the heat, is very healthy, I have known Blacks one hundred and twenty years of Age, and one hundred years old is very common amongst Temperate Livers.

The Mercury in the Barometer stands at about the same heighth and has the same alterations as in *England*, though it change not so often as in *England*.

The Air is here not more Nitrous than in *England*, nor is there any Saltpetre to be had from any natural Earth; but some kind of *Tincal* or *Borax* out of a red Earth, which is improper for the culture of Sugar-Canes. What Saltpetre is to be had here, is from the Earth dug out of Caves where *Indians* were buried, or where Bats, and their Dung, are in great quantities. This I am certain of, because the Duke of *Albemarle* carried several people to *Jamaica* on purpose to try to make Saltpetre, having had a Patent for that Design.

'Tis frequent, riding in the night, to meet with here and there an hot Blast, for some few paces of Earth you ride over; these Blasts, which are also met with in *Egypt*, and other parts of the World, are counted very unhealthy, as are also Norths, which blow clear over the Island for a Month together, about *Christmas*, in which time, in the North-side, no Canes will grow, but if planted, the Worms eat them.

(c)　　　　Butter

The Introduction.

Butter, through the heat of the Air, is so soft here as 'tis when half melted in *England*; and Tallow-Candles here are very often so soft as not to be able to stand upright, without falling and doubling down, which makes the nicest sort of people desire those of Wax.

The Dog-days, and some weeks about that time, are intolerably hot, and unhealthy, few people find themselves then perfectly well and easie, be they either the antient Inhabitants, or new Comers.

The heat of the Air here is endeavour'd to be amended by great Fans in some Parlours, such as are us'd about *Montpelier*, and by lying in *Hamacs*.

There are as many sorts of Water here as in *England*; River-water, Pond-water, Well-water, Spring-water, &c.

Fresh-water is very scarce in dry years, or some parts of the year, in the *Savannas* distant from Rivers, so that many of their Cattle die with driving to water. If the place be near the Sea, or sandy, the Well-water, as at *Port-Royal* is brackish. This brackish water, which is very common in Wells on Sea-shores, is not wholesome, but the cause of Fluxes, and other Diseases in Sailers drinking of it. It may be easily discovered by dropping a solution of fine Silver into it, which according to Mr. *Boyles* computation precipitates a white settlement, if it contain $\frac{1}{1700}$ of Salt in it. I have not yet known, or heard of any method which will clear Water of this brackishness but Destillation.

Pond-water, or River-water here, is more pure and not so much infected with Weeds as in *England*, because of the swiftness of the Currents, or great Inundations, destroying the Water-Herbs at certain Seasons of the year. River-water, because of its great Descents and Precipices, carries with it much Clay and Earth, whereby 'tis muddy and thick; this drunk, has an odd taste, which, in the Town-River, gave occasion to the *Spaniards* to call it *Rio Cobre*, and the *English* to say 'tis not wholesome, and tasts of Copper; whereas on trial of the Sand and other Sediments, there is no Metal found therein. This River-water, if suffered to settle some days in earthen Jars, the Sediments go to the bottom, and the Water is good. It's thought that Pebbles in the bottom of the Jar promote this separation; and Seamen think a lead-holed plate does the same; but 'tis likely these two methods only hinder the subsidence from rising easily. The porous Stones for percolating water is the best remedy for this muddiness; they must be clean'd every day, and sometimes the water put through them twice or thrice. They are brought from the *Canaries* to the *Spanish* Main, and thence to *Jamaica*. They are made into the form of Mortars, the water being put into their Concave side, foul and troubled, passes through them, and is filtred, leaving its filth in the pores of the Stone. Sometimes this

water

water is pass'd through three of these plac'd one under another. *Thames*-water, at Sea, is generally thought to ferment, and to rise to a vinous Spirit, but it is not so; for this is to be ascrib'd mostly to the filth or tincture of the Cask, for the Water itself *in* Jars, does not ferment nor smell; in Cask first it acquires a colour from the filth or Wood, then ferments, smells, and turns vinous, neither is it all so, but only that in some Casks. Brandy, by the Cask, from limpid, turns yellowish in colour, but does not so in Jars or Bottles. Water gather'd off the Ships Decks from Rain, smells and ferments presently, because of Spittle, Dung, &c. *Thames*-water is accounted the best for Ships, though probably passing by so great a City as *London*, it be the foulest in Contents.

Spring water is reckon'd preferable to other kinds; there are fine, large Springs here, many of them as well as Rivers, petrify their own Channels, by which they sometimes stop their own Courses, by a Sediment and Cement uniting the Gravel and Sand in their bottoms. When this petrifying water falls drop by drop, it makes the *Stalactites*. Several Caves have their bottoms and tops united by this Stone, so that they appear Pillars.

Upon the whole matter, the cleansing Water from Clay, Mould, Water-Herbs, and other accidental Impurities it meets with in its Course, seems to be the way of making it good in all parts of the World. In many places the Inhabitants let it purifie itself by quiet and subsidence, in others they do it by help of a sort of Beans, or rather *nux vomica*, as on the Coast of *Coromandel*; in others by percolating through porous Stone-Mortars. I have seen in *London* the muddiest Water made as fine as ever I saw any, by filling a Cistern with Sand, scattering the muddy Water on the upper part of it, which soaking through by a Hole (guarded so as not to be choak'd with mud) at the bottom of the Cistern, left behind all its impurities in the Sand. This Sand is turn'd into a soft Stone, which once a year is taken out, broken with Mallets, cleans'd of its Clay, and put again into the Cistern to serve for the same purpose another year.

A hot Bath or Spring is near *Morant* in the Eastward part of the Island, situated in a Wood, which has been bathed in, and drunk of late years for the Belly-ach with great Success.

A great many Salt-Springs arise in level Ground under Hills, in *Cabbage-Tree bottom*, at about a Mile or two distance from the Sea, which united make what is call'd the Salt-River.

Salt is made here in Ponds, whereinto the Sea or Salt-water comes and by the Heat of the Sun, the moisture being exhal'd, leaves the Salt, which is in great plenty at the Salt-Ponds, about *Old Harbour*, &c. The Salt is not perfectly white, nor in small Grains, but in large lumps, and has an Eye of red in it, as some *sal gemmæ* I have

seen

seen come from *Spain*, or what comes from the Illand called *Salt Tortugas* near the Main of *America*, which is here reckoned the stronger and better Salt.

Lagunas, or great Ponds, are many here, one whereof, *Rio Hoa* Pond, receives a great deal of water by a River, which yet has no visible Rivulet, or Discharge runs from it.

Rivers here in the Mountains rise above, and go under ground again in a great many places, as *Rio d' Oro* falls under, and rises above ground above *Sixteen Miles Walk*, three or four times, and so it is in many others.

At *Abraham's* Plantation in the North-side, is a River which has stop'd it's own Course by letting a settlement fall, and petrifying its own bottom.

It's ordinary to have Cataracts, Cascades or Precipices, in Rivers amongst the Mountains fifty or sixty Foot high. I have heard some people have been in Currents forc'd down these without hurt. The Water making a great Noise down such Precipices, gave the name to the *roaring* River in the North side.

The Rivers, especially that called the *dry* one, (because 'tis sometimes dry) when it rains in the Mountains, come down with great force, carrying along any thing in their way. These Rivers have done dammage to several people by coming down, they being not aware of them, it having rain'd above in the Mountains by their Springs, though not below them in the Plains. Many Fish are in these Rivers, up in the Mountains, especially Cray-fish, wild Hogs feed on them when the Springs are low. The Fish oft are brought down and left in Holes, where *Negroes* take them by intoxicating them with Dogwood-bark.

Many fallen Trees come down the Rivers, and crossing one another make a stop, whereby the Neighbouring grounds suffer great Inundations, this, as it is beneficial to some, so it is hurtful to others, according to the wetness or driness of the Soil.

Milk-River, is so called from the bottom of the River, which being a white Clay, has given it it's Name: it is dangerous fording it, because the Fords remove as the water puts the Sand by it's Current on one side or other.

Rocks of incredible bigness are brought down by the impetuosity of Rivers which sometimes almost stop their ancient Courses making them run another way.

By this steepness of the Hills, and consequently impetuous motion of the Current, are made the very steep and deep Gullies and sides of Rivers, so that the Banks or Brinks of a River are sometimes a great many yards perpendicularly high above the waters surface.

The

The Introduction.

The Tides here are scarce to be taken notice of, there being very little increase or decrease of the water, and that depending mostly, if not altogether, on the Winds, so that the Land-Winds driving off the water of the Island, makes a Foot, two, or more Ebb, which is most apparent in the mornings: in the Harbour of *Port Royal* one may see the Coral-Rocks then sensibly nearer the surface of the water, and all along the Sea-shore, the water is gone for a small space, leaving it dry, and this much more on the South side of the Island, when the Norths blow. On the contrary the Sea-breeze driving the water on the shore of the Island, makes the Flood, so that in the evening it may be said to be high water, especially if a South, or other Wind, blows violently into the Land for some time together, with which the water comes in, and is much higher than ordinary. The Breezes being stronger or weaker according to the Moons Age, it may be thought the Tides or Currents may follow that; but I rather believe they only are the effect of the Winds.

The outward face of the Earth seems to be different here from what I cou'd observe in *Europe*, the Vallies in this Island being very level, with little or no rising Ground, or small Hills, and without Rocks, or Stones. The Mountainous part for the most part is very steep, and furrowed by very deep Gullies on the North and South sides of the highest Hills, on each side of which are very great Precipices. The Ridges left are where the High-ways are made, to pass from one side of the Island to the other. The Gullies are made here by frequent, and often very violent Rains, which every day almost fall on these Mountains, and first making a small Trough or Course for themselves, wash away afterwards whatever comes in their way and make their Channel extraordinary steep.

The greatest part of the high Land of this Island is stony, rocky, or clayey; these sorts of Soil resist the Rains, and so are not carried down violently with them into the Plains, as are the Mould proper for Tillage, and other more friable Earths, either natural to these places, or made of the faln and rotten Leaves, and Trunks of Trees, or Dust carried by Wind and Rain; hence it is that in those mountainous places, one shall have very little or none of such Earths, but either a tenacious Clay, or a Honey-Comb, or other Rock, on which no Earth appears; and this is generally true, unless in some few places where the Rain may carry some of this Earth, and there leave it, the situation of the place being the cause of such an accident, by being a bottom among Hills.

(d)

On the same account that the high Land wants tillable Earth, and is barren, the low Land is very deep of fat and black Mould in some places a great many yards deep, so that the fat, black Earth of *Olaus Rudbeck*, would be no certain Argument of the length of time from the Deluge, there being none in the high Land, and a great deal more than enough in the low. Hence it comes that all low Land, near high, is the most fertile, and all high Land is scarce cultivated, the one being extreamly fertile, the other not.

All the high Land is covered with Woods, which are great high Trees, some of them very good Timber; they are very tall slender, straight, and one would wonder how such Trees could grow in such a barren Soil, so thick together, among the Rocks. The Trees send down their fibrous Roots into the Crannies of the Rocks, where here and there they meet with little Receptacles, or natural Basins, wherein the Rain water is preserv'd not only for the Roots of the Trees to give them Nourishment; but likewise to give Birds and Insects drink, and sometimes Passengers on the Roads. It is likewise ordinary for Rain-water to be kept in the Spurs of Cotton, and other Trees made hollow, and to be drunk by Hunters, and others, sucking it out with a wild Cane.

'Tis a very strange thing to see in how short a time a Plantation formerly clear'd of Trees and Shrubs, will grow foul, which comes from two causes; the one the not stubbing up of the Roots, whence arise young Sprouts, and the other the Fertility of the Soil. The Settlements and Plantations of, not only the *Indians*, but even the *Spaniards*, being quite overgrown with tall Trees, so that there were no Footsteps of such a thing left, were it not for old Palisadoes, Buildings, Orange-Walks, &c. which shew plainly the formerly clear'd places where Plantations have been.

There are the same Layers of Earth one over another, as are to be met with in *Europe*. And the same difference of Soil, appears here, that does in *England*, on digging of Wells, &c.

Most of the *Savannas*, or Plains fit for Pasture, and clear'd of Wood like our Meadow-Land, lie on the South side of the Island, where one may ride a great many Miles without meeting any the least Ascent. Some of these Plains are within Land encircled with Hills, as the *Magotty Savanna*, &c. 'Tis probable, these being void altogether of Trees, that they are either so naturally, or rather have been clear'd by the *Indians*, in order to plant their Maiz there, and other Provisions.

These *Savannas* answer our Meadow-Grounds in *Europe*, and after Seasons, *i. e.* Rain, are very green and pleasant, but after long droughts are very much parch'd and withered.

Low

The Introduction.

Low Land clear'd of Wood is very proper for Hay, which has been here made in very few days, and feeds Horses very well, but the greatest part of the Nourishment of Horses is *Scotch* Grass, and *Indian* Corn.

After Seasons, *i.e.* three or four, or more days Rain, all manner of Provisions, Maiz, *Guinea* Corn, Pease, Patatas, Yams, Plantains, &c. are planted. The Ground, after these Grains and Provisions are gathered, is clear'd, before they expect a new Season, of the remaining Weeds, Stalks and Rubbish, which are put in heaps and burnt.

Their new clear'd Grounds are too rich, those which have been manur'd for a long time need Dung, which now they begin to look after, not burning their Trash as formerly, but keeping it in great heaps to rot, in time to make the better Dung.

The Stalks of *Guinea* Corn and trash, (that is the *marc*, or remaining part of the Sugar Canes, after the Expression of the Juice whereof Sugar is made,) is used in *Barbados*, for fire wood, in three or four of the Stoke holes, where a less degree of Fire is sufficient, and begins to be in use in *Jamaica*, in places where Fire wood is scarce.

In places clear'd, and ruin'd or grown wild again, the several sorts of scandent or climbing Plants, especially *Convolvuli*, do so much abound and prosper, that there is no passing without a Bill to cut ones way; they are so high, as not only to mount up the young Shrubs but even to the tops of the tallest Trees, whence they hang down, and often pull down the Trees with them.

Trees faln in the Path, and turning to Mould or Earth, I have observ'd here to yield just the same sort of Earth for colour, &c. as the Soil is, in which they grew.

The Meat of the Inhabitants of *Jamaica*, is generally such as is in *England*, as Beef, Pork, and Fish, salted and preserved, and sent from hence and *Ireland*, Flour, Pease, salted Mackrels, &c. from these Places, and *New-England*, or *New-York*; on which not only the Masters feed, but also they are oblig'd to furnish their Servants both Whites and Blacks with three Pounds of Salt-Beef, Pork, or Fish, every week, besides Cassada Bread, Yams, and Patatas, which they eat as Bread, and is the natural Product of the Country.

Although there is here in the *Savannas* great plenty of Cattle, yet they cannot keep Beef past some few days, and that salted, otherwise in three or four hours 'tis ready to corrupt. Butchers always kill in the morning just before day, and by seven a Clock the Markets for Flesh-Meat are over.

Their

The Introduction.

Their Beef here is very well tasted, and good, unless when *Guinea* Hen-weed rises in the *Savannas*, which is immediately after Rains, or when they are so parch'd that Cattle can find nothing else to feed on, this having a very deep Root, and being then green. Then their whole Flesh tasts so much of it, that one cannot well eat it, at which time likewise it infects their Milk, and very much their Kidnies. Every thing made of Milk tasts, when the Cows eat it, so strong of it, that there is no using with pleasure any thing made therewith. This is commonly thought to come from their eating *Calabash*-Tree-Leaves, which in that scarcity are fell'd to feed their Cattle; but it comes from this Herb, and not thence as is vulgarly supposed. Cattle feed on the *Calabash* Fruit in dry times. Horses in the Woods are sometimes kill'd by them, the Fruit sticking so fast to their Teeth that they are not able to open their Chaps to feed.

The Butchers remedy the smell of the *Guinea* Hen weed in Cattle, by putting them into other feeding Grounds before they are slaughtered.

Veal is very common, but none thought good but what comes from *Luidas*, where the Calves are very white flesh'd; whether this comes from this places being mountainous, or bleeding and giving them Chalk, as in *Essex*, I cannot tell, but the price of it was so extravagant, that in the Assembly they past an Act that it should not be sold dearer than twelve pence *per* Pound.

A great part of the Food of the best Inhabitants, for their own Tables, of the produce of the Island, is Swines-flesh, and Poultry of their own raising.

Swine are of two sorts, one running wild in the Country amongst the Woods, which feed on the salt Fruits, &c. and are sought out by Hunters with gangs of Dogs, and chiefly found in the more unfrequented, woody, inland parts of the Island. After pursuit, and that they are wearied by the Dogs, when they come to a Bay, they are shot or pierc'd through with Lances, cut open, the Bones taken out, and the Flesh is gash'd on the inside into the Skin, fill'd with Salt and expos'd to the Sun, which is call'd Jirking. It is so brought home to their Masters by the Hunters, and eats much as Bacon, if broil'd on Coals. These Hunters are either Blacks or Whites, and go out with their Dogs, some Salt and Bread, and lye far remote from Houses, in Huts, in the Woods, for several days, in places where Swine come to feed on the Fruits, &c. returning with more or less Meat, according to their good or bad Fortune. The *Indians* are very exquisite at this Game. The same method is used for wild Kine which are now but very few, and those in the Woods in the North side. Wild Goats there are some on the Salt-Pan Hills, not to be seen but in dry Seasons when they come down for water.

Swine

Swine fed at *Crawles* are in very great plenty. These *Crawles*, or Houses and Sties built for feeding and breeding Hogs, are kept by some Whites, *Indians* or Blacks. The Swine come home every night in several hundreds from feeding on the wild Fruits in the neighbouring Woods, on the third sound of a Conch-Shell, where they are fed with some few Ears of *Indian* Corn thrown in amongst them, and let out the next morning, nor to return till night, or that they hear the sound of the Shell. These sort of remote Plantations are very profitable to their Masters, not only in feeding their own Families, but in affording them many Swine to sell for the Market. It was not a small Diversion to me, to see these Swine in the Woods, on the first sound of the Shell, which is like that of a Trumpet, to lift up their Heads from the Ground where they were feeding, and prick up their Ears to hearken for the second, which so soon as ever they heard they would begin to make some movements homewards, and on the third sound they would run with all their speed to the place where the Overseer us'd to throw them Corn. They are call'd home so every night, and also when such of them as are fit for Market are wanted; and seem to be as much, if not more, under Command and Discipline, than any Troops I ever saw.

A Palenque is here a place for bringing up of Poultry, as Turkeys, which here much exceed the *European* and are very good and well tasted, Hens, Ducks, *Muscovy* Ducks, and some very few Geese. *Muscovy* Ducks are here most plentiful, and thrive extreamly, they coming originally from *Guinea*. These Poultry are all fed on *Indian* or *Guinea* Corn, and Ants Nests brought from the Woods, which these Fowls pick up and destroy mightily.

Cattle are penn'd every night, or else they in a short time run wild. These Pens are made of Palisadoes, and are look'd after very carefully by the Planters. The Oxen who have been drawing in their Mills, and are well fed on Sugar-Cane-tops, are reckoned the best Meat, if not too much wrought. They are likewise fatted by *Scotch* Grass.

Turtle, (Tortoises) are of several sorts, those of the Sea call'd green Turtle from their Fats being of that colour, feed on Conches or Shell fish, are very good Victuals, and sustain a great many, especially of the poorer sort of the Island. They are brought in Sloops, as the Season is for breeding or feeding, from the *Caymanes*, or South *Cayos* of *Cuba*, in which forty Sloops, part of one hundred and eighty, belonging to *Port-Royal*, are always imployed. They are worth fifteen Shillings apiece, best when with Egg, and brought and put into Pens, or Palisadoed places in the Harbour of *Port-Royal*, whence they are taken and killed, as occasion requires. They are much better when brought in first, than after languishing in those Pens, for want of Food.

They infect the Blood of those feeding on them, whence their Shirts are yellow, their Skin and Face of the same colour, and their Shirts under the Armpits stained prodigiously. This I believe may be one of the reasons of the Complexion of our *European* Inhabitants, which is chang'd, in some time, from white to that of a yellowish colour, and which proceeds from this, as well as the Jaundies, which is common, Sea Air, &c.

Land-Tortles are counted more delicate Food than those of the Sea, although smaller. They are, as I have been told, on the main Continent of *America*, pen'd and fed with Patata-slips, &c. and drawn out as occasion requires either for victualling the Flota, or for the private expence of their Houses.

All sorts of Sea Tortle, except the green, are reckon'd fishy, and not good Food.

Maniati, is taken in this Island, very often in calm Bays, by the *Indians*. It is reckoned extraordinary good Victuals.

Fish of all sorts are here in great plenty; but care must be taken they be not poysonous, this is known by the places where they use, where if *Mancaneel*-Apples are commonly eaten by them, they are very dangerous.

Salt Mackarel are here a great Provision, especially for *Negros*, who cover them extreamly in Pepper-Pots, or Oglios, &c.

What is used for Bread here, by the Inhabitants, is very different from that in *Europe* : that coming neareft our Bread is *Caſſada*. The Root dug up is separated from its outward, small, thin Skin, then grated on a Wheel, or other Grater. After searcing, the powder is put into a Bag, and its juice squeez'd out; the ends of the Roots are kept for other uses. The searc'd and dry *Farina*, is spread in the Sun to dry further, then put on a Gridiron set on Coals, and there bak'd as Oat-Cakes are in *Scotland*. 'Tis observable, that although it be a Powder when put on the hot Iron, yet presently it sticks together very fast, and becomes one solid Cake, which being bak'd on one side for some few minutes, is turn'd and bak'd on the other almost as long, then put on the side of a House to Sun. The ends of the Roots are made into a coarser Flour, and a Bread is made of a coarser sort, for making a kind of Drink call'd *Perino*. The juice is poisonous, so that any creature drinking of it (after swelling) dies presently. But if Swine be by degrees accustomed to it, 'tis the most fatning Food that is. This juice is whitish, and if let settle, has a Settlement or *Fæculæ* subsiding, which make a very fine Flour, and this fine Flour by some is reckoned the best and most wholesome *Farina*.

This

The Introduction.

This Bread is eaten dry as ours, or dipt in water, on which it immediately swells, and has no very pleasant taste this way, though dry it has none at all. Dipt in sugar'd water this Bread is still more pleasant, and if it be a little tosted afterwards, it eats yet better. If dipt in Wine, it will not swell as if dipt in Water. It will keep a long time without Corruption, so that it is taken as Provision for the Sloops trading to the *Spanish* Main, &c. This Bread is worth about seven Shillings and six pence the hundred weight, sometimes double that, according to its scarcity. People who feed altogether on this, live as long, and in as good Health as they who feed on any other sort of Bread.

Plantains is the next most general support of Life in the Island. They are brought in from the Plantain-Walk, or place where these Trees are planted, a little green; they ripen and turn yellow in the House, when, or before they are eaten. They are usually rosted, after being first clear'd of their outward Skins, under the Coals. They are likewise boil'd in Oglio's or Pepper-Pots, and prepar'd into a Past like Dumplins, and several other ways. A Drink is also made of them.

The next *Succedaneum* for Bread, in this place, are Patatas. They are rosted under the Coals, or boil'd, and are eaten as the former.

Yams are likewise us'd here in lieu of Bread, and are prepar'd as the others, only because they are very large, they are usually cut in pieces.

Grains in use here, are, 1. *Guinea*-Corn. 'Tis prepar'd, and us'd as Rice, and tasts as well, and is as nourishing. It is usually the Food of Poultry and Pigeons.

2. *Indian* Corn or Maiz, either tosted or boil'd, is fed on by the Slaves, especially the young Ears of it, before ripe, are rosted under the Coals and eaten; this is thought by them very delicious, and call'd Mutton; but 'tis most us'd for feeding Cattle and Poultry.

3. Rice is here planted by some *Negros* in their own Plantations, and thrives well, but because it requires much beating, and a particular Art to separate the Grain from the Husk, 'tis thought too troublesom for its price, and so neglected by most Planters.

Pease, Beans, and Pulse of sorts different from those of *Europe*, are here very common. They are eaten when green, as ours of *Europe*, and when dry, boil'd, afford the *Negros* very good and strong Provision.

Flour from *New-York* is counted the best, but this as well as all other Flour, and Bisket, are subject to be spoil'd with Weevils, or small *Scarabæi*, if long kept.

The Introduction.

Chocolate is here us'd by all People, at all times, but chiefly in he morning; it seems by its oiliness chiefly to be nourishing, and by the Eggs mixt with it to be render'd more so. The Custom, and very common usage of drinking it came to us from the *Spaniards*, although ours here is plain, without Spice. I found it in great quantities, nauseous, and hard of digestion, which I suppose came from its great oiliness, and therefore I was very unwilling to allow weak Stomachs the use of it, though Children and Infants drink it here, as commonly as in *England* they feed on Milk. Chocolate colours the Excrements of those feeding on it of a dirty colour.

The common use of this, by all People in several Countries in *America*, proves sufficiently its being a wholesome Food. The drinking of it actually warm, may make it the more Stomachic, for we know by Anatomical preparations, that the tone of the fibres are strengthened by dipping the Stomach in hot water, and that hot Liquors will dissolve what cold will leave unaffected.

Besides these ordinary Provisions, the Racoon, a small Quadruped, is eaten. Rats are likewise sold by the dozen, and when they have been bred amongst the Sugar-Canes, are thought by some discerning people very delicious Victuals. Snakes or Serpents and *Coffi* (a sort of Worms) are eaten by the *Indians* and *Negros*.

As I have formerly observed some wonderful contrivances of Nature, for propagating the Kind, I shall now on this occasion take notice of one very obvious, and yet not regarded for the preservation of the *individuum*. 'Tis the great variety of Foods Mankind is sustained by, not only here but in the several parts of the Earth. Mankind would be at a great loss were they restrain'd by Nature to any certain limited kind of Food. For when they should come to multiply and replenish the Earth, and live in all Climates, where the difference of Air and Soil raises variety of Vegetable and Animal Productions, they would come to want necessary sustenance for Life, were they not fitted by Nature, or rather the All-wise Author of it, to make use of what they find ready for that purpose.

'Tis for this reason Man has cutting and tearing, as well as grinding Teeth, and a natural *Menstruum* or dissolvent in his Stomach and Guts, of great force and power in extracting Nourishment from the great variety of Meats, found and used in the several parts of the World. Chymists have with great industry many years sought after an *Alcahest*, Universal Dissolvent, or *Menstruum*, whereby to open or extract the Quintessence of Bodies, and have not, so far as I can see or learn, been yet able to attain it. We see every day Nature surpass them in this particular, for whereas with them so many kinds of Drugs or Bodies, so many kinds of *Menstrua* are required, the Spittle, or whatever is the *Menstruum* comes from

the

The Introduction.

the Blood into the Stomach and Guts, and is there mix'd with our comminuted Victuals, is able to open and extract from them what *is* good and proper, whether they be Roots, Stalks, Leaves, or Seeds of Vegetables of several kinds; Fat or Lean of the Flesh of Animals, or parts of them, sweet or sower, acid or *Alkali*, 'tis all one, the best parts are kept, and the worst, unuseful, or earthy, thrown off by Excrements. There will be no need of proving this, if we do but consider how many live very well on Vegetables only, thinking it inhuman to kill any thing to eat; others live on Flesh only, most on both Vegetables and Flesh. Many live on the *Irish* Patatas, a sort of *Solanum*, (on which, I have heard, they live in the Mines of *Potosi*, and in *Ireland*) the common Brakes, as in the late Famine in *France*; on the Roots of *Argentina*, called Mascorns, in *Scotland* and the North of *Ireland*, the Stalks of the *Fucus Phasganoides* called *Tangle* in *Scotland*, or on the Roots of *Bulbocastanum* or Pignuts. The greatest part of Mankind have their chief Sustenance from Grains; as Wheat, Rice, Barley, Oats, Maiz, Buck-wheat, *Zea* or *Spelta*, Rye, some from the Seeds of a wild Grass called *Gramen Manna* in *Poland*, or from wild Oats, or *Folle Avoine*, growing in the Lakes of *Canada*, on which the *Indians* feed; or from the Seeds of the several sorts of Millet and *Panicum*. Some in *Barbary* feed on Palm Oil, others on that drawn from Organ or Erguen Nuts, many on Oil Olive, or that from Walnuts or *Sesamum*, which last is much used in *Egypt* and the *East-Indies*. Kine, Goats, Swine and Sheeps Flesh sustain most people in these parts, and so does Camels in *Arabia*, and Horses in *Tartary*. Most in *Groenland* feed on large Draughts of Train Oil; and in *England* the poorer sort have strong Nourishment from Milk-meats, (on which feed the longest Livers) Butter and Cheese. In many parts of the World, as *Lapland*, &c. Fish is their chief subsistence.

Besides these already above mentioned, *Joachimus Struppius*, has written a Book printed *Francof*. 1573. *in quarto*, called *Anchora Famis*,&c. and *Giovanni Battista Segni*, *trattato sopra la Carestia è fame*,&c. *Bol*. 1602. *in quarto*. wherein I find some of the following Vegetable and Animal Productions were made use of in times of Famine, which may be not only curious to consider, but useful in the direction of others in the like necessities, should it please God to inflict the like Calamity. There are likewise other Instances of extraordinary feeding taken from other Books, as Voyages, Sieges, &c. *Petronius de victu Romanorum*, Mundy, Muffet, &c. Roots, not mention'd already, affording Sustenance, are Carrots, Parsneps, Parsly, Navews, Skirrets, Radishes, Onions, Turneps, *Scorzonera*, *Sassafie* or *Tragopogon*, Peony, *Gladiolus*, *Papyrus*, Fennel, *Daucus*, *Asphodil*, Liquorice, Bur-roots, White-thistle-roots, *Alisanders*, *Satyrium*, *Trasi*, *Arachidna*, & *Bambu*.

(f) Though

xxii   *The Introduction.*

Though Stalks and Leaves afford no great Nourishment, yet they have sometimes kept many from starving, they are either such as are tender or succulent of themselves, or made so by earthing. They are Cichory, Selery, Endive, Sorrel, Coleworts, Mallows, of much use in the Kitchin of old, Bugloss, Lettuce, Blite, Cumfry, Spinage, Leaves of Apples, Pears, Beech, Artichokes common and prickly, Mushrooms, Purslain, on which some people have lived in desert Places, *Sium*, Primrose, Cesaglione or the head of the Dwarf-Palm, the Head of the Palm called Cabbage-tree, Asparagus-like, young Shoots of Briony, black and white, Hops, *Ruscus*, *Equisetum*, which is reckon'd unwholesome, and *Ferula*, Sea-weeds, tender Leaves and Sarments of Vines, White-thorn, and Tamarind-tree-leaves. I find also in want, that People have thought of young Oak-Apple, and Pear-tree-bark, as well as Fir-tree-Cones.

Many feed on Pulse, as Pease of the Garden, and wild Beans, Vetches, *Orobi*, Lupins, Kidney-beans, Lentils, *Lotus Ægyptiaca*. Many on the Seeds of *Lolium* or Cockle, which is plentiful when Corn is scarce, and prepar'd by being well water'd, boil'd, dry'd, and made into Flour for Bread, which has been used in scarcity of Food. Sometimes this Bread has been taken out of the Oven, soak'd in Water, and bak'd again to free it of it's bad qualities.

Poppy-seeds were likewise in use of old, and Flax-seed, though thought unwholesome, also Fenugree-seeds, and Hemp-seeds, on which I have known a Woman in *England* feed many Months.

Not to speak of Acorns and Beech-mast, the Food of our Fore-Fathers, Dates, the Food of many People in *Barbary* and *Arabia*, Figs, *Pistachias*, the *Sevennois* in *France* feed on Chesnuts, the Broth or Soupe of which I have heard is very nourishing. *Joachinus Strupius*, in his Book abovementioned, tells us that Bread may be made of Apples, Citrons, Oranges, Pears, Sloes, Cherries, Sorvices, Almonds, Hazelnuts, Heps, Plums, Grapes, Pine-kernels, Bill-berries, Rasberries, Strawberries, Mulberries, Peaches, Cucumbers, Melons and Pompions. The Pulp of the Pods of the Carob, or *Siliqua*-tree, in which the Seed lies, is also eaten in *Italy*, and the Bichy or Buzzee-tree in *Guinea*. The Fruits of *Arbutus*, *mala insana*, *tribulus terrestris*, and Coco, are likewise fed on in the places where they grow, and the latter transported for provision to other places.

All sorts of Skins of Beasts, or Leather, or Things made of them, Girdles, Shoes, Belts, Shields, &c. are soak'd, and eaten, in want of better Food: Mules Flesh, and Buffalos Flesh dry'd and powder'd, Panthers, Bears, Lions, Foxes, Rhinocerots, Bats, large Toads in *New-England*, Wolves, Cats, Otters, Badgers and Beavers, Mice, *Tatou* and *Opossum*; Birds of Prey when young, *Oxygala*, sowr Milk,

or

or Bonyclabber, Fish-bones, Tortoise-Eggs, Crocodiles, Blood of most *Animals* Periwinckles and Frogs, are eaten in scarcity of other Food.

The Natural *Irish*, in scarcity of Grain, live on the Leaves of Water Cresses, Chambrock or *trifolium pratense purpureum*, Dils, Sea Snails, *Patellæ*, and small Fish the Sea-shores afford on the Ebb. The like do the *Scots* in the Western Islands, where they feed on the *Lactuca Marina*, as in the West of *England*, where 'tis called Leaver. The *Scots* also feed on the first Leaves of *Atriplex Sylvestris folio sinuato angusto candicante*, called by them Milds. When I was in the South of *France*, I was inform'd that the *Spanish* Troops in their Neighbourhood in *Catalonia*, in scarcity eat Tallow-Candles. At the Siege of *London-Derry*, the Inhabitants were supported with Tallow. The *English* Sea Pease have sustained many People in cases of Famine, and the Roots of *Astragalus Sylvaticus* have serv'd for the same purpose. In a Famine that spread over *Germany* in 1572. in *Suevia* and *Silesia*, Bread was made of Reed-Roots, of those of *Raporculus*, *iris*, &c, *acorus nostras*; at the same time tender Oak Leaves were boil'd in *Hassia*, and Children were fed with Grass and boil'd Hay. I was told by my good Friend Mr. *Cuningham*, that in *Cochinchina* is a small Bird, which makes its Nest of a sort of round-leav'd short Sea-Plant or *Fucus*, which it dissolves by keeping it in its Crop, and afterwards makes use of in the building its Nest against the Rocks. These Nests are eaten in the *East-Indies*, where they are in great esteem as well as in *Europe*. The Crops of wild Pigeons in the beginning of the Spring, contain the young Buds of Trees and Plants, which after Maceration there, is reckoned one of the most delicate Sallets or Sauces, and so are the Insects contained in the Stomachs and Guts of Woodcocks. The *Hottentots* eat the small Guts of Cattle and Sheep, with their Contents, after being worn as Bracelets about their Necks, and there in part dried. *Mæcenas* was not the only person lov'd Asses Flesh, for in the late War some of the *Vaudois* that made a Descent into *Dauphine*, esteem'd Asses Foals the best Dish they could get. I once knew one cast away on a Desert Island, who liv'd sometime only on Oisters. Snails of several kinds are eaten with much satisfaction in *Italy* and *France*, for which reason the Lord *Arundel*, lest his native Country should want them, brought from *Bourdeaux*, to *Ashted* near *Epsom* in *Surry*, some live ones to breed, whose Posterity remains there in great plenty. In *Silesia*, Dr. *Krieg* has inform'd me, they make places for the breeding of Snails at this day, where they are fed with Turnep Tops, &c. and carefully prepar'd for the Market. The *Romans* took care of them formerly after the same manner, as may be seen by the following passage.

Varro

Varro de re Rustica, lib. 3. cap. 14. gives us an account of the ways of making their *Coclearia,* in the following words. *Nam & idoneus sub dio sumendus locus coclearis, quem totum circum aqua claudas, ne quas ibi posueris ad partum, non liberos earum, sed ipsas quæras. Aqua, inquam, finiendæ, ne fugitivarius sit parandus. Locus is melior, quem & non coquit sol, & tangit ros. Qui si naturalis non est (ut ferè non sunt in aprico loco) neque habeas in opaco, ut facias, ut sunt sub rupibus ac montibus, quorum alluant radices lacus ac fluvii, manu facere oportet roscidum: qui fit, si eduxeris fistulam, & in eam mamillas imposueris tenues, quæ eructent aquam, ita ut in aliquem lapidem incidat, ac late dissipetur. Parvus iis cibus opus est, & is sine ministratore. Et hunc, dum serpit, non solùm in areà reperit, sed etiam si rivus non prohibet, in parietes stantes invenit. Denique ipsa exgruminantes ad propalam vitam diu producunt, cum ad eam rem pauca laurea folia interjiciant, & aspergant furfures non multos. Itaque coquus has vivas an mortuas coquat, plerumque nescit. Genera coclearum sunt plura, ut minutæ albulæ, quæ afferuntur è Reatino, & maximæ quæ de Illyrico apportantur, & mediocres, quæ ex Africa afferuntur. Non quo non in his regionibus quibusdam locis, eæ magnitudinibus non sint dispariles, nam & valde amplæ sunt. Quanquam ex Africa quæ vocantur solitannæ, ita ut in eas 80 quadrantes conjici possunt, & sic in aliis regionibus eædem inter se collatæ & minores sunt ac majores. Hæ in fætura pariunt innumerabilia. Earum semen minutum ac testa molli, diuturnitate obdurescit, magnis insulis in areis factis, magnum bolum deferunt æris. Has quoque saginare solent ita, ut ollam cum foraminibus incrustent sapa & farre, ubi pascantur, quæ foramina habeat, ut intrare aer possit. Vivax enim hæc Natura.*

Not only places were made for Snails by the Romans, as Parks for Deer with us, but also conveniences were made for Rats to feed, and be bred for the Table, as appears by what the same Author tells us in the next Chapter.

Glirarium autem dissimili ratione habetur, quòd non aqua, sed maceria locus sepitur, Tota lævi lapide, aut tectorio intrinsecus incrustatur, ne ex ea erepere possit. In eo arbusculas esse oportet, quæ ferunt glandem; quæ, cum fructum non ferunt, intra maceriam jacere oportet glandem & castaneam unde saturi fiant. Facere his cavos oportet laxiores, ubi pullos parere possint. Aquam esse tenuem, quòd ea non utuntur multum, & aridum locum quærunt. Hi saginantur in doliis, quæ etiam in villis habent multi, quæ figuli faciunt, multo aliter atque alia, quòd in lateribus eorum semitas faciunt, & cavum, ubi cibum constituant. In hoc dolium addunt glandem, aut nuces juglandes, aut castaneam. Quibus in tenebris, cum cumulatim positum est in doliis, fiunt pingues.

All these several and vastly differing Bodies; which, when no other are at hand, must be the Food of Mankind in the places where they are produced, are not only digested by the Artifice of
Nature

The Introduction.

Nature into good Sustenance to repair its Losses, and propagate its Kind, but likewise, however strange to us, are very greedily sought after by those us'd to them. Thus Persons not us'd to eat Whales, Squirrils, or Elephants, would think them a strange Dish; yet those us'd to them, prefer them to other Victuals.

Men and Women, who have not so nice a Smell as some Beasts, nor Faculties to distinguish by their Senses what is wholesome Food so well as they, were infinitely short of them in this particular, were it not for Providence, and the due use of their Reason. It was some Matter of wonder to me, to think how so many People, perhaps one fourth Part of the Inhabitants of the whole Earth, should come to venture to eat Bread, made only by baking the Root of *Cassada*, which is one of the rankest Poisons in the World, both to Man and Beast, when Raw. Though, I must confess, there is an Instance in several *Ranunculi*, common in our Meadows, which when green, Blisters and Ulcerates the Flesh, and are us'd for that purpose by sturdy Beggars, to excite Compassion; these are not touch'd by Cattel, but left standing in the Fields; and yet, (as I am told) fed on greedily by all sorts of Cattel, when only dry'd into Hay. There is an Instance also of this in the Roots or Leaves of *Arum*, of which many kinds, uncommon to *Europe*, are eaten, when dry'd and prepar'd, as *Colocasia, &c.* and even the Roots of the common ones are eaten in *Italy*, when dry'd into a Flower, and made into Bread, though every body knows the great Acrimony they have when Raw. I was somewhat likewise surpriz'd to see Serpents, Rats and Lizards, sold for Food, and that to understanding People, and of a very good and nice Palate. But what of all these things was most unusual, and to my great Admiration, was the great Esteem was set on a sort of *Cossi*, or Timber-Worms, call'd Cotton-Tree-Worms, by the *Negros* and *Indians*; the one the Original Inhabitants of *Africa*, the others of *America*.

The *Negros* and *Indians* are not the only Admirers of these Vermine, for I find the most polite People in the World, the *Romans* living in a Neighbouring Country, accounted them so great a Dainty, as to feed them with Meal, and endeavour breeding them up. That they were commonly known and used, is likely from the word *Cossus*, *Festus* tell us, used to signifie, one lazy or slow, like Worms: and a considerable Family at *Rome*, from the Wrinkles and Furrows in their Face was call'd *Cossutia*.

Pliny, where he speaks of the Diseases of Trees, *lib. 17. cap. 24.* says thus, *Vermiculantur magis minusve quædam, omnes tamen ferè: idque aves cavi corticis sono experiuntur. Jam quidem & in hoc luxuria esse cœpit: prægrandesque roborum delicatiores sunt in cibo: cossos vocant, atque etiam farina saginati hi quoque altiles fiunt.* I cannot find any mention made

made of them in *Apicius de re culinaria*: but *Ludovicus Cælius Rhodiginus*, in his *Antiq. Lectiones*, tells us, that *Phryges ac Pontici, vermes albidos, obesosque, capite nigricante, qui è cariosis premuntur materiis, inter delicias habent, ac xylophagia comedisse luxuria est*. And *Ælian de Animalibus*, in the 14. Book, Chap. 13. p. 817. has to this purpose this passage. *Inter cætera animalium naturæ propria hoc quoque non omiserim, Indorum rex secundis mensis & bellariis non iisdem delectatur quibus Græci, qui palmarum pumilarum fructus expetant. At ille vermem quendam in planta quadam nascentem secundis mensis igne tostum adhibet*, (and so the Indians and Negros in *Jamaica* eat them at this Day) *suavissimum quidem illum in Indi aiunt, & eorum qui gustaverunt nonnulli asserunt, quibus ego fidem habuerim.*

Also, *John de Mandeville* tells us, that in a certain Island call'd *Talache*, were, *vermes similes iis, qui in lignis putridis existunt, hosque principibus in mensis apponi.*

Likewise St. *Hierom* in his second Book against *Jovian* in the second Tome of his Works, *Francfort* Edit. *p.* 53. has this Passage to this purpose; namely, *In Ponto & Phrygia vermes albos & obesos, qui nigello capite sunt, & nascuntur in lignorum carie, pro magnis reditibus paterfamilias exigit. Et quomodo apud nos Attagen & ficedula, mullus & scarus in deliciis computantur, ita apud illos ξυλόφαγοι comedisse luxuria est.*

Aristotle does extremely extol young soft *Cicadæ*, and has been at the Pains in his History of Animals, to tell us, that the time to kill them to the best Advantage, is the Males *ante coitum*, and the Females after, when they are most savoury.

Athenæus speaks of a Marriage-Dinner, where one of the greatest Dishes were *Cicadæ* salted and dried.

As for Locusts or Grashoppers, 'tis most certain, that as they are a Curse to some Places, by devouring the Fruits of the Earth, so they are a great Blessing to others, where the Inhabitants feed on them, and are destitute of other Provisions. They are only dry'd in an Oven, and so kept, or powder'd and mixt with Milk, and, as I have been told, by those us'd to them, they eat like Shrimps.

Lopez de Gomara tells us (*Hist. general de las Indias*, cap. 69.) that after the *West-Indians* (in the Continent about *Santa Martha*) had been overcome by the *Spaniards*, they found many Baskets of Provisions the *Indians* had gathered to Traffick with the People further within Land: the Provisions were these *Cangrejos*, Land-Crabs, which burrough in the Ground like Rabbets, feed on Vegerables, and far exceed ours in taste, *Caracoles sin cascara*, or naked Snails. *Cicadæ*, of which before, *grillos* Crickets, & *langostas de las que destruyen los panes secas y Salados*, Locusts or Grashoppers.

After

The Introduction.

After what has been said, it will seem very strange that the same Author, who has given one of the best Accounts of the *Indies*, in the same Book, Chap. 219. says, that the *Indians* of *America* were made and declared Slaves to the *Spaniards*, for these Reasons that they eat *Piojos*, and *Gusanos* (our very *Cossi* before mentioned from the corrupted word *Cusi*) *Crudos*, that they intoxicated themselves with their kinds of Wines, that is of *Maiz*, &c. and smoak of Tobacco, and that they were without Beards, and if they had any grew, they pluck'd them out. These Reasons, though appearing small, yet were the only Pretences, according to their own Historians, of driving them to Slavery in Mines, where the greatest part of them perished. And if any compassionate Person oppos'd these inhuman Proceedings, such was the Power of Interest, as to bring him speedily to the like end, as appears by the sad Story of *Bartholomeo de las Casas*.

The most common Drink here is Water. 'Tis reckoned the most wholesome Drink by many, amongst whom I am one. 'Tis a common Custom to drink a large Draught of Water in the Morning here, which is thought to prevent the Belly-ach; it may very well be, in that not only it may cool the inflam'd Blood, creating a *Rhumatism*, (very often taken for, and almost always join'd with the true Belly-ach,) but that also it may clear the Guts of some lower, or sharp parts that may lie in them, the Relicks of Lime-juice, or other Heterogeneous or Morbifick Matter lodg'd in any of their Cells, and 'tis the more proper for this, in that 'tis a *Menstruum* very fit to dissolve, as well as dilute all saline and acid Substances, and that it may by its fluidity run into every corner of these Passages through which it must go. And therefore in large quantities drank, it may be affirm'd to be the best Counter Poison.

The Spittle, and Excretions of the *Glandulæ* of the Ventricle and Guts, are known to be the chief dissolvents of our Victuals, to which Water, by its *Analysis*, seems to be next a kin. It dissolves all sorts of Food, making them soft, and into a Gelly (which parts Nature seems to want) whereas Wine is for those *Gelatinas* a very improper, if not impossible *Menstruum*. 'Tis every where ready at hand to all Mankind, and all Sanguineous Animals coming near the Structure of Mankind, make use of no other with their good wills. Water when put into the Stomach, dissolves what ever is nourishable in our Victuals, carries it through the *Vena Lactea* into the Blood, increases the *Lympha*, to dilute the Chyle, and then goes off without leaving any Heterogeneous parts in the Blood or Stomach; whereas Wine or vinous Liquors do not that. *Negros*, *Indians*, *Mahumetans*, and a great part of Mankind know not the use of this Wine or vinous Liquors, and yet look fresher, and are much healthier than we. The

Northern

Northern Nations, *Goths*, and *Vandals*, who by their Numbers and Strength overcame moſt Parts of the World, ended not their Victories 'till by coming over the *Alps* they taſted and drank the Wines, whence they ſtop'd their Conqueſts, became Effeminate, and not Fruitful.

Madera Wine is the next moſt general Drink mixt with Water; 'tis very ſtrong, and a ſort of *Xeres* or Sherry; 'tis of two ſorts, the White uſually the ſtrongeſt, tho' thought not to keep ſo long, and therefore not us'd ſo much, or the Red Wine made of the White, with ſome *tinto* or deep Red Wine put to it, which may preſerve it. The Wines from the Weſtward Iſles are thought unwholeſome, both becauſe of the Grapes and mixture of Lime, Jeſſo, or Plaiſter, in making. The longer or ſhorter time that Wine made of Red Grapes ſtays fermenting with the Huſk, the more or leſs it is ting'd, and the longer or ſhorter time it will keep, and the more or leſs it is Auſtere or Stiptick. The Virgine Wine, which has but a very ſmall time ſtood on the Huſks, ſoon is ready for drinking and fine, and ſoon ſpoils; that which has a greater Tincture keeps longer, as being impregnated with ſome parts of the Huſk, as hop'd Beer keeps longer than Ale; and that ſort of Oil which has ſtood longer with the Rind and Stone of the Olives, keeps much longer than that call'd Virgin Oil which has not been ſalted. *Madera* Wines have this particular to them, different from *French* Wines, and all others coming hither, that it keeps better in a hot Place, and expos'd to the Sun, than in a cool Cellar; whereas the other Wines brought hither muſt be kept cool, and will for all that remain but a ſmall time without being prickt and turning ſower.

Syder, Beer and Ale, do not keep well here, they huff, and fly in this ſtrange Climate, and few Caſks are opened with any of theſe Liquors bottled, where they are not broken at leaſt one third of them; but Mum keeps very well.

The ſmall Beer coming hither is uſually ſaid to be brew'd with *Barbados* Aloes inſtead of Hops, the one being cheaper than the other.

The common Drink of *London*, that is Beer and Ale, alone or mixt, is very much coveted here, the Beer is often ſoweriſh, and the Ale is generally too ſweet and heavy, the one too old, and the other not well wrought (hence, as I have heard, few belonging to a Brewhouſe will taſte any Ale) whence it ſhould ſeem to drink theſe Liquors ſhould be very unhealthy. And yet more to drink ſmall Beer, which is the ſecond or third running off of the Malt, whence it muſt come that a great many Feculencies remain in the Blood. Notwithſtanding all this, 'tis certain our Fore-fathers, who drunk theſe Liquors for common drink, lived to as great an Age with as much Health as we.

There

The Introduction. xxix

There seems to be two great evils following the excessive use of vinous Liquors; one the Diseases they cause in the Head; the other their inflaming of the Blood; besides the *Nausea* the Phlegmatick parts occasion in the Stomach, or the Hiccough there caus'd by their sharpness, always following those who have drunk much. Some of these evils are the effects of the spirituous parts of vinous Liquors, and the others the effects of the *Caput Mortuum*, which remains after Distillation of vinous Liquors in the bottom of the Still without rising, and is a very nauseous substance.

Cool Drink made of Molossus and Water, Perino, Corn Drink, Cane Drink, that made of Sorrel or Pines, are all accounted unwholesom, they turning sower in twelve or twenty four hours, and owing their strength to the Sugar, and Fermentation they are put into. Although I have known some people drink nothing else, and yet have their Health very well.

Acajou Wine, made of the Fruit so called, is very strong, keeps not long, and causes vomiting; 'tis reckoned a good remedy in the Dropsie.

Plantain Drink is stronger than any of the others, except *Acajou* Wine, though subject to grow sower in a short time.

For the better understanding what these Drinks are, it will not be amiss to set down the way of making some of them here, reserving the way of preparing others to their proper places.

To make cool Drink, Take three Gallons of fair water, more than a Pint of Molossus, mix them together in a Jar; it works in twelve hours time sufficiently, put to it a little more Molossus, and immediately Bottle it, in six hours time 'tis ready to drink, and in a day it is turn'd sowr.

To make Perino, a Drink much used here, and in *Gujana*, or *Surinam*, and many other places, Take a Cake of bad *Cassada* Bread, about a Foot over, and half an Inch thick, burnt black on one side, break it to pieces, and put it to steep in two Gallons of water, let it stand open in a Tub twelve hours, then add to it the froth of an Egg, and three Gallons more water, and one pound of Sugar, let it work twelve hours, and Bottle it; it will keep good for a week.

The common fuddling Liquor of the more ordinary sort is Rum-Punch, to the composition of which goes Rum, Water, Lime-juice, Sugar, and a little Nutmeg scrap'd on the top of it. This as 'tis very strong, so 'tis sower, and being made usually of the Sugar-Pot bottoms, is very unhealthy, and because 'tis cheap, Servants, and other of the poorer sort are very easily fuddled with it, when they come from their Masters Plantations; this, as all other vinous Spirits, puts them into a fast Sleep, whereby they fall off their Horses in going home, and lie sometimes whole nights expos'd to the injuries of

(h) the

the Air, whereby they fall in time into Confumptions, Dropfies, &c. if they mifs Apopleƈic Fits.

Rum is made of Cane-juice not fit to make Sugar, being eaten with Worms in a bad Soil, or through any other fault; or of the Skummings of the Coppers in Crop time, or of Moloſſus and water fermented about fourteen days in Cifterns, and then diftill'd off, of which an account will be given hereafter. It ſeems to be much the ſame with Rack, or *Arac* (made in the *Eaft-Indies* of Rice) and other vinous Spirits, the Creatures of Fermentation, and has an unſavoury Empyreumatical ſcent, which is endeavour'd to be taken off by Reƈification, mixing Roſemary with it, or after double Diftilling letting it ftand under Ground in Jars.

They talk of a common Experiment here, that any Animals Liver put into Rum grows ſoft, and not ſo in Brandy, whence they argue this laft leſs wholeſome than that, but their Experiment, if true, proves no ſuch thing. I think it may be ſaid to have all good and bad qualities of Brandy, or any fermented or vinous Spirit.

It is, and may be us'd outwardly, inftead of *Hungary* water, in Aches, Pains, &c. eſpecially that which is double diftill'd.

The better ſort of People lie as in *England*, though more on Quilts, and with few, if any Coverings; they hold here that lying expos'd to the Land Breezes, is very unhealthy, which I do not believe to come ſo much from the qualities of the Air, either manifeft or more obſcure, as from this, that the Air is, when one goes to ſleep here, very hot, the Sun beams having heated it ſo long, it retains this heat for ſome confiderable time in the night, which afterwards wearing away, it grows towards morning very cold, and affeƈs one ſo much as by the coldneſs ſometimes to awake one if ſleeping. This muft of neceſſity check inſenſible tranſpiration, and ſo may be the cauſe of many Diſeaſes. To avoid this, *Negros* and *Indians* ſleep not without a Fire near them.

Hamacas are the common Beds of ordinary white People, they were in uſe amongft the *Indians*, and are much cooler than Beds, ſo cool as not to be lain in without Clothes, eſpecially if ſwung, as is uſually the cuftom here. This ſort of Bed is very proper for People troubled with Rhumatiſms, or inflam'd Blood, as alſo thoſe who have any great defluxion on any part of their Legs or Thighs; alſo for Women with Child in danger of Miſcarriage, the high lying or poſition of the parts, the Heels being up as high as the Head, favouring this. 'Tis a very good way, and eaſie for a ſick Perſon to be carried in one of theſe from place to place by four Men, as is the Cuftom of the better ſort of people in *Guinea*.

Indians

Indians and *Negroes* lie on the Floors, most generally on Mats made of Bull-rushes, ordinary Rushes, Ribs of Plantain Leaves, or the *Spathæ*, or *Vaginæ* of Cabbage-tree-Flowers, with very little or no coverings, and a small Fire near them in their Cottages. Hence they and ordinary white Servants, who lie not in Beds, are not laid to go to Bed, but to go and Sleep: and this Phrase has generally obtain'd all over the Plantations.

Beds are sometimes covered all over with Gauze to hinder the *Mosquitos* or Gnats from buzzing about, biting or awaking those lying in them. This is chiefly after Rain.

It is esteem'd here the wholesomest way to go to Bed early, and rise early.

The heat of the Air exhausting the Spirits, no wonder if some of the edge of Mankind to Venery be taken off; it is thought by some Men, that they are bewitch'd or charm'd by the Air; by others that that desire in Women by this heat is Augmented, but I believe neither; for what I could find by several People this Appetite is the same as in other places, neither are men more bewitch'd or charm'd here than in *Europe*; but I believe People being here more debauch'd than in *England*, the Consequences may be more taken notice of; and I am apt to think that a great many Dropsies may come from this, nothing depauperating the Blood like excessive Venery. I once saw a very great Dropsie fall on a strong young man, occasioned by one nights very excessive debauchery.

Exercises here are not many, because of the heat of the Air; riding in the mornings is the most ordinary, which by its easie moving the *Abdomen*, and so consequently its Contents, and by that means forwarding the depuration of the Blood in the several *Emunctories* there plac'd, has a very great power in keeping a Man in sound Health, as well as recovering a Man when sickly and ill.

The Passions of the Mind have a very great power on Mankind here, especially Hysterical Women, and Hypochondriacal Men. These cannot but have a great share in the cause of several Diseases, some of the People living here being in such Circumstances, as not to be able, to live easily elsewhere: add to this, that there are not wanting some, as every where else, who have been of bad Lives, whereby their minds are disturb'd; and their Diseases, if not render'd Mortal, yet much worse to cure than those who have sedate Minds and clear Consciences. On the same account it is that those who have not their Wills, Minds, and Affairs settled, in Distempers are much worse to be cur'd than other Men. On all which respects the

Indians

Indians, who are not covetous, nor trouble themselves about many things we do, have much advantage of us.

Frost or Snow are never seen in this hot Climate, but sometimes Hail, and that very large, of which during my being here I saw one instance, it comes with very great Norths, which reach with great violence to the South-side, and throw down every thing before them.

The Dews here are so great, as in the morning to drop down from the Leaves of Trees, as if it had rain'd; they fall most in the morning when coolest. One riding in the night perceives the greatness of the Dews, for he will find his Cloths, Hair, &c. very wet in a small time.

There are few, if any Fogs, in the Plains or sandy places near the Sea, only in mornings over moist places, as Rivers, Ponds, &c. there rises a great Fog; but in the Inland parts, as Sixteen-Miles-Walk, Magotty *Savanna*, &c. are great Fogs every morning most part of the year, which are clear'd up as the Sun rises, or Sea-Breeze begins to blow. Although these Fogs are as above, yet the People living there are esteemed healthy.

Rains are here very violent and lasting when they come, the drops are very large, probably from the Clouds breaking; it being observable, that if one pour water out of any Vessel, the higher 'tis from the Ground the smaller will the Drops be. The Rainbow here is as frequent as any where in times of Rain.

According to the different Positions of the places, so the Rains are more or less violent, and come at different times; but generally speaking, the two great rainy Seasons are in *May* and *October*, in which Months, at new or full Moon, they begin, and continue day and night for a whole fortnight with great violence; so that the Earth in all level places is laid under water for some Inches, and it becomes loose for a great many Inches deep, and so consequently the Roads are almost unpassable. In the Town of St. *Jago de la Vega*, in those rainy Seasons, I was forc'd to ride on Horseback, although but from door to door, to visit the Sick. And these Seasons, as they are call'd, from their being fit to Plant in, are generally so over the whole Island, though they are much alter'd in their time and violence of late years, which arises from the clearing much of the Country of Wood.

In the month of *January* is likewise expected a Season or Rain, but this is not so constant nor violent as are the other two, and probably may come from the violent Norths, at that time passing over the Mountains, with part of their Rains with them, for

In

The Introduction.

In the North-side of this Island Rains in that Month are generally very frequent and violent, coming along with great Winds, which nevertheless seldom pass the Hills, or ridge of Mountains running through the middle of the Island, so that very often the Seasons of the one are different from those of the other.

For all the Summer-months, or when the Sun is near, or over their Heads, or indeed almost the whole year round, towards Noon, it rains on some part of the Ridge of Mountains running through the Island, with Thunder and Lightning. These Rains seldom reach two or three Miles into the Plains; wherefore on the account of these Rains, the Valleys lying very near, or amongst the Mountains, have more Seasons, and are more fertile than the Plains farther off, which, if they have any Rain, it is but the Out-skirts of that in the Mountains, and therefore inconsiderable.

At other times of the year, sometimes for three or four days together there may be a Shower about Twelve, or four a Clock in the Afternoon, which only serves to moisten the surface of the Ground without any profit.

It will be necessary for the better understanding of these Things, to give a Journal of the Weather, observed by me at St. Jago de la Vega *in* Jamaica.

May 1688.

2. A Great Sea-Breeze all day, begins in the morning early; no Land-Breeze the night before.

3. A great Sea-Breeze all day, begins about nine or ten in the morning; in the first part of the night a great *Halo* about the Moon.

4. In the morning early a Sea-Breeze pretty great, about Eleven in the morning Thunder in the Mountains with Rain; a little of both came to us, with the Land-Winds, and from thence to night a dead Calm; in the evening a very great Dew.

5. No Breeze in the morning, which begins at one or two by the Clock after noon.

6. The Breeze is great, and lasts till late in the night.

7. The Breeze is moderate.

8. The Breeze is pretty strong.

9. No Breeze till two or three in the afternoon.

10. A great Sea-Breeze, but at Twelve, Rain with Thunder came with the Land-Wind, and no Breeze after.

11. A Sea-Breeze, and about Twelve, Rain from the Mountains

tains with Thunder, after which no Breeze.
12. A fresh Sea-Breeze till night, very hot evenings and mornings; most people are broke out with Pustles in their Skins.
13. A great Sea-Breeze.
14. I went to *Port-Royal*, the Sea-Breeze came in the morning about nine or ten; yesterday by it's violence it had broke a *Guinea* Ships Cable, and set her adrift.
15. A great Breeze and Thunder with Rain in the Mountains.
16. About nine the Breeze rose, it was moderate considering the time of the Moon.
17. A great Sea-Breeze, Thunder with Rain in the Mountains.
18. A moderate Sea-Breeze, towards noon Thunder, and two great Showers.
19. A Sea-Breeze, moderate, but no Rain.
20. Little or no Breeze, great Lightning last night, with Thunder towards the Sea in the morning.
21. A Little Rain in the morning, no Breeze till the afternoon, then a great Sea-Breeze with abundance of Rain, but no Thunder nor Lightning, the Rain lasted till six at night.
22. A great Sea-Breeze, no Rain.
23. A moderate Breeze, Rain towards the evening, it continues all night with great Thunder.
24. A pretty strong Breeze.
25. A very strong Breeze, some Rain in the night.
26. A great Breeze, no Rains; Quotidians, or every day *Agues*, very frequent.
27. A great Breeze, no Rain, but pretty cool weather.
28. A great Breeze. Coming from *Port-Royal* our Boat out-sail'd, or went faster than the Breeze.
29. No Rain, but a great Breeze.
30. A pretty large Breeze, with Rain in the Mountains.
31. A great Sea-Breeze.

June 1688.

1. A Moderate Breeze.
2. A great Sea-Breeze.
3. A great Sea-Breeze.
4. A great Sea-Breeze.
5. A moderate Sea-Breeze, it continued the most part of the night.
6. A great Sea-Breeze, it begins late.
7. A moderate Sea-Breeze, it begins late.
8. A moderate Sea-Breeze.
9. The Breeze begins very late.
10. A very easie Sea-Breeze, if any at all.
11. Rain about four and five this morning, no Breeze all day, towards the evening one from the Land.
12. Little or no Breeze from the Sea in the morning, towards twelve a Clock a very great one, with Rain in the evening.

13. Rain

The Introduction. XXXV

13. Rain this morning, no Breeze, but a Sea Breeze, with Rain towards the evening.
14. The moſt part of laſt night a great Sea Breeze, with ſome little Rain in the morning, about eight a great Rain, it continued all day to rain and blow from the Sea alternatively.
15. This morning fair, but no Breeze till towards Noon, and then very moderate.
16. No Breeze in the morning, but towards Noon a pretty ſtrong one from the Sea.
17. A pretty great Sea Breeze, Rain with Thunder in the Mountains.
18. In the morning Thunder, with Rain from the Sea, a pretty large Breeze towards Noon, fair all day after, with a moderate Breeze from the Sea.
19. A moderate Breeze towards ten, and about Noon it was very ſtrong with Thunder and Rain in the Mountains.
20. A moderate Sea Breeze.
21. A very great Breeze from the Sea.
22. A moderate Sea Breeze till night, then a very violent one, with Rain from the Sea.
23. A very great Sea Breeze, in the evening one from the Land.
24. A very great Sea Breeze.
25. A moderate Sea Breeze.
26. A very ſmall Sea Breeze.
27. A very moderate Sea Breeze.
28. A very great Sea Breeze.
29. A moderate Sea Breeze.
30. A moderate Sea Breeze.

July 1688.

2. A Very great Sea Breeze.
3. A very great Sea Breeze.
4. A very great Sea Breeze till towards two in the Afternoon, then Showers of Rain.
5. A moderate Breeze, Rain about Noon from the Mountains.
6. A moderate Breeze, with Rain towards Noon.
7. A moderate Breeze.
8. A moderate Breeze.
9. A very eaſie Breeze, with Sultry uneaſie weather, towards the evening Thunder with Rain in the Mountains.
10. A very moderate Breeze.
11. A moderate Breeze, Rain in the afternoon from the Sea, with two very heavy Showers in the night.
12. A moderate Breeze, with Thunder and Rain in the Mountains, and the tail of a Shower in Town.
13. A moderate Sea Breeze.
14. A moderate Sea Breeze.
15. A moderate Sea Breeze.
16. A very great Sea Breeze, with Thunder in the Mountains.
17. A great Sea Breeze.
18. A very ſtrong Sea Breeze.
19. A very great Sea Breeze.
20. A very moderate Sea Breeze, with overcaſt cloudy weather.
21. A moderate Sea Breeze, which continues pretty ſtrong till nine at night.

22. Very

22. Very little or no Breeze, this day cloudy, and overcast, towards the evening a Breeze from the Sea which lasts till nine at night.
23. Little or no Breeze, overcast, thick, cloudy and sultry weather.
24. Little or no Breeze, overcast, hot, cloudy weather.
25. Little or no Sea Breeze, overcast sultry weather.
26. Little or no Sea Breeze, overcast sultry weather, extreamly hot and uneasie. I was at this time Sick my self.
27. Little or no Sea Breeze, sultry weather. Fainting fits are very common.
28. Very little Sea Breeze, uneasie weather.
29. Very little Sea Breeze, the same uneasiness.
30. A modeeate Sea Breeze.
31. A great Sea Breeze.

August 1688.

1. A Great Sea Breeze.
2. A Great Sea Breeze, hot between the Sea and Land Breeze, a little Shower.
3. A moderate Sea Breeze.
4. A moderate Sea Breeze.
5. A moderate Sea Breeze, hot between the Sea and the Land Breeze, the Breeze blows at at *Port Royal* all night.
6. A moderate Sea Breeze.
7. A moderate Sea Breeze.
8. A moderate Sea Breeze, very hot, and few people perfectly well, Loosnesses in the night common.
9. A moderate Sea Breeze, which continues till eight or nine at night.
10. A moderate Sea Breeze, blows late.
11. A great Sea Breeze, with one Shower from the Mountains, and another from the Sea in the afternoon.
12. A very easie Sea Breeze, with Rain in the afternoon.
13. A very easie Sea Breeze, with Rain in the afternoon.
14. A very small Sea Breeze, with no Rain, though sultry weather.
15. Little or no Sea Breeze, some, though little Rain, with much Thunder, Plants begin to bud.
16. Extream hot, little or no Breeze from the Sea.
17. Little or no Sea Breeze, very hot, *Quotidians* common still, Thunder with Rain in the afternoon.
18. A moderate Sea Breeze, Thunder with Rain from the mountains in the afternoon.
19. A moderate Sea Breeze, Thunder and Lightning all night, without Rain.
20. A little after day break yesterday morning an Earthquake was taken notice of at Point or *Port Royal*, and *Liguanee*, a moderate Sea Breeze blew this day.
21. A Moderate Sea Breeze, with some Rain in the afternoon.
22. A moderate Sea Breeze, it was cloudy in the afternoon.
23. A great Sea Breeze.
24. A moderate Sea Breeze.

25. A moderate Sea Breeze.
26. A moderate Sea Breeze, in the afternoon some Rain.
27. A great Rain.
28. Fair weather, with a very easie Breeze.
29. A small Sea Breeze, Rain with Thunder.
30. A very great Shower, with Rain in the afternoon.
These last four days I observed the Weather at Mr. *Elletson*'s Plantation in *Liguanee*, and at my return to St. *Jago de la Vega*, I was told it had rained there but one day of the four.

September 1688.

1. Rain at *Liguanee*, with a moderate Breeze from the Sea, and after from the Mountains.
2. A moderate Sea Breeze, no Rain, very much clouded, and sultry weather.
3. An easie Sea Breeze, the weather makes the people somewhat faintish, *Tertians* are frequent.
4. A very great Sea Breeze.
5. An easie Sea Breeze, with some drops of Rain in the afternoon.
6. A great Dew, in the morning early getting on Horseback, after day light, my Periwig and Cloths were throughly wet with it before Sun rising, a moderate Sea Breeze.
7. A moderate Sea Breeze, pretty cool in the night.
8. A very moderate Sea Breeze.
9. A very easie Sea Breeze, in the afternoon a great gust of wind from the North, with some small Showers, the night following there was much Lightning, Thunder, and Rain.
10. About ten or eleven a Clock great Rain, with a Sea Breeze.
11. In the morning a pretty great Sea Breeze after Rain, which came in the night before with Thunder. With the Breeze this day came Rain, which was very violent all day.
12. This morning by four it began to Rain, and continued till eight in the morning, then fair till night.
13. This day small Rains, with dry weather between.
14. Small Rains, with dry weather between the Showers.
15. A moderate Sea Breeze.
16. A very great Sea Breeze, with a Shower in the Mountains about twelve.
17. A moderate Sea Breeze, with Rain from the Mountains about nine of the Clock.
18. A moderate Sea Breeze, with Rain from the Sea about eight or nine in the morning.
19. Last night Showers and Sea Breeze all night, this day the same.
20. Very hard Rain from the Sea all day, with sometimes great Winds.
21. Rain and Showers alternatively.
22. A great Sea Breeze, with Thunder and Rain in the Mountains, and some drops here.
23. In

23. In the mornig Rain, and so continues till three of the Clock.
24. A moderate Sea Breeze.
25. A moderate Sea Breeze, without Rain all day.
26. A very fresh Breeze from the Sea all day.
27. A moderate Sea Breeze without Rain.
28. Little or no Breeze, but a great many Gnats or Mosquitos.
29. A moderate Sea Breeze.
30. A moderate Sea Breeze, and very hot weather.

October 1688.

1. A Moderate Sea Breeze with very hot weather.
2. A moderate Sea Breeze, with Thunder and Rain in the Mountains.
3. A Sea Breeze with Thunder and Rain, very hot.
4. A very easie Breeze.
5. A moderate Sea Breeze, with a Sea Wind, and great Rain about noon.
6. A moderate Sea Breeze, with Rain at noon.
7. A moderate Sea Breeze, with great Rain from the Sea.
8. Great Rain from the Sea.
9. No Sea Breeze, nor from the Land.
10. No Breeze, but fair.
11. No Breeze, but a great deal of Rain all day.
12. A pretty strong Sea Breeze, but fair.
13. A pretty strong Sea Breeze.
14. A strong Sea Breeze.
15. A pretty strong Sea Breeze.
16. Fair weather with a Sea Breeze.
17. A small Sea Breeze with fair weather.
18. Going over the Mountains, between the North and South sides of this Island, at the *Moneque Savanna* I met with a Breeze from the North Sea. I continued in the North side of the Island from this day to the twenty third, where it was fair weather, then returned to the South side.
19. A Breeze from the Sea.
23. A fair day with a small Sea Breeze.
24. A Sea Breeze with some Rain.
25. Fair weather, with a small Sea Breeze.
26. A Sea very easie Breeze.
27. A pretty strong Sea Breeze.
28. A pretty strong Sea Breeze, with a great Shower in the night.
30. A pretty strong Sea Breeze.
31. A pretty strong Sea Breeze, with great Rain after Sun set.

November 1688.

1. A Very strong Sea Breeze.
2. A strong Sea Breeze.
3. A very strong Sea Breeze, with Rain in the Mountains in the afternoon. *Tertians* and *Quotidians* are very common.

The Introduction. xxxix

4. A strong Sea Breeze, Rain from the Mountains in the afternoon.
5. The Weather very hot, in the forenoon scarce any Breeze, but Rain in the afternoon.
6. A great deal of Rain in the morning till about noon, then fair afterwards.
7. Very hot, in the morning no Breeze.
8. A pretty considerable Breeze with fair weather.
9. Little or no Breeze, and very hot.
10. Very hot, with little or no Breeze till the afternoon, then it was strong.
11. A pretty strong Sea Breeze.
12. A pretty strong Sea Breeze.
13. A very strong Sea Breeze.
14. A very strong Sea Breeze, with some Rain in the morning.
15. A very strong Sea Breeze.
16. A very strong Sea Breeze, which lasts till very late in the evening.
17. This day a strong Sea Breeze.
18. A very strong Sea Breeze.
19. A pretty strong Sea Breeze, though not so much as the days before, towards noon some few drops of Rain.
20. Very little Sea Breeze, with some small drops of Rain in the afternoon.
21. A very easie Sea Breeze, and very hot.
22. A very easie Sea Breeze and very hot, Rain in the afternoon.
23. Little or no Sea Breeze, a part of a North, at night some Rain from the Mountains.
24. Little or no Sea Breeze, a great Shower from the Mountains, the Norths, or North Winds, are expected.
25. Little or no Sea Breeze, about ten by the Clock, a North with Rain.
26. A North with Rain.
27. An easie North.
28. A great North.
29. No Breeze in the morning, but towards the afternoon a great North.
30. In the morning no Breeze, but in the afternoon a great North.

December 1688.

1. IN the morning calm, in the afternoon a North pretty fresh.
2. Very calm with Rain, though small and from the North.
3. Very calm, with a smart Shower from the Sea in the afternoon.
4. Calm, with drisling Rain in the morning.
5. Last night great Rain, which continues this day from the Sea.
6. A great Sea Breeze begins early, and continues all day.
7. A great Rain begins in the morning from the Mountains, and continues all day.
8. Little or no Breeze.
9. Fair weather, with a small North. *Tertians* and *Quotidians* are very common.
10. Little or no Sea Breeze, towards the evening a North, which

The Introduction.

which blows very hard all night.
11. A pretty fresh Sea Breeze, and in the afternoon a North continues all night very strong.
12. No Sea Breeze till ten a Clock.
13. An easie North.
14. Little or no Breeze.
15. A moderate Sea Breeze.
16. Little or no Sea Breeze, but very hot.
17. Little or no Breeze, but very hot.
18. A moderate North.
19. A very easie North, with some small drops of Rain in the afternoon.
20. A great North.
21. A North easie all day, in the evening and all night very violent.
22. In the morning very calm, continues so all day.
23. Between *Passage Fort* and *Port Royal* I found a hot or warm Wind before Sun rising, coming from the Salt-Pond-Hills over the Mangroves and Ponds. I ask'd the Watermen if they did observe it, which they told me they did, and found it very evident. Very hot, and sultry weather.
24. In the morning a small Fog in the *Savannas*, and in the Afternoon a Sea Breeze very small, very hot and sultry weather.
25. In the morning no Breeze but very hot, in the afternoon and all night a very strong North.
26. A very strong North all day and the night following.
27. In the morning a small North, which increases all day, and continues the most part of the night.
28. An easie Sea Breeze, at four in the afternoon a North with a little Rain.
29. An easie North.
30. Little or no Breeze.
31. Little or no Breeze.

January 1688.

2. Little or no Breeze in the morning, very hot all day.
3. Little or no Breeze.
4. Little or no Breeze.
5. A pretty fresh North.
6. A small North.
7. A small North.
8. The most part of this day small Rain, a little overcast with a small Sea Breeze, in the evening Rain.
9. In the morning a pretty strong Sea Breeze, in the evening Rain.
10. An easie Sea Breeze.
11. Little or no Sea Breeze.
12. A pretty strong Sea Breeze, a little Rain in the evening.
13. A pretty strong Sea Breeze, with some drops in the afternoon.
14. A pretty strong Sea Breeze, and overcast.
15. A very strong Sea Breeze, and overcast.
16. An easie North.

The Introduction.

17. A pretty strong North with some Rain.
18. A pretty strong Sea Breeze, the North yesterday strong here, (at St. *Jago de la Vega*,) it reached not *Port Royal*, this evening a small Rain.
19. A very strong Sea Breeze.
20. A great North, with a great Shower in the afternoon.
21. A great North, and all day overcast, with some drops of Rain in the afternoon.
22. A great North, with fair Weather.
23. Little or no Breeze, but about four in the afternoon a Land-wind.
24. A pretty easie Sea Breeze, a pretty great Shower towards *Passage Fort*.
25. Little or no Breeze, but very hot.
26. A pretty strong Sea Breeze.
27. A pretty strong Sea Breeze.
28. A very strong Sea Breeze.
29. A moderate Sea Breeze.
30. In the morning at *Guanaboa* a North, in the afternoon a Sea Breeze. Their great Rains (at *Guanaboa*) are in *May*, and continue so till *October* from the Sea: then their Norths come in.
31. A very great Sea Breeze.

February 1688.

1. IN the morning very warm, towards Noon and in the Afternoon, a very strong Sea Breeze.
2. In the morning hot, in the afternoon a strong Sea Breeze which lasts till late in the evening.
3. A moderate Sea Breeze, and very warm.
4. A moderate Sea Breeze, and very hot.
5. A very strong Sea Breeze.
6. A very strong Sea Breeze.
7. A moderate Sea Breeze, in the afternoon very hot, all Plants are burnt up.
8. A very great and strong Sea Breeze, though in the morning about Sun-Rising, great appearance of Rain, with some few drops.
9. All last night a very strong Sea Breeze, this morning it continues very strong, with some drops of Rain, and blows all day very hard.
10. In the morning calm, towards Noon two great Showers with a Sea Breeze.
11. In the morning calm, a Land Breeze all day.
12. A pretty strong easterly Wind.
13. A North or Land-wind this morning, and continues most of the day, about Noon a small Shower.
14. In the morning calm, towards evening a small Shower.
15. A small Sea Breeze.
16. An easie North.
17. An easie North.
18. An easie Sea Breeze.
19. An easie Sea Breeze.
20. In the morning calm, afterwards a small Sea Breeze, it grows pretty violent towards night.
21. A moderate Sea Breeze.

22. A pretty strong Sea Breeze, with Rain at night in *Guanaboa*.
23. A strong Sea Breeze.
24. A strong Sea Breeze.
25. An easie Sea Breeze, with a very great Rain about twelve a Clock from the mountains, which continues till night.
26. An easie Sea Breeze, hot in the morning, at Noon Rain from the Mountains, which continues all day.
27. In the morning hot.
28. In the morning hot, about Noon great Rains, which continue till night.

March 1688.

1. IN the morning overcast, about Noon it begins to Rain, and so continues till night.
2. In the morning about nine it begins to Rain, and continues very heavy till night.
3. In the morning fair, at one in the afternoon Rain, continues till night, great Rains are now in the North side of this Island.
4. In the afternoon little Rain.
5. A strong Sea Breeze.
6. A strong Sea Breeze.
7. A fair morning, in the evening a little Rain.
8. A fair morning and hot, it continues so all day.
9. A fair morning, very hot, and continues so all day.
10. Very hot in the morning, and calm.
11. Fair, but a very great Breeze from the Sea.
12. Fair, and a very great Breeze from the Sea.
13. Hot in the morning, a great Sea Breeze by Noon, which continues late.
14. Very hot in the morning.
15. An easie Sea Breeze.
16. An easie Sea Breeze.
17. An easie Sea Breeze, with a small drisling Rain, coming with the Land Breeze.

This Island being several Degrees within the Tropic, has the Trade Wind continually there, which is on the South side of the Island call'd the Sea Breeze. It comes about eight a Clock in the morning, and increases or freshens till twelve in the day, and then as the Sun grows lower, so it decreases till there is none about four at night. About eight at night begins the Land Breeze, blowing four Leagues into the Sea both in *Jamaica* and the Continent, and continues increasing till twelve at night, and decreases again to four, when there is no more of it. This course generally holds true. The Sea Breeze now and then is more violent than at other times, as at new or full Moon, and incroaches very much on the Land-wind, and the Norths when they reign, *viz.* in the months of *December*, *January*, or *February*, blow over the ridge of Mountains with violence, and hinder the Sea Breeze. Sometimes the Sea Breeze will blow all night, but

this

The Introduction. xliii

this is to be taken notice of, that the Sea-Breeze blows stronger or longer near the Sea, as at *Port-Royal*, or *Passage-Fort*, than it does within Land, as at *St. Jago de la Vega*, or *Spanish* Town. As contrariwise the Land-wind blows harder at the Town than at *Passage-Fort*, or *Port Royal*. These things are evident to any who peruses the preceding Journal, where when the Breeze is mentioned, 'tis to be understood the Sea-Breeze in the day, or if in the night, the Land-Breeze; those Breezes ordinarily succeeding each the other.

As the Trade-Wind, between the Tropics, comes not directly from the East, but varies from North-East to South-East, according to the place and position of the Sun, so the Sea-Breeze here has the like Variation, not coming always from the same Point; on the contrary, the Land-Winds or Breezes, come always from the Ridge of Hills, and from the same point of them, and this holds both on the North and South sides of this Island. In Valleys amongst the Mountains, the Sea-Breeze, or Land one, has seldom any great influence, but the North-Winds very much, prostrating very great Trees, &c.

The Land-wind blowing at night, and Sea-Breeze in the day-time, is the Reason why no Shipping can come into Port, except in the day, nor go out but soon after break of day.

The Norths come in when the Sun is near the Tropic of *Capricorn*, and so farthest off Southerly. Mariners going from *England* meet the Trade-Wind in such or such Latitudes, later if the Sun is farther off, or sooner if nearer to them. This North is a very cold and unhealthy Wind, it is more violent in the night, because it then has the additional force of the Land-Wind with it, and comes through the Gulf of *Florida*, and a great deal farther off, which is evident from no Ships being able to go that way in the Norths Season. This Wind is more violent in the North-side of the Island than South, wherefore it checks the growth of Canes, and all Vegetables, and is hinder'd by the ridge of Mountains from shewing as much of its fury in the South, where it seldom Rains with this Wind.

The South-Winds in the South-side are usually rainy, and the lasting Seasons come in with the Sea-Breeze, it being certain in the South-side that no Rains from the Land are lasting.

As at Sea in the Trade-winds one meets with *Tornados*, so at Land sometimes will be a violent West, directly contrary to the Trade-Wind, for a few hours, with generally violent Rains, but this happens seldom, and is soon over.

The Sea Breeze, when it blows hard, is thought to hinder the Rain from coming to the Plains; it for the most part then raining in the Hills. On this account 'tis that there are in the Mountains many Springs and Rivers, and few or none in the Plains, and this is likewise the cause why there is never want of water in the Rivers coming from

them

them through the Plains, and likewise that sometimes Rivers suffer very great increase and inundations in the Plains, whereas no Rain fell in the places where such inundations appear.

Earthquakes as they are too frequent in *Hispaniola*, where they have formerly thrown down the Town of *Santo Domingo*, so they are too common here also; The Inhabitants expect one every year, and some of them think they follow their great Rains. One happen'd on *Sunday* the 19. of *February*, 1688. about eight in the morning. I found in a Chamber one Story high the Cabinets, and several other Moveables on the Floor to reel, as if People had rased the Foundations of the House. I look'd out at a Window to see what was the matter, and found that the Pigeons and other Birds in an Aviary hard by were on the Wing in as great Astonishment, keeping themselves in that Posture, not knowing where to alight. Wherefore concluding what it was, and the Danger in being in an high Brick-House, I made what haste I could to get out; but before I had passed through two Rooms, and got to the Stair-case, it was over. It came by Shocks; there were three of them, with a little Pause between. It lasted about a Minute of Time in all; and there was a small Noise accompanied it. A pair of Stairs higher it threw down most Things off the Shelves, and had much more visible Effects than below. This was generally felt all over the Island at the same time, or near it; some Houses therein being crack'd and very near ruin'd, others being uncovered of their Tiles, very few escaped some Injury, and the People in them were generally in a great Consternation, seeing them dance. The Ships in the Harbour at *Port-Royal* felt it; and one who was Eastward of the Island coming thither from *Europe*, met with, as he said, at the same time, an Hurricane. One riding on Horseback was not sensible of it. A Gentleman being at that time abroad in his Plantation, told me, he saw the Ground rise like the Sea in a Wave, as the Earthquake passed along, and that it went Northward; for that some small time after he had felt it, he saw by the Motion of the Tops of the Trees on Hills some Miles distant, that it had then reach'd no farther than that place. The *Spaniards* who inhabited this Island, and those neighbouring, built their Houses very low; and they consisted only of Ground-Rooms, their Walls being made of Posts, which were as much buried under ground as they stood above, on purpose to avoid the Danger which attended other manner of Building from Earthquakes. And I have seen in the Mountains afar off bare Spots, which the Inhabitants told me, were the Effects of Earthquakes throwing down part of the Hills, which continued bare and steep. But I will not here enlarge on this, there being several Accounts at large published

by

The Introduction.   xlv

by me of another dreadful Earthquake which happened afterwards in this Island in the *Philosophical Transactions*, Numb. 209. p. 77.

Thunder is here almost every day in the Mountains, with the Rains there, so that any person in the Plains may hear it, as well as see the Rain. It does not so ordinarily accompany those Rains that come from the Sea, although when it does 'tis very violent, and has, on the several substances it meets with, either animate or inanimate, the same effects as follow Thunder in *Europe*.

Lightning for the most part precedes Thunder in this Island, as elsewhere: and if it be fair Weather, especially in the hottest Seasons, it lightens almost all the night over, first in one part of the Sky or Horizon, out of some Clouds, and then out of others opposite to them, as it were answering one another, as it happens often in the Summer in *England*, &c. and gives occasion to people of fancy to foretel strange Wars, &c. when they please, making these Apparitions in the Air, Soldiers in Battalia, &c.

Failing Stars are here as common as elsewhere.

If the Seasons of the year be to be counted from the Spring of Vegetables, the Spring will be after every Season or great Rain, every thing then springing up after much Rain, so that during the time of such Rains is reckoned the Winter time. But they ought rather to follow the nearness or distance of the Sun, and so they will have the same time for the Seasons as in *Europe*; and indeed although the visible effects are not so plain, having the Sun when most distant so near them, yet that those times are more to be counted so than any others, is plain from this, that in the months of *November*, *December* and *January*, the time of the Suns greatest Southern distance from them, a great many Trees shed their Leaves, although they are destitute of them, neither all at a time, nor for many weeks, the warm Sun, even then, when farthest off, cloathing them speedily with new Garments. In the months of *February*, *March* or *April*, is the best time for planting Yams or Perennial Beans, they then, viz. in the Spring, taking better Root, and thriving more than at other times. Besides the Dog-days, or the time near them, are, as other where, very infamous for their sultry heat.

Their Agriculture is but very small, their Soil being as yet so fruitful as not to need manuring or dunging their Land, although they begin to lay by their Dung for future use, they seeing by the example of their Neighbours in *Barbados*, that they may need it. And even they themselves here have in some places fail'd of Sugar-works, as near the *Angels*, where the ground had been cultivated or manur'd before their coming to the Island. It was, and is among several, the custom to burn their Trash, which is the *Marc* or re-

(m)   mainder

mainder of the Sugar-Canes after the juice is squeezed out for the making of Sugar, but now of late 'tis kept in heaps, that so after their new clear'd Land has by Tilling been worn out, they may Dung it with this.

They clear ground, likely to be useful, by felling the Trees as near the Root as they can, the Timber if near their Work, they cut into smaller pieces, split it, and use it in the Stokeholes as Fire-wood to boil up the Sugars, if not they gather the Branches, put them in heaps, and fire them here and there in the Field, wherby the Field is not only clear'd, but made rich with the Ashes. The most part of Fields are not stub'd up, but the Roots of the Trees, with about three, four, or five Foot of the Truncs stand in the Field, and sometimes the fallen Body of the Tree lies along till it decays and rots. A Field being so clear'd, *Negros* with Hoes, make smaller or deeper holes, at nearer or farther distances according to the thing to be planted, and another coming after throws in the Seed, or plants the Root, and covers it with Earth, and so if a good Season has preceded it seldom misses to thrive, and is kept clear of Weeds till it be able of its self to choak them.

Fields which have before been planted several times are before Rain is expected clear'd of the Stalks and Roots of any Plants may have grown there, and they being gather'd in heaps, are burnt ; and so the Fields are planted after a Season, or let grow in Grass for Pasture, as occasion requires.

There is no places after Rain more fruitful than the *Savannas*.

Where the Roots of Trees do not hinder, trial has been made of the Plow, and it has done well for Sugar-Canes, *&c.* to be planted in.

The Inhabitants of *Jamaica* are for the most part *Europeans*, some *Creolians*, born and bred in the Island *Barbados*, the Windward Islands, or *Surinam*, who are the Masters, and *Indians*, *Negros*, *Mulatos*, *Alcatrazes*, *Mestises*, *Quarterons*, &c. who are the Slaves.

The *Indians* are not the Natives of the Island, they being all destroy'd by the *Spaniards*, of which I have said something before, but are usually brought by surprize from the *Musquitos* or *Florida*, or such as were Slaves to the *Spaniards*, and taken from them by the *English*. They are very often very much checquered in their Skin, by Cupping with Calabashes, are of an Olive colour, have long black lank Hair, and are very good Hunters, Fishers, or Fowlers, but are nought at working in the Fields or slavish Work, and if checkt or drub'd are good for nothing, therefore are very gently treated, and well fed.

The *Negros* are of several sorts, from the several places of *Guinea*, which are reckoned the best Slaves; those from the *East-Indies* or *Madagascins*, are reckoned good enough, but too choice in their Diet, being accustomed in their own Countries to Flesh Meat, *&c.* and do not well here, but very often die. Those who are *Creolians*, born in the Island, or taken from the *Spaniards*, are reckoned more worth than others in that they are season'd to the Island.

Clothing of the Island is much as in *England*, especially of the better sort, that of the *Indians* and *Negros* is a little Canvass Jacket and Breeches, given them at *Christmas*. It seems to me the *Europeans* do not well, who coming from a cold Country, continue here to Cloth themselves after the same manner as in *England*, whereas all Inhabitants between the Tropics go even almost naked, and *Negros* and *Indians* live almost so here, their Cloths serving them but a very small part of the year.

When they sleep they unty their Breeches, and loosen their Girdles, finding by experience this Custom healthy, and there is good reason for it, for by that means the Circulation of the blood is not interrupted, and so consequently humours are not deposited in the several parts of the Body, which ever follows such interruption.

The Buildings of the *Spaniards* on this Island were usually one Story high, having a Porch, Parlour, and at each end a Room, with small ones behind for Closets, *&c.* They built with Posts put deep in the ground, on the sides their Houses were plaistered up with Clay on Reeds, or made of the split Trunes of Cabbage-Trees nail'd close to one another, and covered with Tiles, or *Palmetto* Thatch. The Lowness, as well as fixing the Posts deep in the Earth, was for fear their Houses should be ruin'd by Earthquakes, as well as for Coolness.

The Houses built by the *English*, are for the most part Brick, and after the *English* manner, which are neither cool, nor able to endure the shocks of Earthquakes. The Kitchens, or Cook-Rooms here, are always at a small distance from their Houses, because of the heat and smell, which are both noisom and troublesome.

There are no Chimneys or Fire-places in their Houses, but in the Cook-Room, this word is used to signifie their Kitchen, and is a Sea word, as many others of that Country.

The Houses of considerable Planters are usually removed from their Sugar, or other Works, that they may be free from the noise and smells of them, which are very offensive.

The *Negroes* Houses are likewise at a distance from their Masters, and are small, oblong, thatch'd Huts, in which they have all their Moveables or Goods, which are generally a Mat to lie on, a Pot of Earth to boil their Victuals in, either Yams, Plantains, or Potatoes,

tatoes, with a little salt Mackarel, and a Calabash or two for Cups and Spoons.

There are very good Bricks and Pots made here of the Clay of the Country, to the easie making of which the few Rains, as well as plenty of Fire-wood conduces much.

The Air here being so hot and brisk as to corrupt and spoil Mear in four hours after 'tis kill'd; no wonder if a diseased Body must be soon buried. They usually bury twelve hours after death at all times of the day and night.

The burial place at *Port Royal* is a little way out of Town, in a sandy Soil, because in the Town or Church it is thought unhealthy for the living. Planters are very often buried in their Gardens, and have a small Monument erected over them, and yet I never heard of any of them who walk'd after their deaths for being buried out of Consecrated ground.

An ampurated Member buried there, and dug up some days after, was found eaten by the Ants all but the Bones. In the Caves where the *Indians* used to bury, the Ants would eat the whole Flesh off of the Bodies, and would perforate the Bones, and eat up the Marrow, of which I have a proof, having brought with me from thence the Bone of the Arm of an *Indian* so perforated, and its Marrow eaten by them.

The *Negroes* from some Countries think they return to their own Country when they die in *Jamaica*, and therefore regard death but little, imagining they shall change their condition, by that means from servile to free, and so for this reason often cut their own Throats. Whether they die thus, or naturally, their Country people make great lamentations, mournings, and howlings about them expiring, and at their Funeral throw in Rum and Victuals into their Graves, to serve them in the other world. Sometimes they bury it in gourds, at other times spill it on the Graves.

They have every one his Wife, and are very much concern'd if they prove adulterous, but in some measure satisfied if their Masters punish the Man who does them the supposed injury, in any of his Hogs, or other small Wealth. The care of the Masters and Overseers about their Wives, is what keeps their Plantations chiefly in good order, whence they ever buy Wives in proportion to their Men, lest the Men should wander to neighbouring Plantations, and neglect to serve them. The *Negros* are much given to Venery, and although hard wrought, will at nights, or on Feast days Dance and Sing; their Songs are all bawdy, and leading that way. They have several sorts of Instruments in imitation of Lutes, made of small Gourds fitted with Necks, strung with Horse hairs, or the peeled stalks of climbing Plants or Withs. These Instruments are sometimes

The Introduction. xlix

times made of hollow'd Timber covered with Parchment or other Skin wetted, having a Bow for its Neck, the Strings ty'd longer or shorter, as they would alter their sounds. The Figures of some of these Instruments are hereafter graved. They have likewise in their Dances Rattles ty'd to their Legs and Wrists, and in their Hands, with which they make a noise, keeping time with one who makes a sound answering it on the mouth of an empty Gourd or Jar with his Hand. Their Dances consist in great activity and strength of Body, and keeping time, if it can be. They very often tie Cows Tails to their Rumps, and add such other odd things to their Bodies in several places, as gives them a very extraordinary appearance.

The Introduction.

Upon one of their Festivals when a great many of the Negro Musicians were gathered together, I desired Mr. *Baptiste*, the best Musician there to take the Words they sung and set them to Musick, which follows.

You must clap Hands when the Base is plaid, and cry, *Alla, Alla.*

The Introduction. li

Koromanti.

Loud. Soft.

Meri Bonbo mich langa meri wa langa.

They are fruitful, and go after the birth of their Children to work in the Field, with their little ones ty'd to their Backs, in a Cloth on purpose, one Leg on one side, and the other on the other of their Mother, whence their Noses are a little flatted against the Mothers Back, which amongst them is a Beauty. The same is the reason of the broadness of their and *Indians* Faces. The Mother when she suckles her young, having no Cloths to keep her Breasts from falling down, they hang very lank ever after, like those of Goats.

Their unskilful cutting the Navel-String, does occasion that swelling which usually appears in their Navels, and makes their Bellies prominent. Their Children call'd *Piganinnies* or rather *Pequenos Ninnos*, go naked till they are fit to be put to clean the Paths, bring Fire-wood to the Kitchen, &c. when a Boy Overseer, with his Wand or white Rod, is set over them as their Task-Master.

They are rais'd to work so soon as the day is light, or sometimes two hours before by the sound of a *Conche-Shell*, and their Overseers noise, or in better Plantations by a Bell. They are suffered to go to Dinner at Twelve when they bring Wood, &c. one burden lest they should come idle out of the Field home, return to the Field at One, and come home at night.

When a Plantation has many Men or Women, 'tis said to be well handed, or in case of few, it is said to be bad handed, or to want Hands. This expression comes, as some others, from the Planters of *Jamaica*, coming a long Voyage at Sea, whereby they get some of the Sea Phrases. At Sea a Man is call'd a Hand, because his Hands are chiefly useful there. Whence all Hands up to Prayers, is as much as to say, let all Men come and Pray, or send a Hand to do this or that, is as much as let a Man do this or the other thing.

They have *Saturdays* in the Afternoon, and *Sundays*, with *Christmas* Holidays; *Easter* call'd little or *Pigganinny*, *Christmas*, and some other great Feasts allow'd them for the Culture of their own Plantations to feed themselves from Potatos, Yams, and Plantanes, &c. which they Plant in Ground allow'd them by their Masters, besides a small Plantain-Walk they have by themselves.

They formerly on their Festivals were allowed the use of Trumpets after their Fashion, and Drums made of a piece of a hollow Tree, covered on one end with any green Skin, and stretched with Thouls or Pins. But making use of these in their Wars at home in *Africa*, it was thought too much inciting them to Rebellion, and so they were prohibited by the Customs of the Island.

They

The Introduction. liii

Their little ones are not black, but reddish brown when first born. From several Countries they are of a deeper or paler colour, when the same persons are paler than ordinary 'tis a sign of sickness. Their colour is heightened by anointing with Palm, or any other Oil. When a *Guinea* Ship comes near *Jamaica* with Blacks to sell, there is great care taken that the *Negros* should be shav'd, trim'd, and their Bodies and Hair anointed all over with Palm-Oil, which adds a great beauty to them. The Planters choose their *Negros* by the Country from whence they come, and their look. The Blacks from the *East-Indies* are fed on Flesh and Fish at Home, and therefore are not coveted, because troublesome to nourish, and those from *Angola* run away from their Masters, and fancy on their deaths they are going Home again, which is no lucriferous Experiment, for on hard usage they kill themselves.

When I was at *Liguanee* I was told that at the Plantation of Captain *Hudson* there was a young Woman white all over, born of a black Mother. I had the Curiosity to go and see her, and Mrs. *Hudson* did me the favour to send for her. She was twelve years old, and perfectly white all over, middle siz'd, broad fac'd, flat nos'd, ill favour'd, and countenanc'd like a Black. Her Hair was fair and white, but not lank like ours, or half lank, half woolly like those of *Mulattos*, but short, woolly, and curled like those of the Blacks in *Guinea*. Her Mother had been bought by Captain *Hudson*, on her landing in *Jamaica*, about eleven weeks before her delivery of this Daughter. Her Mother was perfectly black, and related that once before in *Guinea*, her own Country, she had been brought to Bed of a white Boy, by a black Father. 'Tis very likely the Mother might have been sold on this occasion, for as *Juan dos Santos* tells us that these white Children, born of black Parents, are worshipped in those parts of *Ethiopia* he lived in, as the Off-spring of the Gods, so in others, if my memory does not fail me, they are put to death for being reputed the Children of the Devil. I was told that in *Nieves* two such were born, and saw my self in *England* a Black, a Servant of Mr. *Birds*, which was mottel'd or spotted with white spots in several parts of his Body and *Penis*. The Skins of such are whiter than ours, and their Hair is also whiter. This is common to almost all Creatures, that the colour of the Hair is black or white as the Skin is on which it grows.

Their Physick consists for the most part in Cupping with *Calabashes* on the pain'd place. They first apply the *Calabash* with some Chips or Combustible matter burning in it, when that is pull'd off they cut the place with Scarifications, and then apply the Cupping-

glasses

liv   *The Introduction.*

glasses or *Calabashes* again. Their Lancet is a sharp Knife, with which they cut through the Flesh held between their Fingers.

This, instead of relieving, sometimes seems rather to add more pain to the place, by making a Flux of Blood that way. There are few *Negros* on whom one may not see a great many Cicatrices or Scars, the remains of these Scarrifications, for Diseases or Ornament, on all their Faces and Bodies, and these Scarifications are common to them in their own Countries, and the *Cicatrices* thought to add beauty to them. * The *Negros* called *Papas* have most of these Scarifications. Other *Negros* take great pleasure in having their woolly curled Hair, cut into Lanes or Walks as the *Parterre* of a Garden, and this I have seen them do, for want of a better Instrument, with a broken piece of a Glass Bottle.

* *John Lok. ap. Hakl. pag. 33.*

Another very general Remedy in almost every Disease, is mixing Clay and Water, and plaistering over either some part, or all the Body in the warm Sun; but as this must of necessity stop the insensible transpiration, so it rarely misses to add a Cough to the Patients Malady, and always, by what I saw, fails of the Cure of the Distemper. Although I will not say but that in some Diseases it may avail.

They use very few Decoctions of Herbs, no Distillations, nor Infusions, but usually take the Herbs in substance. For instance, in a Clap, they grind the Roots of Fingrigo and Lime-Tree, between two Stones, and stir them into Lime-Juice till it be pretty thick, and so make the Patient take it evening and morning for some time. This is the same method of preparing Medicines, with what in the *East-Indies* is practised, for I have seen many Simples from thence, and all, or most, are to be ground on a Stone with some simple Liquor, and so given the Patient.

Besides Simples, the *Negros* use very much bleeding in the Nose with a Lancet for the Head-ach. They thrust up the Lancet into the tip of the nose, after tying a Ligature about the Neck, and some drops of Blood follow, whence they think themselves relieved in Colds, with Hoarseness and stuffed Noses.

Bathing is very much used by them. They boil Bay-Leaves, Wild-Sage, &c. in water, in one of their Pots, when boil'd they tye a *Fasciculus* of these Plants up together, and by putting that into the Decoction sprinkle their Bodies all over with it as fast as they can, they being naked.

The *Negros* and *Indians* use to Bath themselves in fair water every day, as often as conveniently they can.

I have heard a great deal of their great Feats in curing several Diseases, but could never find them any way reasonable, nor successful in any, and that little they know of Simples here, seems to

come

The Introduction. lv

come from the *Indians*, they being us'd for the same Diseases in *Mexico* and *Brazile*, as *Piso*, *Marcgrave*, *Hernandez*, *Ximenes*, and others relate.

One of the greatest remedies of the Planters living here to prevent Diseases, or the ill effect of what they call ill Fumes or Vapours, is an infusion of the Seeds of that kind of *Aristolochia* called *Contra Yerva*. The Seeds steep'd in Wine afford a clammy, yellow substance which sticks close to them, and their infusion in Wine is used in a morning in sickly times, to prevent Infection. The Seeds infused in Spirit of Wine, and distill'd, afford a Spirit very good to promote Sweating in Colds and Fevers. This Plant is used for much the same purposes, both in the Continent of *America* and the *East-Indies*. Dr. *Smallwood* an *English* Physician, who liv'd in *Guatemala*, (having been taken Prisoner by the *Spaniards*) told me that the *Spaniards* made great use of this Herb in all poisonous and Malignant Diseases. The *Indians* about *Guiana* had first discovered its Vertue to the *Spaniards*, for the Doctor being pursued by them and wounded by one of their poison'd Arrows, to find out his Cure, they took one of their *Indian* Prisoners, and tying him to a Post threatned to wound him with one of their own venemous Arrows, if immediately he did not declare their Cure for that Disease, upon which the *Indian* immediately chaw'd some of this *Contra Yerva*, and put it into the wound, and it healed. The Inhabitants here use also with great success, Vomits of the infusion of *Crocus Metallorum* in Wine, in Fevers and other Diseases.

The Trade of *Jamaica* is either with *Europe* or *America*. That of *Europe* consists in bringing thither Flower, Bisket, Beef, Pork, all manner of Clothing for Masters and Servants, as Osnabrigs, blew Cloth, Liquors of all sorts, &c. *Madera* Wine is also imported in great quantities from the Island of that name, by Vessels sent from *England* on purpose, on all which the Merchant is supposed to Gain generally 50 *per cent*. Profit. The Goods sent back again, or Exported from the Island, are Sugars, most part *Muscavados*, Indico, Cotton-wool, Ginger, Pimento All-Spice or *Jamaica*-Pepper, Fustick-wood, Prince-wood, *Lignum Vitæ*, *Arnotto*, Log-wood, and the several Commodities they have from the *Spaniards* of the *West-Indies*, (with whom they have a private Trade,) as *Sarsaparilla*, Cacao-Nuts, Cochineel, &c. on which they get considerable Profit. There is about 20 *per cent*. in Exchange between *Spanish* Money and Gold in *Jamaica*, and *English* Money paid in *England*.

Their

The Introduction.

Their Trade among the *Spaniards* privately in *America* manag'd chiefly by Sloops, is with all those things mention'd to come from *Europe*, especially Clothing, as Serges, &c. on which they have either in Truck or Money 55. *per cent*. Gain, one moiety whereof goes to the Masters and Owners of the Sloops, the other to the Merchant Adventurer. There are also many *Negros* sold this way to the *Spaniards*, who are either brought lately from *Guinea*, or bad Servants, or Mutinous in Plantations. They are sold to very good profit; but if they have many *Cicatrices*, or Scars on them, the marks of their severe Corrections, they are not very saleable.

The Commodities the *English* have in return, besides money, most usually are *Cacao*, *Sarsaparilla*, Pearls, Emeralds, Cochineel, Hides, &c.

The Trade of *Jamaica* with the *Dutch* at *Corasol* is chiefly for Provisions which are wanted very much on that Island. The Island of *Corasol* is very small, and very little Provision grows on it. The chief advantage the *Dutch* have of it, is, that 'tis a place whereto Goods are brought to Trade with the *Spaniards* privately on the Continent of *America*, for which purpose 'tis very advantageously seated.

The Turtlers who furnish the Island with Turtle, may be reckoned among the trading Sloops.

There is likewise a Trade with this Island from *New-England*, and *New York*. It consists usually in an exchange of Rum, Molossus, Sugar, and Money, for Horses, Beef, Pork, Flower or Rusk, 'tis manag'd by Brigantines, or small Craft, who now and then touch at the *Bahama* Islands, and kill Seals, or Whales for the Train-Oil, or *Sperma Ceti*.

When the Trade of the *Assiento* for furnishing the *Spanish West-Indies* with *Negros* was in this Island, it was not only very beneficial to the *African* Company and their Factors, but to the Governours of this Island, as well as the Captains of the Frigats who convey'd them to *Porto Belo*, and on their delivery there had immediately paid them the Money agreed on by the Head.

The Religion of those of the Island, either *Europeans*, or descended from them *Creolians*, is as in *England*, and the same proportion of Dissenters are there as in *England*.

The *Indians* and *Negros* have no manner of Religion by what I could observe of them. 'Tis true they have several Ceremonies, as Dances, Playing, &c. but these for the most part are so far from being Acts of Adoration of a God, that they are for the most part mixt with a great deal of Bawdry and Lewdness.

The *Negros* are usually thought to be haters of their own Children, and therefore 'tis believ'd that they sell and dispose of them to Strangers for Money, but this is not true, for the *Negros* of *Guinea*

being

The Introduction.

being divided into several Captainships, as well as the *Indians* of *America*, have Wars, and besides those slain in Battles many Prisoners are taken, who are sold for Slaves, and brought hither. But the Parents here, altho their Children are Slaves for ever, yet have so great a love for them, that no Master dare sell or give away one of their little ones, unless they care not whether their Parents hang themselves or no.

Many of the *Negros*, being Slaves, and their Posterity after them in *Guinea*, they are more easily treated by the *English* here, than by their own Country-People, wherefore they would not often willingly change Masters.

The Punishments for Crimes of Slaves, are usually for Rebellions burning them, by nailing them down on the ground with crooked Sticks on every Limb, and then applying the Fire by degrees from the Feet and Hands, burning them gradually up to the Head, whereby their pains are extravagant. For Crimes of a lesser nature Gelding, or chopping off half of the Foot with an Ax. These Punishments are suffered by them with great Constancy.

For running away they put Iron Rings of great weight on their Ankles, or Pottocks about their Necks, which are Iron Rings with two long Necks rivetted to them, or a Spur in the Mouth.

For Negligence, they are usually whipt by the Overseers with Lance-wood Switches, till they be bloody, and several of the Switches broken, being first tied up by their Hands in the Mill-Houses. Beating with *Manati* Straps is thought too cruel, and therefore prohibited by the Customs of the Country. The Cicatrices are visible on their Skins for ever after, and a Slave, the more he have of those, is the less valu'd.

After they are whip'd till they are Raw, some put on their Skins Pepper and Salt to make them smart; at other times their Masters will drop melted Wax on their Skins, and use several very exquisite Torments. These Punishments are sometimes merited by the Blacks, who are a very perverse Generation of People, and though they appear harsh, yet are scarce equal to some of their Crimes, and inferior to what Punishments other *European* Nations inflict on their Slaves in the *East-Indies*, as may be seen by *Moquet*, and other Travellers.

The Horses here are very fine, small, and for the most part well turn'd and swift, though very weak; they are of the *Spanish* breed, but very much degenerated, the *English* taking no care of them, but letting them breed in the *Savannas*, in the mean while destroying their best and strongest Horses in their Mills for grinding their Canes. They are very smooth Skin'd, and by that easily di-

(q)   stinguish'd

The Introduction.

stinguish'd from *New England*, or other Horses, which have rough Coats or Skins. A great many are wild in the Woods, who are taken with Cords, and tam'd by Horse-Catchers.

At the time of the first taking of the Island by the *English*, there was great plenty of wild Cattle in all the *Savannas*; and they were caught the most part by cutting the great Tendons of their Legs behind, whereby they were disabled to run, and were for the most part destroyed by the Soldiers who had little else to feed on. This way of taking wild Black Cattle by hunting, cutting their Tendons or Lancing, is what is used by the *Spaniards* in their Islands and Continent, and by the Privateers or Bucaniers; but in *Jamaica* there remain very few wild Cattle to be taken, and those are in the North side of the Island, in the less frequented parts. The manner by which the *Spaniards* and *English* kill'd these Cattle, besides the wild Dogs who used of themselves to hunt and kill them, was with a Lance or Halberd, on the end of which was an Iron sharpned, and made in the shape of a Crescent or Half-moon. These wild Cattle are said much to exceed the others in taste.

When I was in *Jamaica*, the Town of *Port-Royal* was situated on a sandy Point; at the farther end whereof, towards *Liguanee* runs a narrow Neck of sandy ground about three Miles long. This Town had a very good and secure Harbour, defended from all Winds by the Land, and by a Reef running out a great way beyond the *Cayos*. In this Harbour there was so deep water, that the greatest Ship might lie with her side to the Keys. The Town or Point with violent Sea-Breezes, had suffered some small Inundations of the Sea, and some small diminution; but by hindering People from taking Stones from the Reefs, and barricadoing the Town against it, no such thing had happened lately. The Town or Point was fortified very well with a Fort, and several Batteries both to the Sea and Land; and on the sandy Neck were likewise planted prickly Pears to hinder the march of any Enemy that way. The Winds have sometimes by making several Currents in the Water, forc'd the Sand into some parts of the Channel where Ships used to come in, wherefore 'tis now become straiter, and some Ships have been put on ground. The Town consisted of about Fifteen hundred Houses, which at first were of Wood, but lately were of Brick for the most part. It was built here for a Conveniency for Seamen and Soldiers on the *English* Fleets being in this Harbour, when they took the Island, and afterwards it came by degrees to what it lately was. It was in its Infancy design'd to have been removed to the Salt-Pans, which is just opposite to it on the firm Land of the Island, that so there might be a readier Communication at any time between

this

this Town and the Island, but I know not what hindered its being put in Execution. The greatest want in this Town is fresh water, their Wells affording only brackish; therefore 'tis the business of several Men to send great Canoes in the morning with the Sea Breeze to the River, and to bring thence many Casks of water to the Town with the Land Breeze at night.

This Point, which was called Point-*Cagway*, where *Port Royal* stood, was never built upon by the *Spaniards* while they remained Possessors of the Island, for two Reasons, as I was informed. The first was the frequency of Earthquakes, which, when considerable, would certainly overturn it. This was found true in a few years after I came from thence, for it was all destroyed by the Earthquake which happened in the year 1692. whereof there is a large account Communicated by me in the *Philosophical Transactions*, Number 209. p. 77. For the whole Neck of Land being sandy (excepting the Fort, which was built on a Rock and stood) on which the Town was built, and the Sand kept up by Palisadoes and Wharfs, under which was deep water, when the Sand tumbled upon the shaking of the Earth, into the Sea, it covered the Anchors of Ships riding by the Wharfs, and the Foundations yielding, the greatest part of the Town fell, great numbers of people were lost, and a good part of the Neck of Land where the Town stood was three Fathoms covered with water. The second cause of the aversion of the *Spaniards* to this place, was its being liable to be wash'd off by the violent Sea-Breezes or Souths. Upon the great Earthquake many are removed and settled in *Liguanee* at *Kingston*, and since a great Fire which happened there lately, more are removed to the same place, so that very few remain at present at *Port Royal*.

It will not be amiss to give here a small description of some places and things, such as I took in Journies when in the Island.

Passage-Fort is made up of some few Houses, Store-houses, and others; it is the place from whence is ship'd the Sugars, and other Commodities of the Island from Sixteen-Miles-Walk, and other Plantations. It was a small Fortification in the time of the *Spaniards*, whence its name, and I suppose was a defence to St. *Jago de la Vega*, which was the chief Town possessed by that Nation.

Old Harbour is a place on the Sea side, about six Miles from St. *Jago de la Vega*, consisting of some Houses, and Store-houses. In the time the *Spaniards* possessed this Island, they loaded their Galeons here. These large Vessels rode at Anchor under a *Cayo* or Rock. Pigeon Island, a small Island, lies off of this Harbour.

It belonged to Major *Ballard*, who told me that formerly there used to resort great numbers of Pigeons and breed there; but the Inhabitants going thither and loading Boats with their young, they were disturbed, and left the place. In the Plains or *Savannas*, about *Old Harbour*, grows that fine Flower which I found in St. *Christophers*, and is hereafter described, and called commonly in *Jamaica* White Lillies.

Near *Old Harbour* is a place called the *Canoes*, which is the chief place for Fishing hereabout.

A little Westward of *Old Harbour* are Woods, and some few Hills, beyond which is a large *Savanna* or Plain, call'd *Palmetto Savanna*, from the great number of Palm-Trees growing in it. From *Palmetto Savanna* I went into the Seven Plantations, where at going into the place I found a very bad smell and Air, which is occasion'd, as I was told, by the River call'd the Dry River. This River had at bottom great numbers of large round Stones, and was dry when I saw it, as it is often, but it is full of running water at some Seasons, and it then brings along with it Fishes of several sorts. When the River does not run these Fishes are left in holes in its bottom, where they are either taken by the Inhabitants, eat up by the Herns, or other Fowl feeding on them, or dye and corrupt the Water and Air. I was sensible of this corrupted Air when I was here. This dry River is swallowed up into the Earth, and rises again in some parts of it. It comes down from the Mountains when it Rains violently on them, so suddenly and impetuously, that I was told it had drowned a Boy, and six Horses he was watering, though no signs of its swelling appeared beforehand at the place where the accident happen'd. This place is cooler than the Town of St. *Jago de la Vega*, and Sir *Francis Watson*, who lived here, used to be more troubled with the *Asthma* then when in Town. For this purpose he had made a Chimney in one of the Rooms of his House, which was the only one I ever saw in this Island, except in Kitchens.

Sir *Francis Watson* had made here a Refining-house for Sugars, which serv'd most part of the Island with fine Sugar, and that candied, little Refin'd being Exported. They dissolve the *Moscovado* or coarse Sugar in water, which they call melting, then they mix a strong *Lixivium* of quick Lime with it, and clarifie this mixture in a Furnace with Whites of Eggs over a Fire, then they strain it through a Blanket plac'd in a Basket over a Cistern, whence it is carried into Brass Coolers, and then is put into Pots. The Surface of it is evened and leveled with a Trouel in these Pots, and then 'tis cover'd with moist Clay, by which in seven Weeks, 'tis purg'd, knock'd out, and put into a Stove to be perfectly

The Introduction.

fectly cured. The Clay used for Sugar is ordinary pale Clay expos'd to the Air, then mixt with Water to the consistence of a Syrup, afterwards it is strained through a Colender, and powred on the Pots where it stands till it sinks the Sugar in them pretty low, sometimes half in half. No quick Lime is used in double Refined Sugar the second time. The Molossus dropt from once Refined Sugar is called Bastard, it is boil'd up again, and clay'd to make it white. Four Gallons of Molossus yield three of Rum, but in *England* four, because of the Fermentation, which in *Jamaica* being brisker from the heat of the Air, evaporates more of the Spirits. Three Loaves of once refined Sugar, make two of double refined. The small quantity that is sent into *England* is beaten to pieces in a wooden Trough. Claying Sugar, as they report here, was first found out in *Brazil*, a Hen having her Feet dirty, going over a Pot of Sugar by accident, it was found under her tread to be whiter than elsewhere. A Refining-House is worth six thousand Pounds, of which there are but two in the whole Island, one at the *Angels* and this here. The Stoves are best arched. Pots for refining Sugar are made at *Liguanee*, though more brittle and dearer than when brought from *England*, but they are made here to supply the present need of the Planters; the Clay of which they are made, is dug up near the place.

I have seen Sugar made at several Plantations; they make it by bruising the Canes between Iron Rollers, in a Mill drawn by Oxen, the Figure whereof is to be seen in *Piso*, and several Authors. The juice is conveyed into the Boiling house, where in a Cistern is mixt about two handfuls of Lime, with One hundred and fifty Gallons of juice, and then both are let into six Coppers one after another, where it is boiled and scumm'd. The Scum is conveyed to the Still-house, only that of the fifth Copper is put into a Jar, that it may be again boiled, in the first Copper, because it is purer than the rest, and so will yield Sugar. In the sixth, with a little Oil or Grease, to lay its huffing and boiling over, it is boil'd up to Sugar, and so cool'd in Troughs, and carried into Pots, where, by a stick run through it, a hole is made, whereby the Molossus is drain'd from it, and leaves the Sugar white. This Molossus mix'd with Water, as well as scum or juice from bad Canes, is carried into the Distilling-house; where, after Fermentation, when it begins to subside, they in the night time distil it till thrown into the Fire it burns not: this in the day time is Re-distilled, and from Low-Wines is call'd high Wines or Rum.

Every several Soil requires a several Temper, as a Lye of Ashes with Lime or Lime-water, &c. which is mixt in the fourth Copper. For this reason the Overseer always gives notice to the Sugar-
boilers

The Introduction.

boilers when he begins a new piece of Ground, that they may be ready to remedy any inconvenience from the variety of Soils.

One Acre of Canes yields sometimes four thousand of Sugar, commonly two thousand.

Sugar-Canes grow well within a Foot of water, and near the Sea they are large, though sometimes brackish.

Good Sugar is known by those used to making it, by its smell before it is made.

I have seen some try to boil Cane-juice to Sugar in an ordinary Skillet, with and without Temper, but both, especially the first, was naught, being black and glewy, the reason given me was the slowness of the fire. The Sugar-boilers always observing to make a very violent fire under their Coppers.

Hogs Grease, or any Oil is put a drop, or the bigness of a Pea, into the Tach, sixth, or last Copper, to lay the boiling over of it, and for no other end: it is immediately quiet upon its being dropt in.

Mountains, and Hill-Canes make bad Sugar, being black: they burn the Coppers, there not being moisture enough, therefore 'tis the best way to mix Mountain and Valley Canes; one bunch of the first, to two of the last, which makes good Sugar, or the Planters mix water with the Cane-juice to hinder the Coppers from being burnt, or the Sugar made black. Likewise too much moisture makes ill Sugar, for 'tis observ'd that after Rains the Sugar is brown, because the water makes it be the longer on the Fire before the superfluous moisture is boil'd off.

Out of Sugars are made great quantities of cool Drinks by Fermentation, for I was informed that *Rap*, is what in *Nieves* they call cool Drink, *viz.* Molossus and Water, and that Sugar Drink is made with Sugar-Canes bruised in a Mortar, or Hand-Mill, and then boil'd with water, and wrought in a Cask: it is clear like water. Locust-Ale is Cane-Juice clarified, mix'd with Rum. Molossus Drink is called *Cawvaw*. Upon drinking the Molossus Drink of Penils, or very bad Sugar, the Belly-ach came to *Barbados*. *Perino*, before described is the wholesomest of all cool Drinks.

Out of the *French* Sorrel, in this and other places, which came from *Surinam*, is made a fine cooling Conserve and Syrup. This Sorrel is described hereafter.

Formerly this place of the Island was famous, in the time the *Spaniards* possessed it, for Tobacco. They now Plant some of it with their *Indico*, but they think their best Seed is lost or degenerated, being it is not so good as formerly. What they Plant here is that sort with long Leaves, and is only cultivated for the use of the Island, and not for Exportation. They take off the tops and side

Buds

Buds of each Plant, that the Leaves may be the better, they gather the Leaves when they are at their full growth, and in vigour, and dry them by hanging them up in the shade. The best Tobacco made by the *Spaniards* in their Plantations is pick'd Leaf by Leaf, with great care that none faulty may be found amongst it. That from the *Nuevo Reyno de Granada* (corruptly called *Verinas*, or *Tabac de Verine*) is reckon'd the best.

Tobacco is likewise planted in other places of the Island, and is of several sorts, from several Seeds: that with the broad Leaves is call'd Bulls Face. There are two sorts from *Oronoque*, and two *Spanish* kinds. It is sown in Beds; when the Leaves are about two Inches long, the Plants are drawn, and planted at four Foot distance one way, and three and an half another, then they are kept clean, and when grown about a Foot high, and going to shoot out their Stalks or Tops, the top of the Stalk or Bud is snipt off. That day seven night the Buds rising *ex alis foliorum* on the sides, are snipt off likewise, and seven days thence the other Under-buds. It stands some time longer, and then the Stalks and Leaves are cut off, hang'd up in a Shed, and if wet weather come, a Fire is made in it to hinder the Corruption of the Tobacco. Some time after the Leaves are stript off and preserv'd in great heaps from the injuries of the Air till 'tis made fit for the Market. It has been made here worth twelve Pence a Pound in *England*.

The Head of the River *Mino* is far up from this place, and there is a Lake where I have been told are to be seen great heaps of Snakes roll'd together, who leap into the Water. Abundance of Fish comes from thence. There were a great variety of Water-Melons here in the *Spaniards* time. The Cows eat them, and dunging, their Seeds, (at this Dry River, where they came to water,) there grew, and were preserv'd, till perfect neglect lost all their sorts.

Very good Fullers Earth was taken out of a *Stratum* of the Earth, in sinking a Well here. I did not observe any difference in the Layers of Earth taken out of this Well, from those in *England*, neither could I find any Shells or Petrefactions amongst the Stones, Clay, Sand, *&c.* brought up. But I was inform'd, upon enquiry, by Colonel *Nedham*, an Eye-Witness, that in *Barbados* at Sir *John Colleton*'s Plantation they dug for a Well, and at forty seven Foot had water, but in dry weather it went away: they dug to fifty, and had water a second time, which dryed away again; after a third digging they came to some Shells, and then into a River, and taking up water they brought up Fish with it. After this their water never decay'd; tho' there wer etwelve Men perpetually drawing

ing it, six at a time for Cattle, &c. it was in the Cliffs of *Barbados*.

It is pretty strange that sometimes at great depths in the Bowels of the Earth, these Substances that have belong'd to real Shell-Fish should be found. They are common in most Counties of *England*. Mr. *Middleton* brought some of them dug up in *Barbados*. I have by me many *Astroites*, (a sort of Coral,) taken up in the in-land parts of *England*, in as great plenty near the Surface of the Earth, as I have seen them in the neighbouring Sea to *Jamaica*, their Native place. At *Richmond* in *Surrey* they find in digging the Clay-Pits for making Tiles, many back-Bones of Sharks, and Shells; and I have in my possession several of the ordinary real *Nautili* that are now only brought to us from the *East-Indies*, that have been dug up there.

At Sixteen-Mile-Walk, or St. *Thomas* in the Vale; the Fog, which is every morning, except in rainy Seasons, lasts till about eight or nine, and then is dispell'd by the Sun. This Fog is not counted unwholesome. The Road thither is by the Water-side, or along the Banks of the *Rio Cobre*, where there is a Stone under which one passes, as under an Arch. There is a Hill, or rather a Rock, on the left side going up, which is at least two hundred yards perpendicular heighth having Bushes here and there on it, down which a wild Boar being hunted precipitated himself, and was at the bottom reduced to Mash. The River is sometimes filled with great Stones, which come rowling down from the neighbouring Hills, and sometimes with Timber rotten and faln into it. The Wood here is Tall, and the Woodbines on them very long. The Rain at Sixteen-Mile-Walk is so furious as sometimes to wash out of the ground the Roots of all the Plants set in it.

I was here told by Eye-Witnesses, that one Dr. *Foster*, at Sixteen-Mile-Walk, had tam'd a great Snake or Serpent, and kept it about him within his Shirt; it would wind it self fast about his Arm, and drink out of his Mouth, and leap at a Call on the Table, to eat Crums of *Cassada* Bread. It was killed by one *Coffin*, after sixteen months being tame: it was about the bigness of ones Wrist.

When the Potatos here, and at other Plantations in this Island are full grown, they hough up the Roots, cleanse them of their Fibres, keep them for use, and give the Stalks and Leaves to their Hogs for Food.

The next Town, in bigness to *Port-Royal*, on the Island, is St. *Jago de la Vega*, or St. *James* of the Plain, a Town improving
every

The Introduction. lxv

every day, 'tis the place where the Governour usually resides, and where the Courts of Justice and Records of the Island are kept. It was very great in the *Spaniards* time, and then consisted of Two thousand Houses built all in good order, every Street running parallel to or else piercing the others at right Angles, being broad and very long. It had four Churches and a Monastery. 'Tis situated on the Banks of the *Rio Cobre*, and has Plains on each side of it for several Miles. Here the Assembly and Supream Courts reside, which must make this place in some time very considerable. There were here some few Palisadoed Houses defended with Guns, but now they are ruin'd. When the Island was taken, the Soldiers burnt many of the Houses, neglected the rest, and made it so much below what it was formerly, that now they reckon its straggling Houses to be reduced to three hundred. Either this Place or *Old Harbour*, were called formerly *Oristan*. Here lived formerly the *Spanish* Governor, who had about One thousand, or Two thousand Pieces of Eight Income, more or less, according to his Profits by Trade. The Duke of *Veragnas*, who was descended from *Columbus*, and whose Ancestors had been Proprietors of the Island from the time of *Fernand* and *Isabella*, had for Anchorage, and other dues, about three or four thousand Pieces of Eight yearly Revenue from this Governor and Island, as the *Spaniards* inform'd Sir *Thomas Lynch*.

Besides these places, there are several other, as *Morant, Withy-Wood*, &c. but neither are they strong, nor very considerable.

I was resolved to go to the North-side of the Island, and visit the Mountains between it and the South side, to see what they brought forth. Wherefore I got some Gentlemen of the Country, one who drew in *Crayons*, a very good Guide, and a sure-footed Horse, and set out. Having passed Sixteen Miles Walk beforementioned, where are some of the best and securest Plantations of the Island, I came to the *Magotty*, a large *Savanna* or Plain. I met here, growing in great plenty, a sort of small and low sensible Plant, not describ'd by any person. If any one mov'd a Switch or Whip over it, as a Pen on Paper, the forms of the Letters remain'd legible for some time after: this is describ'd hereafter. I went on towards Mount *Diablo*, at the bottom of which, being benighted, I lay. The Horses of our Company were tied with *Withs*, and fed upon such Grass as they could reach. My Company and I went into a Hunters Hut, and lay on Plantain and Palm-Leaves all night. Our Sleep was very much interrupted by the Croaking of a sort of Tree-Frogs, described hereafter, the singing of Grashoppers, and noise of night *Animals*. We got up early in the

the morning, cross'd the Mountain, on which I saw those wonderful Ferns described hereafter, and observ'd the Trees cover'd with the *Phaseoli*, called *Cocoons*, of which Snuff-Boxes are made. Going over the *Moneque Savanna* I gather'd the sensible Plant, and came to St. *Anns*.

I observed the Ruins of the Town called *Sevilla*, among which a Church built by *Peter Martyr* of *Angleria*, of a sort of Freestone (to be had near this City) and Bricks. A Pavement was found two Miles from this Church, the City was so large, it had a fortified Castle, the Walls of Pebbles and Brick, four Foot thick; it was and is a good Port. There was formerly here one great Sugarwork at a pretty distance, the Mill whereof went by Water, which was brought some Miles thither. The Axletree of this is to be seen intire at this day. This Town is now Captain *Hemmings*'s Plantation. The Church was not finished, it was twenty Paces broad, and thirty Paces long: there were two rows of Pillars within, over the place where the Altar was to be, were some Carvings under the ends of the Arches. It was built of a sort of Stone, between Freestone and Marble taken out of a Quarry about a Mile up in the Hills; the Houses and Foundations stand for several Miles along and the ground towards the Country is rising. Captain *Hemmings* told me, he sometimes found Pavements under his Canes, three Foot covered with Earth, and several times Wells, and sometimes Burial-Stones finely Cut. There are the beginnings of a great House call'd a Monastery, but I suppose the House was design'd for the Governour. There were two Coats of Arms lay by, not set up; a Ducal one, and that of a Count, I suppose belonging to *Columbus* his Family, the Proprietors of the Island. There had been raised a Tower, part Brick, and part Hewn Stone, as also several Battlements on it, and other lower Buildings not finished. At the Church lie several arched Stones to compleat it, which had never been put up, but lay among the Canes. The rows of Pillars within were for the most part plain. In the time of the *Spaniards* it was thought the *Europeans* had been cut off by the *Indians*, and so the Church left unfinished. When the *English* took the Island, the ruins of this City were so overgrown with Wood, that they were all turn'd black, nay; I saw a Mammee, or Bastard Mammee-Tree grow within the Walls of the Tower, so high as that it must have been a very large Gun could kill a Bird on the top of it, and the most part of the Timber fell'd off this place, when it was planted, was sixty Foot or more long. A great many Wells are on this ground. In the Bay, under this, is a very fine Harbour, made by a Reef running

ning out a great way, able to hold a great many very good Ships. The West-Gate of the Church was very fine Work, and stands very entire, it was seven Foot wide, and as high before the Arch began. Over the Door in the middle was our Saviour's Head with a Crown of Thorns between two Angels, on the right side a small round figure of some Saint with a Knife struck into his Head, on the left a Virgin *Mary* or *Madonna*, her Arm tied in three places, *Spanish* Fashion. Over the Gate, under a Coat of Arms, this Inscription,

PETRUS. MARTIR. AB. ANGLERIA. ITALUS. CIVIS. MEDIOLANEN. PROTHON. APOS. HVIVS. INSVLE. ABBAS. SENATVS. INDICI. CONSILIARIVS. LIGNEAM. PRIUS. ÆDEM. HANC. BIS. IGNE. CONSVMPTAM. LATERICIO. ET. QUADRATO. LAPIDE. PRIMVS. A. FUNDAMENTIS. EXTRUXIT.

The words are entire, except *Mediolanensis*, which I have supplied because this *Peter Martir*, a famous Man, wrote himself of *Milan*. He was Author of the *Decads*, *Epistles*, and some other Books, which gave him great Reputation in the World.

I went from St. *Anns* towards St. *Georges*, where I crossed the River called *Rio Nuevo*. I saw the old *Spanish* Fortifications, whither the *Spaniards* retreated, and kept themselves till they were carried to *Cuba*, where they, for the most part, settled about a place called St. *Jago*. Colonel *Ballard*, who was present at the taking of the Island, assured me that the *Spaniards* (who Inhabited the Island to the number of Five thousand, with as many Blacks) retired to the North side, where Seven hundred fortified themselves very well, but were beat in their Forts by so many *English*. The Governour was an old decrepid Man, who was brought to them in an *Hamaca*, his Name was *Don Juan Ramires de Arellano Cavallero del Habito de S. Jago*. They held it out in this North-side for some time.

The same Person likewise told me that when the *Spaniards* were retreated into their Fortifications, at the first coming of the *English* hither, the *Spanish* Dogs went all wild, and that they were almost as big as *Irish* Grey-Hounds. They used to hunt of themselves the Cattle which were in the *Savannas* and Woods. One day Colonel *Ballard* assured me he saw a little reddish one, call'd a *Busc*, howl, and was answer'd by the others in the Woods, who came from all quarters to him, and then went orderly about to take their Supper. The *Soldiers* used to follow the wild Cattle on Horseback, and take them in the manner before related. The wild Dogs, who

not

The Introduction.

not only devour'd and took at Bay the Cows, but Asses, Colts, &c. did much mischief in the night. On the first Discovery of the *West Indies*, Dogs were very much used by the *Spaniards* to hunt the poor *Indians*, who had escap'd them over Rivers or into Woods; and their Voyages or Relations tell us there was a certain share of Booty due to the Master of the Dog, upon such Excursions, I think about half of what was given to a Man.

Ants are said to have killed the *Spanish* Children by eating their Eyes when they were left in their Cradles in this part of the Island: this is given as one Reason why the *Spaniards* left this part of the Country, where they had first settled, and built the Towns of *Sevilla* and *Melilla*. Sir *Thomas Lynch*, when Governour of *Jamaica*, sent to the old *Spanish* Inhabitants of it on *Cuba*, to know what Reason they had to leave it, and go to the South-side; the answer they made was, that they left it because their Children died there, that there were abundance of Ants, that there was no good Port, and that it was out of the Road for the Trade of *Cartagena*, and *Santo Domingo*. How troublesome Ants may be to Men and Women, much more to Children, may be seen in the Relations of *Africa*, particularly by *Denys* and *Carli*, who tell us that when the Ants set upon a House, the Inhabitants are forc'd to run for it. I once went to visit Mr. *Rowe*, a sick Person at St. *Jago de la Vega* in *Jamaica* in a morning, and found him more than ordinarily discompos'd, for that the Ants by eating in the night some of the joints of his Bedstead, his Bed of a sudden had fallen to the ground; but of this and their wonderful Actions, I shall have occasion to Discourse hereafter. In the Northern side one tried to kill them with a Train of Gun-Powder, but could not. If you thrust an *Animals* Thigh Bone into one of their Nests, they will be all kill'd by the Wood-Ants for love of the Bone.

Ginger is planted in this North-side of the Island in holes four Inches deep, made with Houghs in clear'd Ground, six Inches asunder one Root from another. They put into each hole a small piece of a Root, and cover it with Earth, in twelve Months it covers the Ground, so that a Hough cannot be put where the Races or Roots are not. At twelve months end, when the Stalks and Leaves are withered, 'tis Hough'd up, clear'd of its Fibrils, Stalks and Strings, by a Knife, or the Hand, then wash'd in fair water, put in a Basket about a hundred Pound at a time, and boil'd in a Kettle for a quarter of an hour, then expos'd to the Sun and dried. Though Rain comes, it hurts it not; this is the Black Ginger. Fresh Roots must be boil'd in fresh water.

The Introduction.

The white sort of Ginger is made by only scraping the fresh Root clean of its outward Membrane, and exposing it in the Sun till it be dry. This white Ginger is easily spoil'd by Worms. Some say that in the Windward Islands or *Caribes*, 'tis only dried on a sandy Bay.

There is a white sort prepar'd as the black, only it is boil'd in Lime-water, which makes it as they say, not so wholesome. This Root is very often injured by Worms as well as *China* Roots, which are apt to breed a small light brown *Scarabæus*, like that bred in Rhubarb.

Preserv'd wet Ginger, is made by drawing up the Ginger while 'tis young, about three months old. After cleansing, 'tis soak'd in water for a day, then boil'd in fair water shifted six or seven times, it giving each water a very biting tast, then 'tis clear'd of its outward Membrane, soak'd again in fair water, and put into a Syrup made of fine Sugar. It draws the Sugar, say they, and leaves the water behind it to be thrown away, then 'tis put into fresh Syrup, and so several times, and not boil'd up till the last shifting, after which 'tis kept for use. Dry preserv'd Ginger is only this expos'd to the Suns Beams till dry.

Guinea Corn, and great *Indian* Corn, are ripe in three Months or a little more, from their respective plantings, Patato's in four, and Yams in twelve months.

The way to make Cane Drink.

Take six or seven long Sugar-Canes, cut them to pieces, beat them in a Mortar, put them into a Kettle, with about three Gallons of water, boil them for a pretty while, then put as many fresh Canes, and about a Gallon of water more; Boil them again. When 'tis cool strain your Drink, set it in a Jar, and put to it the white of an Egg beat to froth, to which some of the Liquor is added. Let it work twelve hours, then Bottle it, it looks very clear.

Bonano and Plantain Drinks are severally made by mashing of either of these ripe Fruits with water, till it comes to be pretty well mix'd with the Fruits, then they let it stand in a Trough twelve hours, and draw it off.

These Liquors are very much us'd for ordinary Drink in the remote Plantations, and North parts of this Island.

I was assured here, that in this part of the Island, at a place called *Wague Water*, Horses still remain wild in the Hills among the Woods, and that whensoever they are taken and forc'd to stay in the *Savannas*, or are openly expos'd to the Sun, they dye in some time.

The Introduction.

In the North-side of this Island, in the month of *January* when the *Norths* blow, they have great Rains, so that the Roads are scarce passable at that time.

I returned from the North-side to the South-side, by a Road on the ridge of Hills called *Archers-Ridge*, and by the Banks of *Orange-River*. Afterwards I came to *Rio d' Oro*, which I saw sink into the Earth, and rise up again out of it in two or three several places.

I went to *Guanaboa*, where are large Settlements and Plantations, and observed that tract of Ground called the Red Hills between *Guanaboa* and the Town. The dust of these Hills is red, apt to fly, and sticks fast to Travellers, the Soil being claiy and barren. *Guanaboa*, or as the *Spaniards* wrote it, *Guainabo*, is a fine Valley, very well planted, having rain almost every day.

On these Red Hills, four Miles from Town, lived Mr. *Barnes* a Carpenter, who used to cut and bring Wood to the Town. He told me that prickly yellow Wood which grows in great plenty here, and is describ'd hereafter, was good for nothing but to burn. He likewise assured me that the day before I was there (which was *Sept. 9. 1688.* when at Town I observed a North-Wind) at his Plantation here, there happened a thing extraordinary, *viz.* with the North came Hail and fell about his House, as big as Pullets Eggs, of various shapes, some corner'd like cut Diamonds, some shap'd like a Heart, &c. he tried to keep them in Flower, but they soon melted. This Hail beat down his high *Cassada* to the Roots, his other things were laid flat by it, it also beat down *Orange*-Trees. He carried me half a Mile up his Plantation, shew'd me the Woods wherein the *Spaniards* had usually planted their *Cassada* for the Town, after felling of the Woods. The Trees were grown, from the time the *Spaniards* had quitted the Island, to the time I saw them, to be at least forty or fifty Foot high, long small, and straight. They often in those Woods meet with Palisadoes, Orange-Walks, Limes, and other marks of formerly planted Ground. He once, ten years ago, in the Mountains in a natural hole in a Rock, found a Coffin partly corrupted, with a Body in it, he suppos'd it to have been some *Spaniard* thrown in there in hast.

Half a Mile from his Plantation, ten years ago, he found a Cave in which lay a human Body's Bones all in order, the Body having been eaten by the Ants. The Ants Nests we found there, the rest of the Cave was fill'd with Pots or Urns, wherein were Bones of Men and Children, the Pots were Oval, large, of a redish dirty colour. On the upper part of the Rim or Ledge there stood out an Ear, on which were made some Lines, the Ears were not over an Inch square, towards the top it had two parallel Lines went round, being

grosely

grosly cut in the Edges near. The *Negroes* had remov'd most of these Pots to boil their Meat in. The Cave was about eight or nine Foot Diameter, roundish, and about five Foot high, it was on a sufficiently high Precipice, of nine Foot steep Ascent before one came at it. It was before opening curiously shut in on all sides with thin, flat Stones. The Ants had eat one Carcass to the Bones, and had made holes in their ends, whereat they enter'd, I suppose, to eat the Marrow.

At *Guanaboa*, in the time of the *Spaniards*, were great Chocolate or Cacao-Nut-Walks; but after that Tree was blasted, most of them were ruin'd. The Trees wild in these ruin'd Plantations, are grown in so short time monstrously, being some of them seventy Foot of good Timber. Surveyors know all the Trees by their Barks. Those that grow low and bushy in the Commons, grow high and tall in the Woods.

At *Guanaboa* the great Rains are in *May*, and continue so till *October* from the Sea, then their Norths come in. I saw here in the Gully, a Rock upwards of sixty Foot perpendicular heighth, being the side of a Hill, which towards the Gully was steep, it was call'd the end of the World. There is another Rock much more than twice as high, towards sixteen-Mile-Walk, in the Road from the Town near the hollow Rock.

Here, on the barren sides of small rocky Hills, I saw great variety of Gourds. They serve the Island instead of Bottles, Pails, Ladles, small hoop'd Vessels, Coopers, Turners, and Glass Wares. They are of several Shapes and Sizes, from small ones, of which are made Snuff-Boxes, to such as will hold four or five Gallons. All of them, except the sweet one, which is us'd for Preserves and Sweet-Meats, are purgative. The Leaves in Clysters are given in the Belly-ach. The Bottle, and other Gourds are clean'd either by lying in the Field till dry, when by cutting off the Top, the Seeds come out, or by putting in Water, which by moistning brings the Pulp and Seeds out. If one drinks the Water that has stood in a green Gourd, 'tis very purging, but 'tis not so in one long us'd.

I saw them likewise here Preserve, or Pickle Green *Indian*-Bell-Pepper. Before it turns red, this *Capsicum* is cut and cleansed from its Seeds, then has a gentle Boil in Water, and so is put into a Pickle of Lime Juice, Salt and Water, and kept for use.

To make *China*-Drink. Take four or five handfuls of the Root cut in pieces, boil it in so many Gallons of Water, take it off the Fire, let it cool, and put two Pounds of clarify'd Sugar to it, let it stand, and after twelve Hours bottle it. It is of a red Colour, and a very pleasant Drink.

The Introduction.

In time of Sugar-making, two Quarts of clear *Moloſſes* is thought better than the Sugar.

Beyond *Guanaboa* in the Mountains, were several *Cacao*-Walks, or Plantations in the time of the *Spaniards*, but now they are ruined, only some few *Cacao*-Trees stand here and there in the Woods, as there is of Oranges, &c. the Wood about them is likewise here grown since the *Spaniards* left the Island, as high as any of the Island, being seventy Foot high of Timber.

I went to *Liguanee*, and crossed from *Paſſage-Fort*, the Arm of the Sea which comes in by *Port-Royal*.

The greatest part of the Shore of this Island, and particularly of this Bay, are full of a Tree call'd *Mangrove*, of which I shall speak hereafter. In the mean time, I think fit only to take notice that Oisters grow or stick to these Trees, not upon them like Fruit, as is vulgarly conceived, but only to so much of the Root of the *Mangrove*-Tree, as is under Water: the Tree-Oisters stick and fasten themselves, and afterwards several of them stick together, the lower down they are the bigger; so that at low Water the best is taken. They cause the Flux and Fevers when eat in excess, and taste somewhat like ours. When through any Accident these Oisters die, they corrupt, stink, and infect the Air and Wind, and are noisom to the places about them, on this account the Land-Winds are thought to bring *Port-Royal* no good Air.

Sloops may, if they know the Paſſage or Canal, go to *Paſſage-Fort* from *Port-Royal*, otherwise they cannot for the Shoals. *Men of-War* Birds, so call'd, appear in this Bay, they fly like Kites, look black, are very large wing'd in proportion to the Body, they fight with Sea-Gulls, (which are to be found here, and are like ours,) for their Prey.

Pelicans fish in this Bay, likewise in blowing Weather, when they cannot fish abroad, and in the calm Mornings they dive after their Prey. *Spaniſh* Mackarel are taken in this Bay in plenty. They are like ours, only made like a *Boneto*. I here observed a small Shoal of small Fishes to leap out of the Water, being pursued by greater Fishes.

The whole Shoals between *Port-Royal* and *Paſſage-Fort* are cover'd with Coral of several sorts, and *Alga anguſtifolia Vitrariorum* or Sea-Grass. There are also Star-Fishes of several sorts, large and five-pointed, as well as small, and several sorts of the *Echinus Marinus*. Allegators are often drawn on Shoar in the *Senne*-Nets by the Fisher-men, whose Nets are generally broken by them. These Allegators are so call'd from the word *Alagarta*, in *Spaniſh*, signifying a Lizard, of which this is an amphibious sort. When I was in *Jamaica*, there was one

of

The Introduction.

of these used to do abundance of mischief to the Peoples Cattle in the Neighbourhood of this Bay, having his regular courses to look for Prey. One of the Inhabitants there, as I was told, tied a long Cord to his Bedstead, and to the other end of the Cord fastened a piece of Wood and a Dog, so that the Allegator swallowing the Dog and piece of Wood, the latter came cross his Throat, as it was design'd, and after pulling the Bedstead to the Window, and awaking the Person in Bed, he was caught. Allegators love Dogs extreamly, but prey also on Cattle. This Allegator was nineteen feet long.

There are also Sharks to be found in the Sea hereabouts. A Man bathing in the Sea by *Port-Royal* had part of the Flesh of his Arm and Breast at one mouthful torn off by a Shark, of which he immediately died. I was told that one *Rockey* a Privateer used to go and fight with them in the Water, and so do some *Divers*, killing them with Bodkins run into their Bellies, while they turn themselves to Prey.

I saw in this Harbour and Bay a Ship come from *Guinea*, loaded with Blacks to sell. The Ship was very nasty with so many People on Board. I was assured that the *Negroes* feed on Pindals, or *Indian* Earth-Nuts, a sort of Pea or Bean producing its Pods under ground. Coming from *Guinea* hither, they are fed on these Nuts, or *Indian*-Corn boil'd whole twice a day, at eight a Clock, and four in the Afternoon, each having a Pint of Water allow'd him. The *Negroes* from *Angola* and *Gamba*, are not troubled with Worms, but those from the Gold Coast very much.

I was inform'd here that Ewes bring forth twice in fifteen months, without any regard to the time of the year; but Cows bring forth their young according to the Seasons of *Europe*.

I saw some *Guinea*-Sheep, they were brought by a Ship from that Country, being provided by the Commander to eat at Sea, but when the Ship arrived they were presented to a Planter in *Liguanee*. They are like Goats in every respect, having for the most part black and white short Hair, like that of a six weeks or a month old Calf. They are much less than Goats, multiply very fast, and are very sweet Meat.

At some Plantations bordering on this Bay many Whites die, as believed by the ill Air, some of them lying in bottoms, bordering on Marshes near the Sea. On the other hand, Plantations that are seated high are very healthy, and the People are not sickly. Colonel *Barry*'s House all galleried round, was formerly, when the *Spaniards* possess'd the Island, the only place in *Liguance* inhabited. A very rich Widow had here a Sugar-work, and abundance of Cattle in the *Savanna's*, near forty thousand.

(t) The

The Spaniards thought that in *Liguanee*, was to be found good Gold and Copper Oar, for when Sir *Thomas Lynch* sent to know of the old Inhabitants of *Jamaica* at *Cuba*, where they had found Mines in *Jamaica*, they answered in *Liguanee*, but that they never had wrought them.

When I was at *Liguanee*, I was inform'd that there was a Plantation in the Mountains belonging to Captain *Harrison*, where was a Garden the best furnished of any in the Island with *European* Garden-Plants; such as are either used for Physick, for the Kitchen, or for Ornament. The high situation of the place made it fitter for the Production of these Vegetables, because the higher the cooler, and that generally there are more Rains and Showers on Mountains than in the Valleys. Here follows a List of the *European* Plants I met with in this Garden, and of those which I observ'd to grow in other parts of the Island. They all thrive almost as well as in *Europe*, save Wheat, Oats, and Apples.

Apium hortense seu petroselinum vulgo. C. B. pin. p. 153. Common Parsley.

Absynthium Ponticum, seu Romanum officinarum, seu Dioscoridis. C. B. pin. p. 138. Common-Wormwoood.

Artemisia vulgaris major. C. B. pin. p. 137. Common-Mugwort.

Melissa hortensis. C. B. p. 229. Bawm.

Lavendula major sive vulgaris. Park. p. 73. Lavender.

Thymus vulgaris folio tenuiore. C. B. pin. p. 219. Thyme.

Satureia hortensis. Park. p. 4. Savory.

Hyssopus sativus vulgaris. Lugd. p. 933. Hyssop.

Pulegium latifolium. C. B. pin. p. 222. Penny-Royal.

Tanacetum vulgare. Park. 462. Tansie.

Carduus Benedictus, Park. p. 530. Carduus.

Horminum sclarea dictum. C. B. pin. p. 238. Clary.

Borago. Cam. hort. p. 29. Borage.

Buglossum. Park. p. 239. Buglofs.

Pimpinella sanguisorba minor hirsuta. C. B. pin. p. 160. Burnet.

Melo vulgaris. C. B. p. 310. Musk-Melons.

Asparagus sativus. Ger. Sparagus.

Cinara hortensis foliis non aculeatis. C. B. pin. p. 383. Artichokes.

Pastinaca sativa sive carota lutea & alba. J. B. Tom. 3. lib. 27. p. 60.

Pastinaca sativa sive carota rubra. Ejusd. 16. Yellow, White, and Red Carrots.

Nasturtium hortense vulgatum. C. B. pin. p. 130. Common-Garden-Cresses

Mentha

The Introduction. lxxv

Mentha prima sive rubra. Park. par. p. 480. Garden-Mint.
Fragaria vulgaris. C. B. pin. p. 526. Strawberry.
Smilax hortensis sive phaseolus major. C B. pin. p. 339. Kidney-Beans or French-Beans.
Raphanus minor oblongus. C. B. p. 96. Garden-Radish.
— — *Rusticanus.* C. B. pin. p. 96. Horse-Radish.
Brassica vulgaris sativa. Lob. obf. p. 122. Icon. p. 243. Coleworts.
Lactuca sativa. C. B. pin. p. 122. Garden-Lettice.
Glycyrrhiza vulgaris. Dod. p. 341. Liquorice.
Ruta hortensis major latifolia. Morif. hift. p. 507. Rue.
Calendula sativa. Tab. tom. 2. p. 49. Marygold.
Cucumis sativus vulgaris. C. B. pin. p. 310. Cucumbers.
Rosa Provincialis sive Damascena. Ger. Province Roses. These grow very well in *Barbados.*
Rosmarinus. Riv. p. 10. Rosemary.
Vitis Venifera. Lac. p. 502. The Vine.
Pomus sive malum. C. B. pin. p. 432. The Apple-Tree.
Salvia major vulgaris. Park. p. 40. Sage.
Brassica capitata alba. C. B. pin. p. 111. Cabbage.
Ficus communis. C. B. pin. p. 557. The Fig-Tree.
Acetosa pratensis. C. B. pin. p. 114. Sorrel.
Marrubium nigrum fœtidum ballote Dioscoridis. C. B. pin. p. 23.
Avena vulgaris sive alba. C. B. pin. p. 23. Oats.
Rapa sativa rotunda. C. B. pin. p. 89. Round-Turnep.
— — *Oblonga.* Ejusd. p. 90. Long-Turnep.
Pisum arvense. C. B. p. 342. Field-Pease.
Cepe vulgaris. C. B. pin. 71. Onion.
Sinapi sativum. Ger. Mustard.
Triticum. Adv. p. 5. Wheat.
Rubus Idæus. Rasberries. This was the same with that of *Europe*, only more creeping.

Flowers.

Tanacetum Africanum majus flore pleno. C. B.
— — *Minus flore simplici.* C. B.
Amaranthus simplici panicula. C. B. pin. p. 121. Amaranth.
— — *Panicula incurva holosericea.* Ger.
Hyacinthus Indicus tuberosa radice. Clus.rar.pl.hift.p.176. Tuberoses.

I went to Colonel *Crew's* Plantation, *May*, 19. they were setting after a days Rain, some clay-colour'd and red Pease. One or more of the *Negroes* went with a How, and scratch'd up some Earth, and a little after followed another Black, who put into the hole three Pease, and then with his Foot cover'd them with the mould scratch'd

off

off by the former. One or two of the Peafe is fuppofed to be eaten by Vermin. They were planted at about two Foot diftance, and are ripe in two months, gathered when dry, and boiled as Peafe in *England*. They are good nourifhment for *Negroes*. They are uncertain in Seafons, or Rains here, therefore plant no Cotton.

Indian Corn cannot be fet to grow but after Rain. It is beft to be hung up in its Membranes to be hardened and kept from *Weevils*.

There is a fort of White *Caffada* not poyfonous, which boil'd as Yams are, may be eat like them.

I was inform'd here that Snails Calcin'd, and a water made of them like Lime-water, it is a good Remedy in Bloody-Fluxes.

For the better underftanding of feveral matters in the *Weft-Indies*, I think it proper to fubjoin fome accounts I received from feveral Credible Perfons who had lived, and made Voyages to feveral parts in the Neighbourhood of *Jamaica*. Thefe follow without any other order than that of the time they were told me, and enter'd in my Journal which was generally when the Perfons came upon their firft arrival to wait on the Duke of *Albemarle* as Governour of the Ifland.

One King *Jeremy* came from the *Mofquitos* (an *Indian* People near the Provinces of *Nicaragua*, *Honduras*, and *Cofta Rica*) he pretended to be a King there, and came from the others of his Country, to beg of the Duke of *Albemarle*, Governor of *Jamaica*, his Protection, and that he would fend a Governour thither, with a power to War on the *Spaniards*, and Pirats. This he alleged to be due to his Country from the Crown of *England*, who had in the Reign of King *Charles* I. fubmitted itfelf to him. The Duke of *Albemarle* did nothing in this matter, being afraid it might be a trick of fome people to fet up a Government for *Bucaniers* or Pirats. This King *Jeremy*, in coming to Town, asking many queftions about the Ifland, and not receiving as he thought, a fatisfactory account, he pull'd off his *European* Cloaths his Friends had put on, and climb'd to the top of a Tree, to take a view of the Country. The Memorial, and fubftance of what he, and the people with him, reprefented to the Duke of *Albemarle*; was, That in the Reign of King *Charles* I. of ever Bleffed Memory, the Earl of *Warwick* (by virtue of Letters of Reprizal granted by his faid Majefty for Damages received from the Subjects of his Catholick Majefty) did poffefs himfelf of feveral Iflands in the *Weft-Indies*, particularly that of *Providence*, (fince called by the *Spaniards* St. *Catalina*,) which is fituate in 13 deg. 10 m. N° Lat. lying Eaft from Cape *Gratias de Dios*, (vulgarly known by the name of the *Muskitos*) between Thirty and Forty Leagues; which put the faid Earl upon trying all ways and means of future

Corre-

Correspondence with the Natives of the said Cape and neigbouring Country, and in some little time was so successful as to gain that Point, and farther prevail'd with them so far, as to persuade them to send home the King's Son, leaving one of his People as Hostage for him, which was Colonel *Morris*, now living at *New York*. The *Indian* Prince going home with the said Earl, staid in *England* three years, in which time the *Indian* King died, and the said Natives having in that time had intercourse of Friendship and Commerce with those of *Providence*, were soon made sensible of the Grandeur of his Majesty of *Great Britain*, and how necessary his Protection was to them. Upon the return of the said *Indian* Prince, they persuaded him to resign up his Authority and Power over them, and (with them) unanimously declare themselves the Subjects of his said Majesty of *Great Britain*, in which Opinion they have ever since persisted, and do own no other Supream Command over them. As to the Fertility of their Soil, 'tis a very great Level, free from any Mountains for several Leagues from the Sea, the Soil black Mold mixt with Sand, where otherwise, the Land is covered with Pine Trees, of the nature of *New England* Fir, well watered with great Rivers and Rivulets. Their chief Commerce being managed in Canoes; for Harbours it hath but two, and they both barr'd, and so not capable of receiving any Vessel that draws above Eleven or Twelve Foot Water, but when within, able to receive Ships of the greatest Burthen, that may lay their Sides to the Shore, and Careen safely. In these Harbours are Banks or rather Rocks of Oisters, from the bigness of Horse-shoes, to those of *Colchester*, or less. As to the Nature of the Inhabitants, they are Affable and Courteous, very Hospitable, and ready to relieve all People in Distress, but more especially endear'd to those of the *English* Nation. The Men generally speak broken *English*, there is nothing more hateful to them than breach of Promise, or telling an Untruth, their Words being inviolable. They are always on their Guard for Fear of their neighbouring Enemies, and for the most part get their Living by Fishing. Their most usual Arms are six hand Lances slight ones, and one other very large and strong, with which they nimbly avoid those that are darted from their Enemy, the lesser ones they handle so dextrously, that they dart small Fishes, about the bigness of a Salmon Trout, Thirty Yards, and though their Country have great plenty of Deer, Pecary, and Wild-Fowl; yet they get most of their Provisions out of the Sea. Their manner of living now is Patriarchal, their Families being numerous, they allowing plurality of Wives, of which only the old Women and Children perform the Work of all the rest of the Family, but every individual Person pays great Duty and Respect to the *Paterfamilias*, who is absolute Judge in all Cases, and from

(u) whom

whom there is no Appeal. All they produce from the Earth, of Provisions (which they fell to the Men of War, and other Veſſels trading on the Coaſt) is equally divided among the whole Family, only the Chief and his Wife have each two Shares. Were ſome Perſons induſtrious to ſettle among them, and encourage them to plant, the richneſs of the Soil would eaſily, and advantageouſly produce any Commodity, eſpecially Indico, but they will not permit any other Nation to ſettle among them but the *Engliſh*; they have ſome propenſity to the *Dutch*, but the *French* they mortally hate for their wanton behaviour towards their Wives. As to their Number, no ſure account can be given, they being ſettled at ſuch great diſtance, and uncapable to give a true Eſtimate of themſelves, being wholly unlearned, only ſome that have been at *Providence*, have learned the Lords Prayer, the Creed and Ten Commandments, which they repeat with great Devotion.

One Captain *Gough*, who had lived there, told me they had thereabout much Cochineel-Tree planted both by the *Indians* and *Spaniards*, ſometimes in Fields of Fifty or Sixty Acres of Ground: that they keep theſe *Opuntia* or Trees very clean, that the Inſects may breed on them; that this ſort grows very high like prickly Pears, only has no Prickles or very few: that the Inſects come from another Tree, and that they appear on the Surface of theſe Plants, in form of little Bladders, which they ſweep down into an Iron Pan, which afterwards being ſet on the Fire, leaves ſomething like a Spiders Web. Afterwards they put this Cochineel into Cheſts as cured; if it be not enough dried, it takes life and flies away. I ſhall have occaſion to ſpeak more of this hereafter.

He told me alſo *Vaniglias* grew here, and are cured by taking them off the Vine (which runs very far) at a certain ripeneſs, dipping them into hot Water, and drying them in the ſhade. If they be pulled too young they break, and are brittle; if too old they open, which they do of themſelves on the Trees. Another Perſon told me the *Vaniglias* were cured after the following manner, and I am the more willing to publiſh it, becauſe they are ſaid to grow in *Jamaica*, and that they cannot be ſent from thence to *Europe*, becauſe of their Ignorance of the way of curing them.

Another Way to Cure Vaniglias.

Gather them when full ripe, prepare a Liquor, or Brine of Water and Salt, ſo ſtrong as to bear an Egg, then put to it a fourth part of Chamber-ly, and a reaſonable quantity of unſlak'd Lime, and when that is incorporated, boil all together about half an Hour, then take it off, and put the *Vaniglias* into the Liquor, and let them remain

main there till they are throughly scalded or parboil'd; then take them out, and dry them in the shade where no Sun may come at them.

The same Person spoke of a Gum or Balsam, called *China*-Balsam, growing in the aforesaid *Mosquitos* Country, which is procured by applying Fire to one side of the Tree, and gashing the other, at which gashes a black Balsam sweats out, very proper for Wounds.

They make use of Plantain or *Musa*-Leaves for Table Cloths, and Napkins.

The Women live very much in awe and submission to their Husbands in this Country.

Towards the *Havana*, in *Cuba*, there are abundance of Spouts to be seen, more than in any other part of the *West Indies*. It is all plain, level, and very fertile Ground thereabouts. I was told that the *English* Prisoners taken, as Pirates or Traders, and kept by the *Spaniards*, by their several Artifices, and Skill in Mechanics, get a very good Subsistence.

In the Lake of *Maracaybo*, Sir *Henry Morgan* told me, afar off he once saw a thick Cloud, which when he came near, he found to be *Mosquitos* or Gnats. The Country thereabout is so marish and wet, as that the Inhabitants are forc'd to build their Houses on the Trees, as Ants do in many places for the same reason.

I perused here at *Jamaica*, a Journal of Sir *William Phipps*, which gave an account of the first finding of the great Plate-Wreck to the North-East of *Hispaniola*. After Sir *William Phipps* had been at *Samana*, on the North side of *Hispaniola*, he went with one *Rogers*, Master of a small Ship to *Porto Plata*, and there discharging three Guns to get the *Spaniards* to Trade, they came down, and the *English* sold small Bables, and Searges for Hides, and jirked Hogs taken by the Hunters there. In the mean time *Rogers* had been on the Wreck, discover'd it by means of a Sea-Feather, growing on the Planks of the Ship lying under water, and brought from thence the news of its being found. They went thither, found it grown over with Coral, and *Lapis Astroites*, and took up Silver as the Weather and their Divers held out, some days more, and some less. The small Ship went near, the great one rode afar off. At last they got in Bullion 22196 *l.* in Coin 30326. of which were Sows, and great Bars 336. After they sail'd for *Turks* Islands for Salt, and going thither, after several hours sailing, had almost been a-ground and wreck'd on the Handkercher Shoal. They about the Wreck were sometimes in seven Fathom Water, and immediately almost out of reach of the bottom by founding.

This

The Introduction.

This Wreck had been a *Spanish* Galeon lost on these Shoals, near the *Abreojos* or *Handkercher* Shoals, to the North-East of *Hispaniola*, about the year 1659. bound for *Spain*. The Inhabitants of *Hispaniola*, who used to Trade with Sir *William Phipps*, had acquainted him with it. He proposed the taking up of this Silver to the Duke of *Albemarle*, who together with Sir *James Hayes*, Mr. *Nicholson*, and others, set out two Ships, a greater and a lesser, laden with Goods to Trade with *Hispaniola*, and the *Spaniards* in the *West Indies*, in case they failed of the Wreck. They found this Wreck, as is above related, and wrought on it till the Ships Crew grew scarce of Provisions, when they had taken up about Twenty six Tuns of Silver. A Sloop from *Bermudas* came to their help: when they fail'd for *England* the Sloop return'd to *Bermudas*, and there disclosed the Matter, which soon went to the other Islands. From these parts, and *Jamaica*, Sloops and *Divers* were sent, who took up a vast quantity more of Plate and Money, so that before a second Fleet came from *England*, the greatest part of what Silver remain'd unfish'd was taken up. Not only the *English* from the Plantations and *England*, but the Prince of *Orange*, afterwards King *William*, from *Holland*, equip'd a Ship which was sent thither, but they came too late. Those who commanded the *Dutch* Ship, and Sir *John Narborough*, who was in the *English*, return'd without any considerable Cargoes of Silver. It happened so not only to the first Patentees, but to many other People, who by the example of this Project (where the Duke of *Albemarle* received Fifty thousand Pound for Eight hundred, and others in proportion) hoping for the same Success, took out Patents for Wrecks lying at the bottom of the Seas in all places, especially in the *West Indies*, where any Traffick is used, not considering that though there have been lost divers Ships laden with Money, on many Shoals of the *West-Indies*, such as the *Serranillas* between *Jamaica* and the Continent, the *Bahama* Shoals, &c. yet in most parts there is such a Vegetation of Coralline matter out of the Sea-water, as that the bottom of the Sea is incrustated with it, and the Wrecks hid by them. The Pieces of Eight in the Silver-Wreck above mention'd, that was lost in 1659. were covered with this Matter about a quarter of an Inch thick, and I have a piece of the Timber of the Ship, with an Iron Bolt in it, grown over with the *Corallium asperum candicans adulterinum* J. B. and some of the Pieces of Eight incrustated, others almost covered with *Astroites*. Those underneath were corroded with the Sea-water, so that many of them stuck together. These things I have caused, at least some of them, to be graved. It is not only from this, but also Sand driving by the Winds and Currents, or Earthquakes that happen at the bottom of the Sea, that Wrecks may be cover'd, and past finding out. I remember

an

an *African* Ship, laden with Elephants Teeth, wreck'd on the Coast of *Sussex*, which Mr. *Halley* told me was in a very short time almost covered with Sand and Oase, so that the Project of recovering the Teeth, was frustrated, though by the help of a Diving-Bell, contriv'd by his extraordinary Skill, they had gone to the bottom of the Sea, and into the Ship, where they had a perfect view of the Ship, and all about it. Though the Money brought into *England* from the first Wreck was very considerable, yet much more was lost on Projects of the same nature. For every silly Story of a rich Ship lost, was credited, a Patent taken out, *Divers*, who are us'd to Pearl-fishing, &c. and can stay under Water some Minutes, bought or hir'd at great Rates, and a Ship set out for bringing home Silver. There was one Ship lost amongst the rest, said to be very rich, near *Bermudas*, which was divided into Shares and sold. It was said to be in the Possession of the Devil, and they told Stories how he kept it. I do not find the People, who spent their Money, on this, or any of these Projects, excepting the first, got any thing by them.

Colonel *Nedham*, who had liv'd some time in *Teneriff*, told me, that in the Year 1649. Locusts destroy'd all the Product of that Island; they saw them come off from the Coast of *Barbary*, the Wind being a Levant from thence; they flew so far as they could, then one lighted in the Sea, and another on it, so that one after another they made a heap as big as the greatest Ship above Water, and were esteemed almost as many under. Those above Water, next Day, after the Sun's refreshing them, took flight again, and came in Clouds to the Island, from whence they had perceiv'd them in the Air, and had by their Military Officers gathered all the Soldiers of the Island and *La Laguna* together, being seven or eight thousand Men, who laying aside their Arms, some took Bags, some Spades, and having notice by their Scouts from the Hills, where they alighted, they went strait thither, made Trenches, and brought their Bags full, and cover'd them with Mould. This did not do, for some of the Locusts escap'd, or being cast on the Shoar, were reviv'd by the Sun, and flew about and destroyed all the Vineyards and Trees. They eat the Leaves, and even the Bark of the Vines where they alighted. After two Months fruitless management of them so, the Ecclesiasticks took them in hand by Penances, with Swords tied to their Arms, voluntary Whippings, &c. by Excommunications by Bell, Book and Candle, by sprinkling with Holy-Water, Processions and Crosses; amongst the rest one voluntary Penance was rolling round the Hands and Arms, Feet, Legs, and Body with hard Ropes; but all would not do, the Locusts staid there four Months;

(x)

Cattle

Cattle eat them and died, and so did several Men, and others struck out in Botches. The other Canary Islands were so troubled also, they were forced to bury their Provisions for fear of being oblig'd to relieve their common Necessity with what was to produce for their own Families. They were troubled forty Years before with the like Plague.

A very credible Person, on the Agreement of the *Assiento* or *Spanish West-India* Company, with the Merchants call'd *Grillo's* of *Genoa*, and of them with the Royal *African* Company, went to *Cartagena*, in a *Spanish* Ship, with Five hundred *Negroes*, he was sent from thence to *Porto-Bello*, where they could not get Liberty to go on Shoar, but received their Money and good Entertainments in their Ship. They return'd to *Cartagena*, where buying *Spanish* Habits, they went about the Town. It is, as he told me, twice as big as *Port-Royal*, stands almost encompass'd by the Sea, is wall'd with Stone, and has several Forts or Castles in it. Its Houses are built of Stone or Brick two Stories high, with Balconies and Roofs jetting over them. Things here are twice as cheap as at *Porto-Bello*, because of the vent of Commodities from this last place to *Panama*.

When the *English*, under the Command of Sir *Henry Morgan*, came from *Porto Bello*, and *Panama*, (which places they had taken) after certain contagious Fevers, they, for the most part, fell into the Yellow-Jaundice, grew worse and died of it, after languishing a great while in the greatest Degree of it, which is commonly call'd the Black-Jaundice. They look'd with the Jaundice like *Indians*, and were, when remediable, chiefly cured by the Infusion of Goose-Dung.

A Sea-man related that he washing his blue Jacket on the Fore-castle, coming hither, the Ship having fresh way he lost it, but two Days after, having been becalm'd, they took a Shark, and found in his Belly the blue Jacket, not otherwise alter'd than by the holes of his Teeth in chawing.

Several Persons who used the Logwood Trade, or who were imploy'd in cutting that Wood, otherwise call'd *Campeche*-Wood, used by Dyers, inform'd me, that at about fifteen Leagues from the Town of *Campeche*, are two Creeks, the Eastern and Northern, in which last they cut Logwood. This is call'd the *Logwood*-River, the Inhabitants live in Huts on each side of this narrow Creek, near Two hundred *English*; and are ready on the appearance of any Enemy to hinder their landing by firing on them on each side, every

one

The Introduction. lxxxiii

one having his Firelock and other Arms ready. It is a knotted crooked Wood, growing in Marshes, three or four together up the two Creeks, or *Lagunas*, about eight Leagues from where the Shipping Rides; it is very hard, and bears a small Leaf like a Heart. They Saw it down, then cut pieces of it about four or five Foot long, then cleave it. It is of a dark or purple, near a black colour. The *English*, who have lived there many years, Cut and Sell it to the Sloops for about Three Pound *per* Tun, for which the Sloops bring them Cloathing, Victuals, Rum, Sugar, &c. The Sloops carry this Wood, and sell it at *Port-Royal* for about Six Pound *per* Tun; the half of the Profit going to the Master of the Sloop. When any of the *English* at *Campeche* resolve to come away, they having got Logwood, it may be Thirty or Forty Tun, they embark it and themselves in a Sloop for *Jamaica*, where the half Profits go to themselves, and the half to the Master, otherwise they send it, and paying the Fraight, *viz.* the half Profits, their Money is return'd them. The *Indians* of this place us'd formerly to Trade with them, but the *English* not keeping their Faith, but taking and selling them, they are retired up into the Country several Leagues. There are on an Island near this, wild Cows and Bulls in abundance; there are also wild Deer near this River. The *Spaniards* who are offended at this settlement equipp'd some *Periaguas* and Hulks against them; but before they were ready they were burnt by the *English*, since they only lie out at Sea off this place cruising on their Sloops and Merchant Men. The *English* have a place stronger than their Huts for their Provision, and when a Strength much greater than theirs comes against them, they retire to the Woods. They have been cut off several times by the *Spaniards* in this place, and yet have settled here again. This usage of the *Spaniards* is somewhat harsh, if what Sir *Henry Morgan* has often told me be true, that this Logwood River was in the Possession of the *English* at the time of the Treaties being sign'd at *Madrid* concerning the *West-Indies*. The Ships lie Eight Leagues from the cutting place, and the Wood is carried to them by Long-Boats and Sloops.

I was told that the Pearl-Fishing of the *Spaniards* at *Margarita* was fail'd, but that there was a fishing at *Rio de la Hacha* not far of. The *Indians* Dive and bring up the Pearl-Oisters, they dare not take any Rake or Iron to Drudge them up, for fear of destroying the young breed, under pain of High Treason; the biggest lies in deepest water, they Dive on the Banks in Nine Fathom water. The *Indians* String the firm part of the Oisters on Lines, and dry them against a Wall, and when they are dry and transparent, they are
eaten

eaten by the *Spaniards*. The Pearls are fold by the *Indians* to the *Spaniards* by the Shell full. They are dark more or lefs, and are not of fo clear a colour as the Oriental Pearls are. When the *Indians* find very large Pearls, they keep and hide them till they find a better Merchant, which are often others than the *Spaniards*. This Pearl fifhery was farmed of the *Spaniards* by *Englifh* Merchants of *Jamaica* when I was there.

New-England Horfes are frequently brought to *Jamaica*, they are bought for Five Pound apiece in *New-England*, and kept by the way on Bran, they ufually are fold in *Jamaica* at Fifteen Pounds, they are rougher than the Horfes in the Ifland ufually Pace, and lofe their Hair at firft coming.

The Inhabitants of *New-England* Pickle Pork and Beef either dry or wet, the firft is done in bad Cask, the other in good Cask, and is much the better, the firft proving ordinarily rufty.

Fifh preferv'd and cur'd, both dry and wet, come from thence.

One half Barrel of *Irifh* Pork is worth about Twenty two Shillings, and a Barrel of Beef as much, which is in Provifion, to be diftributed for the fupport of Servants and Slaves, reckoned equivalent to it.

Salt wet Mackerel comes from *New-England*, and is much ufed in this Ifland for the fupport of Slaves and Servants.

The true way of fatting Cattle, as I was inform'd by the Grafiers of *Jamaica*, is by bleeding them in the Jugular Vein (which will ftop of its felf) and then purging them with Aloe or *Sempervive* Leaves clear'd of their outward Skins, and thruft down by Gobbets till a whole Leaf is fwallowed.

The fame has been effectual in a Man, in reftoring the tone of his Stomach loft by drinking It purges Cattle and Men of Worms, and may make them fat that way.

The lefs Nourifhment the Grafs affords, the greater the Paunch of the Beaft feeding on it, fo that the Bellies of Cattle are fo large in dry times in hot Countries, as if they were big with young.

On *Hifpaniola*, at *Samana* are many *French* Hunters. They go out Twelve in a Company, for fear of the *Spaniards*. The Hogs they take have fometimes Stones in their Bladders; one of thefe Stones I had brought me thence was long, of divers Tunicles, the outward white, thofe within it reddifh, and fmelling very ftrong of Urine. I had two others bigger than any Peafe, round, and fet about with protuberant Prickles they were taken out of the *Urethra* of Hogs. They were fhining and Cryftalline, and exactly refembled thofe taken out of Human Bodies. At

At *Samana* the *French* setled several Families, who were cut off by the *Spaniards*, of a Town called *Isabella*, of three or four hundred Inhabitants. They have Horses and Mules in the *Savannas* about it. The *English* at *Jamaica* buy most of their *Mules* from the *Spaniards*. It is not far from *Porto-Plata* where the *Spaniards* are also settled. The *Moiati* Stones are brought from thence, and taken from behind the Ears of that Animal, each Ear having one.

One Doctor *Fritz* a *German* Chirurgeon and Chymist, who had been in the South-Seas with Captain *Townley*, told me he had seen of the *Cascarilla*-Trees, or those on which grows the *Peru* or Jesuits Bark: that they grow near the Sea side and are very large Trees; that they cut a piece of the Bark round the Trunk of the Tree near the ground; that a while after the Bark withers on the whole Tree, and falls off; that they had taken a Ship named the *Cascarilla*, and had thrown most of that Bark, her loading, into the Sea, to make room for Provisions. That the Leaves generally found in the Bags of the Bark, was of that Tree, and Plenty of the Trees is to be met with in the South-Sea.

A *Bristol* Ship, coming towards *Jamaica*, struck on a Rock two Miles from without the Town of *Port-Royal*, but this Ship when lighted of the Goods was got off again. This is very ordinary, for the Rocks and Shoals being here-about covered with Coral and Coralline Substances, the Ships coming upon them, are not often pierc'd nor bulg'd, but bruise these Coralline Substances to Sand, and very often get off again without much damage.

Captain *Groves* told me he had left a while ago on the Island of *Tobago*, seventy *Curlanders* in a Fort on that Island, that their Food was *European* Bread, which they had provided for several Months, wild Hogs, Hogs with their Navels on their Backs, *Armadillos* an excellent Food, and *Racoons*, with very great plenty of Fish. Their Trade was as yet chiefly cutting of very large Mastick, Cedar and other Trees for Timber, to be sent to the Island of *Barbados*, with which by Licence from the King, they might Trade. This Island is, I think, held by the *Curlanders* of the Crown of *England*.

They have there a sort of Pleasant, very good Meat.

He told me that having been often in the *Mediterranean*, at certain Seasons some sort of wild Pigeons were so plentiful on the Isle of *Zante*, as the Inhabitants sold them for half pence apiece, that they came from *Egypt*, and that on the Island *Caprea*, near *Naples* Quails are sold at a certain Season at the same rate. Pigeons are at *Jamaica* very plentiful at some Seasons. Ground-Doves are sold for a Bitt, or Real a Dozen. They are taken with *Clavanues*, and wild

(y) *Cassada*

The Introduction.

Caſſada Seeds for Bait. The Cane-Rats are numerous, of a gray colour, cheap, large, and very good Victuals.

I was inform'd that *Sarſaparilla* is very frequent and cheap up *Rio San Pedro*, in the Bay of *Honduras*, where are several *Indian* Towns. There is brought into *Jamaica* great Quantities of *Sarſaparilla*, by Trade with the Bay of *Honduras*, *New Spain* and *Peru*. It grows in all theſe places on the Banks of the Rivers, and in moiſt Grounds: The *Spaniards* think it makes the Water of thoſe Rivers, where it grows wholeſome. It is a ſort of *Smilax aſpera*, and comes very near to that common in *Spain*, *France* and *Italy*, though it differ from it. It is alſo very near akin to *China*, the Strings or *Sarments* of the Roots of *Sarſaparilla*, taking their Original from a knobby *Tuberous* Root, like that of *China*, and going very deep into the Ground. It is moderately warm, uſed to promote an eaſie Sweat, and open Obſtructions. There is an Account and Figure of it publiſhed in a Book printed at *Mexico* in the year 1570. in *Latin*, written by *Franciſcus Brayus*, a Phyſician, who lived there, whither I refer the Curious.

I went from *Port-Royal* in a Boat to *Houſe Key* and *Gun-Key*, or *Cayo*, ſo called from the Tryal of Guns from the Fort; they ſhooting thence at a Cask ſet up here. They were defended by Coral and Aſtroites Rocks to the South, and were very ſmall Iſlands, with ſome few Buſhes on them. Such places are called by the *Spaniards Cayos*, whence by corruption comes the *Engliſh* word *Keys*, uſed to ſignifie ſuch places in thoſe parts.

A Maſter of a Veſſel from *Barbados*, bound to ſeek a new Wreck, came into *Jamaica*, and told me he had been in *Crab-Iſland* near *Porto Rico*, ſo called by our Seamen, from the great number of Land-Crabs on it: in the Charts 'tis named *Borrinquem*. A little before his being there, two Ships appear'd before the Iſland with *Engliſh* Colours, but coming nearer put out *Spaniſh* Colours, and ſent five *Engliſhmen* on Shore with a Flag of Truce, who enticed the Commander in Chief on board the Ship. When there they forc'd him to write on Shore for the Inhabitants he had left, to tell them of his civil Treatment, who all went on board likewiſe, and were carried to *Santo Domingo* Priſoners; they were reported to be one hundred and fifty in all, Men, Women, and ſome few *Negros*. They burnt all the Houſes, which were of Timber. A *Negro* lying in the Woods, made his eſcape to the North ſide of that Iſland, and embarq'd in a Sloop which lay there for the Windward Iſlands, where he gave this Relation; the like eſcape had five other white Men in a Sloop, that lay in another place, where they were cutting Timber to carry to the Windward Iſlands.

The Inhabitants had been there ſeveral years, and had come from the Leeward Iſlands, chiefly *Anguilla*, to ſettle *Borriquem*. *Anguilla* is rocky

rocky and barren, and this fruitful and rich in Soil, and easily to be clear'd, and of a very rich Mould, well water'd with Rivers, and Springs to be had for digging a few Feet deep. They had planted *Caſſada*, *Yams*, *Patatas*, *Indian*-Corn, and other Proviſions, with ſome few Sugar Canes. Since this the *Engliſh* have again poſſeſſed this Iſland, and I am told have now Captain *Sharp*, formerly an *Engliſh Commander* in the South Seas, for their *Commander*. The Governor of *Anguilla* had beat ſome *Spaniards* off with the loſs of ſome of their Men, with the help of only fifty Soldiers, and had deſir'd leave of the Governour of the Windward Iſlands to ſettle it again. The *Spaniards* in this Attempt were thought to have had *Bear* a Pirat with them, and to be ſent out by the Governour of *Santo Domingo*. The *Spaniards* are very barbarous to all Nations in theſe parts where they are ſuperior. They think they have the only right to the *Weſt-Indies*, and it was a long time ere they would hear of any Treaty with *European* Nations. This was the firſt riſe of the Privateers, Bucaniers or Freebooters, who conſiſted of all Nations except *Spaniards*, from whom they often took great Riches, and as eaſily parted with them to the people of their own Nations.

A Maſter of a Sloop from *Anguilla*, related to me that this *Bear* a notorious *Engliſh* Pirat, under a *Spaniſh* Commiſſion, had made a deſcent on that Iſland, and barbarouſly handled threeſcore Families of *Engliſh* which lived: in it ſuch treatment is very ordinary in theſe parts of the world where the Inhabitants are not able to defend themſelves, and ſmall Iſlands often on this account change Maſters. This Iſland is reckoned amongſt the *Caribe* or *Leeward* Iſlands, and is under the Government of their Captain General or Governor.

Some Turtlers being at the South *Cayos* off of *Cuba*, had been robb'd by a *Periagua*, ſent by the Governor of *Havana*; they were ſtript, and the *Spaniards* talk'd of putting them all into one Sloop, and burning them in it. The Turtle-fiſhery there, and at the *Caymanes* were thought by Sir *Henry Morgan* to be ours by right, ſeeing it could be prov'd by ſeveral at *Jamaica*, that thoſe were in the Poſſeſſion of the *Engliſh* at the time when the Treaty between *Spain* and *England* concerning the *Weſt-Indies* was ended at *Madrid*. The ſame is to be ſaid of the Turtle-fiſhing at the Iſland *Vaches*, off of *Hiſpaniola*, pretended to by the *French* of the Iſland *Tortugas*. This Iſland *Tortugas* or *Tortue*, lies a little off the North-Weſt ſide of *Hiſpaniola*. It was planted by the *French* ſome years ago, and from that ſmall Iſland they have ſpread themſelves over a great part of the North-ſide of that large Iſland, where they have a Governor ſtiled *Gouverneur des Cotes de St. Domingue*. They have pretended lately to a right to the Iſle *des Vaches* or *Aſh*.

There

The Introduction.

There is a fort of Loggerhead Turtle or Tortoise at *Jamaica*, very little differing from the common fort, only in every part leſs, and having the Breſt of a yellowiſh white colour.

The Turtle or Tortoiſes come to *Caymanes* two ſmall Iſles Weſt of *Jamaica*, once a year to lay their Eggs in the Sand, to be hatch'd by the Sun, and at that time the Turtlers take them in great numbers. At other times they go to the South-Cayos off of *Cuba*; there to feed on the Sea Graſs growing under water: wherefore the Turtlers go thither in queſt of them, and it may be four Men in a Sloop may bring in thirty, forty, or fifty Turtles, worth ſeventeen or eighteen Shillings apiece, more or leſs, according to their goodneſs. The Female with Egg is reckoned the beſt. They ſometimes get their Loading in a day, but are uſually ſix weeks in making their Voyages. The Turtlers feed on Turtle, Bisket-Bread and Salt. They catch the Turtle with Nets made of Yam larger than Whipcord. When they come home to *Port-Royal* they put them into the Sea, in Fourſquare-Paliſadoed places, where they keep them alive till there be occaſion to kill them, which will be very long ſometimes, though the ſooner they are killed after taking, they are the fatter. The Callepee, or under part of the Breſt or Belly, bak'd, is reckon'd the beſt piece, the Liver and Fat are counted Delicacies. Thoſe who feed much on them ſweat out a yellow *Serum*, eſpecially under their Armpits. Their Fat is yellow, taſts like Marrow, and gives the Skin a yellow Hue or Tincture.

Saltertudos, is corruptly called ſo, it being, properly ſpeaking, the Iſland *Tortuga*, near *Margarita*: 'tis uninhabited, but has ſeveral Salt-Ponds, filled with Salt, reputed very good, it being large grained. 'Tis always to be found there in great quantities, except about *July*, when the Rains moiſten and diſſolve it. The *New England* Veſſels carry *Lumber* and Fiſh to *Barbados*, and from thence go for this Iſland, and ſtretch it thence again ſtreight home. They carry with them Wheelbarrows, and Bags to load this Salt. It is reckon'd much better than the Salt of *Jamaica* commonly ſold here, and looks reddiſh. The Governour of *Barbados* takes Salt *Tertudos* to be under his Dominion.

The *Bahama* Iſlands are fill'd with Seals, ſometimes Fiſhers will catch one hundred in a night. They try or melt them, and bring off their Oil for Lamps to theſe Iſlands.

There are Iſlands lying North off of *Hiſpaniola*, where are many Salt-Ponds; but becauſe of the Neighbourhood of the *French* and *Spaniards*, they are accounted more dangerous than the other. They are called the *Turks* Iſland.

A Ship came into *Jamaica* from *Carolina* with Beef and Pork. A curious Person on board related to me, that that Country abounds in every thing for Food. That they chiefly plant *Indian*-Corn. That he had travell'd from *Palatzo* or the *Apalathean* Mountains, to St. *Augustin* in the *Spanish* Dominions, and that the whole Country was level, most sandy and barren, except about five Miles extent near the Rivers, where it was planted with *Indian*-Corn, and some Wheat for the Priests. That the Fort of St. *Augustin* had Twenty four Guns in it, and that the *Indians* paid yearly to the *Spaniards* a piece of Eight a Head Tribute-Money. That in *Carolina* Pines and Walnuts were the commonest Trees, with some Oaks bearing Acorns, on which the Swine feed, as well as on some other wild Fruits abroad, and on *Indian* Corn within. He said, Fevers and Agues were there common and mortal. That he had come through the *Bahama*-Islands, and stretch'd it between *Cuba* and *Hispaniola*, and so came to *Port-Royal.* The Duke of *Albemarle* once shew'd me a very rich piece of Silver Oar which his Father had sent him from the *Apalathean* Mountains on the Confines of *Carolina.*

One from *Tortuga* and *Petit-Guaves*, told me that at this last place the *French* have about Thirty Inhabitants keeping always good Guard for fear of the *Spaniards*. They have no Sugar-Works, but Indico. Tobacco and Hides were their chief Commodities, the last they get by hunting; but that fresh wild Beef is scarce, for they must go a great way from their Habitations to find it, and that in Companies. That *Petit-Guaves* is in the middle way between the Isles *de Vacas* and *Tortuga.*

The Introduction.

Of the Diseases I observed in Jamaica, *and the Method by which I used to Cure them.*

BEfore I conclude this Introduction, I think it necessary to give an Account of the Diseases of *Jamaica*, and how I endeavour'd to relieve them. This may be useful to some, and I am sure would have been to me, had I been so fortunate before my going thither, as to have met with any such *Observations*. I was told that the Diseases of this place were all different from what they are in *Europe*, and to be treated in a differing Method. This made me very uneasie, lest by ignorance I should kill instead of curing, and put me on trying with the utmost caution the Remedies and Methods I had known effectual in *Europe*, which in a very little time, I found to have great success on the Diseases there. My Medicines had the better operation, because people had a belief I could help them, and submitted to the taking Remedies in the order they were prescribed without changing the Medicines, altering the Method, or judging harshly in case the Person died. Indeed, at first, the Inhabitants would scarce trust me in the management of the least Distemper, till their observation of the good effects the *European* method had in the Duke of *Albemarle*'s numerous Family, in the same Diseases, brought them to make trial of what I could do with some of the meaner sort, accounted in desperate Conditions. I shall give some of these Observations both in the Voyage thither, and during my abode there, in as few words as I can, chiefly relating Matters of Fact, whereby, abating some very few Diseases, Symptoms, *&c.* from the diversity of the Air, Meat, Drink, *&c.* any Person who has seen many sick People, will find the same Diseases here as in *Europe*, and the same Method of Cure. For this reason I have put down some very ordinary Observations and Methods, that this matter may be very plain. For my own part I never saw a Disease in *Jamaica*, which I had not met with in *Europe*, and that in People who never had been in either *Indies*, excepting one or two; and such Instances happen to People practising Physick in *England*, or any where else, that they may meet, amongst great numbers, with a singular Disease, that they had never seen before, nor perhaps meet after with a parallel instance.

Of a *Cholera Morbus*, want of Appetite, *&c.*

Captain *Nowel* aged about forty, Cholerick, who had drunk very hard, and was very thin of Flesh, sent to me, he was ill of a *Cholera Morbus*, Vomiting, and going often to Stool. I found him weak, not able to bear any farther Evacuation, I gave him there-

fore

The Introduction. xci

fore about 15 Drops of *Laud. Liquid. Cydoniat.* in a convenient *Vehicle.* His Vomiting being stop'd, I gave him for his Loosness the *Decoct. alb.* for his ordinary Drink, and order'd him Rice Milk, and Milk-Meats for his Food, which, with the help of the *Laud.* repeated, soon cur'd him of that Indisposition. He continuing to drink hard, weakned his Stomach, so that he vomited almost every Morning, lost his Appetite, and complain'd of a great pain in his Breast under the *Sternum,* which I ghess'd to be some small Inflammation in the *Mediastinum,* or other Membranes of the *Thorax.* I try'd by bitter Wine, and other Stomachics, to rectifie the Stomach; by Milk-Diet, Diet-Drinks, Steel Course, and Bleeding to Cure the latter, but without success, the Reason I ghess was his drinking Drams in the morning, chiefly Brandy and Sugar. He reduc'd by these means, his Stomach to that weakness, that at last, since I came from *Jamaica* I have been told he could keep nothing therein but the Milk of a *Negro* Woman he suck'd.

Mr. *Rhadish* was seiz'd with a *Tertian* Ague at *Plimouth,* and cur'd by the *Cortex Peruv.* given as usually, without any return. *Of a Tertian.*

Mr. *Mark Collet,* in the thirtieth year of his Age, at the same place and time, was seiz'd after the same manner, only there was very little Intermission, and he was delirous for several hours. Upon the taking of this Bark given by the Ship-Doctor, by my Directions, he was perfectly cur'd. He had before my seeing him, taken in vain abundance of Juleps and Cordials, been Blister'd, Bled, &c. It was a while before I, who was generally in another Ship, could find out the Intermission, the Fever being scarce ever off. *Of an intermitting Fever, with little intermission, Delirio, &c.*

In *February* 1688. he was taken in *Jamaica* after the very same manner again, I had him Bled and Blister'd, he being delirous. I gave him, for his cold Sweats and weak Pulse, about ten Drops of Spirit of Hartshorn every four or five hours, and after a full discovery of the Intermission, I cur'd him with the *Cortex.*

Mr. *Anthony Gamble,* aged about Forty five, a Cook, given to Drink, had, some years before I saw him, in an Engagement with some *Turkish* Ships, a great part of the Flesh of his right *Hypochondre* shot away with a Cannon-Bullet. He fell into very great pains in his Belly, which was bound. I gave him some *Extractum Rudii,* to loosen his Belly, which not succeeding he had Glisters, Suppositories, Decoctions, Bolus's of *Ther. Venet.* Draughts with some Drops of liquid *Laud.* in proper *Vehicles,* Juniper, and other Cordial Waters, outward Fomentations, and Bags Emollient and Anodyne of all sorts contriv'd to procure a Stool, as well as ease the Pain, but the Disease was too violent to yield to any of those Medicines, before
several

Of a Colick.

several days of intolerable Pain were over, when by an easy ordinary Glister he was relieved, and escaped that time. Drinking very hard, some time after, he fell into the *Hemorrhoids* with intolerable pain, and at the same time had a Flux and Fever, the Flux being a Crisis of the latter. He sent for a Chirurgeon, who gave him at night, as I suppose, a Bolus of some Opiat Medicines, which stop'd the Looseness, but increas'd the Fever to that extremity, that he was in great danger. I, on all these accounts, order'd him to be Bled to ten Ounces, gave him cooling Juleps, and directed the *Anus* or *Hemorrhoids* to be easily anointed with *Ung. Comitiss.* and *Popul.* In some time he recover'd by degrees his former state of Health, but was very often subject to violent Colicks, which I judg'd might be occasion'd by some part of the Guts adhering to the *Cicatrix* of the great Wound in his *Hypochondre*, and by that means occasioning some small stop or obstructions to the passage of the Excrements in that place, as it happens, for another Reason, to those troubl'd with Ruptures; but in some time (which was requisite for the Guts to do their Office, the Excrements to be moistned, and pass this stop) it usually went off. He was very much troubl'd with the *Hemorrhoids*, and inflam'd swell'd Eyes, the first I cur'd with Bleeding, and the Ointments before mention'd, the latter with Bleeding, Purging and Blistering, according to the greatness or stubborness of the Disease.

Of a Consumption or Hectic.

A Seaman, aged about Forty, had a quick Feverish Pulse, especially towards the Evening; a very troublesome Cough, which had been his Companion for some Months. I order'd him to take some Pectoral Medicines, and at night an Anodyne Draught, by which he found some ease. I chang'd, after a while, his Medicines, and gave him *Locatelli*'s Balsam; but he grew weary, and went to change the Air. He came, in about nine Months after, to me, very much emaciated, with his former complaints, only in every respect worse, with a great Looseness, for which I gave him every Evening about ʒj. of *Ther. Ven.* with gr. j. *Laud. Lond.* which stop'd his Looseness and other Symptoms. I know not what became of him afterwards, he not coming near me, but by the common course of such Distempers, 'tis likely he died soon after.

One *Saturday* evening, when we were in hot Weather, a Hog being kill'd, and the Blood sav'd (to make Puddings) till *Monday* morning, they prov'd very hurtful, for although some, who had eat of them, complain'd not, yet several others were taken violently ill; some Vomiting with great pain, and others Vomiting and going to Stool with great Anxieties. Being call'd, and asking if they had eaten or drunk any thing to occasion such great disorders, I concluded the

Puddings,

The Introduction. xciii

Puddings to be the Cause, and whereas 'twas advis'd to stop the Vo- Of several miting, I thought it most proper to forward it; for that it seem'd People sick to be the readiest and easiest way to follow the motion of Na- Hogs- ture, and to discharge the Cause, especially considering the Puddings Puddings were scarce yet out of the Stomach. I therefore gave small Beer of which and warm Water with a little *Infus. Croc. Metall.* and help'd them they were up, and after a sufficient Evacuation, gave some Cordial Draughts made ha- of *Conf. Alkermes*, Cinamon water, and *Syr. Caryophyll.* and all the kept too persons were in a little time very well. The best way of managing long. most Persons Poison'd or Surfeited, is by Vomits, if the matter remain in the Stomach; Purging if it be in the Guts, and Diluting.

One —— who had had a *Gonorrhæa* often, and a pretty while before he complain'd, had many Symptoms of the Pox, which threatned his Life, or at least the present flatting of his Nose. The question was whether being at Sea he might be flux'd, I told them I saw nothing to hinder it in such an urgent Case, and therefore advis'd one, who pretended to understand Salivation well, to do it by Unction, as the surest way to Root out the Distemper. He put him into a Of a Sali- very close Cabbin, anointed him, and the Flux rose very well, and manag'd. the Symptoms ceas'd; I concluded all would go on as usually they do in such Cases, and gave the Person who had the care of him general Directions how to behave himself. But it prov'd otherwise, for he was ignorant of the Method of treating in a Salivation, spar'd his Medicines, substituted others in the Places of those I order'd, alledging amongst his Comrades, he knew better. He likewise kept his Cabbin too hot, as well by burning two Candles always there, and never suffering a vent to the Steams, as by giving a great many Cordials, or hot Sudorifics. By these several means unknown to me, he inverted the Course of Nature, and threw what ought to have come by Spitting, through the Pores of the Skin, wherefore in about Fourteen days his Spittle thickned, the *Serum* being thrown out another way, and he was choak'd and died, notwithstanding what could be done for him.

A Gentleman, aged about Forty, of a Sanguine Complexion, much given to Drinking and Venery, fell ill of the Gout, for which he following some Emperics Advice, plaister'd all his affected Joints Of the with Tar, whereby in some time he fell into a Quinsie, there be- bad effects ing a translation of the Matter from the Joints to the Throat. I of tampe- had him immediately Bled to a good quantity, for fear of Suf- deavour- focation, and gave him a Dose or two of *Extr. Rud.* which ing to cure working well, he was freed of his sore Throat. He, afterwards, the Gout. by the use of Bitter-Wine, and *Elixir Proprietatis*, recover'd his Appetite, but drinking several months afterwards to a great ex-

(a a) cess

xciv *The Introduction.*

cess in Syder and Punch, he was taken after a shivering Fit and Fever, with pains in his Side, for which he was Bled, and thereby reliev'd. The Gout coming again, he could not be perfuaded, or kept from tampering it with Cows Dung and Vinegar fried, and applied as a Poultefs; the Cows Dung I thought might be an innocent Anodyne; but the Vinegar as being a diffolver and thinner of the Blood, I oppos'd. On the use of it he fell into a Loofnefs, and sometimes Vomiting, which continued, notwithstanding the *Decoct. Alb.* cafe Opiates, and whatever I could think of, till he died. His Stomach was always out of order, becaufe of his exceffive drinking, efpecially Brandy and Sugar, by way of Dram in a morning, to fettle, as he thought, his Stomach.

Of spitting Blood.
Colonel *Walker*, aged about Forty five, Plethoric; upon drinking, used always to be troubled with Rheumatick and Gouty pains through all his Joints, after an exceffive manner, of which by bleeding he was ftill reliev'd, though fometimes he was forc'd to fly to Opiats. Once he fell inftead of his pains, into a spitting of Blood, which came up in large quantities without pain. Going to the Palifados in a hot day to drink Milk, he fpit or vomited up half a pint, for which he was Bled, and took an Opiat at night, with other Aftringents. I advis'd repeating of the Bleeding, continuing in the ufe of Opiats, great Quiet, Iffues in the Shoulders, *&c.* with which, Rice Milk, and other cooling, thickning, *&c.* Medicines for the Blood, he was perfectly cur'd. Upon his return to *England*, he fell into a Relapfe, with the same Symptoms, and I have heard died Confumptive.

Of a common continual Fever.
Mr. *Rayney*, of about Seventeen years, fell into a Fever, from which he was freed by bleeding, cooling Juleps of Barley-water with Syrup of Lemons, and other things of that kind.

Of the effects of hot Weather on the Body by bringing out Puftles.
When we came into hot Weather, it was a very ordinary complaint in every ones Mouth, that they were fo troubl'd with an itch from fmall red Puftles or Wheals, that they knew not what to do to be eafie. They came out ufually on the Back, along the Spine, though fometimes they cover'd the whole Body. I told them I thought this Diftemper was the greateft advantage they could have, it being a great Purger of the Blood from hot and fharp parts, and therefore was fo far from complying with their defires of curing them, that I ufually gave fomething to forward the eruptions, as *Flos Sulph.* or fome other innocent Diaphoretick; but if their impatience was to be complied with, Bleeding, and Purging after it, was an infallible Remedy. I concluded the alteration of the Climate was the

occafion

The Introduction.

occasion of this Disease, by putting the Blood into a brisker motion, and perhaps putting into it some fiery Particles Nature threw out this way. I was not much troubled with this, but in lieu of it had a small Carbuncle came out on my right Wrist.

Afterwards at *Jamaica* in hot Weather, the same Disease was more troublesom to us New-comers, and even sometimes, though rarely to those had been a long time acquainted with the Climate, my Answers to such Complainants and Remedies were the same. Something more will be said of this hereafter.

Mr. *B.* aged about Forty, of a Sanguine Complexion, and Plethorick, at coming ashore on *Port-Royal*, fell into a Fever, he had an high quick Pulse, an inclination to Vomit, and uneasiness all over. I ordered him to be Bled, gave him the next morning a Vomit of ʒvi. of *Vin. Emet.* with half an Ounce of *Oxym Squill.* which with the help of Watergruel wrought very well, but remov'd not his indisposition. He was forbid the tasting any Wine or Flesh, and whereas about Twelve a Clock at night he usually had a large Stool or two, about Ten at night I ordered him to take about Ten gr. of *Extr. Rud.* thereby to forward that motion of Nature, by endeavouring to help to throw out the Morbifick Matter by Stool, which it accordingly did, and clear'd him of his Disease. *Of a Fever with great weakness and drouth.*

He afterwards grew very faint and weary, and for that finding relief in drinking *Madera*-Wine and water for the present, he made use of it too often, whereby he became usually, the more he drank, the more dry, so that after a small time he was necessitated to drink again. By the Air without, and the Wine within, his Spirits and Moisture were exhausted. Once or twice in the evening and night he was a little incoherent in his Discourse, wherefore I immediately had him Bled, and gave him a Vomit, and in a while, he keeping an orderly Diet, was well. This Gentleman, after his arrival in *England*, fell into a Fever, in the latter end of which he voided several Pints of Blood, and soon after died.

I was sent for at St. *Jago de la Vega*, to a Child of Collonel *Fullers*. It was a Boy about twelve years of Age, had been in a Fever for some time, and was then in Convulsions, cold Sweats, &c. his Pulse quick and low. I advis'd he should be Cupt, with Scarification in the Shoulders, taking away some Blood. He had Cordials with *Confect. Alkerm. aq. Cinamom. Syr. Caryophill.* &c. and Blisters with *Sp. C. C.* and *Ol. succin.* some Drops of which last were inwardly given him, but in some hours, his weakness increasing, he died. *Of a Fever with Convulsions.*

His

xcvi   *The Introduction.*

Of *Hysterick* Fits.
His Mother falling into violent Hysterick Fits upon his death, I gave her twenty Drops of *Sp. Sal. Armon.* and order'd her to smell to a Bottle, wherein the Volatile Salt of it was enclos'd. The Salt was impregnated with some Particles of *Castor*, with which it had been sublim'd, the Bottle was only half full, that thereby there being Particles ready to issue out in plenty, the Sensories might be the more irritated, and the Fits taken off. I shall say more of this Distemper hereafter.

Of a very *Epidemic* continual violent Fever.
About the month of *January* 1688. most of his Graces Family were taken very ill of continual Fevers, one after another being seiz'd, till it went round the whole House, some very few only excepted. It usually invaded them without any apparent Cause. I had it my self, and could not assign any cause, if not being a little uncover'd in the night by the Sheets falling off. The Symptoms were a great pain in the Head, and Back about the Loins, a Reaching to Vomit to no purpose, a very great pain in the Limbs, and all over the Body, as in a Rheumatism, which seem'd to be from the violent heat and boiling of the Blood in the Vessels and Membranes. It usually ended in twelve or eighteen hours, and the Remedies I us'd were these. If call'd at first seizure, I immediately order'd bleeding *Ex Vena maxime tumida*, to ten or twelve Ounces, and if there was an inclination to Vomit, I gave them *Infus. Croc. Metal.* with *Oxym. Scill.* according to their strength, and these two Remedies timely given, would check it presently in the very bud, the Vomit working usually well, and the bleeding giving immediate ease. If it had been on them some time, then it was necessary to cool very very much with Barley-water and Syr. of Lemons for their ordinary Drink, forbidding the use of Wine and Flesh, or Broath in any degree, and in case of any Symptom of a *Delirium*, blistering the Neck, Arms, and Ankles, and to remedy cold Sweats (very familiar here) 'twas now and then necessary to give some Drops of *Sp. C.C.* in any potulent Liquor. By this Regimen I thank God none committed to my care miscarried, but those who would not observe Rules, or were treated after another manner, usually were in danger, as you may see by the two following Instances.

Of one in great danger from drinking Wine in a Fever.
Mr. *Lane*, aged Twenty five, or thereabouts, being seiz'd, and the Fever running high, he being Plethorick and hot, was Bled and Blister'd, *&c.* and by this Course his Fever was abated, and almost at an end. About eighteen hours after, coming to him I found him very much disorder'd, and almost as bad as at the beginning, wherefore having repeated my Orders about him to his Nurse, he yet very hardly escap'd. He afterwards told me the Reason of his Relapse, which was his privately drinking White *Madera* Wine contrary to direction.   One

The Introduction. xcvii

One *Richard*, a white Servant, belonging to Colonel *Ballard*, about thirty years of age, was taken with this before mentioned *Epidemic* Fever. After he had been treated with *Bolus*'s of *Diascor-* Of a Fe-*dium*, and Cordials to Sweat him, he grew worse and worse. I ver made found him at Twelve a Clock at night in a vast Agony, as every worse by one thought a dying. He had a mighty oppression and anxiety on over-heat-him, a very great difficulty in breathing, and could scarce speak. I ing with told them I believ'd this Disease was partly forc'd, and therefore and took off his superfluous cloathing, set by the Cordials (as they call'd Cloaths. them) and sent him a Bottle or two of cooling Julep made of *Aqua Font.* (which I use, in Bills to Apothecaries, to call very justly *Aqua Cordialis frigida*) acidulated with *Ol. Vitriol.* and sweetned with *Syr. Caryophyll.* I desired him to drink plentifully of this, and in two or three days time, without any other considerable Remedy, he was well. Though I did not my self see any dye of this Fever, yet I heard with hot treatment some persons died.

Most part of people who had been troubled with this Fever, fell afterwards into very great weakness, so that although this Disease, Of great with good management, lasted not past twelve or eighteen hours, yet weakness their weakness was as great as if they had been under a Distemper this Fever. for several Months. This was, I think, peculiar to this Fever, though at first I suspected it was to all Diseases here, by reason of the hot Climate, but I found all other Diseases accompanied with the same Symptoms as if in *Europe*, and therefore look on this Symptom as a thing particular to this Fever, and such uncommon Symptoms now and then attend *Epidemic* Diseases every where. For this the best Remedies were procuring a good Appetite, and a regulation of Diet.

I know not whether this last very great weakness and faintness might not come from another Distemper, very ordinarily follow'd this Fever, which was the Jaundice, for about some few days after Of the this Fever was over, the Jaundice very often began to shew its self Jaundice by great slothfulness. Afterwards the yellow Face and Eyes, as well this Fever. as thick yellow Urin discover'd it plain. This Distemper usually was cur'd by an easie Vomit, or Purgers, first gentle of *Pil. Ruff.* and then stronger of *Extr. Rud.* with *Curcum. Milleped.* Saffron, *Elixir Proprietatis, Castile-Sope*, and such other easie Medicines given between Evacuations, though sometimes 'twas so difficult as not to yield to Courses of those Medicines taken every day for a Month together. I remember the *Serum* was so discolour'd in some, that all the Pustles rose on the Skin, were fill'd with an *Ichor* as yellow as the infusion of *Curcuma* or Saffron in water. Perhaps the weaknesses, hindering people from going about, or Exercise, might be in some measure the occasion of this Jaundice.

(b b) A great

A great many were of opinion that this Fever was what is call'd the Seasoning, that is to say, that every New-comer before they be accustomed to the Climate and Constitution of the Air in *Jamaica*, are to have an acute Disease, which is thought to be very dangerous, and that after this is over, their Bodies are made more fit to live there, with less hazard than before; and this is not only thought so in that Island, but in *Guinea*, and all over the remote Eastern parts of the World. That this Fever was not so, is manifest in that not only we New-comers were taken with it, but likewise many of the ancient Inhabitants of the place, as several of the Family of Colonel *Ivy*, *&c.* and that a great many of us who were lately arrived, escap'd this and all acute diseases whatsoever. If there be any such thing as Seasoning, the Itch or Pustles formerly mention'd must be it, the alteration from cold to heat being by degrees done by the way, and that Symptom appearing on increase of the heat.

Of the Seasoning. That this Fever was not the seasoning.

Sir *H. M.* aged about Forty five, Lean, sallow coloured, his Eyes a little yellowish, and Belly a little jetting out or prominent, complained to me of want of Appetit to Victuals, he had a kecking or reaching to Vomit every morning, and generally a small Looseness attending him, and withal was much given to drinking and sitting up late, which I supposed had been the original cause of his present Indisposition. I was afraid of a beginning Dropsie, and advised him to an easie Vomit of *Oxymel. Scill.* with the help of a Feather, and thin Watergruel, fearing *Vin. Emet.* might disorder him too much by putting him into a Looseness, or too great Evacuation. After that I gave him some *Madera*-Wine, in which the Roots of Gentian, Tops of Centaury, *&c.* had been infused, with which Vomit, it working easily, and the bitter Wine taken every morning for some days, he recovered his Stomach, and continued very well for a considerable time. Not being able to abstain from Company, he sate up late, drinking too much, whereby he not only had a return of his first Symptoms, but complain'd he could not make water freely. His water was thick and very red, and his Legs swell'd a little. When these Symptoms appeared, Doctor *Rose* and I being join'd, we ordered him an Electuary of *Cassia*, Oil of Juniper, *Cremor. Tart.* and other things to purge easily the watery Humours, enjoyn'd Temperance, and desired the continuance of his former Medicines. This Course did very well with him, but making but very little water, and being much troubled with Belchings, and a Cough in the night, he sent to another Doctor, who, when he came, was of opinion that his Disease was a Timpany, and that the swelling of his Belly came only from wind, according to *Hippocrates*, and that he was troubled with neither the beginning of a Dropsie,

Of a Dropsie in a very bad Constitution.

The Introduction. xcix

Dropsie, nor had Gravel (which is not unusual in this Case, and he had been always troubled with) I told him later Observations upon the Dissection of deceased *Morbid* Bodies, had discovered the Bellies of People dying of supposed Timpanies, to be distended with water, and no more Wind than what is supposed to be the effect of Phlegm, and Crude Humours lying in the Stomac and Guts. I desir'd him that we should put off talking of the Theory, and come to the Practice, that perhaps we might very well agree in the Medicines he should take, as it very often happens to Physitians, who may disagree in the Theory, and yet agree in the Practice. I waited on Sir *H.* and told him Dr. *Rose's* and my Opinion, which agreeing, he was satisfied therewith. We gave him all manner of Diuretics, and easie Purgers we could find in *Jamaica*, Linseed and Juniper-Berries infus'd in Rhenish Wine, *Milleped. ppd.* in Powder, Juniper-water, advis'd him to eat Juniper-Berries, us'd *Oil* of Scorpion, with *Ung. Dialth.* outwardly, by which means he recovered again. On intemperance he fell into a great Looseness, threatning his Life, which by an Opiat, &c. at night we stopt, and he enjoy'd his Health for some time longer very well. Falling afterwards into his old Course of life, and not taking well any Advice to the contrary, his Belly swell'd so as not to be contained in his Coat, on which I warn'd him of his very great danger, because he being very weak, and subject to a Looseness, there was no room for purging Medicines, which seem'd to be the greatest Remedies for his Dropsie, threatning his Life, seeing Diureticks did not now produce the desired Effect. On this alarm he sent for three or four other Physitians, who, as I was told, said he had no Dropsie, because his Legs did not swell, the Reason of which was, because he lay in a *Hamac* with his Legs up, and us'd very little exercise. They advised him to a *Cataplasm* of Vervain of this Country, &c. for his swell'd Belly, and would have given him a Vomit next morning, but that it was an unlucky day, as indeed it had in all likelihood been to him, if he had taken it, for he fell naturally by only the *Cataplasm* into a very dangerous Looseness, which had almost carried him off ; so the thoughts of this proceeding was put off. He chang'd soon his Physicians, and had first a Black, who gave him Clysters of Urine, and plaister'd him all over with Clay and Water, and by it augmented his Cough. He left his Black Doctor, and sent for another, who promis'd his Cure, but he languished, and his Cough augmenting died soon after.

Mrs *Barret*, about Forty years of age, of a spare Body, fell into a Tertian, which naturally, or by Medicines, was very violent, there being scarce any intermission. Her Tongue was very black ; and she delirous for the most part. She had by her several Cordials, Of a very violent *Tertian.*

as *Bolus's* of *Diascord*, &c. which I suppos'd had in part brought her to this. I told them I hop'd the best, and prepar'd her some *Cortex Peru*, with which, and the use of cooling diluting Drinks she entirely recover'd, although she was by every Body thought to be in a desperate condition.

Of a Lethargy in a Child.
Mr. *Fletcher's* Child, about a year and an half old, was taken with a sleepy Disease. It lay with the Eyes always shut, and asleep. I advis'd the Mother to give it a little *Manna* immediately, and to Blister its Neck, which being done, and the Physick working well, the Child recovered entirely its Health.

Of a Lethargy in a Woman.
One R. a Tavern-keeper's Wife, about Forty years of age, Fat and Phlegmatic, was upon excessive drinking of Brandy, taken with a Lethargy, inclining to an apoplectick Fit. She would on very violent irritations lift up her Eye-Lids, but would not speak. I immediately order'd bleeding, blistering in the Neck and Arms, gave her ℈ij. of *Diagridium* in a Glass of Water, with some Drops of *Sp. Sal. Armon.* Ordered one to hold to her Nose the volatile Salt of the same in a Bottle, and a Snuff for her of *Majorane*, Betony, and White Hellebore, which being put to her Nose, she snuft up very often. By the help of these Medicines she first went to Stool in the Bed. Her Blisters rose, and then on the use of the Snuff she Snees'd. She was plied hard with them two days, then look'd up more, could say a word or two, and call for the Pot to make water. I continued them two days longer, and she grew better, but being morose would take nothing, and shut her Eyes. I told the standers by, to frighten her, that I would get a Pan of Coals and burn her with them on the Head, which so alarm'd her, that she took things, and was well above a year. But then, I suppose, on the like occasion, fell into an apoplectick Fit, and being sent for, before I came she was dead.

Of a *Tertian* in a Black Boy.
A little black Boy, of a year and an half old, belonging to *Tho. Rowland* was taken violently ill of a *Tertian*. I gave him the Cortex prepd. as usually, which being forc'd down, the little one was well without relapse.

Of a Dropsie.
One *Stephen Lego*, a Wheel-wright, aged about Forty five, Phlegmatic, sent for me. He was sitting in a Chair, with his Legs swell'd like Posts, on a Stool before him. He could not lie down, nor so much as lean down his Head, for an *Orthopnæa* He had likewise a very violent Cough molesting him at all times. One would have thought he could not have liv'd three hours

in

The Introduction.

in that Agony. I order'd him immediately a *Linctus* made of *Syr. de Succo Heder. Terrestr. Diacodium*, Sugar-Candy, and *Flor. Sulph.* which I bid him lick every now and then, from the Point of a Knife. This reliev'd him extremely, so that every thing seem'd to be better with the continuance of this Medicine. He slept lying, his Legs were not swell'd so much, and his Cough gone. I gave him some *Sp.* of Hartshorn for his Weakness, some Pill of *Extr. Rud.* for his swell'd Legs, and some *Locatelli's* Balsam for his Lungs, to hinder Putrefaction in them. These Remedies succeeded very well, so that in a few Weeks time he went abroad, riding about the Town every Morning. Having formerly been troubled with *Erysipelas's* on his Legs, the depending Posture of them in riding brought down an *Erysipelas*, which being very painful, and mightily inflam'd, hinder'd him of Sleep, took away his Stomach, and brought to his Legs a great Defluxion of serous Humours. The parts affected were bath'd with a *Lixivium*, in which were boil'd Wormwood, Rosemary, Thyme, Bay-Leaves, Orange-Leaves, *&c.* with a Bottle of Wine added to it at the latter end. With this the parts were often bath'd to evaporate the Humours, and hinder a Gangreen; but every thing growing worse, they ask'd my Opinion whether he would live. I told them, I believ'd he would not live many Days. They consulted the Astrologers, (who were much esteemed in *Jamaica*) who told them, that if he surviv'd the next Day's Noon, the Aspects of the Planets positively agreed to save his Life. He liv'd three Days after the time, and yet when he died, these same People said they had by the Stars exactly foretold the Minute of his Death. He had before his Death a Gangrene appear'd in *Perinæo*.

One —— —, aged about Fifty, came from his Plantation, where he had been under the Care of several Physicians without Relief. He complain'd of a great Pain in his right *Hypochonder*, Thinking his Liver obstructed, by reason of a Tumour there, I gave him such things as usually avail in such Cases, Hot-Gum-Plaisters, *&c.* Finding this Course did not work the desired effect, but that he rather grew worse, and that he began to find some Difficulty in making Water, I began to doubt a Caruncle, and the Pox to be the chief of his complicated Diseases, and questioning him very hard about that matter, he at length confess'd it, and that he had several times had a *Gonorrhœa*; whereupon I alter'd my Course, and he being so very weak as not to endure any manner of Salivation, I gave him some *Merc. Dulc.* with an easie Medicine to work it off, and some things for his Cough and swell'd Legs, which had been on him a great while. He fell afterwards into a Flux, which could not be stopt by Opiats, nor any other Remedy I could give, and so he died.

Of a complicated Disease of the Dropsie, Consumption and Pox.

Mrs.

The Introduction

Of a Woman with Child in danger of Miscarriage, which was prevented notwithstanding she had regularly her Catamenia.

Mrs. *Fletcher* had been with Child four or five Months, and had a great pain in her Back and Loins, as if ready to bring forth, with a Flux of Blood, or an appearance of the *Menstruæ Purgationes*. I immediately had her Bled for fear of Abortion, enjoyn'd her to keep her Bed, lying very still, with her Heels high, and a Pillow or two under her Loins, and gave her an easie *Hypnotick*, *viz.* about 15 *gtt* of *Laud. Liq.* in a Draught of water, wherein was dissolv'd some *Eleosacch. Cinamomi*. With this she rested well, and by its continuance all the Symptoms were quell'd, but she, during all the time of being with Child, had her *Menses* as regularly as when well, for all the Medicines and Directions I could give her. She notwithstanding went out her time, and brought forth her Child very well.

According to the notions of some Ancient Physicians, there was some reason to be apprehensive the Child would not be healthy, being defrauded of its Nutriment while in the Belly; yet, contrary to this Opinion, it continu'd as lively and brisk as any, till it was five months old, or thereabouts. It was then emaciated very much, did not sleep, and was always froward and crying. I found its *Of the Child brought forth by this Woman.* Belly, *&c.* very well, which is generally swell'd in Children emaciated from Knots, or *Scrophulous* Tumours in their Mesenteries. Being apprehensive that the Nurses Milk did not agree with the Child, she was chang'd, but notwithstanding that, and all the innocent Medicines I durst use, the Child languish'd more and more, and died. I believe a great cause of the variety observed in this case, may come from the *Plethoric*, or other Constitution of the Mother.

Of a considerable Ptyalism stopt.

Mr. *Byndlosse*, aged about Twenty four, was for several months troubled with a great spitting, on which he look'd very ill, he did not Cough, but wasted strangely. I was apprehensive that this might bring him in time to a Consumption, and therefore ordered him to take thrice a day about seven Drops of *Opobalsamum* in Sugar, drinking after it a draught of Diet-drink made of *Sarsa, China, Sassafras, ras. C. C. eboris, &c.* made with Raisins bruised, to give it a good taste, and made fresh every day, left it should ferment and spoil. By these Medicines he in a while grew very well. I was doubtful whether this Distemper might not be an easie *Ptyalism* from some Mercurial Medicine taken unknown to me, or perhaps to himself, some Physicians being very fond of giving Mercurial Remedies without any urgent cause, in which I think they are to blame, they having an uncertain Operation, and being sometimes exhibited not without danger.

One

The Introduction.   ciii

One *Prince*, a lusty *Negro*, had been ill of the *Yaws* (of which I shall have occasion to say more hereafter) and flux'd for it in one of the Chirurgeons Hot Houses at Town, where being kept extremely hot, and abridg'd of Victuals, he, either being mad, or extremely uneasie, broke open the Door, and ran home in a very great Breeze of Wind. Upon this his Spittle thickn'd, and his spitting stop'd, it running by Stool, and griping him very much. Major *Bragg* sent for me to him, I order'd a little place in a corner of the House to be made moderately warm for him, and gave him as much Watergruel as he could eat or drink, one Scruple of *Merc. Dulc.* in Conserve of Roses several times, and to stop the Looseness, some drops of *Laudan. Liq.* were put to it. By these means his Salivation rose again, and all the Symptoms ceas'd, only on the upper part of his Foot was one Sore, not yet dry'd, for which towards the latter end of his Spitting, I gave him 7 gr. of *Turbith. Min.* in a *Bolus* of *Consery. Ros.* which working well upwards and downwards, it dry'd. He continued well without Relapse.

Of a Black who being mad or uneasie, ran out of the Hot House in the height of a Salivation.

Mrs. *Duke*, aged about thirty five, was always at the usual time of the *Menstruæ Purgationes*, extremely troubled with intolerable pains in her Belly and Loins, with a great press downwards, so that sometimes she had a Suppression of her *Menses*, and at other times a *Procidentia uteri*. I endeavour'd to remedy these Accidents by all manner of *Menses moventia*, bleeding and purging, Steel-Courses, *Pulegium* Decoctions, which prov'd to no purpose for some Months. Then I endeavour'd to Cure this Distemper by Bleeding, Purging, &c. just before the usual time of the coming of the *Catamenia*, but she found very little Amendment. Afterwards I grew a little cautious, lest she might be with Child, and proceeded no farther. 'Tis very ordinary to have before, or at the beginning of the *Catamenia*, these Symptoms, especially when the sick Persons are out of order, have receiv'd any injury in Childbed, or are troubl'd with the *Fluor albus*. I have seen many methods of Chalybeat and Bath-Waters, tried for this Disease in several persons without effect. The most ease they find is by having Children, the Vessels about the *Uterus* being thereby distended, and afterwards their Pains are less.

Of a Woman who had great Pains, &c. at the time of the Catamenia.

Robert Nichols, aged Thirty or thereabouts, usually drunk with Brandy, fell into a violent *Hemorrhage* at the Nose, it running out in great quantity. After a while I was call'd, and order'd him to be bled at the Arm ten Ounces, and blew up through a Quill a Powder made

Of a violent Hæmorrhage at the Nose.

civ   *The Introduction.*

made of equal parts of Alum, *Vitriol* and *Bole Armeniac*, which stopt the *Hæmorrhage* for some time. It returning several times, the same Medicines being repeated with bleeding, great Abstinence, Cooling, and a cold Bath, he was entirely cured.

Of blindness.
Dr. *Rooks*'s Wife, aged about Thirty five, of a Phlegmatic Constitution, lost entirely the sight of one of her Eyes, and with the other could very hardly perceive any thing, and distinguish nothing. The Pupil of the one stood always wide open, and that of the other on looking at distant or near Objects, scarce alter'd, contracted, or dilated its self, which is a sign of a very bad sight. The Doctor told me that he came to me to satisfie her Relations, but that he despair'd of a Cure. He had given her *Pil. Lucis*, whereby she had had some Stools, and had made a Seaton in her Neck. Enquiring concerning the *Mensium Fluxus*, I was told, she had been out of order that way for some months. I encouraged them all I could, told them there was room for hope, and took my Indication from the Obstruction, knowing what wonderful effects, and how many Diseases in Women, come from thence. I order'd her to be Bled by Cupping with Scarification in the Shoulders, to be blister'd in the Neck, to be purg'd with *Pil. Lucis* sharpned with *Diagrid.* to some Grains, twice a Week, and in the intermediate days to take a Steel-Electuary made up with Cephalicks, *viz. Limat. Chalyb. Subtiliss.* trit. made up with *Conserv. Flor. Rorismarin. &c.* and Chymical Oil of Thyme to some Drops. I advis'd her likewise against Sneezing Powders, and to take after her Electuary, and twice a day besides, a good draught of a Decoction of Sage and Rosemary, into which an *Eleosacch.* of Rosemary was dissolved. I likewise desired her to take *Millepedes* alive, to one hundred in a morning, rising to that number by degrees on the days when she took nothing else. By these means persisted in, she first felt some relief, by degrees recovered the sight of one Eye, and then of the other, so that she could at last read Bibles of the smallest Print, and was entirely cured.

Of an Epidemic Chincough.
In *Jan.* and *Feb.* 1688. after some hard Breezes and Norths, (*Winds*) most part of the poor Children who lay in the *Savanna* Houses (which were Huts made of Palisadoes or Reeds, and cover'd with Palm-Leaves) expos'd on every side to the Winds, and not strong enough to keep them out, were taken with Chincoughs, which was very Epidemical, and contagious among them, scarce any escaping. After trial of several things, I could not find any relief till the violence of the Distemper forc'd me to *Laud. Liq. & Lond.* both of which cautiously given cur'd them all, but I had a great care of the Dose. I have given it to many Children at the Breast, dissolv'd in

the

The Introduction.

the Mothers or Nurses Milk, with very wonderful success, and I thank God had never any bad Accident follow'd its use, although I have given it to hundreds of Infants. The same Remedy (*Mutatis mutandis*) never misses the other degrees of this Distemper in other Ages, if administred as it ought to be.

Mr. *E. H.* aged about Forty five years, much given to drinking *Rum*-Punch, had several times fallen into the Belly-ach, by which he had lost the use of his Limbs. He came to me complaining of want of Appetite, had likewise a squeamishness or inclination to Vomit, a very great *Paralytick* shaking all over him, and was very weak. I gave him a Vomit of *Oxymel Scill.* which increased, during the time of working, his *Tremor* to such a heighth, as one would have thought him Expiring, but he telling me it was usually so with him in Vomiting, I wrought it off with thin Watergruel, and after the use of Bitter-Wine, *Sp. C. C. &c.* For some time he seem'd to be very well recovered, so that he was able to go about his Business, his Stomach was good, and he eat his Victuals very heartily, and grew stronger every day. *Of the Belly-ach.*

He rode out one morning about seven Miles, and drank the Milk or inward Juice of three *Coco* Nuts, which being too great a Load for so weak a Stomach and Body, he fell presently into violent Vomiting and Looseness. This last continued with him, for which I ordered him the *Decoct. Alb.* for ordinary Drink, I gave him Cordials of *Confect De Hyacinth.* made up into Draughts with Cinamon-water, and *Syr. Caryophill.* to which was now and then as occasion requir'd, added either Opiates to stop, Bezoar-Powder, or *Sp. C. C.* to some Drops, but all in vain, for every Stool weakned him more and more, so that in a very few days he died.

His Wife, recovering of a Fever, turn'd yellowish in her Complexion, and had a bitter tast in her Mouth. I gave her a Vomit of *Infus. Croc. Metall.* about ʒvj. with as much *Syr. Cariophyll.* as made it palatable. She vomited, and was reliev'd entirely by taking some Pills of *Extr. Rudii.* She was some Months after taken ill, much after the same manner, I repeated the Vomit, and gave her three Pills made of *Pil. Ruff.* and finding they agreed with her, I gave her a Box of them, and some *Elixir Proprietatis*, to remove the Jaundice, (which seem'd for the most part to lurk about her,) when ever she should find occasion again. She came by this method to a perfect Health. *Of a Cachexy.*

(dd) A

The Introduction.

A *Negro* Woman, belonging to Mr. *Forwood*, was brought to me, she had a great many Ulcers in the Extremities of the Fingers and Toes, and about the Joints. There was also several Bladders fill'd with *Serum* on several of her Joints, as if *Cantharides* had been applied there to raise a Blister. These Bladders or *Cuticula* fill'd with serous Matter, came either on her Fingers or Toes, every Full and New Moon, and in process of time each of these Bladders brought an Ulcer, leaving the Flesh raw, and sometimes deeper, sometimes shallower corroded, so that the longer the Bladders had been rais'd, the deeper were the Ulcerations. The virulency of the Humour was such, as that after it had eaten into the Bone, the joints of the Fingers and Toes would drop off, and they die, as I have been assur'd by those who have lost several *Negros* of this Disease, I was assured was peculiar to Blacks. Her Master told me she had been in the Hands of a great many Physicians, who had Bled, Purg'd, Sweared, &c. her to the greatest degree, without any success. I told him, I thought Fluxing, or Salivation, bid fairest for the Cure of this Disease, and having got a corner of an Out-house ready, she was therein flux'd by Unction. After a while she was not only so well that all the Symptoms of Bladders formerly rising on Full and Change of the Moon, did not appear as usually, but the Ulcers all over her extremities dry'd up, and were cicatriz'd, so that I did not doubt but all was perfectly well, Salivation being a great Remedy in Diseases where the *Serum* of the Blood is Peccant, either as to quantity or quality. I was very much disappointed, when her Master told me about three Months after, that her Distemper was again, on Full and Change, return'd on her. I concluded that the Salivation had not been prosecuted to the heighth, by my judging her Disease cured, and therefore order'd her to be shut up, and seen rub her self as directed. This second Salivation was very copious, and she well again. I, notwithstanding, towards the latter end gave her a Vomit of *Turpeth. Min.* and continued her Spitting for several days with *Merc. Dulc.* and afterwards order'd her a Diet Drink made of the Woods *Sarsa*, &c. boil'd in Lime-water. This preserv'd her as formerly, for some time, but did not secure her from a Relapse. So soon as this Disease again appear'd, I thought, that perhaps, this was proper to Blacks, and so might come from some peculiar indisposition of their black Skin. Knowing nothing more effectual in Cutaneous Diseases of this Nature than *Sulph. Vivum.* I made an Ointment of this with, *Ung. ex Oxylapath.* and order'd her the use of this Liniment on all the diseased parts. This being done, all was seemingly relieved, but not without a return as violent as ever of her Distemper. Her Master being displeas'd with the loathsome sight of her

Of a strange Disease in a Black Woman rotting her Fingers and Toes, following the Full and Change of the Moon.

The Introduction.

her about his House, remov'd her from thence to his Plantation, whereby I had no opportunity of further trial of Skill with this Distemper. This was a very strange Disease not only in its self, but that it followed very regularly the Full and New Moon. I have seen more Diseases than this, come exactly at those times; but they have generally been *Epilepsies*, or other Diseases of the Head, and have not been so visible as were these Bladders of water before-mentioned.

One night, very late, I was sent for to the *Crawle* Plantation to a Girl of about Twelve years of Age, of Mr. *Bozles*. She had a very great swelling in her Throat with pain, difficulty of swallowing and breathing, join'd with a Fever. I immediately took Ten Ounces of Blood out of her right Arm, gave her about one Scruple of *Pil. Coch. Min.* dissolv'd in water, for her more easie swallowing her Medicine, order'd her cooling Drinks and Juleps, made a *Linctus* of fair Water, Whites of Eggs beat to water, and Sugar-candy, which she was desired to swallow down easily. With the use of these Remedies, I furthered desired, that in case, in my absence, her difficulty of breathing should augment, they should immediately bleed her again. By the use of these directions she entirely recovered. *Of a Quinsie.*

The same Girl, some months after, was at *Port-Royal* taken ill of an intermitting Fever, which hanging very long about her, discolour'd her Face. She look'd very pale, had likewise a swelling in the right *Hypochondre* on the Region of the Liver, and was very *Cachectic*. I told them I thought this Disease was the effect of the Fevers long continuance, and lurking about her, without being cur'd by some effectual Remedy. I thought it convenient to carry off some of the *Morbific* Matter to purge easily with *Extr. Rud.* to give her a Diet Drink of *Sarsa, China, &c.* for ordinary Drink, and to put her in a small time into a Steel Course, by which means in some Weeks her Belly grew lank, she well colour'd, and perfectly rid of her *Cachexy*. *Of a Cachexy.*

Two of her Brothers were then troubled with *Quotidians*, the Fits lasted twelve hours, and they were treated by Physicians of several Notions, with several methods for removing them. They were not at all relieved, but grew *Cachectic* as their Sister had been, their Bellies began to swell, and they to look pale. I advised without any delay, the use of the *Cortex Peruv.* which given as I us'd to do, cur'd them; but their Fevers had continu'd so long about them, that they being *Cachectic*, I put them into the Steel Course before-mentioned, whereby they were perfectly recovered. *Of a Quotidian.*

Dr. *Cooper*,

Of the Belly-ach.
Dr. *Cooper*, aged about Forty five years, of a yellowish swarthy Complexion, was a great Drinker of Rum-Punch, and told me that he had had Twenty five several violent Fits of the Belly-ach, with drinking that sort of Liquor. He had been ill in the Country several days, and had had some Convulsion Fits. He had an intollerable pain about the Region of his Navel. For this he had taken several Clisters and Purgers, which, although they wrought well, yet cur'd him not, but gave only a small momentary relief. I advis'd him to a Fomentation all over his Belly, with a Decoction of Roots of *Althæa*, Leaves of the same, and Mallows, Fenugreek, and Linseed, with Camomile-Flowers, Juniper-Berries, and Cummin-Seed. With these he had a small relief, but grew presently as bad as ever. I gave him *Extr. Rud.* which purg'd him very well, but remov'd his pain but for a moment. I then gave him an easie Opiat, mixt with a Purge, but it had no effect. In the use of a Clister, in which was boild Gourd Leaves, he found a moment's relief, during the time of its working, but soon was taken with Convulsion Fits. For these Fits I gave him *Sp. C.C.* and Volatile Salt of *Sal Armoniack* to smell to, as well as *Ol. Succin.* dropt on Sugar, in a convenient Vehicle. He was now very weak, had cold Sweats, a weak Pulse, not able to endure Physick, every Stool endangering his Life, and the use of his Limbs was almost wholly taken from him. I chang'd the Medicines as occasion required, sometimes he took one Ounce of *Sena* boil'd in Chicken-Broath, and drank after it a Gallon or two in a day, to endeavour the washing away, of any sharp or sower Humour lying and corroding the Coats of the Guts. For the same purpose he would at my desire drink huge Draughts, and often of thin Watergruel, left the Chicken Broath should inflame too much, his water being thick and high colour'd, and he complaining of Erratick pains, (as it is with most others in this Disease) but all to the same purpose. Sometimes on vomiting he found great relief, there being always an inclination that way, but this relief was but of a short continuance. He being desirous of new Medicines, after I had given him an agreeable Cathartick, sent for another Physician, who gave him some strong Opiat, as I think of *Diacodium*, by which the Operation was stopt of the Cathartick. After sleeping, as he thought, he was considerably reliev'd, wherefore he persisted in the use of that; but in a few days he fell into a strong Convulsion Fit, and died. A third Physician coming to him, propos'd Wild-Liquorish-Leaves, boil'd in water, and the Decoction drank, as likewise a Decoction of Purslane; but they both seem'd very unequal to remove so great Diseases, especially this last, which we in substance eat every day to great quantities in Sallets, without any sensible alteration. Mrs. *Cook*

The Introduction.

Mrs. *Cook* sent for me to a Child of hers, which, when I came, was just out of a Convulsion Fit. It was a year and a half old, and was not breeding Teeth. I gave it some drops of *Sp. C. C.* and a drop or two of *Ol. Succin.* in Sugar, dissolv'd in water, and ordered three or four Grains of *Cinnabar* to be given in any Liquor, or what way they pleas'd. The Child falling into another Fit, on repeating these Medicines had immediate relief. The same Child fell into an intermiting Fever, and in the time of the Paroxysm had Convulsion Fits. I immediately gave the *Cortex Peruv.* in Chocolate, hinder'd the next *Paroxysm* of the Fever, and cur'd the Child.

Of Convulsion Fits.

Mrs. *Fuller*, aged about Forty years, of a Sanguine Complexion, had a swimming in her Head, could not Sleep, but was in a manner lightheaded in the nights, having a great many incoherent and troublesome Fancies and *Chimæra's* in her thoughts. She had likewise some pains in her Back, sometimes in one place, and sometimes in another, and, to be short, told me she had no free part about her, but every where was troubled with one Ail or other. She had likewise every day several loose Stools, and had had several Physicians, who had gone on several methods; but mostly on evacuations by Stool. She grew always worse and worse on these Courses, was extremely apprehensive of her Life, saw every thing of her Distemper through a magnifying Glass, and upon any sudden fear or danger, she fell into violent Fits, of which you may see an instance before on her Sons death. She was very earnest for Evacuations, but I told her that was not the way. Another Physician, being consulted, told her that she could not be relieved, and that it was Fits. I easily assented to his Opinion of the Disease, and proceeding a contrary way towards her cure, would not as yet suffer any Evacuation, either by Bleeding, Stool or Vomit, but strove by all means to stop her habitual Looseness by *Decoct. Alb.* for her ordinary Drink, Papers of *Creta*, or Chalk powdered, with which I mixt some *Castor*, and gave her over night about Two Scruples of this, with a very gentle Opiat. She had very many other *Cephalicks* given her, as *Sp. C. C. Ol. Succin.* a Bottle of Volatile Salt to smell to. She had also Juleps of *Aq. Puleg.* and other Hysterics, a Decoction after the manner of Tea made of Sage and Rosemary, and several other things of this Nature, by which she did not seem to be very much reliev'd, only the Looseness stopt. Her Head was still very light, and full of Fancies. I durst not after the first nights tryal adventure on any Opiats, but persisted in the other Remedies to stop her loose Stools, and in the Hysterics. So soon as the Looseness was stopt by this method, I put her into a Steel

Of Hysteric Fits, with a Looseness, and incoherent Fancies.

Steel Course, which after it was patiently gone through with great promises of relief, she keeping a very good Diet, and exercising on Horseback and Foot, was brought to her perfect Health. Complaining to me very much of a giddiness in her Head, and that she had not regularly her *Catamenia*, I had her bled in the Foot, once or twice, which did her a great deal of good, and she was ordered to take *Millepedes* every morning in a Glass of water, with which she was very well.

About two months after she was well of this Disease, she fell into a Colick and Constipation of her Belly. She had not been at Stool in some time, had very violent and extraordinary Colical pains in her Belly, and was shivering as one in a Fit of an Ague. I gave her immediately about *gr.* 1. of *Laud. Lond.* with 15. *gr.* of *Extr. Rud.* that after her pains were asswaged by the Opiat, she might have a Stool by the *Extr. Rud.* She fell into most violent convulsive motions, and foam'd at Mouth, but being brought to herself by burnt Feathers, Volatile Salts, Spirits, &c. she complain'd of intolerable pains, till the Opiat made her easie. After she had ease, the Physick began to work, it gave her some very good Stools, and remov'd the Distemper in her Belly, by voiding some hard, round Balls or Pellets, like Sheeps Dung. She was feverish, and inclining to Sweat. I gave her some *Sp. CC.* in a Decoction of Sage and Rosemary to forward that motion of Nature, to try thereby to root out that which I was afraid might in time show its self, *viz.* an intermitting Fever. This accordingly happen'd, and it return'd in about Forty hours, with the same Symptoms, only no Belly-ach, which had been accidental to the first Fit. This second Fit began with inclination, and reachings to Vomit, which I promoted by thin Watergruel, and a Feather put into her Throat. By this proceeding there was a present alleviation of Symptoms for a moment. After these Vomitings came the cold Fit, with which I struggled with *Sp. C C.* and such things, she having Convulsions in it, which were strong. After the Convulsions came the hot Fit, very burning, in which was a great Thirst, and delirous Discourse, for these I gave her as many cooling Liquors as she desired to drink. When the Sweat began to appear, then I ply'd her with hot Sudorificks, by which she sweated very plentifully. After the Fit was over, considering its violence, and her Constitution, I told her I apprehended some danger, if she suffered any more Fits, and therefore advised her to cure this Fever immediately with the *Cortex Peruv.* which I gave her. She mist all signs of a Fit after she had taken it. I gave her *Milleped.* to one hundred in a morning alive, to hinder Obstructions, and forward the *Menstruæ Purgationes*. She found relief in them, and was perfectly well. People who have the Belly-ach have generally rhumatick or gouty

pains

Of the Belly-ach, with an intermitting Fever and Convulsions.

The Introduction. cxi

pains in their Joints or Limbs, and a settlement in their water like Brick-dust.

Major *Thomas Ballard*, Plethoric, of a Sanguine Complexion, aged about Thirty five, much given to extravagant drinking, watching, and sitting up late, sometimes for several nights together without Sleep, was, after a Debauch in Brandy for some days and nights without Rest taken extremely ill. He sent for me, I found him complaining very much of a giddiness in his Head, Palpitation, or as he call'd it, a fluttering at his Heart, very great Faintings, cold Fits, and great cold Sweats. These Symptoms were not always on him, but were sometimes very much abated, and would return several times in a day or night with great violence, insomuch that there seemed to be great hazard of his Life, and that he could not be brought to Sleep. I was at first very apprehensive of an Apoplexy, and therefore had him bled, and for his fainting Fits he took now and then some Cordials, as they are call'd, made of *Aq. Ceraſ. Nigr. Pæon. Comp.* with some Bezoar or *Gaſcoyne-*Powder in them. By these means the cold Sweats were taken off, and he at present relieved. He had likewise some Volatile Salt of *Sal Armoniac* to smell to, some *Ol. Succini* to take some times, and some *Sp. C. C.* at others. He was also blistered in the Neck and Arms, and had an Issue cut in his Left Arm. Notwithstanding these Medicines, his Faintings and Palpitation continued. I gave him Betony, Sage, and Rosemary, to make of them a Decoction to be drank after the manner of Tea, many Preparations of *Caſtor*, and a little *Gaſcoyne-*Powder several times in his Cordial Juleps. He was, likewise, at the instance of some, Bled in the Foot once or twice, but I could not find any of these things to relieve very manifestly, only the Disease seem'd to go off insensibly by degrees, and was in some weeks, by the help of Infusions of *Hierapicra*, and Purges of *Diagridium*, &c. carried off. His Distemper nevertheless lurk'd so about him, that drinking very much, riding in the Sun, or any thing heating the Blood, immediately brought a *Paroxiſm*, which did not easily yield to any Medicines, till it wrought off by degrees of its self by temperate living. I was very apprehensive that these Symptoms proceeded from a *Polypus*, lodg'd somewhere in the great Vessels near the Heart, and advis'd him, on that score, to great Temperance, and a Steel Course. It is very plain this Disease must have come from some great disorder about the Heart, upon the Blood its coming thither in quantity, and not being able to be discharged from thence; but by Palpitations of reiterated Pulsations and efforts. I have seen such a Distemper more than once come from bony Excrescencies about the *Aorta*, great Artery, or Valves of the Heart: perhaps such excrescencies might

Of an extraordinary Palpitation of the Heart.

might be produced the sooner by intemperate living, for Spirit of Wine turns the fibrous part of the Blood, of which *Polypi* are made into a hard *Cattilaginous* or bony Substance. This Gentleman enjoy'd after this Sickness a perfect Health.

Of a Dropsie.
A Servant of his, one *Charles*, a white Man, came to Town from his Plantation. He was about Thirty five years of age, was quite discoloured all over his Body, looking pale, his Face was bloated or swell'd extremely, so were his Legs like Posts, and all his Body, but especially his Belly and *Scrotum*. He made water very little, if any, that which he made was with very great difficulty, complaining of great sharpness and heat. His *Scrotum* was so swell'd with serous Matter, as that it was much bigger than his Head, yet almost transparent, and hiding of his *Penis*, so that very little of it appear'd. This Disease was thought by every one to be an incurable Pox, but I told them I thought there was no Symptoms here but those of a Dropsie, and gave him presently some Jalap in Powder, to about half a Dram, which wrought with him very well by Urin and Stool, insomuch that he found himself much better. I continued it every other, or third day, sometimes changing it for *Extr. Rud.* and on the intermitting days, wherein he took no Purges, I order'd him a *Decoctum ex Lignis*, &c. in a small time, with the help of a *Crocus Metall.* Vomit, now and then given him, his Belly, *Scrotum*, and whole Body were lank. He made water well, and was in perfect Health. To confirm this I gave him a Steel Electuary to carry into the Country, to hinder a Relapse, ordering him to Exercise much, he being a very Lazy Fellow. He not obeying directions, was taken ill again, after the same manner as at first, and came not to Town till the swelling all over was so great, as that he could not stir off his Back. He had likewise a great Cough. I began as formerly with him with Purgers, with which he was very much eas'd, and brought to go about again, but he could not eat, he fell likewise into a Looseness, with which, although I struggled all I could by Opiats, &c. yet in some days he died.

Of a Consumption.
George Thrieves, a Bricklayer, about Thirty five years of age, had labour'd under a Cough for several months, by which he could not sleep in the night. He was also very much troubl'd with it in the day. He had a very quick and Feverish Pulse, especially towards the Evening, and was very much emaciated and weak, that he could very hardly stir. I gave him a *Linctus* of *Ol. Amigd. Dulc. Syr. Capill. Ven. Diacodium*, and Sugar-Candy, and besides in the evening he had an *Hypnotick* Draught, and some Drops of *Liq. Laudan.* with which he was very much better, and although he had been left as desperately

The Introduction. cxiii

desperately ill; yet by these, and some Pectoral Decoctions, he was above two months in a condition to ride abroad, and very hearty, so that 'twas thought he would have intirely recovered. The constant use of the Opiats stopt him so much that I was forc'd now and then to give a Pill of *Extr. Rud.* to procure him a Stool. He took some Balsam of Sulphur, and *Locatelli's* Balsam; but he fell into a Looseness, and so died, notwithstanding *Confect. de Hyacinth.* and other things of that Nature given to stop it.

His Wife, after his death, was taken very ill of an intermitting Fever, which was very violent. She had been treated with several sorts of *Diaphoreticks*, with which she grew worse, but sending for me, I perswaded her, with much ado, to take the *Cortex Peruv.* by which in a little time she was cured. *Of an intermitting Fever.*

Mrs. *L.* aged about Forty years, on drinking too much Wine, fell into a *Cholera Morbus.* She sent for me, was vomiting very frequently, and going to Stool often. I order'd her to drink a great deal of very thin Chicken Broath, and so helpt her to Vomit much more than she did before, and towards the evening when I saw a sufficient Evacuation both ways, and that she was not well able to bear much more, I gave her some *Confect. de Hyacinth.* in Cinamon-water, into which was dropt fifteen Drops of *Laud. Liq. Cydon.* with which a little fair water being added, and *Syrup. Gariophyll. ad grat. Sapor.* she was perfectly cured. *Of a Cholera sicerbus.*

About six months after she fell into a very great Pleurisie. She had a most extraordinary violent pain in her Side especially on breathing, Sighing or Coughing, all which she was troubled with. I had her Bled in the Arm to Ten Ounces, gave her some powdered Crabs Eyes, some Linseed-Oil and Sugar-Candy. These did not relieve her, but she grew worse and worse, had cold Sweats, &c. I gave her then some Cordials made of several Vinous Spirits, made her take some *Sp. C. C.* to keep her alive, and had her bled to a good quantity, five times in her Foot and Arm in Twelve hours. I had likewise her side rub'd with *Ung. Dialth. & Ol. Lumbric.* as also a hot Bag of parch'd Salt put warm to it. In this she felt some small relief, but she found more advantage in the bleeding than any thing else, though it seemed very excessive. By these methods she recovered her perfect Health. *Of a very violent Pleurisie cur'd by large bleeding.*

Her Husband had been sick and weakly for many years, his Skin yellow. I gave him some *Extr. Rudii* by which he was cured, and rid abroad. He sent, on another Fit of the same Disease, for another Physician, what was done I know not, but when I was sent for again he had an extraordinary yellow Skin and Eyes, a great weakness, scarce able to stir, a vomiting of very filthy mucous disco- *Of a Cachexy and Singultus.*

(ii)

discolour'd Matter by Mouth fulls, and a perpetual Hiccough. I thought his Distemper now scarce curable, but advis'd him to take some Drops of *Elix. Propr.* with some Tincture of *Arnotto* or *Achiotle*, which he us'd to take for the Stone in Rhenish Wine. This Tincture, (which is a fam'd Diuretic in these parts,) he took, but his *Singultus* perpetually following him, except when he had a little Sleep, which was very short, he died.

Of intermitting Fevers.   Mr. *Fletcher*, twice in two Epidemical Constitutions, fell into intermitting Fevers, which by the help of the *Cortex Peruv.* were both cured without returns.

Of an intermitting Fevers, and a Cough, resembling an *Hectic*.   A Man Servant of his likewise very many Symptoms of a begun Consumption and Hectic, as great coughing, especially in the night, difficulty of breathing, &c. with these he had an intermitting *Tertian* Fever. I gave him the *Cort. Peruv.* as usually, and he was not only freed of his Ague, but perfectly recovered of his other Symptoms, which I was apprehensive would have been much more troublesome than the Fever. Thus I have seen very often that seeming Hectics have been cured by the Bark.

Of Madness.   A *Negro* Woman of his called *Rose*, who us'd to be about the House, and attend Children, grew Melancholy, Morose, Taciturn, and by degrees fell into a perfect Mopishness or stupidity. She would not speak to any Body, would not eat nor drink, except when forc'd, and if she were bid to do any thing she was wont to do, before she had gone about it, she would forget what her Commands were. If one brought her out to set her about any thing, she would stand in the Posture she was left, looking down on the Ground, and if one further, as for instance, put a Broom in her Hands to sweep the House, there she stood with it, looking on the ground very pensive and melancholy. She had fallen into this the Full Moon before I saw her, and had afterwards her Exacerbations always on Full or Change. I had her Cupt and Scarified in the Neck, ordered her a very strong Purge of *Extr. Rud*, to be forc'd down her Throat. This did not work. I gave her six Ounces of *Vin. Emet.* telling her it was a Dram, which wrought pretty well with her. I gave her also, several days very strong Doses of *Diagridium*, or Jalap amongst her Victuals, which sometimes wrought none at all, and at other times would work pretty well. In a months time there was much alteration for the better, so that she was concluded not to be bewitched by her own Country people, which was the Opinion of most saw her. This happens very often in Diseases of the Head, Nerves or Spirits, when the Symptoms of them are extraordinary, or not understood, to be attributed by the common People to Witchcraft, or the Power of the Devil. *Assa fœtida* is used in Exorcisms,

which

The Introduction.

which I take to be more proper for Hysterick or Nervous disorders of those to be Exorcis'd, than to offend the Nostrils of the Devil. I blistered her Neck, gave her now and then *Vin. Emet.* to six Ounces, or *Merc. Vit.* to eight or nine Grains, she being very hard, as all mad People are, to work on. She had some white Pustles rose all over her Skin, and by the use of these Medicines alternatively, she came to her self, went about her business, and was well.

Worms of all sorts are very common amongst all kinds of People here, especially the Blacks and ordinary Servants. They are very often obliged to eat the Country corrupt Fruits, Roots, and other Meats apt to breed many kinds of Vermin in the Guts. Sometimes these Worms cause Fevers, which run very high with great intermissions and exacerbations, sometimes Convulsion-Fits, very often great pains in the Belly and Stomach, now and then bloody Excrements, and at other times persons Vomit up Worms of divers shapes and magnitudes. It is often very hard to find out the cause of these Symptoms. I used to give immediately some *Diagridium* and *Merc. Dulcis* mixt together, which usually brought some of them away, or quieted the Symptoms. I us'd to allow a Grain of *Diagridium* to every year of the Childs Age, and about half the quantity of *Merc. Dulc.* Sometimes in aged People I gave *Pil. Coch. Min.* or *Extr. Rud.* mixt with *Calomelanos*. I have seen eighteen Worms come away in a day or two with these Medicines. Sometimes I would give Cinnabar to Children this way diseased, if I apprehended their Head to be affected, or if other Medicines took no place. I often gave *Corallina* in very great quantities, which I never saw do any great matter, but by Wormseed given I have found great success, as also by Oil. Very often the Worms here are proof to all these Medicines, and carry off abundance of People, which chiefly happens when they have eat through the Guts, or are in so great numbers, or lodg'd in such corners and recesses of the Guts; that although there be a plentiful Evacuation of them, yet so many remain as to be mortal, or lie in such places as the most effectual Medicines come not near them. 'Tis usual here to give Children easie Purges, or Wormseed, at the Change or Full of the Moon. These Worms in Children come very often from sucking Sugar-Canes Raw, and make the Children look very pale and wan.

Of Worms.

One *Harris*, a Joyners Wife, came to me with a Child of about seven years of Age in her Arms, which had the Face strangely swell'd, especially the Lips, which were tumified prodigiously, and made the Child look very deformed. The Nose was also all over red, and swell'd out in lumps which were very much inflam'd, his Throat was also somewhat affected. I ask'd her whether or no she

The Introduction.

Of a Child in danger of Life by Salivation on taking Mercurius Dulcis.

She had given any thing to this Child for the Worms. She told me she had, and that one for that purpose had given her a Powder. Desiring to see it, she shew'd me a Paper wherein was a great deal of *Merc. Dulc.* Lest the Child should be choakt by the stopping of the Salivation, I order'd it to be kept moderately warm, and to be bled to six Ounces, and after that to take some *Diagridium* to make a Revulsion, and carry the humours off by Stool. I order'd her afterwards to apply *Empl. Diachyl. cum gum.* to all the lumps to ripen them. I desir'd they should have a care of opening them on the outside by Incision, lest Scars might follow, and make the Face all over deform'd and ugly, for in such case the more the Child grew, the greater would be the Scars. If small Tumours break without a Knife or Caustic, there is not so much danger of marks. With *Diagridium* Purges now and then given, the Face lessened, came in shape, and was pretty well, and by the application of the Plaister, the lumps ripened and broke one after another, and the Child was well without so much as the appearance of a Scar. If *Merc. Dulc.* be mix'd in a larger quantity with Jalap, or other purging Powders, after the Powder has been stir'd, and some Doses taken out, the *Merc. Dulc.* being heavier, goes to the bottom, and so consequently, after some time, is taken in larger proportion than the Purgers, and occasions such Accidents as Salivation, &c. This preparation of Mercury likewise acquires a bad Corrosive quality by lying long in the Air, and so does *Antimonium Diaphoreticum* get an *Emetic* quality.

Of a giddiness in the Head.

A lean spare Woman, aged about Fifty, complain'd very much of a giddiness in her Head. She told me that when she was in the Fortieth Year of her Age her *Menstruæ Purgationes* had left her, and that then she began to be out of order. She had had the *Catamenia* first at Eleven. I gave her some *Sp. C. C.* thrice a day, and would have bled her, but that her weakness was a *Contra-Indication.* With that Spirit she recovered.

Of a Fever and Cholera Morbus.

Mrs. *Pain*, aged about Thirty five years, was taken very ill of a Fever, with which she had perpetual Vomitings and Stools, without any respite. She being in some time very much weakned by it, I gave her some *Laudanum* at night in a *Bolus* with *Theriac. Androm.* but it had no success. I order'd her to drink great quantities of Watergruel to wash away the Cause, and then gave her again some easie Opiat, which, nevertheless, succeeded not. I then gave her great quantities of *Sal Prunell.* by which she was much reliev'd of her vomiting, and was in some time by keeping a cool Regimen very well recovered.

The Introduction. cxvii

One *Pavey*, of about Fifty five years of age, complain'd to me of a great Oppression, or Lump at her Stomach, that she could not swallow nor eat, she had likewise a pain there. I durst not give her a Vomit for her weakness and age, but I ordered her about Fifteen gr. of *Pil. Coch. Min.* in two Pills, with which she was gently purg'd and well. Of an oppression at the Stomach.

A Cooper had a blow on the *Sternum*, with a Horses Foot, which healing, had a knot or *Callus* visible on it, he complain'd of a great and constant pain at his Stomach, which had been proof to several methods us'd by several Physicians. I suspected it might be a depression of the *Cartilago ensiformis*, and order'd him to be Cupt on it, to endeavour drawing of it to its place. I also order'd a sticking Plaister to be drawn violently off of his Breast, with which he found relief, but not so much as when at Sea. I doubted Whether the *Compressio in congressu Venereo* might not depress the *Sternum* and make him worse, he being worse every morning when at home, and better when he was at Sea absent from his Wife. Of a depression of the Sternum and Cartilago Ensiformis.

Mr. *F—* aged about Twenty four, extremely Corpulent and Fat, us'd to eat very heartily, and drink very hard without any great prejudice. One evening he made a Challenge to another, who thought himself able to bear more drink than he, desiring him before the present Company to come to a fair tryal in that matter. They had drank, by computation, about a Quart and a half before Supper, and at Supper, in about three quarters of an hours time, drank to, and pledg'd one another in six Draughts of *Madera* Wine, drunk out of six Calabashes or Cups, holding each a Quart by measure. The drinking so hard, and in so short a time, seiz'd this Gentleman all of an instant, that his Eyes turn'd in his Head, stood fix'd, and he began to sink down in his Chair. He was carried out of the Room, and plac'd in a great arm'd Chair, where he immediately fell into an extraordinary deep Sleep. Nature struggling, and making now and then an effort in the Stomach to disgorge what was unwelcome to it both in quantity and quality, was always check't through his fast Sleep. I thought I could hear a begun Vomit in the bottom of the Stomach towards the *Oesophagus*, but being so fast asleep, that he could not give way to it, it was stopt there, and could get no farther. He was likewise in danger by his Head hanging in several Postures, whereby in some the *Aspera Arteria* might be compressed, or the Jugular Veins, that he might be strangled. To avoid this I set one or two to watch and keep his Head in such a position as might hinder those Accidents, and forward as much as might be Vomiting, which I likewise did by thrusting Feathers as long as I could get into Of two Persons who drank a great quantity of Wine.

(g g) his

his Throat. This help'd the coming away of a great deal of discolour'd mucous matter, which lying behind might have choak'd him, unless he were help'd a little, by keeping his Mouth open with a great Key thrust between his Teeth, with the Wards turn'd uppermost, so that his Mouth could not be quite fill'd with that Matter. He foam'd a little at Mouth, and breath'd very high and uneasily, wherefore apprehending some imminent and present danger of some Apoplectic Distemper, I had him very largely bled at the Arm, which very much calm'd all the Symptoms, especially those threatning his Life every minute. He was narrowly watch'd all night. I would have given him a Purge if he could have got it down, but he was so fast asleep that it could not be done. He was carried in a double Sheet, after five or six hours Sleep, to his Lodgings, and laid abed, his Head high, and taken great care of lest he should be strangled. Finding him still in great danger of a Head Disease, being speechless, not to be awak'd, I gave order that he should be forthwith bled again in the Arm to some ten or twelve Ounces. After this he could bring out half a word, and then by degrees came to speak a whole word, than two or three, and so a Sentence. Then I gave him some *Pil. Coch. Min.* about two Scruples dissolv'd in water, to clear him of his inflam'd and swell'd Throat, as well as of the disorders of his Head and Stomach. After this had wrought he was very sensible of his escape, Penitent, and with Tears in his Eyes express'd his concern. He was scarce yet able to speak intelligibly, but by another Purge next morning, in some few days he recovered his Health.

The other, who pledg'd him, had drank less by three Pints, and before he fell asleep had vomited a very great quantity of the Wine he had drank. He slept till morning, and continued three or four days very much flustered or hot headed, without any further mischief. Both these Gentlemen died since in *England*, and, as I have been told, shortened their Lives by such Actions.

Of the Belly-ach and Rheumatism.

Mrs. —— aged about Thirty five years, of a yellowish colour, had been in an intermitting Fever, for several Weeks, which ended in the Belly-ach. She had a very great pain about the Region of the Navil, Constriction of the Belly, high colour'd thick Water, and frequent pains all over the Body, like those of Rheumatic People. She had also a *Nausea* or inclination to Vomit, which when forwarded (which I generally did in this Disease) would ease her somewhat by emptying the Stomach. She had gone through all Courses ordered by several Physicians without success. I had her bled several times, by which she found no great relief, neither did she by easie Purgers, which although they wrought well, and gave some immediate

mediate ease, yet they were far from taking away the Distemper. She was extraordinary weak, insomuch that I was forc'd to allow her Cordial vinous Spirits, *Sp. C. C.* and several things of that kind, which by their inflaming qualities seem'd to be very prejudicial to her, although absolutely necessary in respect of her faintness. She had also Emulsions of all sorts, and because she could not Sleep a great while together, I gave her an easie Opiat without any success. I likewise gave her Purgers with *Merc. Dulc.* and order'd her Clysters of all sorts. Her Diet was Broaths of all kinds, Watergruel, *&c.* I was told she drank much Brandy and strong Liquors, which I was inclinable to believe, because these cooling Remedies did not at all relieve her. I told her a cooling Regimen was much the best, but could not hinder her drinking strong Liquors, whereby she continued under her Distemper several months. I was told there was something Venereal in it, wherefore I gave her some Mercurial Purges, sweated her at night, and order'd her a Diet-Drink of *Sarsa, China, &c.* by which she found some, but not much relief. She went into the Country after she had lost the use of her Limbs. She recovered them in some measure, by degrees, with the help of the *Green With,* and some Salves, and came to her perfect Health. By this I apprehend she was clear of all Venereal infection, but that strong Liquors had been the occasion of the long continuance of this Malady.

A Turner belonging to Colonel *Nedham,* of about Forty years of age, was taken with a great pain about his Navel, he could not go to Stool, and had a great *Nausea,* or squeamishness at his Stomach, which made him Vomit sometimes a small quantity of mucous matter. I gave him some *Pil. Coch. Min.* about fifteen Grains, to try whether it would work it down, it did not soon, wherefore being in pain he sent for another Physician, who gave him a Clyster, which did not at all ease him. He sent to me again the next day, I gave him some four or five Pills made up of about two Scruples of *Pil. Coch. Min.* and order'd him to drink much Watergruel. It wrought very well, and he was very much eas'd, but his Disease return'd in a small time (as it usually does) and he was in the same condition. After some hard Balls of Excrement or Pellets had come away by Stool, with the same Medicines he was perfectly cured. Of the Belly-ach.

A Tailor of Colonel *Nedhams* was soon after taken ill after the same manner, as were likewise most of the *Indians* and Blacks about his House, some whereof fell into this Distemper by drinking Rum-Punch, others by other Causes. I found the aforesaid method with *Extr. Rud. Pil. Coch. Min.* Jalap, *Diagrid,* or any Purgers to be very effectual, Of the Belly-ach.

effectual, and Clysters to be seldom beneficial, except sometimes they were very strong, made of Gourd or Tobacco Leaves. These Clysters are sometimes so violent as to cause very great disorders, and to bring after Convulsions death, which has happened to several so affected. 'Tis very ordinary to have in this Distemper Relapses, for several times after one another in some hours, and at each return, after the working of purging Physick, there are voided Pellets like Sheeps Dung, as hard as Stones, after which comes ease, and then violent pains, whence a necessity of taking Physick, and then the like Balls or Pellets again. Opiats in this Distemper seldom relieve, and are very hurtful in that they stop up the Belly, and give no great immediate ease unless sometimes when they are mixt with Purgers. Balsamics are very proper in this Disease. I us'd to prevent a Relapse in this Distemper, to give the Leaves of *Sena* to about one or two Drams, to be boild in thin Watergruel, or Chicken Broath, which keeping the Belly open, they have been reliev'd. This word (the Belly-ach) is given to several Diseases, where there are great pains in the *Abdomen*, as Stone, &c. and always to the Rheumatism, and for the most part this last is join'd with what one may call the true Belly-ach. I think the Belly-ach consists of such variety of Symptoms, that there is no curing of it but by several Medicines us'd in a right method, and persisted in for some time.

Of an inflam'd Eye from the Milk of the *Mansanillo*-Tree.

The Turner before-mentioned, in felling a *Manfanillo*-Tree in the Woods, some of the Milk spurted into his Eye, whereby it was extremely sore and inflam'd, and in a nights time the Eyelids were so swell'd and glu'd together by the gumminess of the Milk, that he could not open them. I order'd him to be immediately bled, gave him a strong Purge of *Extractum Rud.* and order'd him to wet his Eye very often in cold water, keeping a wet brown cold Paper to his Eyes. When one Paper was hot a fresh cold wet one was put on, to hinder a defluxion on the Eyes, and to cool and take away the acrimony of the Humours came that way, and occasion'd great pain, heat, and restlessness. These things being done, in three days he was cured.

Of the *Lues Venerea*.

One sent for me to her Daughter, about Fourteen years of age, she was strangely distorted in her Fingers, and at every joint there was a white swelling round about it. She had likewise several Ulcers on her Feet, and was so Lame that she was forc'd to use Crutches, and could scarce stir with them. I was told by her Mother, that these Disasters had come on her after the Small-pox, she not being purg'd, she seem'd likewise to have some snuffling in her nose. It was thought she came into the World, when her Mother was tainted with some Species of the *Lues Venerea*, but her Mother would

The Introduction. cxxi

would not hear of Fluxing. I gave her about Fifteen Grains of *Pil. Coch. Min.* which she took every Week twice, and about Fifteen Drops of *Sp. C. C.* every intermediate day. I continued this Course for several Weeks, only in lieu of *Pil. Coch.* she took some powdered Jalap, and by this means, in some time, she by degrees recovered her Health, and was able to go about her Business, climb Trees, and throw away her Crutches. The Tumours on the Fingers, and Joints subsiding, she grew every where well, except one small Sore in the bottom of her Foot, which was not skin'd. I order'd her for that Ulcer to be purg'd, and a *Decoctum Chinæ*, but no Chirurgeon taking care of it, neither she keeping a good Diet, but eating Pepper, &c. she grew worse. In this case I left her, neither did I believe she would recover without fluxing, this Disease being, as I thought, a Species of the *Lues Venerea*.

Her Mother was very weak, and complain'd to me of great pains, that she could not Sleep in the night, and could scarce walk about. She was old, very weak and Paralytic. I gave her some easie Purges, and had her bled without any relief. She was not capable of enduring a Flux, for she could not be patient. I gave her some *Ol. Chym. Rorismar.* she found no relief. I ordered her about fifteen Grains of Jalap, to be taken next morning. I went then to see her, and found she had vomited, was in cold Sweats, and speechless. I gave her some *Sp. C. C.* and burnt Wine, but she grew worse. I would have given her an Opiat, but she could not swallow it. Notwithstanding she recovered, and Spit as if she had been salivated for a great while after. I suspected, because she had formerly taken the same Dose of Jalap several times without any such effects, that either designedly to do her good, or maliciously to Poison her, her *Negro* Woman had chang'd the powdered Jalap for some *Merc. Sublimat. Corrosiv.* or some such other violent Medicine. About a month after she died, but of what, or how, I know not. *Of a Salivation.*

An Overseer belonging to Colonel *Ryves*, aged about forty years, had been several times troubl'd with the Belly-ach, after curing of which, for some considerable time, he was usually blind. This blindness had now been on him, after this Fit, for some months, and he was very much discolour'd in his Face and Skin. I advis'd him to Bleed, and next day Purge with *Pil. Coch. Min.* which he did. I also gave him an *Eleosacch. Rorismarin.* & *Sp. C. C.* in great quantities thrice a day. He took likewise some *Ol. Succini*, designing by these Remedies to remove the Obstruction of the Optic nerve, and envigorate his Spirits, his Eyes having no outward visible Disease. I blister'd his Neck, and although I continued some days in this course, *Of blindness after the Belly-ach.*

(h h)

course, yet no success follow'd these Medicines. I order'd him to take about fifty live *Millepedes* in a Glass of Water twice a day. He did this some days, found his Eyes much strengthened, but would not stay any longer. I gave him directions when he went into the Country, but know not what became of him afterwards. I have seen total blindness come many times in the Belly-ach, both in *Jamaica*, on the way thither, and in *England*. There is no blemish to be seen in the Eye, but it seems the *Morbific* Matter is translated up to the Head. I never saw any but what recovered their sight afterwards by proper Remedies. Convulsions are likewise ordinary in this Disease, but I think they are much a worse sign than blindness.

Of a bad sight from excessive Venery.

One *Henry*, a *Negro*, Overseer of Colonel *Ballards*, much given to Venery, fell into a blindness by degrees, so that he could see very little at any distance, nor well near at Hand. I look'd very earnestly on his Eyes, but could not see any blemish. I advised him to be very Chaste for some time, and had him cup'd and scarified in the Shoulders, blister'd in the Neck several times (which I account more effectual then a *Seton* because there is a sudden great Evacuation of serous Matter in the one, and but a slow and habitual small discharge in the other) gave him great quantities of *Sp. C. C.* and *Millepedes* without any relief. After a great many Weeks persisting in this Course, and use of several Cephalic Oils, by way of *Eleosacchara*, I gave him an Electuary made of Steel, &c. and order'd him a Regimen proper for a Steel Course. By this in some time he by degrees recover'd his Eye-sight, and found them very much strengthened every way. He was sent into the Country to mind the Plantation business as formerly, whither he went provided with a quantity of his Electuary, and I never heard he had a Relapse, which in all likelihood I should have done had his Distemper return'd; for Planters give a great deal of Money for good Servants, both black and white, and take great care of them for that Reason, when they come to be in danger of being disabled or of Death.

Of Fluxes, *Diarrhæas*, and *Dysenteries*.

Fluxes and *Diarrhæas* of all kinds, as well as Dysenteries or bloody Fluxes, are at all times here very common to all manner of People. As for Fluxes, provided they be moderate and within bounds, I always avoided stopping them, but rather if I saw that they went on easily, *Cum bona ægri tolerantiâ*, gave some innocent Remedy, or some easie Medicine to help it forward. This is one of the most usual and Salutary ways Nature disburthens its self of Morbific Matter, which otherwise might occasion great Disorders. But if a great Fever be join'd, or if there be so great an Evacuation that the Person is grown weak, I us'd to order the Person to be immediately bled for

the

the Fever. Very often in this, as well as in the Belly-ach, there is an inflammation in the Guts, which occasions a Gangreen if not timely remedied. This appears frequently upon the Diffections of diseased Bodies. I have not only seen this in Men, Women and Children, but in Horses. I used to order Rice to be boil'd in Water for ordinary Drink, and the Rice eat with Milk, as also frequently to give *Decoctum Album* for the ordinary drink of the Patient, or some *Creta albissima*, or fine Chalk powdered, and made the same way into a Drink as the Hartshorn Calcin'd in the *Decoctum Album* is wont to be us'd. I would put to it some times *Bolus Armen.* and likewise give half an Ounce of these Powders twice or thrice a day, and usually in the Evenings a *Bolus* of *Diascordium* or *Ther. Andr.* with an easie Opiat of liquid or solid *Laud* according to the Age of the Patient. If the Looseness continued long, it usually washed away the *Mucus intestinalis* corroded the Guts, and ended in a Dysentery, for which I give, after bleeding several times, the same Medicines, as for a *Diarrhæa*. It is very ordinary after eating Shell-fish, as *Conches*, Oisters or Crabs, by people thrown on Cayos, Desert Islands or Rocks by Shipwrack, and feeding on these for their only suftenance, to fall into Fluxes and Loosenesses, greater or lesser according to the time they have continued on such places. I had one under my care, who had been Shipwrackt so on some Cayos on the North side of *Hispaniola*, going to the great Plate Wreck, whom I could very scarcely recover with all the abovesaid methods several months persisted in, and with Bees Wax inwardly given. He took also Rhubarb in Powder without success. On taking of the Wax form'd into Pills it came away by Stool the same way that it was taken in, without much alteration. I saw once in *Jamaica* in the latter end of a *Phthisis* one Dram of Rhubarb, with five of *Terra Sigillata*, some *Confect. de Hyacinth.* and Cinamon-water, do very well when nothing else could stop a dangerous *Diarrhæa*, but the vertue of it only continued for some small time. I have known in Epidemic Dysenteries Flower boil'd in Milk, with some Wax scrap'd into it, do very great Cures. But by the abovesaid Medicines, some, or all of them, I have cur'd hundreds in *Jamaica* of these Distempers. Paper boil'd in Milk was us'd in *France*, in *Diarrhæas* infesting the Army, with very great success. This I was assur'd by an Officer in the French Army at *Toloufe*,

Mrs. *Halstead*, aged about forty years, of a clear Complexion, was very much troubled with flushings in her Face, and small lumps, which by drinking Water, or cool Drinks, which she thought would remedy them, she grew worse. I order'd her to drink as much Wine as usual, and to be bled, after which she was purged with *Extr. Rud.* Of flushing in the Face.

and

and then took an Emulsion of the cold Seeds, with which she was most violently purged. After bleeding, purging, and the use of some easie Diaphoreticks, I gave her a mixture of Allum and *Sulph. Viv.* powdered. These mix'd and ty'd up in a Linnen Cloath, I ordered her to dip in water, and then to rub on her Face several times a day. This Medicine being continued for a great while, she was perfectly well of that troublesome Distemper.

Of the Itch.

— — — A Laundrey Maid, was troubled very much with the *Pruritus* or Itch, it rose in small little whales, all over her Body, especially between the Fingers, and was uneasie both by its Itching and unseemliness. I bled and purg'd her, ordered her for three mornings and nights, to take one Dram of *Flor. Sulph.* in any Vehicle, and then to anoint herself with *Ung. ex Oxylapath.* in which powder'd *Sulph. Viv.* is mixt with some Drops of Chymical Oil of *Saſſafras*, to take away the smell. Every night before she rub'd, she took a little *Flor. Sulph.* inwardly. With these things she was cured.

Of *Chegos*, and the Consequences of them.

I found an uneasiness, soreness, or pain in one of my Toes, as if a small Inflammation or Tumour had been there rais'd by the pressure of some part of my Shoe. I had a *Negro*, famous for her ability in such cases, to look upon it, who told me it was a *Chego*. She (who had been a Queen in her own Country) open'd the Skin with a Pin above the swelling, and carefully separated the Tumour from the Skin, and then pull'd it out, putting into the Cavity whence it came, some Tobacco Ashes which were burnt in a Pipe she was smoaking. After a very small smarting it was cured.

This Tumour is accounted of two sorts, either poison'd or not poison'd, both are about the bigness of a small Field Pea, being almost round. They have a few Fibres, by which they are fastened to the Flesh as by a Root. That call'd poison'd has a black spot in it, and is accounted worse than the other to cure. They contain, within a thick Skin, a great number of small Eggs or Nits, white, and crackling when bruis'd. These are the Spawn of a small blackish sort of Louse or Flea, which harbours its self and lives in Dusty or unclean places. The Mother, I apprehend, puts and insinuates these Eggs under the Skin of Men and Women, as other Insects do their Eggs into the Barks or Leaves of Trees. They infest the Feet of most People under the Nails of the Toes, or any where about the Heel. If these insects be left to themselves they will Spawn and multiply in the Feet to great numbers, and bring bad Accidents, which I think come rather from the depending of the part than any poisonous quality of this Insect or Air.

A

A very neat Lady had one of these Bags bred in one of her Toes, part of it was by a Black taken out with a Pin, but it seems not the whole Bag, (as it ought to have been) She complain'd of some pain therewith, which by her walking about inflam'd very much. She shew'd it me, I advis'd her to keep her Bed, and to dress it with a little *Ung. Basilicum* and *Precipitate*, putting over this Liniment some *Diapalma* Plaister. By these in some Weeks time the festering was gone, and she with simple *Basilicum* cur'd with keeping her Bed, or Leg up, for on the least hanging down, it would inflame again. Although she was well in her Toes, yet she had a swelling rose in her Knee about the *Patella*, it was very painful and red, as if Wind were under the Skin. I thought, it being a dangerous Place for an Ulcer, it was best to Bleed and Purge, thereby to hinder a defluxion of Humours to the part. This being not my proper business, I committed her to the care of a Chirurgeon, who applying Poultesses, &c. to it, it broke and kept running for a long time, after which it cicatriz'd not without great trouble.

A little *Negro* Boy, by leaping off a high place at Twelve years of Age, strain'd his Knee, whence came a puffing, soft, painful Tumour, red and large. This being an ill place for a swelling, because of the *Patella*, &c. and its being a depending part, I order'd him to be bled presently, to be purg'd the next day, and to keep up the Knee. By another Purge or two his Knee healed, the Swelling falling by degrees without any sign of an Apostem. *Of a Swelling in the Knee.*

An old Woman of Seventy years of Age, complained to me she had not made water in a fortnight, her Belly was very much swell'd. I gave her a Dose of *Jalap*, by which she not only went to Stool, but made Water very plentifully. I order'd her a Bath of fair Water and *Culilu*, to sit in it lukewarm. She continued well for a while, but then, as I heard, died, I suppose, with Age, and weakness. The like Operation as is above related I have found several times on the giving of Jalap, and sometimes I have observed a contrary effect, though from a different Cause, which was, that it had by great purging so drain'd the *Serum* from the Blood by Stool, that none remained in a Day or more, to come by Urine. *Of the great Effects of Jalap upon the Serum of the Blood.*

I was call'd to one who made extraordinary complaints of very great pain in the bottom of her Belly; she could not make water, nor had in several days. I gave her some *Sal Prunel.* and put her into a Bath of *Culilu* and fair Water lukewarm. She vomited, and had all the Symptoms of the Stone in the Bladder, although she *Of the Stone in the Bladder, taken for the Belly-ach.*

(ii) took

took it for the Belly-ach, and call'd it so. Her ordinary Physician returning to take care of her, I went no more near her, nor know I what became of her.

<small>Of a Worm in the muscular Flesh of *Negros* coming from some parts of *Guinea*.</small>

From some parts of *Guinea*, as is before related, come *Negros* troubled with Worms, they are flat and long, something like the *Lumbricus Terrestris*, and lodge amongst and in the muscular Flesh. I saw one who had one of these Worms in his Thigh, half an Inch of the end of it was hanging out, which was flat and blackish, and there issued out a thin Ichor by its sides. I was told that the only Remedy for this Distemper was to draw it out by degrees every day some upon a round piece of Wood, as a piece of Tape or Ribbond. After they have pull'd it all out, they apply a Plaister to the part. I was assured that if any part of this Worm, which is tender and very long, and requires great care in the management of it, should chance to break within the Skin, that there follows an incurable Ulcer.

<small>Of the *Yaws*.</small>

A *Negro* lusty Fellow, was taken ill of the *Yaws*, he had not been long from *Guinea*, and was all broke out into hard whitish swellings, some greater, some lesser, from the bigness of a Bean to that of a Pins Head, of which last size there were many which appear'd like the Glands of the Skin swell'd and white. When these Tumours are large, they are usually white at top, from some of the *Cuticula* and Humours dried lying in Scales over it, and sometimes they weep out an Ichor. At other times the Ulcers are much larger. They likewise complain sometimes of great pains in the Bones, and this Fellow, whom I cur'd, was broke out very much about the *Penis, Scrotum*, and Elbows. I flux'd him by Unction in the Corner of an Out-House, feeding him with as much Watergruel as he could eat or drink. The Flux proceeding as it ought to do, he was quite clear'd of this filthy Distemper, only on his Elbow he had one swelling, not quite dry, to which I applied calcin'd Vitriol, which made the Scales fall off, and heal as the rest.

This Distemper is thought to be contagious, and to be communicated from one to another, from Blacks to Whites, and from Parents to Children, but I cou'd n't observe it to be more or less contagious than the Pox. There are few Plantations without several of these Diseas'd Persons, who are usually cur'd as above. Though 'tis commonly thought that fluxing does not cure without Relapse, yet I, by what I could observe, find it does, and do believe the return of this Disease comes from their not being thoroughly flux'd by anointing, or by being kept too warm, or wrong treated afterwards, whence some Remains of it staying behind in the Body, these Dregs by Degrees bring the same Distemper again.

It

The Introduction. cxxvii

It is commonly thought that this Distemper is curable without fluxing by Purgers, but I could never find it so, or that there were what is pretended by many, *Negros* who understand by some Specific Herbs to root it out. I believe that Purgers, &c. may, as in the Pox, take away some of the Symptoms for a while, so that sometimes they may not appear in a considerable time, and afterwards may shew themselves the same as at first. Some sorts of this Distemper seem to me to be the *Elephantiasis* or true Leprosie of the Antient and *Arabian* Physicians. Others said to have this Disease were plainly *Scrophulous*, or had the Kings-Evil; and most said to have it, had the *Lues Venerea*. Though this Disease is thought to be propagated by ordinary Conversation, or trampling with the bare Feet on the Spittle of those affected with it, yet it is most certain, that it is mostly communicated to one another by Copulation, as some other contagious Diseases are.

Several *Negros* belonging to Mr. *Batchelor*, had after Rain gather'd Mushrooms, and eat them plentifully in their Pepper-pots, or *Oglio's*. whence they all fell into a vomiting and purging with great anxiety. They after a while recover'd, all except one. He had, besides the aforesaid Symptoms, such as are common to a Rheumatism from inflam'd Blood. His Head was very much affected, having a *Vertigo*, &c. on which accounts, as well as the others, he was immediately bled. I order'd him a Contemperating cooling Diet, to take off the Heat and Acrimony of the Blood, as drinking much water, &c. and in a while, he was perfectly well of his pains as well as *Diarrhœa*, which had continued on him a great while after his eating the Mushrooms. Mackarels salted, and beat up with Soot and Salt into a Poultess applied to the Feet, were used in this case, and are reckon'd to draw very powerfully from the Head. Several Gentlemen, some years before, were all very dangerously ill after eating Mushrooms, and one died: the Symptoms all shewing (as I was informed) great Inflammation of the Blood and inward parts. *Of the bad Effects of Mushrooms.*

A lusty strong *Negro* Woman belonging to Captain *Halstead*, had a very unseemly superficial Ulcer on her Wrist. She had likewise one on her Forehead, and pains all over her Body. She also spoke a little through her Nose, and brought thence very often some mucous Matter. I salivated her in the corner of an Out-house, by Unction, and desired she might be fed with as much Watergruel as she could eat, whereby in a Months Time she was quite cured. She had been in the Hands of several Physicians, who did not apprehend this to be the Pox, and therefore had in vain given her several sorts of Physick. *Of the Lues Venerea.*

Wellington

Wellington, a Boy of seventeen years of Age, had a *Bubo* rose in his Groin. I advis'd immediately the ripening of it, as the means to preserve him from the Pox, which not being done, he broke out all over. I could not get any to receive him into their House, because of the Disease, nor a Nurse to look after him, but a *Mulatta*, who after his Flux was rais'd, neglected him in not giving him sufficient Drink, &c. Notwithstanding his Salivation went pretty well on, and he was clear'd all over. Being neglected by his Nurse, some body brought him Tarts, made of unripe Fruit, which he eat. By this means and coming into the Air, his Flux stopt in some measure, but with care his Salivation returned, and he came abroad well. He drank Wine presently to such excess that he made himself Mad. A *Tertian* Fever, which was very Epidemical, likewise seiz'd him. To the other Symptoms of this Fever in him, was join'd a swell'd Throat, which increas'd to that degree in this person, who had by occasion of his late Salivation, a disposition to the Humours coming that way, that he fell into a Quinsie, and so died.

Of a Salivation, after which followed a *Tertian* and *Quinsie*.

Gonorrhæas of all sorts amongst Men and Women are very common here, especially in Plantations amongst *Negroes*. They complain first of the great heat of their water, &c. and have the same Symptoms as in *Europe*. I us'd to purge with *Merc. Dul.* and *Pil. Coch. Min.* &c. which took away most of the Symptoms. If these Medicines did not take place I gave Vomits of *Infuf. Croc. Metall.* or *Turpeth. Mineral.* which in some time never failed the Cure of any, either Man or Woman, with the use of Emulsions. I never gave Adstringents, because there is no certainty in this method when the Cure is perfect. It is generally believed in *Europe*, that *Gonorrhæus* and the *Pox*, are with more ease, and sooner, cured in *Jamaica* and hot Countries, than in *Europe*. I was of the opinion of the generality of the World when I went to *Jamaica*, but found as the Disease was propagated there the same way, and had the same Symptoms and Course amongst *Europeans*, *Indians* and *Negroes*, so it requir'd the same Remedies and time to be cur'd.

Of *Gonorrhæas*.

A black Man, of about forty Years of Age, told me he had great pains about his Navel, so that he could not sleep, he was in a cold Sweat, in great pain, and had not been at Stool in four days. I gave him immediately about fifteen Grains of *Pil. Coch. Min.* on which he found sudden relief, for it wrought four or five times in the night, he taking it pretty late about six a Clock at night. The next day I gave him about six Grains of the same Pill to preserve him from a Relapse, and in three or four days observing that method he was well.

Of the Belly-ach.

One

The Introduction.

One *Mountague*, a Shoemaker, aged forty five, complain'd of a great oppression at his Stomach. I gave him *Infuſ. croci Metal.* an Ounce, and *Syr. Cariophyll. Q. S. ad gratiam.* He took it, it wrought very well upwards and downwards, and he was cured.

Of an oppression at the Stomach.

I was sent for to a Servant of about Twenty years of Age. His Mistress had given him a Vomit of an *Infuſ. Croc. Metal.* It had wrought him upwards and downwards, till he was seized with the Cramp, and violent pains in his Hands and Feet. He was in a cold Sweat, and his Pulse faint. The Mistress, for her vindication, ordered her Servant to shew me how much she had given, who pour'd the Infusion out muddy. I perceiv'd the cause of the Super-Purgation to come from the Powder, or the substance of the *Crocus Metallorum*, having been taken instead of the clear Infusion. I order'd him burnt Wine, and design'd to have given him an Opiat, but he was well without it.

Of an Hyper-Catharſis occaſion'd by the muddy Infuſion of Crocus Metallorum.

Mr. *Thomas Rowe*, about Forty five years of age, sent for me in *Feb.* 1688. I was told he had the Belly-ach. He complain'd very much of a great oppression at his Stomach, and of a load there, he vomited every thing he took, whether Liquids or Solids, had a pale yellow look, and the whites of his Eyes were yellow, he had been sick a month. I concluded it to be the Jaundice in a great measure, and gave him about fifteen Grains of *Pil. Coch. Min.* made into two Pills, to try whether by that means the Stomach might not be emptied of its Load, an irritation at the same time made in the Guts to solicite the excrements downwards, contrary to that motion in him from the Stomach upwards, which was preternatural. Immediately the Pills came up. I therefore ordered him next morning to have thin Watergruel ready to work off a Vomit easily, which accordingly was done very early, because of the heat. His Vomit was *Infuſ. Croc. Metal.* Six Drams, *Syr. Cariophyl. Q. S. ad gratum ſaporem.* This wrought very well, first upwards, then downwards, and he found himself very much reliev'd and eas'd. I follow'd the stroak, and gave him next morning Fifteen Grains of *Pil. Coch. Min.* in two Pills, they wrought him pretty severely, but he found himself reliev'd of all his Diseases except weakness. I wish'd him to get this off by taking an exact care of his Diet, that it were easie of Digestion and pleasant to his Palate, and that he should very carefully avoid taking away any Blood or making use of Physick till further occasion.

Of an oppression at the Stomach and Vomiting.

Mr. *Ridley* a Painter, sent for me, he complain'd of great pains in all his Muscles and Flesh, he look'd very ill, and was yellow, especially

(k k)

especially the whites of his Eyes, vomited often, and went seldom to Stool. He told me he had pains in his Sides, and that he had been bled twice for them. I gave him presently some *Sal Prunellæ* about half a Dram in a Glass of water, he took it, had a Stool, and was eas'd of his pains. His inclinations to Vomit continued violent, wherefore I gave him the Vomit prescribed in the last Observation. This immediately reliev'd him of his vomiting, and after it had wrought well both upwards and downwards, he lost altogether the use of his Hands and Feet, but was very much at ease. I desir'd him to take now and then some *Pil. Coch. Min.* about Seven Grains in one Pill, to keep his Body open for fear of a Relapse, which he did. He complain'd of fainting Fits, and cold Sweats, especially on striving to get up to Stool. I gave him a Cordial to take now and then a spoonful or two of, when faint or cold. It was made of *Aq. Epidem. font.* of each Four Ounces, *Syr. Cariophyll. Q. S. ad gratum saporem.* He recovered by these Medicines his health and strength very well.

Of an oppression at the Stomach, and loss of the use of the Limbs.

I was on the eighteenth of *February* 1688. in the evening consulted for a Child of a year old. It was breeding Teeth, had six or seven Convulsions the same day and had been cup'd. I gave two Spoonfuls of the following mixture to the Child, and order'd it to be repeated every fourth hour. *Recipe Ol. Succin. Opt. gutt. iv. Sacchar. alb. Q. S. fiat Eleosaccharum. Sp. C. C. gtt. iv. Cinnab. gr. iv. aq. font. uncias quatuor M.* I left a Bottle of the Spirit of Hartshorn for the Child to smell to, if it had any more Fits, and ordered blistering in the Neck. They neglected this, and pretended that the Fits came so fast on the Child that it could not be done. On the nineteenth in the morning they sent for another Physician, and about noon the Child died. The Fathers name was *Green.*

Of Convulsion Fits in a Child.

A Child about Ten years of age was brought to me by the Mother which had had for some days so great a *Vertigo* that it could not go alone for reeling. I immediately ordered it to be bled, to Seven Ounces, out of any appearing Vein, and the next morning I gave the Purge following. *Recipe Mass. Pil. Coch. Min. gr. 15. f. Pil. ii. si deglutire pilulas non possit dissolve in S. Q. aq. font. & f. potio.* After this the Child was well.

Of a Vertigo.

One *Isaac*, belonging to the *Crawle* Plantation, was taken very ill, he had a Vomiting and Looseness, which had been violent on him for some time, and had weakned him considerably. I gave him some *Liq. Laud.* which stopt the Vomiting, and some *Sp. C. C.* to strengthen him. By the use and repetition of these Medicines on

occasi-

The Introduction. cxxxi

occasion he recover'd very well, but would not be persuaded to rise out of Bed. He thought himself very ill, that he should not live, but certainly dye of this Illness, his mind being very much sunk within him, I advis'd the People about him to chear him as much as possible, to ease his mind, and get him up out of Bed. He died being very morose and seem'd to have no Distemper on him but Sullenness and Melancholy, and though I took much pains to examine him nicely, I could find no Disease, but only he said he was sure, say what I could, that he would not recover. The Passions of the Mind, both Hope and Fear, have a very great influence on the Body.

Of one who dyed of an ill opinion of his Health and Melancholy.

A Lady about Thirty five years of Age, little, of a Sanguine Constitution, the Mother of many Children, at the Birth of each of them had so copious a Flux of the *Lochia* that it endanger'd her life. She had about a month of her Reckoning yet to come, when, without any extraordinary occasion, she was taken with great pains, as if she were to be deliv'r'd; they were soon follow'd by a very copious Flux of Blood. I was sent for, and took with me all manner of Adstringents, as well as Forcers, to endeavour the Birth of the Child, for in such a case when the *Fœtus* keeps the capacity of the *Uterus* distended, 'tis impossible to stop the Flux of Blood till the Woman is deliv'r'd. Therefore if the flooding be not easily stopt, the best way is to force it away, whereby the Vessels of the *Uterus* being corrugated, the Blood by degrees stops of its self. This must be endeavour'd by all means without delay, for in the Blood is the Life. Though I hastened all I could she was expiring when I came, and had suffered a very extraordinary Flux of Blood, was delivered three quarters of an hour before she died, and had been bled some hours before. If this dangerous Distemper had been timely remedied, by forcing the Child away, the Mother might in all human probability have been sav'd. It is ordinary that the *Menstruæ Purgationes* here, are both less in quantity, and continue for a shorter time than they do in *Europe*.

Of flooding.

Mrs. *Aylmer*, aged about Thirty five years, a spare lean Woman, giving suck to one of her Children Thirteen months old, was taken very ill in an Epidemic Constitution of an intermitting Fever: The *Paroxisms* returned sometimes every day, and at other times every other day. She had gone through several Febrifuge Courses of Vomits, Gentian-Roots, Centaury the lesser tops, *Carduus Bened.* Sweaters, &c. without relief. She complain'd of her feverishness and a pain on the Region of her Spleen, which was somewhat swell'd. The Fits of her Fever decreased upon taking the *Cort. Peru*. I advis'd

Of an Intermitting Fever.

vis'd her to wean her Child, both on her own, and the Childs account, she being weak, and her Milk perhaps not very healthy, but she would not. I advis'd her to a gentle Vomit, bitter Draughts, and what I most relied on, a Steel Course, with which she recovered perfectly her Health. She was sometime after taken Ill of a Quinsie, which by Bleeding and Purging was taken off, the swelling was not very dangerous, being not so much in the Throat, as Mandibles and Cheeks.

Of a sickness at Stomach.

Loveney, a very sensible *Negro* Woman of Colonel *Ballards,* complain'd very much of a great illness, at and about her Stomach, with which she was always out of order. I gave her a Vomit of *Inf. uf Croc. Metal.* sweetned with *Syr. Cariophyll.* which working very well, she was cur'd of all her Indispositions had lurk'd about her for several years. When tough and phlegmatic humours oppress the Stomach, 'tis hard to remove them but by a Vomit. The reason is, because the way out of the Stomach by the *Oesophagus* is short and straight, by the Guts it is long, and these humours are apt to stick in their many turnings and Cells.

Of Inflammation, and other Diseases of the Eyes.

Sore Eyes, inflam'd, and painful, are very ordinary here. I was always the more fearful of an inflammation of the Eyes, in that I have observ'd most Diseases of the Eyes, and even sometimes a Cataract to begin with an Inflammation there. I therefore immediately order'd such persons to be bled, and purg'd so often as seem'd requisite, which with blistering in the Neck seldom miss'd to cure any of them, unless the distemper came by much Venery, which was not to be cur'd any other way but by these Medicines and Abstinence from it. I used outwardly to drop into the Eyes a little Rose-water, into which is put the subtile powder of *Lapis Calaminaris, & Tutia.* These Powders are made fine by mixing them with Rose-water, and letting the gross powder subside, the fine and impalpable remaining in the Body of the water. Whites of Eggs beat up with Alum is good applied outwardly, so is Lime-juice and water. But a Mucilage of *Psyllium*-Seeds, Quince-Seeds, Saffron, and Lin-Seed, is much more Anodine in hurts than any other. Any Powder in such a case grates the tender Eye, is very painful, and increases the Inflammation. Whites of Eggs stick so fast to the Eyelids, that the force used in pulling them off sometimes injures the Eye.

Of a Tertian with Worms, cured by the Cortex Peru.

A Black Boy of *John Youngs,* about twelve years of Age, was very ill of an Epidemic *Tertian,* there was very little intermission. I gave him the *Cortex Peruv.* as usually. He voided a great quantity of Worms by Stool, and was perfectly cured.

Mrs. *Bal-*

The Introduction. cxxxiii

A young Gentlewoman, about Twelve years of Age, had for several months a few Pustles broke out on the hairy Scalp; they were red, and when dry'd turn'd into small Scabs or Scurf. I thought it was best to Cure them cautiously, and had her first bled to about Six Ounces. She was the next day purg'd with Jalap in Chocolate, and afterwards took for some Weeks *Flos Sulph.* in the morning and evening, drinking after it a good draught of Diet Drink. I thought it safe after these Medicines to anoint the eruptions with *Ung. ex Oxylapath.* made thick with *Flos Sulph.* There was drop'd some Oil of *Rhodium* into it, to take off the ill smell. This Ointment was rub'd first all over the Head, then on any place where any spot appear'd, observing the eruptions carefully for a while. She was perfectly cured. *Of a scabby or scall'd Head.*

I was desir'd to look on a Servant of Mrs. *Copes*, he had been very weary, and by advice of some ignorant Person wash'd his Feet and Legs in a Decoction of Physick-Nut-Tree-Leaves, whereby the whole Feet and Legs were rais'd into inflam'd Blisters, some of which were turn'd into superficial Ulcers. I advis'd the voiding the hot *Serum* by clipping open the Bladders, applying some *Basilicum*, and at the same time to Purge with Jalap several times. To hinder the Defluxion of Humours into the part, I advised keeping of it up, for which purpose lying in *Hamaca* is very effectual. He was cured. *Of a great inflammation of the Legs and Feet by washing them in a Decoction of Physick-Nut-Tree-Leaves.*

A Gentleman, aged about Forty years, had been very much given to Venery and Drinking. His Face was yellowish, his Belly very much swell'd, he could not sleep nor make water, had no Stomach, and complain'd of great uneasiness, especially in his Back. I purg'd him with Jalap, which was too weak. I therefore gave him some *Diagridium* about Two Scruples, which wrought very plentifully, but did no good, his Belly continuing as much swell'd as ever. He took all manner of Clysters and Diureticks, but without any success. I was for persisting in purging Medicines, and after the watery humours had been voided, I had resolved to have given him Steel Medicines, not forgetting bitter Stomachics and Exercise. He was perswaded out of this method to take Steel, and some Alterative *Arcana's*, which stopt up his Belly, so that in a great many days he had not been at Stool nor slept. I gave him in these Circumstances a Dose of Jalap and *Diagridium* mixt, which wrought very well, and at night a very easie Opiat, with which he slept very well. He was perswaded to alter my Medicines, and took a strong Opiat, as I believe, for he came to Town dos'd, had slept all the way, and could scarce be *Of a Dropsie.*

(11) awak'd

awak'd till he died. I had him bliftered, gave him ftrong Purgers, held irritating Medicines to his Nofe, &c. and did all things I could think of to raife him out of his Sleep, but to no purpofe.

Intermitting Fevers of all kinds, were very Epidemic all over the Ifland when I was there, fo that the third part of Mankind were taken ill of them, from Children at the Breaft to old aged People. They were generally very violent, fo that idle talking, light-headed- nefs, &c. were ordinary, and they very feldom yielded to any of the common febrifuges, but generally grew worfe on the ufe of them, and lafted a very great while. In thefe Fevers if the Per- fon died not by the violence of the Fever, but recovered, they were often very much difcolour'd, fallow, Cachectic, and ufually had fwellings in the left Sides, called Ague Cakes, which were very painful, and in time kill'd them. At other times a Dropfie follow'd, which rarely mifs'd, but certainly brought death. Thefe intermitting Fevers, and drinking extravagantly, I look on to be the reafon of Dropfies being fo common here. I us'd, if fent for in the Fit, to give fome Medicine forwarding the motion of Nature, as if I found the Patient Vomiting, I would help it forward, by warm water, a Feather, &c. If hot and thirfty, I fuffer'd him to drink cool Drinks as much as he pleas'd, and if I found the Fit going off by Sweat, I gave fome *Sp. C. C.* in *Decocto Salviæ vel rorifmarini*, to forward that. If the Fever affected the Head very much, I gave order for Bleeding, Blifters, Cupping, &c. So foon as the Fit was over, I immediately gave the *Cortex Peru.* in powder about one Dram every four hours in a Glafs of water, till they had taken Two Ounces of it. If there was fufficient time in the day before the next return to give Six Drams of the Bark, then I troubled them not in the night, but if the Symptoms were dangerous, and little intermif- fion, I order'd it to be given as well in the night as in the day. If it purg'd violently I gave it in a *Bolus* with *Conferv. Rof.* into which was dropt a very little quantity of *Laudanum* to take of its purging quality, which infringes in fome meafure its vertue. If it purg'd only the firft or fecond Dofe eafily, there was no need of any thing, for very often afterwards it would bind up the Belly. According to the prefent Circumftances of the Conftitution of a Man, or Seafon of the Year, it purges or purges not with him. I advifed Chil- dren fhould take it in Chocolate well fweetned. Sometimes I gave it in Pills made up with Gum *Tragacanth. Mucilage.* The beft and eafieft way of giving it is in fair water fweetned with *Syr. Cariophyll.* and aromatiz'd with Cinamon water. I ufually after the Fit was off begun to give it them, or fo foon as they found themfelves a little eafie. If I found the cafe urgent I gave it at any

Of inter- mitting Fevers.

giving

The Introduction.

time in the intermission, and although some times I have observ'd on giving it about four hours before the Fit, that the Fever came more violently on the Patient than before, yet I never could find the giving of it then dangerous. This I was first taught by giving it accidentally to one whose Fit anticipated its time, or came sooner than he expected, coming from a *Tertian* to a double *Tertian* or *Quotidian.* Several Chymists and Apothecaries in *England* and *France* pretend to an Extract or Tincture of this Bark of equal vertue with it, which because of the unpleasant tast of the Powder, in substance were very valuable, but on giving both Tincture and Extract in the Doses prescribed, they are found far less effectual than the bark in substance. Although where the Powder in substance cannot be taken, these are the next best Remedies, yet they are not to be depended upon. 'Tis most certain that the Stomach and Guts, with their juices, surpass all Artificial Vessels and *Menstruums* in Extracting what is beneficial from this Simple, and that no Art is able to make so effectual a Medicine out of it, as the Bark its self, as it happens in many other cases. This Powder, when I went to *Jamaica*, was in very great disrepute here, insomuch that it was charg'd with the death of several People, whereas the ignorant way of giving it was the cause of its disgrace. The Inhabitants used to Purge after it to take away the Relicks of the Powder, which never misses bringing a Relapse. They likewise faild in not giving so great a quantity as was sufficient, or as it ought to be given. They used only just before the coming on of the Fit, to give so much as might prevent it, without any farther regard to the Fever. Avoiding these Rocks, with the blessing of God I never mist the cure of *Quotidian, Tertian,* or *Quartan,* in whatever Age or Sex, and although I have given it to many hundreds, yet I never knew any bad consequence on its use, which with justice I could ascribe to it, but always a perfect recovery. Indeed it will not cure all Diseases, neither will it touch upon a continual Fever, but in this last, I think, if any prejudice arise to the Patient by it, 'tis only that it must be taken in large quantities, and hinder perhaps the taking of more proper Remedies. It very often works by insensible Perspiration, and sometimes by Sweat, the Patient having a breathing, and sometimes more copious Sweat every night after taking it. Intermitting Fevers here, are call'd Fevers and Agues. A Physician who had practis'd many years in *Barbados,* told me there was no such Distemper there, and that from the *Leeward* Islands (where it was common) they came thither, and were cur'd by that Air. The cause of the great frequency of the Agues in *Jamaica,* at some times of the years, are the Rains which fall so violently, and continue so long as that the low Plains are for some time covered with water. Hither *Aquatic* Birds, and those that are

Waders

Waders, or live *Circa aquas*, resort at those Seasons, and from these waters, I take it, proceed the intermitting Fevers in *Jamaica*, as they come in the Hundreds of *Essex*, and other fenny and marshy Countries of *England*. These situations are coveted, because in the Neighbourhood of such Marshes generally the Lands are very Fertile, such waters inriching the Soil very much. It is very common for this Distemper, when Epidemic, to have all other Diseases run into it, as you may see by the following instances.

<small>Of the Belly-ach, and intermitting Fevers.</small>

Captain *B.* aged about Thirty years, Sanguine in his Constitution, his Stomach being out of order, whereby he eat little, was taken very ill with all the Symptoms of the Belly-ach, *viz.* great pains about his Navel, Vomiting, thick muddy Urin, &c. He had given him Purgers of all sorts, Clysters, &c. by which he was reliev'd. He had a Relapse again in a while, as usual in this Distemper. He had some easie Opiats, &c. but these Medicines, and whatever else he took, though sometimes alleviating the Distemper, yet rooted it not out till I observ'd it had form'd its self into a *Tertian* then reigning. This Fever run very high, affecting his Head very much, but by the *Cortex Peru.* given him as it ought to be, came to be very well.

<small>Of an Asthma, and intermitting Fever.</small>

The Lady *Watson,* aged about Fifty years, very fat, was taken ill of a great Cold, she was somewhat feverish, and had a very great wheezing Cough, and difficulty of breathing. Her Fever increasing with her wheesing, I had her bled, made her ingredients of Amber, Rosemary, Betony and Sage, to smoke as Tobacco is taken, with which she found some relief at present. I likewise gave her a *Bolus* of a small quantity of *Laud. Lond.* in a little *Theriac. Andr.* by which she was reliev'd. Notwithstanding these Remedies she had very great and dangerous returns of it, wherein I repeated the same Medicines, and gave her some Diet-Drink, or *Decoct. ex Lign.* &c. All these avail'd not, till I observ'd the Disease had form'd its self into a *Tertian,* when by the giving of the *Cortex* she was perfectly cur'd.

<small>Of the same.</small>

Sir *Francis Watson,* aged Fifty five years, had been for many years troubled with wheesing and an Asthma. He had it not in *England,* and in *Jamaica* usually slept in a *Hamaca,* wherein he was swung in the evening for some hours, and then rose wheesing. He drank Wine for it very liberally, which usually gave a present relief, but he was restless, and hot all night after, with his Tongue furr'd in the morning, had no Stomach to eat, and was very dry. I gave him an easie Vomit, and afterwards a bitter Wine, and advis'd him against sleeping in the *Hamaca*; and drinking Wine, but he would

continue

The Introduction. cxxxvii

continue them. I gave him frequently *Bolus's* with *Ther. Andr.* and a little *Laud.* but they dos'd him too much, although they cur'd him always when ill of a cold. At *Port-Royal*, on drinking hard he fell very ill of a great Cold and Fever, which although I endeavour'd, by all means I could think of, to remedy, he being dangerously ill, yet I could not, till I found it form its self into a *Quartane*, when giving him the *Cortex*, he recover'd intirely.

For his Asthma the best Remedy I found was a Diet Drink of *China, Sarsa, &c.* made fresh every night, and continued for three or four months, every day, by which I thought him extremely reliev'd.

It very often falls out in intermitting Fevers, that during the time of the Fits or Paroxysms, there will happen very dangerous Symptoms as *Deliria*, Convulsions, Asthmas, *&c.* according to the Constitution of the Person or year. I have observ'd all of them if they return periodically to be cur'd by the Bark, notwithstanding the foolish opinions of some who pretend that that Remedy locks up or binds the humours in the Blood, Head or Breast. That Opinion has been the cause of the Death of many Persons, deterring Physicians from giving what was proper.

One *Cornwall's* Daughter, about Twenty five years of age, giving suck, was taken very ill of an intermitting Fever. I advis'd her to wean her Child, or get another Nurse, she would do neither, nor would she take any effectual Medicines her self. The Fever by neglect lurke about her a long time before she was well. The Child then grew very dangerously ill, but by the help of *pulvis de gutteta*, given frequently, the Child recovered. *Of one giving Suck, who was taken with an intermitting Fever, and the effects her Milk had on the Child.*

A Lusty Woman was taken with an intermitting Fever, and all the Symptoms of a beginning Consumption, as a very violent Cough and Looseness, *Atrophie, &c.* I gave her some *Creta* made into a white Drink like *Decoct. Alb.* and *Laudanum* for the Looseness. I gave her the *Cortex Peru.* for the Fever and Ague, which she had every night. The Fever being by this Remedy taken off, the other Symptoms, which seemed dangerous, presently vanished. *Of an intermitting Fever with a Cough, Looseness, and Watling.*

A Servant of Mr. *Fletchers* was very ill with all the Symptoms of a beginning Consumption. He had Exacerbations and intermissions, for which reason I gave him the *Cortex*, whereby the Fever being taken off, the other Symptoms left him. *Of the same.*

It is very ordinary towards the latter end of a Consumption for the Patient in the evening to be cold, shiver and quake, as one in the beginning of the *Paroxysme* of an intermitting Fever, which nevertheless

(m m)

The Introduction.

vertheless goes on till death, notwithstanding the giving the *Cortex*, or any other Remedies. This I have seen happen in many, in several parts of the World.

Of an *Asthma*.
Mr. *Nich. Philpot*, aged about Forty five years, living at *Rio d' Oro* above Sixteen Mile-Walk towards the North side, in the time of Rains in the North parts, and a North Wind withal, was taken very ill of a great Asthma. He could not sleep lying, but in a Chair sitting straight up, and even very little this way, having a very great wheesing and tickling Cough. Considering it threatned his life every minute, he was brought to Town, and committed to my care. I order'd him immediately to be bled to Ten Ounces, and a blistering Plaister to be put to his Neck. I gave him a *Bolus* wherein there was a small Dose of *Laud. Lond.* By the help of this he slept in a more declining Posture, and easier than before. I gave him in the morning some *Bals. Sulph.* and some *Sp. C. C. Ol. Tereb.* or *Opobals.* changing these Medicines now and then one for another. He fell ill again, I repeated what I had done, and after the violence of the Disease was over, I gave him a *Decoctum ex lignis*, by which he was cured.

Of the same.
Mrs. *Thoroughgood* was taken ill after the same manner of this *Orthopnæa*. I bled her, and gave her Diet-Drink, and she was well.

Of Hypochondriac Melancholy.
One *Barret*, of about Twenty years of age, thin, and of a swarthy Complexion, complain'd to me of fainting Fits, and a great many indispositions afflicted him. I took them all to be *Hypochondriacal*, and order'd him a Steel Course for them, which so soon as it took place entirely cured him. It is not very ordinary (though it sometimes happens) to find labouring Men troubled with this Distemper as this Patient was, who was very diligent, and wrought hard about a Pen of Cattle, and small Plantation he had some few Miles out of Town. He had taken several Medicines and Courses before, amongst the rest Steel, which not being given as it ought to be, had no ways reliev'd him. I suppose it had been given with Purgers.

Of a Pleurisie.
Mr. *Molines*, aged about Twenty six years, of a Sanguine Complexion, and Plethoric Body, sent for me on *March* 18. 1688. He we was on board a Ship bound for *England*, and could scarce speak or breath, he had been sick for two days, complaining extreamly of a great pain in his Shoulder, or rather inside of the *Pleura* answering that part, which increas'd on breathing high, sighing or coughing. He had likewise a short Cough. It was taken by all for Sea-sickness, but

The Introduction. cxxxix

but I told them I thought they were deceived. I forthwith ordered him to be bled in the Arm to about Ten Ounces, and gave him a *Linctus* of Sugar-Candy and Oil, and a Pectoral Decoction of Barly, Liquorish, Raisins, &c. He immediately found himself much better. I ordered him to continue this, and to take of Crabs-Eyes, and *Sal Prunellæ*, of each half a Dram, and to swallow morning and evening the half on't, drinking afterwards a Pectoral Draught, and in case of a Relapse I ordered him to be bled. The Ship Chirurgeon, contrary to my desire, gave him a Vomit, he himself knowing nothing of it till it was down. I was sent for after it had wrought five or six times upwards and downwards, and found him in very great anxieties, with a small Pulse, Cramps or Convulsions, cold Sweat, cold Hands, Arms, Feet and Legs. I gave him presently some *Aq. Cinam.* with *Syr. Cariophyll.* Some Fifteen drops of *Sp. C. C.* order'd him to smell to it, gave him some burnt Wine, and about half a Grain of *Laudanum*, after which he was much better, had a small quiet Sleep, and in some hours time seem'd much reliev'd. His pains afterwards returned, though not so violent as at first. I bled him twice on two several days, and with *Decoctum hordei*, Sugar-Candy and Sallet-Oil beat up together, he was cured. A while after he spit up very much purulent matter, but an easie Opiat, and these Remedies perfected the cure.

One aged about sixty years, in the Dog days had been so much troubled with faintness, as even several times to fall into *Syncopes*. I advis'd him to stir as little as possible, to eat good Gelly-broths, and to take about Ten drops of *Sp. CC.* every morning and evening in a Glass of fair water, by which means he grew stronger every day, and escap'd those Fits he had been troubled withal. To take away his faintness, he had prescribed, as I apprehend, too much Wine and spirituous Liquors, which may give present relief, but in the end destroy. He having us'd this too much in the evenings, appear'd for the most part somewhat flustred. On March 22. 1688. he fell into a Lethargic Distemper. I suppose the vinous Liquors, by little and little, had made so many attempts on his Brain, that at last it was overcome. He talk'd incoherently, when ask'd a Question would answer, though not able to speak many words, his Pulse was quick and low, and his under Lip had convulsive motions. I ordered him to be bled to Six Ounces, that blistering Plaisters should be applied to the nape of his Neck and Wrists, that he should be cupt on the Shoulders, that he should have Fifteen drops of *Sp. CC.* given him, and that if his Senses were not very well awaked with these methods, but his sleepiness increased on him he should have some Snuff made of White Hellebore

Of a Lethargic, and Apoplectic Distemper from too much Wine drunk for faintness.

cxl *The Introduction.*

lebore put up his Nose. He fell soon into foaming at Mouth, and immediately died in Convulsions.

Of the Belly-ach. A Gentlewoman, aged about Fifty years, complained to me extreamly of the Belly-ach, she had pains about the Region of the Navel, and all over her Limbs and Body, and had been with all the Physicians of the Island. She had the use of neither Hands nor Feet, although her Pulse went very well, and was strong. I gave her some Emulsions of the cold Seeds, and would have had her bled but she was so affraid of it she durst not, and would not take any thing else. She recovered of that Fit, but fell ill afterwards much after the same manner. I gave her as occasion required some *Pil. Coch. Min.* to Fifteen Grains to give her a Stool, which kept her Body open. I had her bled several times, and desir'd she would altogether abstain from Wine. This last I insisted much upon, and gave her much water to drink with Steel, but neither it nor any thing else did succeed. I attributed this Disease to Wine, Punch, and Vinous Liquors, but she would not abstain, alledging that her Stomach was cold, and needed something to warm it.

Of danger of miscarriage. One *Evans*, very big with Child, about Eight Months, was taken with great pains in her Belly and Loins, as if ready to bring forth. She had likewise great pains in her Body all over. I advis'd her forthwith to be bled, to hinder Abortion, and to asswage her pains, prescrib'd here a contemperating cooling Course, as is usual in Rhumatisms, and desir'd her to abstain from Flesh and Vinous Liquors for some time, with which in a few days she was well. I order'd her to keep her Bed, or be very quiet till it should please God she were deliver'd, she being extraordinary big, which she did, and was safely delivered at the due time. She had about a year before been delivered of three Children at one Birth.

Of one who supposed her self to be with Child, but was not. One aged about Thirty five years, concluded her self for some time to be with Child. She found her self in much pain, and after a *Fluxus Mensium*, which continued on her somewhat violent, several Physicians were consulted, who advised her to *Trochisci de Myrrha*, and other very forcing Medicines, and continu'd their use for some time, although she had her *Catamenia* in due time. I was consulted, and advised her to take the bitter Wine for her Stomach, that being very much out of order, not to take any other Medicine, but patiently to expect the event, she did so, and found in some time that she was better, not with Child, and pretty well.

Colonel

The Introduction. cxli

Colonel *Fuller*, aged about Forty five years, very much troubled with the Gout, on taking a Dose of Jalap in Powder, which wrought copiously, fell into a Nausea, or great inclination to Vomit, and into a Looseness. He was troubled with this endeavour to Vomit chiefly after Meat. I thought the best Remedy was to give him some *Elixir Proprietatis*, by that means to remedy his Stomach, and at the same time by making an easie irritation downwards to stop the motion to vomit after Meals. This in some time took place, and he was well. Once in a violent Fit of the Gout he was taken ill of a Looseness, in which I ordered him to take some easie Opiats, and he was cured. *Of a Squeamishness, and Looseness after taking Jalap.*

One *Hercules*, a lusty Black *Negro* Overseer, and Doctor, not only famous amongst the Blacks in his Master Colonel *Fuller's* Plantation, but amongst the Whites in the Neighborhood, for curing several Diseases, and particularly *Gonorrhœas*. He had been three years before troubled with that Distemper, which he thought by the Country Simples he had cur'd, but came to me, complaining of a very great heat in making water with intolerable pain, and scalding. Looking upon the part affected, I found he had neglected his Clap, and that Caruncles had grown up and stopt almost quite the passage of Urin or *Urethra*, wherefore Nature had by a Tumour and Apostemation made a passage for the Urin in *Perinæo*. This passage had callous Lips. I order'd him some Mercurial Medicines, and would have try'd several other Remedies for his Cure, had I not soon after left the Island. There are many such *Indian* and Black Doctors, who pretend, and are supposed to understand, and cure several Distempers, but by what I could see by their practice, (which because of the great effects of the Jesuits Bark, found out by them, I look'd into as much as I could) they do not perform what they pretend, unless in the vertues of some few Simples. Their ignorance of Anatomy, Diseases, Method, *&c*. renders even that knowledge of the vertues of Herbs, not only useless, but even sometimes hurtful to those who imploy them. *Of a Negro Doctor famous for curing Gonorrhœas, who was so far from being able by Specifics to cure that Disease, that he was very ill of it himself.*

Emanuel, a lusty *Negro* Footman, was ordered over night to get himself ready against next morning to be a Guide on Foot for about an hundred Miles through Woods, to a place of the Island, to seize Pirats, who, as the Duke of *Albemarle* was informed, had there unladed great quantities of Silver, to Careen their Ship. About Twelve a Clock in the night he pretended himself to be extraordinary sick, he lay straight along, would not speak, and dissembled himself in a great Agony, by groaning, *&c*. His Pulse beat well, neither had he any foaming at Mouth, or difficulty in breathing. The Europeans *Of a Negro who dissembled a great Sickness,*

(n n)

peans who stood by thought him dead, Blacks thought him bewitch'd, and others were of opinion that he was poyson'd. I examin'd matters as nicely as I could, concluded that this was a new strange Disease, such as I had never seen, or was not mention'd by any Author I had read, or that he Counterfeited it. Being confirm'd that it was this latter, and that he could speak very well if hepleas'd, to frighten him out of it, I told the Standers by, that in such a desperate condition as this 'twas usual to apply a Frying-Pan with burning Coals to the crown of the Head, in order to awake them throughly, and to draw from the Head, and that it was likewise an ordinary method to put Candles lighted to their Hands and Feet, that when the flame came to burn them they might be awaked. I sent two several People in all haste to get ready these things, in the mean time leaving him, that he might have time to consider and recover out of this fit of Dissimulation, which in a quarter of an hour he did, so, that he came to speak. I question'd him about his pain, he told me 'twas very great in his Back. I told him in short that he was a Dissembler, bid him go and do his business without any more ado, or else he should have due Correction, which was the best Remedy I knew for him, he went about his Errand immediately, and perform'd it well, though he came too late for the Pirats.

Of the same.

I was call'd to a Carpenter, a lusty Rogue, who pretended himself sick of the Belly-ach, he had got a Blanket about his middle, and made wry Faces, bemoaning himself very much. He told me, upon examining of him, that he went to Stool very well every day, and did not Vomit, &c. I told him that I believ'd he dissembled, and that if he were well chastis'd it would be his best Cure, he seem'd not to be of that mind, but very soon recover'd without any Physick.

Of the same.

'Tis very ordinary for Servants, both Whites and Blacks, to pretend, or dissemble sickness of several sorts, but they are very easily with attention found out by Physicians, who are used to converse with Diseases, for the Symptoms do not answer one another, and they may, by proper questions be discovered as Forgeries, Perjuries, or Lyes. In people who pretend sickness, and have none, I us'd, in order to be rid of them, (they never growing better,) to order harsh, yet innocent Remedies, as blistering, taking bitter Medicines, &c. Thus I used to be free'd of their trouble. Sometimes they pretend to have a Cough, but that is easie to be seen, if feigned, by a more leisurely great inspiration than those really troubled with that Disease can admit off.

In

In case Women, whom I suspected to be with Child, pretended themselves ill, coming in the name of others, sometimes bringing their own water, dissembling pains in their Heads, Sides, Obstructions, &c. thereby cunningly, as they think, designing to make the Physician cause Abortion by the Medicines he may order for their Cure. In such a case I used either to put them off with no Medicines at all, telling them Nature in time might relieve them without Remedies, or I put them off with Medicines that will signifie nothing either one way or other, till I be further satisfied about their Malady. 'Tis a very hard matter for a Physician in these cases to be certain, but after taking what care he can to inform himself, he must use his discretion. If Women knew how dangerous a thing it is to cause Abortion, they would never attempt it on any account whatever. I know but one case beforementioned, which is flooding, wherein 'tis necessary, and then 'tis best done by the Hand. One may as easily expect to shake off unripe Fruit from a Tree, without injury or violence to the Tree, as endeavour to procure Abortion without danger to the Mother. This is a most certain truth, and I have seen it confirm'd by the sad experience of such, who, upon political considerations, to avoid scandal, having too many Children, or the like, had endeavoured, without effect, to procure Abortion, and instead thereof had brought themselves near their ends. *Of the danger of using Medicines which cause Abortion.*

One *Booker*, a Woman of about Thirty five years of Age, was taken very ill with a malignant Fever, she had been blistered, and treated after the Cordial way, that is, by giving her great quantities of *Diascordium, Ther. Andr.* and other Diaphoreticks, she was kept hot, and not suffered to drink any thing that was cool. I gave her some easie Medicines, as a very little *Confectio Alkermes,* order'd her a cooling method, to drink as much cool Drinks as she pleas'd, and by them she was well. *Of a dangerous Fever cured by a cool Regimen.*

Her Husband was ill at the same time of a violent Vomiting and Looseness, which had been on him for many days, by drinking to a very great heighth in Canary. He was very much weakned, dry, and troubled chiefly with the Vomiting. I gave him, considering his weakness, some drops of *Laud. Liq.* and some other things of that nature, to stop both Evacuations, but he being sometimes morose and ill-natur'd, and at other times Phrenetic, and so not taking his Medicines as he ought, and withal, his Stomach being so mightily disorder'd that nothing could stay there, in some time he died. *Of Vomiting and Looseness from excessive drinking of Canary.*

John

John Parker, about Thirty five years of age, a lusty full-blooded Fellow, was much given to drink. He had been taken ill' of the Epidemic continual Fever, reigning at first when I came to the Island, and recovered, as others out of it, of which before. Soon after he committed a great debauch in Rum Punch, after it lying on a cold Marble Floor. He fell from these causes into a *Mania*, so that he was observ'd to speak and act very incoherently, and to get up in the night, &c. His rage increas'd to a very high degree, and he died in a very few days, notwithstanding all the methods usually followed in these cases.

Of a Mania from excessive drinking of Rum-Punch.

Roger Flower, a Baker, a strong Man, of about Forty five years of age, of a Sanguine Complexion, and Plethoric, was much given to drink Sengury, or Wine, Sugar and Water in the morning early, continuing till night, thereby endeavouring, as he thouht, to quench his thirst, and relieve his Spirits. He was taken very ill of a *Cholera Morbus*, in which his vomitings were very violent. After a sufficient Evacuation by Vomit and Stool, help'd on by thin Water-gruel, and Chicken-Broath, I endeavour'd to stop them with *Laudanum*, when I thought his Stomach and Guts sufficiently wash'd. This reliev'd for some small time, as it never missed the Cure of many others so Diseased, but he after some hours fell ill again with Vomiting and a Phrensie. I endeavour'd what I could to remedy both the one and the other, but he grew more outragious, and notwithstanding blistering, &c. died in a few hours.

Of a Cholera Morbus from the same cause.

Dick a Postillion, Plethoric, Choleric, much given to drinking Rum-Punch, and strong Liquors, fell into a Fever, which chiefly seiz'd his Head, so that he was in a very great rage. I treated him after the cool Regimen, had him bled and blistered, but notwithstanding this found him still worse. At length, I learn'd his Nurse gave him much Wine and Flesh, contrary to instructions. I order'd the contrary, and by the continuance in this course, When the Aspect of the Sun and Moon chang'd, on that very minute, from great rage he came to himself, and recovered quite of his Distemper. I have in several persons observed the same, but these Aspects which I was sure to have any effects, were only the Fulls and Changes, or Oppositions and Conjunctions of the Sun and Moon. I have seen their effects, principally on Persons used to excessive drinking, and that chiefly of Brandy, which after some time turned them maniacal, with very great Fits, for some days before and sometimes after these Aspects. I have not seen so much of the effects of the Sun and Moon's Aspects in *England* and *Europe*, as *Jamaica*,

Of a Mania which was occasioned by excessive drinking, and had Fits which follow'd the Full and Change of the Moon.

The Introduction. cxlv

maica, an instance of which is before related. I have been able in this Disease, by considering the strength of the Person, and the time of the Full or Change, ere it was likely such Fit would end, to foretel whether in probability they would out-live the Fits or not, for I could very seldom bring this sort of Distemper under by Medicines till that time was over.

One aged about Fifty years, little, very Cholerick and Hot, much given to drinking, had been troubled very much with an Asthma, for which I gave him a Diet-drink, made of *Sarsa*, &c. which he continued to take for some time, and found great relief. He was taken ill of a *Gonorrhæa*. I gave him some purging Mercurial Pills, and Emulsions. He was afterwards seized with an Epidemic Fever, being a *Tertian*, then reigning, for which he being afraid, and not daring, lest the *Gonorrhæa* should be stopt, to take the *Cortex Peru*. on the intermitting day I gave him a Vomit of *Oxymel. Scillit.* and infusion of *Crocus Metall*, whereby in a small time it working very well, he was cured of both Distempers. Of an intermitting Fever and Gonorrhæa.

A month after he fell ill of great pains in his Reins and Back, and being usually troubled with Fits of the Stone at that time of the year, he took those Medicines he us'd to find relief in, without any success. I gave him all manner of Diuretics, as *Sal Prunell. oc. cancr. ol. terebinth.* a decoct. of the Roots of *Althæa*, Linseed, &c. Anodine Fomentations, &c. as well as Ointments and Oils, without any relief except some ease for a day or two. He was then advis'd to stir much, I thought that hazardous, because it might throw down the Stone, and so occasion a Paroxysm, but he would do it, and was worse. I endeavoured to remedy these Symptoms by all the ways I could, without success, wherefore I conjectured some *Abscess* to be in the Back, or one of the Kidnies. I bled often, purg'd easily with *Cassia*, &c. gave all manner of Diuretics, but his pain continued. He grew feverish and weak, I endeavour'd to remedy these Symptoms by more cool Medicines, without success. He sent for another, who gave him Diaphoretics and Cordials. He had an Issue in his Arm, which being neglected, the Arm inflam'd and swell'd, to which were applied, Adstringents, and the humour was repelled, whereby he grew delirous, and had a very small intermitting Pulse. I order'd him some Diaphoretics to force the repell'd matter out again, by which the swelling appeared and he was reliev'd. It turn'd to an Apostem, and complaining very much of it some days after, I had it opened. It appear'd to be of the nature of an *Erysipelas*, he was drest by the Chirurgeon, and in a day it appear'd blackish about the Issue, and discolour'd in several places. There appeared some discolour'd Pustles or Whales where- Of an Ulcer in the Kidnies, and Gangrene in the Arm.

(o o)

wherefore to avoid a Gangrene, he was scarified with deep Incisions, and had Pledgets of *Ægyptiacum* dissolv'd in *Sp. V.* applied, but although with this, and a Fomentation of Wormwood, &c. in a Lye, and *Sp. V.* it seem'd to be at a stand, yet it sphacelated more and more, and he being so weak as not to endure the Amputation of his Arm, he died.

Of a very great *Vertigo.*
The Reverend Mr. *Leming*, of a Plethoric and Sanguine Constitution, aged about Forty five years, by walking in the heat of the day, exposed to the Sun-beams, was taken ill with a very great *Vertigo,* so that he was reeling every step, and could not see to Read or do any thing else, neither could he walk without the assistance of several people under his Arms to support and guide him. He had been two days in this condition, neither had he taken any thing but a Clyster. I wondered very much that in so long time he had not been Apoplectic, and ordered him immediately to be bled in the Arm to Ten Ounces, the next morning to take a pretty strong Dose of *Extr. Rud.* in Pills, that he should be blistered presently in the Neck, afterwards cup'd with Scarification in the Shoulders, and that he should take some alternative Medicines, as Sage and Rosemary made into a Drink after the manner of Tea, and drink a quantity of it several times every day at convenient Seasons. I order'd some drops of *Sp. C C.* to be put into it, and gave him directions to change and repeat these Medicines till he was well, which accordingly he did. He found immediate relief on bleeding, and by the use of the rest of these Medicines recovered in some days his perfect health.

Of a Consumption from straining the Lungs.
The same Gentleman, in preaching used to strain his Lungs so much, that he became obnoxious to several Coughs and defluxions. I advis'd him to remove from his Parish, where he used his Lungs too much, to a place where no opportunity should be given of exercising them so much. By this method he was relieved. I heard since, that returning to his Parish he fell into a spitting of blood, turn'd Consumptive and died. I have seen the like happen several times to people who have in their several Professions used their Lungs too much. The ancient Physicians, and Philosophers prescribed, Reading aloud, Disputing, &c. as necessary for exercising the Lungs, which they thought as proper for the keeping of them sound, as other Exercises for other parts of the Body. But I have often observed that the immoderate straining of them by Singing, Hunting, Trumpeting, inviting People to Shows, &c. have by degrees brought Hoarsnesses, Coughs, Consumptions, great pains, Ulcers under the the *Sternum,* and Death.

A

The Introduction.

A Child of about a year and an half old, was taken ill of an Epidemic *Tertian* Fever, with the Symptoms of which it had a very great swell'd upper Jaw. The Mother told me, that the Child breeding Teeth very hardly, the Jaw had been formerly cut, where the swelling now was. I concluded the cutting had fowl'd, or hurt the upper part of the Tooth, and that part of the Morbific Matter was sent to that place as weak, during the Paroxysm of the Fever. I gave it the *Cortex*, and it was in some few days cured of both Fever and swell'd Jaws. *Of a Tertian, and swell'd Jaws in a Child.*

Face-Cloaths, or Linnen to be pin'd over the Face of New-born Children, are never used in *Jamaica*, it being hot, and thought there very unhealthy. Cradles are not us'd very much, but *Hamacas* for Children to be laid in, wherein they are toss'd or swung as if they were rock'd in a Cradle. They make an Engine of Wood as long as the Child, a little broader, and a Foot and an half high, arch'd at top. The sides and top are covered with Gauze to hinder the Gnats or *Mosquitos* molesting the Child lying under it. *Of the management of Children in Jamaica.*

A great many White Women, all *Indians* and *Negros*, keep not their Beds over a Week, after having brought forth, when they return to their ordinary Business. Sometimes through a contrary Custom this may be prejudicial, as I knew a Mother of many Children, who getting up so much earlier than she used to do, fell into great pains in her Sides, after having some diminution of her *Lochia*. In this case I ordered bleeding, and the use of Sage-Tea, by which she recovered in some time. *Of the treating Women in Childbed.*

A Woman being got with Child, endeavoured to hide it, and took violent Medicines designing Abortion. They had not the desired effect, for although she took Mercury Sublimate in Broath, yet she went out her Time, and after violent Vomitings, and great Spitting for some time, she was privately delivered, and the Child buried in a Field. It was discovered by Birds, which feed on corrupting Flesh, are a sort of Vulture, and call'd Carrion Crows. The Child thus found, being brought to Town, a search was made, and the supposed Mother carried to Prison. She there fell into a *Delirium*, with other feverish Symptoms very high. She was blister'd, and took *Sp. C. C.* and other things, by which she recovered, and was seemingly well, but I heard she died a Week after. *Of Mercury Sublimate taken inwardly.*

Blacks

The Introduction.

Of Black Nurses. Blacks are as often taken for Nurses as Whites, being much easier to be had. They are not coveted by Planters, for fear of infecting their Children with some of their ill Customs, as Thieving, &c. I never saw any such Consequences, and am sure a Blacks Milk comes much nearer the Mothers than that of a Cow, and yet in *Jamaica* some Children are bred up by the Hand very well.

Of Childrens Diseases. Some Women being here very debauch'd as to drinking, &c. when they are Nurses, can scarce abstain from it, and thereby infect Children very often with Pustles, breakings out, &c. I us'd in such cases to persuade changing the Nurse, or bringing up the Child by Hand, tampering with Physick too much with Children, where the Disease is not plain, being not safe, they not being able to inform the Physician of their Malady, but by frowardness and crying.

Of Chocolate given to Children. Chocolate is given to young Children here, almost the first Meat they take except the Mothers Milk, and is found to agree with them as well as Milk-Meats in *England*.

Of the Ringworm, Impetigo or Lepra Græcorum. Mr. *William Kayes*, aged about Forty years, complain'd to me he had been several years troubl'd with Ringworms on his *Abdomen*. I desir'd to see them, he show'd me a spot or two on his Belly, about the Circumference of a six pence, in which was a superficial ulceration of, as it would seem, the *Cuticula*, with some scales about the edges. It was of a dark brown colour, and there seem'd to issue out a small Ichor. He had in this place a most intolerable uneasie itching which was very troblesome. He had not infected his Wife with it, although most Men are thought to communicate it to their Wives. I order'd him to be bled and purg'd, gave him inwardly, after twice purging, a great many Doses of *Flor. Sulph.* in the morning and night, drinking after it Diet-Drink, for some weeks. After this preparation an Ointment was made with *Sulph. Viv.* powdered and mixt with *Ung. ex Oxylapath.* and scented with Oil of *Rhodium*, with which he was to anoint it, but by the use of the first Medicines he was well, and continued so. This is a very ordinary Disease here, and in most parts of the World, continuing many years, and fixes its self in several places of the Body. It seems to come near the *Lepra Græcorum*, and is the *Impetigo* mention'd by *Piso* to be in *Brazile*. It uses always to be cur'd by the abovesaid Medicines, though now and then, on the Aspects of the Moon it would return again, and be cur'd on use of the Medicines as before. It had been tried to be cur'd by most of the Physicians of the Island without success.

Mr. *Hem-*

A Gentleman, aged about Forty five, looking very black in the Face, or of a livid Aspect, had been very much given to Venery, and intemperance in Drinking. He had always after a debauch some bruised places about him, which were hurt by Accidents and Falls. For these Casualties it was proper to bleed him very often to prevent his death, as on hurting his Sides, came Pleurisies and Inflammations, for which there was a necessity to give him several Medicines, and more especially to use Phlebotomy. He complain'd to me one day he thought his Belly swell'd, and that he made a small quantity of Urine. I told him I thought 'twas very dangerous and advis'd Jalap, and other purgers of water to carry off the humours, which threatned a Dropsie, as also Chalybeats and Diuretics, with Exercise. Although these Medicines wrought well, yet his Belly swell'd more and more, whereby I was almost sure he would not live long, and so it prov'd, for in about three weeks time his Belly swell'd most prodigiously, made his breathing uneasie, the watery humour overwhelmed his Brain, made him Delirous, and also seiz'd his Lungs, so that he had a great Cough and died. He us'd to drink two Bottles of burnt Wine every night when well, in the night time, to support, as he thought, his Spirits. *Of a Dropsie from intemperance in Wine and Venery.*

One *Lambert*, a young Man, complain'd to me of a great pain in one of his Kidnies, with Vomiting, he us'd to be troubl'd with the Stone there. I gave him about one Grain of *Laud. Lond.* in about seven Grains of *Extr. Rud.* he had ease presently, a Stool some while after, and was perfectly cur'd of that Paroxysm. The like I have known in a great many others. Easing of the pain takes of the constriction of the Ureters and Membranes, and then the Sand, Gravel, or Stone come away. *Of the Stone.*

A Woman of about Fifty years of age, was taken with an Epidemic *Tertian*. I gave her the *Cortex*, by which she was cur'd, and continu'd well some weeks. On *Christmas* Eve, she took a great cold, and fell into an Epidemic Pleurisie, which was then frequently join'd with Rhumatic pains, and mortal. I advis'd her forthwith to bleed plentifully, and gave her some *Sal Prunell.* and Crabs-Eyes in large quantities. She sent for some ignorant Fellow who could not bleed her, and neglected that Remedy two days, in which time her pains increas'd. I sent for a Chirurgeon, who bled her, directed her a Clister, gave her inwardly Antipleuritics, order'd her side to be rub'd with *Ung. Dialth.* I advised her to repeat bleeding, which was done five times in two days, without any success, for her pains increased in a small time after bleeding, *Of a Pleurisie.*

(P p) and

The Introduction.

and She died in Convulsions. There is no Remedy in a Pleurisie so effectual as bleeding. This is not done, either through obstinacy of the sick Person, ignorance, or aversion of the Physician to that Remedy, or unskilfulness of the Chirurgeon, who generally pretend, on missing the Vein, that the Blood is too thick, as if it could Circulate through the capillary Vessels, and yet be so thick as not to come out of a hole made in the side of the Vein. If sufficient bleeding is neglected in a Pleurisie, on any of these accounts, the extravasated Blood increases, difficulty of breathing follows, and occasions either a Suffocation by the great quantity of Blood stagnating in the Heart, and great Vessels, a Gangrene, or at least an Apostem, whence follows an *Empyema* and Consumption.

Of an Erysipelas. Her Daughter was taken very ill at the same time of a great pain in her Arm, after which came out an *Erysipelas*, which I advis'd to be treated with a Fomentation of Wormwood, Sage, &c. in Lye and Wine. With this Remedy, after breaking out, and going in several times, it came out in several Boils and hard Lumps on the Hand, which breaking, and running, cur'd her. I have seen many troubl'd after the same manner, with great pains in the Arm, after which comes out the *Erysipelas*, then it goes in again, and the pains return till it again breaks out, which is chiefly effected by *Ol. Lumbr. & Cham.* with which some *Petrol.* is mixt. They ought to be bled and purg'd, &c. and yet often recover not in half a year. The Skin of these people is usually afterwards very much discolour'd.

Of the Fluor Albus. A great many Women are here troubl'd with the *Fluor Albus*, with which Distemper they usually have very great pains in their Back. I usually gave them a Medicated Wine, with the Roots of *Angelica, Imperatoria, Bistort. Tormentill. &c.* infus'd in *Madera*. I advis'd them to drink for some time of it, several times aday, forbidding all manner of Evacuation. This method usually cur'd both Whites and Blacks.

Of bruises. On outward or inward bruises, I us'd, in danger, to bleed immediately, and in the outward to embrocate with fair water, by applying Papers or Cloaths dipt in water, and repeating them, when they grew hot, by which the humour was repell'd. I have seen this method do great matters, as much or more than Bole or Astringents *cum Album. Ovor. Sperm. Cet.* and Crabs-Eyes. I us'd to give inwardly, *Sp. C.C.* and to order a *Montagany* Plaister to be applied.

Of an Ulcer in the Kidnies. A lusty blind Fellow of about Fifty years of age, had been languishing a long time with pains in the Region of his Kidnies, as

well

The Introduction.

well as all over his Body; he now made and had done for some time past white purulent smelling and thick water. I concluded it to be an Abscess of the Reins, and told him I was apprehensive it was incurable. However I had him bled twice or thrice, purg'd several times a week, with *Pil. Coch. Min.* he took on the intermediate days, *Ol. Tereb. Bals. Sulph.* and a Decoction of Barley. By which means contrary to expectation he was quite cured.

Mrs. *Purifie*, aged Thirty five years, complain'd of great pains in her Kidnies, had a great heat there, and had gone through several Courses without success. I put her into the same method as above, but the Oil was too hot for her. I gave her some *Vitriolum Martis*, making artificial Mineral waters, as well for that as a swell'd Spleen. After she had gone through this Course, she recovered very well, which she attributed more to the outward application of Orange Leaves then any thing else. She had for her Spleen a *Galbanum* Plaister, and one *Ex Cicuta cum Gum. Ammoniaco*. I have known this last Plaister do great matters, with Jalap inwardly given at the same time. They drain a great quantity of blackish *Serum*, which fills and swells the capacity of the *Abdomen*, as well as Liver and Spleen to that rate, as that they meet almost about the Navel. One told me she found on the use of this Plaister over the aforesaid *Viscera* as if some body had squeez'd them with their hand, and the Belly at the same time fell several Inches in a week by measure. Of the same, and a swell'd Spleen.

One *Devons* Wife brought me to see her Husband, who had been very melancholly for several months, was morose, would scarce speak, but was always drowsie and sleepy. I order'd him to be bled, gave him a Purge, some *Sp C. C.* and had his Neck blistered by which in a few days he was quite well. Of a sleepy Disease.

She fell ill of an Epidemic intermitting Fever, which in one of the Fits run so high, as that after long and Phrenetic discourse, notwithstanding blistering, &c. she, after falling into cold Sweats, died. Of an intermitting Fever which was mortal.

I had heard very much of a dry Dropsie, a Distemper that was said to be very Mortal to many of this Island, both Whites and Blacks, and was pretended to be a very strange Disease (as it would have been, had it answer'd its Name) and proper to this Climate. At length one was brought to me from Colonel *Nedham*'s Plantation, where he had been a Labourer at the Stokeholes. He had in this Employment been extremely heated, and sweated, and by taking Colds thereon, he had contracted a great Cough, which had continued for some time. He was wasted in his Body, was hot and Feverish, and had his Legs *œdematous*, puff'd up, and a little swell'd Of an Hectic, call'd in *Jamaica* a Dry Dropsie.

swell'd. I found this to be the beginning of a Consumption, and perfect *Febris Hectica*, having no Symptoms, but such as are common to our *English*, and all other Hectics. I therefore order'd him, *Bals. Sulph.* to preserve his Lungs from being tainted, order'd him a Diet Drink of *Sarsa, &c.* with Barley and Raisins, and gave him in the Evenings to keep off his Cough, an easie Opiate, with which, and good Diet he was cured. A Steel Course is much commended here in this Distemper, and a furr'd white Tongue is reckoned one of the chief Pathognomonics of this Disease, which is nothing but what often accompanies the Hectic in *England*, and every where else.

Of the Kings-Evil.
A Boy of about Thirteen years of age, had a great and ugly swelling on his Leg, which yielded great quantities of an Ichorous Sanies from the Bone. He had some marks of Ulcers remaining about his Eyes, which shew'd him to be scrophulous, or troubled with the Kings Evil. He had been flux'd in several hot Houses, and proceeded with after several Methods, by several Physicians, who ordered the matter so that this Tumour had been almost cur'd several times, but by neglect it return'd. He found great relief in washing and bathing in salt-water. I ordered him to be well purged thrice a week, with *Pil. Coch. Min.* and *Merc. Dulc.* Sometimes I chang'd this for a Vomit of *Infus. Croc. Metal.* and gave him a Diet-Drink, made of the Woods, Roots, *Sarsa, &c.* boil'd in Lime-water. He put to the Ulcer some Basilicon with Precipitate, and sometimes without, with which he grew much better. I left the Island before he was quite well, but order'd them to prosecute the cure the same way.

Of Burns.
The *Negros* and *Indians* of Plantations usually have Fires near the places where they and their Children sleep. They make these Fires both for their Healths sake, and to keep themselves from Gnats, *Mosquitos*, or Flies, which would be troublesom, were they not kill'd by the smoak. The Slaves are usually so well wrought in the day, and sleep so fast at night, that they do not easily awake. Several of their young ones fall into these Fires, whereby their Arms or Legs are sometimes burnt off. I always found a Cataplasm of Onions, Salt, and white Soap beaten together, to do very great matters in the cure of such Accidents, and these ingredients are almost every where to be had.

Of a Rupture.
One about three years before she advis'd with me, had been troubl'd with a long and tedious Delivery, in which she found in the Childs coming away a crack as if something had broken in her Groin. She complain'd to me of a great pain there, in which was

a small

a small lump, which was about the bigness of a Pigeons Egg. I took it for a Rupture, ordered her to lye with her Heels and lower parts high, to endeavour by an Anodyne Carminative and Discutient Fomentation to put up the Gut, and gave her an easie Opiat in the Evening. This did well, and she was better, but still had Pain. I desir'd her to take great care to be quiet, and to repeat these Medicines as she found occasion, upon which the Pain went away, and she was better.

One came to me complaining he was troubled very much with Itch about the *Os Pubis*, which proceeded from Lice or *Ascarides*. I advised him to rub and wash all the part over with Sope and Water, but that did not kill them. I then ordered him to beat some of the Seeds of *Staphisacre*, and strew it on the part, which kill'd them all in a very short time. It is an almost certain Remedy for any Vermin of this kind.

Of *Ascarides*.

One of about Fifty five years of Age, given to good Fellowship and Drinking of Drams, had been very ill of the Belly-ach, several times, on which he had lost the use of his Limbs. He had not long before I saw him been taken with a very severe Fit, and was recovered out of it by the help of Ginger in Cyder and Wine mixt, and heated with Sugar. After the violence of the Fit was over, because he was very weak, he had suckt two *Negro* Womens Milk, by which he was perfectly recovered. He seem'd to be very Hypochondriac, was Melancholly, and look'd Yellow in the Skin and Eyes. Being consulted for his Health I advis'd him, because he had no Stomach nor Appetite to Victuals, to take an infusion of Gentian Roots, Centaury Tops, &c. in *Madera* Wine, and a Diet-Drink of *Sarsa*, *China*, &c. mix'd with an equal quantity of Cow's Milk every Morning. Now and then as his Belly was bound up, I order'd him a Pill of *Extract. Rud.* whereby he was kept Soluble. By these Medicines his Health was preserved without any great Sickness, but coming on Board in order to a Voyage for *England*, and drinking Punch more than ordinarily, he first fell into an unusual Weakness of his Hands, and afterwards into Pains all over his Body. I would have remedied these Symptoms by Bleeding and Purging, but that his Weakness put a stop to any such Courses. He grew worse, having had no Stool for some Days (for he had, besides his Colick, a Rupture) and complained very much of Pains every where, especially in his Belly. I gave him *Extract. Rud. gr.* 15. which did not work, wherefore in some hours more I gave him thirty Grains of *Pil. Ruffi* for his Jaundice and bound Belly, which with *Sena* boil'd in Chicken-Broath, or great Quantites of thin Water-

Of the Belly-ach.

gruel

gruel did not relieve, but he grew phrentic and idle in his Talk. I gave him, after a while, a very strong Decoction of *Sena* in Water, which took effect, working four or five times, but it was so far from alleviating his Distemper, that he grew worse. He lost his Sight quite, although his Eyes look'd well and without blemish, for which I order'd him to be bled. I blister'd him in the Neck likewise, and thought his Distemper uncurable, if at Full Moon, or two or three Days after, he did not recover. He slept not, but had strange Persuasions or Imaginations in his Head, and dos'd at first, but afterwards fell into a perfect Lethargy. When his Sleepiness was over, he awak'd, but seem'd to be pensive at some strange things in his Mind. I apply'd Blisters to his Wrists, and on the Day of the Full Moon, he came to see somewhat, and at the same time recover'd some of his Understanding; yet, a great many things were blotted out of his Memory, so that the Remembrance of things past, not only during his Sickness, but likewise before, were lost, and some Imaginations and Fancies, were so fast imprinted in his Mind, during the time of his being not *Compos Mentis*, that afterwards, when he discours'd and reason'd very well, there was need to take Pains with him to undeceive him, and make him sensible of his Mistakes, but in a few Days that Reason and Experience had taught him to judge rightly, he was well. When he was recovering, it was very hard for him to bring out some Words at first, which, I suppose, might proceed from his forgetfulness of them. This blindness is not a very common Symptom with the Cholic, or Belly-ach, but yet appears now and then. I have my self seen several Instances of it. There appears no Blemish in the Eye in this Case, and they are struck Blind unknown to themselves or the by-standers, till they come to try their Eyes upon any occasion. This sort of *Gutta Serena* goes off in some Days, and they recover their Sight, at least as many as I have seen or read of, recover'd it by the use of Bleeding, Purging, Blistering and Cephalics.

The End of the Introduction.

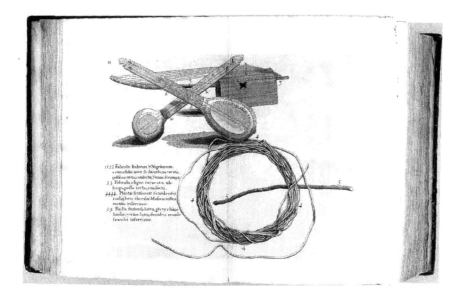

1,2,2. Fidicula Indorum & Nigritarum, concurbita inter se diversis, ex ovatis pellibus ficu confectis, Strum Strump.
3,3. Fidicula e ligno excavato, oblonga pelle tecta, confecta.
4,4,4. Planta fruticosa Scandentis cuius loro chordae Musica instrumentis inserviunt.
5,5. Radix fruticosa lutea, gtveyrhinae fundis, exortus fupra dentibus mundificandis inserviens.

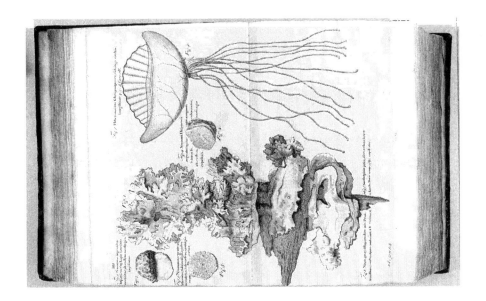

A VOYAGE TO JAMAICA.

WHEN our intended Voyage to *Jamaica* was drawing near, I was desir'd by several who were to go the same Voyage, to give them my Advice what Physick would be best to prevent their being Sick at Sea, and receiving injury thereby, and by the change of the several Climates we were to pass through; to which my Answer was, that I thought the best Counsel I could give, was, to eat and drink what was fitting, and to use Exercise, and the other six non-naturals with that moderation, that their Bodies might be kept in a healthful state, and made strong and able to endure any Disease should through unavoidable contingences attack them; for that when one is well, if Physick be taken, it must either make no alteration at all on the Body, and then it will not deserve the name of Physick, but be a Chip, and so consequently a needless trouble and good for nothing; or it must make an alteration on those, who being supposed perfectly well in Health, must by it be changed and made Sick. Some of those who would take Physick notwithstanding this warning, felt this true to their Cost, being by Purges thrown into Gripes, and other troublesome Distempers from which they were not so easily freed. The same Argument might be urg'd against those, who when perfectly well will take preventive Physick, who if they escape Death (which the famous *Machiavel* did not) or Sickness, will at least by Custom, which will become at last necessary, make themselves Slaves to bleeding and purging every Fall and Spring to prevent Diseases, which are much rather brought by these means than hindered. And it seems as reasonable to me that a Soldier should before a Fight come to a Chirurgeon to ask a remedy to prevent his being Shot, as when one is altogether Well, to a Physician for a Remedy against Sickness. And to confirm this necessary Caution a little further, I have seen more than once in Seasons for Epidemical Diseases, as the Small-pox, &c. that those who have been over-wise, in either taking Medicines or Journies to shun the common Distempers, have, by the agitation they put their Bodies into, been taken with what perhaps otherwise they might have avoided.

Preventive Physick consider'd.

On *Monday* the 12. day of *September* 1687. I went on Board the *Assistance* Frigat, one of the King's Ships, of forty four Guns, and two hundred Men, Commanded by Captain *Laurence Wright*, lying at Anchor at *Spithead* near *Portsmouth*. She had in Company two large Merchant Ships, and the Dukes Yacht, carrying His Graces Provisions and Servants. We weighed that

A Voyage to JAMAICA.

The Common Sea-sickness consider'd.

Afternoon, the Wind being North-West Northerly; but next Morning the Wind coming South South-East, and blowing hard with Rain, we came back to St. *Helens* Road, where we lay in eleven Fathom water, Oosie Ground, and about three Miles from the Land, till the 19. the Wind being South-westerly all this while, with rainy dirty Weather. We who had not been accustomed to the Sea, at first setting Sail, or even on first going aboard, by the Heaving and Setting, as Seamen term it, or the Motion of the Ship by the Waves, were all of us almost Sea-Sick. This first appears by a great uneasiness and load about the Stomach, disorder and aching in the Head, high colour'd Urine, and Vomiting at first what has been lately eat or drunk, then a great quantity of serous Matter insipid to the Tast, and mixt with Ropy Phlegm. Then if the Vomiting continues, comes the Yellow bitter Bile or Choler, pump'd up by the inverted motion of the *Duodenum* out of the Gall-Bladder, as well as the subacid Juices from the *Pancreas* and neighbouring Glands, which give a greenish Tincture to whatever comes up, and sometimes a sour Tast; and after these Liquors vomited up, as after a Natural or Artificial Vomit, the Persons generally from desponding and not caring what happens to them, come to be very easie. There are great Controversies among learned Men concerning the

Acosta's Opinion concerning it refuted.
* *Martens, cap. 6.*

Cause of this Sickness. *Joseph Acosta* ascribes it to the Sea-Air, but I think without any manner of Reason, for it seems only to proceed from the unaccustomed motion of the Ship *, which is sometimes higher or lower, or on one side or t'other, according to the Waves on the top of which she swims, or the Sail she carries; few People are Sick when the Waves are small, and the motion of the Ship inconsiderable, and those unaccustom'd to the Sea are always out of order more or less in proportion to the motion of the same, and are extremely sick in great Storms, when even the Seamen themselves, who have used that life many years, are not free from it, as many have related to me. It seems to be a greater or lesser Vomit, according to the Weather, and that, as Artificial Vomits, they sometimes work easily on some Tempers, and on other People of a different Constitution not at all. That a very small unaccustom'd motion of Man's Body will produce Vomiting, is very plain from those not used to Coaches, or to ride backwards in them, which altho it is not a very extraordinary motion; yet in some will produce the same effects as if they were at Sea. And that a greater will seldom miss, is evident from the Vomitings which mostly follow violent or Consumptive Coughs, which seem only to affect the Stomach in making it move irregularly with its Contents, by the great motion of the *Diaphragma* and Muscles of the Belly: and this is more plain by that sort of punishment used in *Switzerland*, and some other Countries, Malefactors being put into a sort of Cage, which being turn'd round makes them so Sick as to Vomit with uneasiness. As to the two Arguments used by *Acosta* to prove his Position, they do not move me; for his first, that the farther off Land, one is the sicker, is not true, for People are generally sickest when they go first on Board, and although the motion of the Ship, because of the largeness of the Waves then is greater, yet by Custom they become not so Sick: and for that other Reason of his, that he was once at Land sick on a great Sea-breeze, I believe his sickness was accidental; otherwise those who live on small Islands far remote from Continents, as St. *Helena*, would have a sickly time on't. Why this unaccustom'd motion should produce this effect, is beyond my giving any tolerable account of, only this is obvious, that Water in any open Vessel, if not lash'd so as to remain Horizontal in whatever Position the Ship be, will on alterations of the posture of the Ship overflow; even so the Contents of the Stomach, if Liquid, may run impetuously several ways, touch or make an irritation on some parts of the Body or Stomach it did not touch before, and how sensible those nervous parts are, need not be told any who have seen Vivisections, where the least, but superficial

A Voyage to JAMAICA. 3

ficial eafie touches, even when the Guts and Stomach are in fome Animals out of the *Abdomen*, will caufe a fenfible and vigorous Periftaltick motion.

As to the fo much celebrated Salutary effects laid to enfue Sea-Sicknefs in all difeafed Perfons, and its no lefs famed ones in preventing Sicknefs in thofe that are Sound; I confefs my felf to diffent from thofe Opinions; for although in feveral Diftempers very frequent now adays from Intemperance, &c. cleanfing the Stomach by Vomit be a very good remedy, yet I think it neither univerfal in every Diftemper, nor this by Sea-ficknefs to be able to do more than a good ordinary Vomit, having likewife this very great Inconvenience, that it is not in the power of the Phyfician to ftop this, or moderate the working of it, as he may a Vomit, no more than he can command the Waves, or ftop the raging of the Sea; whence many People by the lafting of this perpetual Vomiting, in long Voyages, have been brought to even Death itfelf, and others extremely weakened thereby. For all thofe Arguments from the flothinefs of the Subftances vomited up are very frivolous; for I queftion not but many perfectly found People vomit them up, who would not have any Sicknefs, if they had not fo vomited them, and to the Oeconomy of whofe Bodies they were abfolutely neceffary; the Serous Matter or Spittle to be a Menftruum to their Victuals, the Phlegm to defend the Tunicles of the Stomach and Guts, and the Gall as well as fubacid Juices from the Glands to help the Digeftion of our Victuals, and forward the Secretion of the ufeful Juices and Excretion of the Excrements, which fo neceffary Subftances, either fimple in their own Colours, or mixt together, and fo making a new Colour, are ufually fo much and undefervedly blamed.

The Effects thereof;

For Remedies for this Sicknefs (which not one in near fifty efcapes) I know none perfectly fo; but keeping in a quiet pofture, in a place where is the leaft motion without any noifome fmell or fight, is the beft. I ufually at firft defired them to drink quantities of warm Water, or Small-Beer to make all come up eafily, on which they found Relief, and in fome days they grew better. Or with the Ships motion this Sicknefs abates, and then when there is only Squeamifhnefs, eating of Victuals relieves rather than hurts. Marmalade of Quinces, Candy'd Citron-Peel, burnt Wine with Spices, and generally all Cordial hot Waters, gave me no eafe but fometimes made me ficker; having been, notwithftanding tryal of all thefe things, troubled with this Sicknefs for a Month together in going to the *Weft-Indies*, and fix Weeks in coming Home; upon every the leaft puff of Wind extraordinary. At the worft a Foot on Land always fet me free from all thefe Symptoms and Maladies immediately, altho my Head would turn round and feem'd to have an odd motion within it for fome time after.

And probable Remedies. * Rauwolf, cap. 1.

'Tis very ordinary for Sea-Sick People to be very much bound in their Bellies, becaufe of the Periftaltick motion's being inverted, the Gall or unnatural Glifter voided by the Mouth, and that they have no Appetite for Solids, neither digeft fo much Food as ufually. To remedy which I ufed to give gr v. extr. Rud. a little Manna, or fome eafie Medicine to Stimulate Nature, and to hinder the baking of the Excrements in the Cells of the Colon, and by that means hinder a Colick. But one need not be over-folicitous about this matter; becaufe if they Eat not as they ufe to do at Land, they cannot expect the fame quantity of Excrements. This confideration may be of fome ufe likewife in the practice of Phyfick, where Phyficians for the fame reafon in difeafed Bodies ought not to expect Excrements as in Health.

The 19. of *September* we Weighed Anchor, the Wind being at North-Eaft in the Morning, it chang'd at eleven at Night, when we Tackt, and was variable from South-Eaft to South-Weft, and for the moft part a frefh Gale. On *Wednefday* the 21. the Wind being at South and South by Weft, thick Weather, much Rain, and like to blow hard, we went into *Plymouth* Sound, where we lay

We arrive before Plymouth.

4 *A Voyage to* JAMAICA.

lay in seven Fathom water, and struck our Topmasts, being a hard Gale and much Rain.

We lay here at Anchor moor'd till *Wednesday* the 5. of *October*, nothing memorable passing, the Wind being for the most part South-Westerly, and blowing hard, with dirty rainy Weather and great Seas rowling in upon us from the Ocean, which made me (who was obliged to visit Sick People on Shore) several times wish it were customary for Men of War to go nearer the Land; where we had been well defended by Hills from the injuries of the Wind; but by lengthening the Cables and our good Tackle the Anchors held firm, and we received no injury. It deserves here to be noted that from our coming on Board to this time, twenty two days, we have had, except one day, almost always South-West Winds, which for the most part reign here this time of the year, making it very hard for outward-bound Ships to get out of the Channel; but there will be occasion to speak more of this hereafter.

We Weighed Anchor from *Plimouth* on *Wednesday* the 5. of *October*, the Citadel saluting his Grace with twenty one Guns, and the Island of St. *Nicolas* with nineteen, the Wind being at East South-East, a fresh Gale, we steered to get clear of the Land, and the next day we got thirty or forty Leagues to Sea, sailing more Westerly than our Course required, the Seamen desiring to clear themselves of that as much as conveniently they can, that when a contrary Wind comes to cross them they may have room to Tack, and not be afraid of a Lee-Shore, which is so formidable to them, that they rather chuse in such a case to take a Harbour than to lie at Sea, where if the Head Sea be not very considerable, by tacking they always get something forward be the Wind never so contrary. I observed at about fifty Leagues or more, of the Lands end, at Sea, many of a kind of *Larus*, or Gull, very like to that describ'd by Mr.*Willughby* in his History of Birds by the name of *Hirundo Marina*, or Sea-Swallow, only larger, for which Reason I shall call it *Hirundo Marina major*:

Hirundo Marina major described.

This Bird was bigger than a Pigeon, it was of a dark grey colour on the Back, and white below, it was sharp-winged, and did not fly high, but very close to the Surface of the Water, which it dextrously shun'd touching, after the manner of Swallows, and closer to it than the before-mentioned common *Hirundo Marina*; being very intent on its Prey, and getting what Fish it could spy there; so as one would admire how they can avoid being wetted by the Seas. When they had wearied themselves on the Wing, they would usually in Companies together, set themselves on the Surface of the Water, especially towards the Evening, which made me inclinable to think that they seldom went to Land; if so, it is not easie to imagine, if it be necessary for them, how they provide themselves with Fresh-Water, for as to Meat, the breaking Waves discover to them sufficient quantities of smaller Fish, which they greedily devour. The further off Shore we went, we had the fewer of these Birds, altho we never were entirely for several days without seeing some of them; but most when we were about the *Madera* Islands, there being some uninhabited Islands and Rocks between that place and the *Canary* Isles, known by the name of *Salvages*, where they multiply and increase yearly in prodigious numbers, not being interrupted by mankind. I am very apt to believe these

*Coron.*l.14. c. 1. Birds to be what *Oviedo** calls *Patines*. What *Denis* † calls *Croiseurs*, who
Ramnus. p. 261. were at top of the Water to take small Fish, having met them most of the way to *Canada*: and what *Ligon* meant when p. 5. he speaks of a Bird not
‡ *Tom.* 2. p. 5. much bigger than a *Castril* turning about every Wave. I am in doubt whether this be the *Gargaio* of *Colon.* f. 29? or *Avis à Lusitanis Garayos dicta Aldrov.**
§ *Ornith. Tom.* 3. p. 546. or *Willockes* of *Purchas* p. 556? and whether these be the same with the Mews mentioned by *Martens* of *Spitzberg*, that stay so long as the Sun, then fly away in Flocks, and rest on the Water by the way. When we could find

no

A Voyage to JAMAICA.   5

no more any Ground, or were got clear of the Soundings, that is commonly counted about One hundred Leagues from the Coast of *England* towards the South-West; the Flag was hoisted on the Main-top-mast-head, and several Huzzas, and Guns discharged at drinking his Grace the Duke of *Albemarle's* Health, who was then Vice-Admiral of those Seas. The Ships in Company likewise shew'd their Respect, in discharging their Guns, by way of Salute to the Flag; the like was done by some Ships Bound for *Guinea*, who kept Company with the Frigat to avoid hazard from the *Sally* Men of War, who are very busie about these parts: one who was some Leagues a Head, hal'd up his Sails, and lay by till the Frigat was past, when he likewise, by a Salute of all his Guns, paid the respect was due to the *English* Commander of those Seas.

On *Friday* the 7th. of *Octob*. 1687. we had a *Grampus*, or small Whale, follow'd us: This sort of Whale which is the least, and of the same kind with the Dolphin or Porpesses, is about forty Foot long, and proportionably broad, is smooth, and of a dark brown colour in the Back, and has two Channels in the Head, through which he spouts out Water in two small Streams when he rises from the bottom above the Surface of the Sea to breath, which he does very frequently, blowing and so going down again. This sort is very frequent in the great Western Ocean. The Sailors commonly shoot at them with small Shot, but without any seeming hurt to the Creatures. They are very devouring, and chase into the Shore the greater Whales, and are frequently thrown on Shore in *Scotland*, where they are about Twenty five Foot long. It is called *Balæna minor in utroque Maxilla dentata qua Orca vocatur*. *Sibbald. obss. de balænis*, p. 6. *Orca Rondeletii*, p. 483. *& Bellonii*, p. 16. *Raii*, p. 40. *Granpasses*. *Purchas*, 569. where it was observed in 73°. *N. lat*. *Graunpisce* of *Terry*, p. 8. *Baleine de Rochef*. 195. *D'Abbeville*, p. 30. *Grampose* of *Pool*. *Purchas*, p. 707. The *Grampus* of *Smith, Virg*. p. 28. *New-England*, p. 227.

The Grampus describ-ed.

We had also Porpesses came about us. There are multitudes of these (which are well described and figured by Mr. *Ray*, p. 31. Tab. A. 1. Fig. 2.) in the great Western Ocean; they go generally many together, and when they appear about a Ship they are counted to Presage a Storm, and that the Wind will blow that way whither they go: they are not so frequent between the Tropicks as in the Northern parts. They are much swifter in Swimming in the Water than a Ship under Sail with a good Gale of Wind, they are taken by Harping-Irons, and the Fat being taken off (especially Fins and Tail) are eaten by some Seamen, but are no way delicate Food, having a very noisome fishy, and rancid Tast.

Porpesses described.

I believe this, and the *Delphinus* to be much the same; they leap up about a Foot out of the Water, (tumultuating and being uneasie or sick by the weather, *) sometimes more, and sometimes less, and are so quick in returning that they seem to be crooked, whence the Painters took their Figures.

* *Martens*.

These are mentioned by the following Writers by the names of *Porpisses* of *Smith, New-England*, p.227. *Porpisce* of *Ligon*, p. 3. *Marsovin appelle pourfille de Deus, Tom.* 2. p. 258. *Gnaperva Marogr*. *Marsovins de Rochef*. 191. *du Pyrard de la Val*. p. 6. *de Cauche*. p. 141. *D. Abbeville*, p. 30. *Marsovin*, called by the *Portuguese Tonans* of *Mandelslo*, p. 197. *Phocæna Rond*. p. 473. *Gesn*. *Raii*, p.31. Tab. A. 1. Fig. 2. *An Botos & Toninas* of *Anonym. Portug. Purchas*, 1314? *Phocæna seu Tursio Bellon*. p. 15. *Scalig*. Porpisces or Hog-Fish. *Terry*, p. 11. *An Ambizeangulo seu Porcus aquatilis*. *Pigafet*. p. 10? Porpasse, *Butskopf*, or Places-Head of *Martens. cap*. 2. The Porpesse of *Escarbot*. *Nova Francia*, p. 69. (From which Authors a larger account may be had of them.)

C   I had

I had very often heard, but never observ'd before, the sparkling light of Sea-Water, which appears thus: In a dark Night (the darker the better will you observe it) if you look attentively on the Surface of the Sea, you shall see now and then a little sparkling light sometimes broader, and at other times narrower, which presently vanishes. If you row in the same, you see it very plain where the Oars touch the Water. On a part of the Sea where a Wave breaks or curls you see it much plainer, and by the Ships side, or Bow, where the Water is more broken, you see it most of all. Sometimes you shall see as it were a Spark of Fire leap up into the Air as if a Flint and Steel were struck together, which nevertheless vanishes very soon, though sometimes I have seen a sparkle left by the Water on the entring Ladder of a Ships side, which has continu'd there shining for some half a Minutes time, like the icy *Noctiluca* or *Phosphorus*, the light of this being as to colour, &c. like that of the other:

Concerning the sparkling light of the Sea-water.
the Seamen told me that they were more ordinarily to be seen in Southerly Winds than any other, how true I know not, but am sure the more the Sea is broken or white, the more you see of them: I endeavoured with a Swab several times dipt into the Water to pull some of those Sparkles up, but could not, for they would not stick to it, wherefore I had a Bucket of Water drawn, and by moving it up and down with my Hand saw some of them appear now and then on its Surface; but once had the good luck to move it in such a manner that one of those Sparkles hit on the Bucket-Rope, and sticking there gave me the opportunity of squatting it with my Thumb, and making it by that means give a larger light, which it did for some small time, and then went out. I did not observe that it had any actual heat on touch. *Nicolas Papin*, who wrote a Treatise in *French* about this, giving it the Title of *Mer lumineuse ou traité de la lumiere de la Mer*, tells us, that agitation without Froth produces it even at bottom, *p.* 129. how true I cannot tell.

Apud Cluf. in Cur. poster.
This strange and surprising *Phenomenon* is by several People taken notice of; *Vanderhagen* speaks of it as extraordinary in some places, so that it was like a Lanthorn giving light, *Martens cap.* 2. tells us that the Sea shines like the lustre of a Diamond, and foretels South or West Winds. *Ligon* when he takes notice of it *p.* 7. thinks it comes from the Saltness of the Sea and the hard Boat striking Fire, and that it would be Fire if not quenched. Not to recite the several Opinions about this, I am very inclinable to believe that it may proceed from some of the smaller Particles of Fishes floating in the Water, although so small as to fly the quickest sight, for I could not observe any difference between that and the most limpid Sea-Water I ever

The reason of this Appearance.
saw. I am the more inclinable to believe this, in that I have seen on the Sands, left uncovered by the ebbing of the Tide, several Portions of fishy matter shining after that manner, only larger; the same is to be seen in Oysters, Lobsters, &c. And I see no improbability in supposing small corrupted parts of Fishes to roul up and down in the Sea-water, and when they come to its Surface either by themselves or in Bubbles, whereby they are more exposed to the Air (broken-water being nothing but a heap of Bubbles) they shine in the same manner that a Piece of rotten Fish will shine in the dark, and the relation of Seamen may well enough agree with this, the South-Winds being warmer, and more promoting of Putrefaction, or aiding to Fermentation than any other Wind whatever. If it be objected, that it seems unreasonable to believe that all parts of the Sea should be furnished with them, I answer, that it is certain, most parts of the Sea are very full of them, and not to mention the relations of several Seamen, who have told me, that both in the North and South Seas they have sailed a great many hours through Fishes Spawn; I my self have done so for more than two days. It seems to me that such Fish-Spawn or fishy Matter, if in great quantities,

A Voyage to JAMAICA. 7

discolours variously the Sea-water in the day-time, and makes it, if dark, appear more or less shining, as it is more or less in quantity. *Lopez de Gomara* tells us, that about *Cubagna* the Sea is at some times Red by the Oysters Spawn or Purgation; and *Terry* takes notice of Sea-water white as Milk not near Land. *Hatch* in his Voyage published by *Purchas* speaks of Water white as Whey, yet no Ground, and *Haynes* of White Water no Ground, in 4°. South Lat. *Downtown* tells us of Water muddy and thick, with spots of clear, near *India*. And *Weymouth* of Water as black as Puddle, and clear again, yet without bottom, at 120 Fathoms. And *Hall* Sailed in Black Puddle Water for three Hours. But to come nearer our matter, in proving this Conjecture. *Bottell* in the Red-Sea falls in the Night on whitish Spots, raising and casting Flames like Lightning, he wondering at it, took in the Sails presently, believing they were on some Banks or Shoals, and commanded to cast the Lead, and found Twenty six Fathoms, Pilots of the Country not fearing went on again, *ib*. *Saris* met with Cuttle-Fish in sailing from the Red-Sea to *India* in 8°. 12's the Wind was at West-South-West, *Sep.* 22. at midnight very dark, saw shining Water, strange and fearful, so as to discern a Letter in the Book thereby, he sailed in it half an Hour, not without fear of Rocks, but he tells us it was from Cuttle-Fish; and *Cancho*, *p.* 20. tells us that the Worms that eat the Vessels, shine in the Night.

Hist. Gener. cap. 8.
Pag. 56.
Pag. 6. 8.
Ib. p. 632.
Ib. p. 310.
Ib. p. 810.
Ib. p. 811.
Ib. p. 815.
Ib. p. 1129.

Apud Purchas, p. 352.

When we were about Forty Leagues off the Land, we had a Lark which had come, or been driven too far from Land by a Storm, perch'd on the Ships Rigging; it was so tir'd that it suffer'd any Body to come within Arms length of it before it would flir, and would have permitted us to have caught it rather than have gone into the Sea, had it not espied the other Ships, to some of which it went for resting itself when 'twas scar'd from ours. 'Tis very ordinary for Land-Birds thus to be driven off to Sea, and to light on Ships, being lean and wearied on the Wing: so *Martens, cap.* 2. tells us that Blackbirds, Starlings, and all small Birds lose their way in a Storm, and cannot recover it but either drop and are drowned, or sit on Ships; and that a Crow he saw (at *Spitsberg*) by mistake, had thus come astray.

A Lark driven forty Leagues to Sea.

On *Tuesday* 11. When we were in about Forty six Degrees of Northern Latitude I first saw what the Seamen call a *Caravel* or *Portuguese* Man of War, which seems to be a *Zoophyton*, or of a middle Nature between a Plant and an Animal; it is of that kind of the soft Fishes called *Urtica* from their Stinging quality, and to me seems different from any describ'd by any natural Historian. I shall call it *Urtica Marina, soluta, purpurea, oblonga, cirrhis longissimis.*

A Caravel describ'd.

It is taken notice of by *Stevens apud Hakluyt, p.* 99. where it is called a Ship of *Guinea*, and by *de Lery, p.* 399. under the name of *Immondicites Ranges*. *Martens* calls it the other sort of Sea-Nettle in the *Spanish* Seas that weighs several Pounds, of a Blue, Purple, Yellowish, and White colour, that burn more violently than those of the North-Sea, they do suck themselves so close to the Skin that they did raise Blisters, and cause sometimes St. *Antony*'s Fire. He says further that one sort of this is called Sea-Spider, and is the Food of Whales, which may, by the way, explain a Passage of *Peyrere* in his *Anonymous* Book, called *Relation de Groenland*, where the Author tells us, that Whales feed on *Araneos du Mer. Ligon* calls it *Carvile*, and observ'd it Five hundred Leagues from Land, and they are named *Grandes Urtica* by *de Laet*, who takes notice of them in *Brasil*.

Pag. 6.
Lib. 15, *c.* 11
p. 573.

This floated on the Surface of the Water, and consisted of two parts, the one was an oblong Cylindrical Bladder not so big as a Turkey-Egg, it was as it were blown up, and full of Wind, almost like the Swim of a Fish, widest at

A Voyage to JAMAICA.

at bottom, and grew straiter or narrower to its top, where round about was a corrugated or curled Ledge or Band, something like a Cocks-Comb, Convex on one side, and Concave on the other, which Seamen said was for its more convenient sailing; all this part of it was of a purple and bluish colour and Pellucid: the other part was a great number of blackish and Red Fibres, Strings or *Cirrhi*; they were long and White, here and there Purple, having several Knots like Nits on it, taking their Original from the bottom of this Bladder, which if stretched were several Feet long, but if curled up were very short, stinging much worse than Nettles, whence it is by some reckoned Poisonous. They are very often to be met with at Sea, and Seamen do affirm that they have great skill in sailing, managing their Bladder or Sail with judgment, as may be most for their purpose, according to their different Winds and Courses; allowing them more Reason, than I, at present, am willing to do of Life, there appearing to me no other parts than the Bladder and *Cirrhi* abovementioned.

On *Tuesday* the 18. at Night we lay by, because we were afraid of running upon the Island of *Madera*, or Rocks about it in the Night when dark, being in its Latitude, or very near it by observation, and by the Dead-Reckonings, Corrected by Observations, it appearing that we should be so far West as those Islands are placed. But after sailing several Leagues, next Morning we could not make any Land, and several of our Seamen being of Opinion that we were to the Westward of it; it was ev'n almost resolv'd that we should hold on for *Barbadoes* the nearest way, seeing with this Wind we could not easily get back thither, and that it might retard us considerably the going to *Palma*, or any of the *Canaries*; although Fresh Water, Provisions, and Wine were a great inducement to our going to one of them. One of the Captains who had made many Voyages to the *Canaries*, being call'd, and coming on Board with the others, (who according to their Instructions and Signals, were to come to consult for the Publick welfare) told the others, that having made this Island usually in his Voyage to the *Canaries*, he knew that as yet we were not come so far West, but if the Wind held we might be there to Morrow; and that the Reason to him seemed this, that either the Island ought to be placed a Degree more West; or that we, as every Body else, sailing through this Sea at this time of the year had met with Westerly Winds, which bringing great Seas, and making a Current which always goes with Wind; * and there being in the Ocean, we had sail'd through a great Current of Water to the Eastward, he thought we had had more Lee Way than we had allowed for, and been more Easterly carried than we computed, and therefore advised that we should persist till Night, then lie by for fear of the worst; which on his positive affirmation we did, and on *Thursday* the 20th. about Twelve at Noon we made the Desarts, which are three small Islands or great Rocks, lying on the East side of the Island of *Madera* about Three Leagues from the Land. Being about Ten Leagues from it, we came in sight of *Porto Santo*, an Island belonging to the *Portuguese*, Three Leagues long, and one and a half broad, in 33°. North Lat. Twelve Leagues to the North-East of *Madera*. It had Five hundred Inhabitants, and yet was taken by *Preston* with Sixty Men, as *Davies* tells us, *Purchas*, 579. It was first Discover'd according to *Jo. de Barros*, by *Giovan Consalvo Larco*, and *Tristan Vaz*, who were sent out to discover *Guinea* beyond Cape *Bajader*, and were carried against their intention by a Tempest to *Porto Santo*, which they called so for their being saved thereby from Shipwreck. They return'd to *Portugal*, and gave an account of it, and went thither again with Three Barks, one *Bartolomeo Perestrello* (whose Widow *Christopher Columbus* married) joyning with them: they carry'd Fruits

and

Purchas p. 10. Towerson Voy. 1. to Guinea ap. Hackuyt, pag. 3 l.

Decad.2,c.2.

A Voyage to JAMAICA.   9

and Seeds, with a Rabbit big with Young, which multiplied so fast as in two years its Offspring destroy'd every thing was planted in the Island. *Barros cap. 2. Dec. 1.*

When *Perestrello* was return'd to *Portugal*, *Tristan Vaz*, and *Giovanni Consalvo Zareo*, discover'd *Madera* from *Porto Santo*, it appearing as a Cloud. It was almost every where full of Wood (as most uninhabited Countries are) and peopled in 1420. *Consalvo* to free it of Wood set Fire to it, it burnt and destroy'd so the Woods that the Inhabitants soon wanted for making their Sugars, and were commanded to Plant. *Ib. cap. 3.* *Of the first discovery of Porto Santo and Madera.*

We lay by all Night, and the next Morning being *Friday, October* 21. and Sixteen days from *Plymouth*, we came to Anchor in the *Madera* Road, the Castle bearing North, distant one Mile, in Forty one Fathom Water.

The Island *Madera*, was so called from its being all Woody that word in *Portuguese* signifying Wood. When the Trees were fired, they burnt so impetuously, that the People were forc'd to go into the Water up to their Necks. *Cadam*, 105. *Ja. Barr. Dec. 1.* It was discovered by *Macham* in 1344. which was before that of *Giovanni Consalvo*, according to *Galvanos Portugal* Chronicle, *Purchas, p. 1672.* It is Situated in 32°. 30′. of Northern Latitude, 9°. some odd Minutes West of the *Lizard*, is high Land, very rocky and steep, six Leagues broad, and eighteen long; inhabited by *Portuguese*, and Populous, having about, at present, Eighty thousand Inhabitants, whereof Fifty thousand are Communicants, that is above Eight years old. It has a healthful Air, some People living here to an Hundred years of Age. It is fruitful in Cattle, viz. small Cows, Swine and Sheep, the latter being Lean, and having long Wooll, almost like to that of Goats, not curled as our Sheep. The Air here is very Temperate, refreshed for Nine Months of the Year by a Sea-breeze in the Day, from Eight in the Morning till Four in the Afternoon, and a Land-breeze in the Night, from Eight at Night to Four the next Morning; between which Breezes generally there is a Calm: from the latter end of *November* till the beginning of *March*, the Wind is at between South and West, and then the Weather is stormy, making great Shipwrecks in the Harbour of the Principal Town *Funchal*, it being exposed to those Winds, and only secure in one place, where a Rock Perpendicular and high, keeps off the force of the Wind from Ships Riding between it and the Shore. Some few Years before I was here, the most part of the Ships in this Harbour had suffered Shipwreck, the Winds being violent, and the Water so deep, that the Cables cannot so firmly hold as in other Roads, whence Ships are forc'd to put to Sea on any extraordinary puffs of South-Westerly Winds, the danger of which frighting Seamen, does not a little hinder the Trade of this Island. The Winters are here so Temperate, as that usually no Snow lies, except on the top of the Mountains, neither is Hail very common, though the Winter in the Year 1683. which was so extraordinary hard to *Europe*, reached this Place likewise, the Inhabitants assuring me they had not felt the like. So I find by *Smith*, that in 1607. there was an extraordinary Frost in *Europe*, and that it was as extreme in *Virginia*, *P.* 21. And so it happens when there is very hard or extraordinary Weather in one place, it generally is so in others, contrary to what one would think. They have Apples, Pears, Walnuts, Chesnuts, Mulberries and Figs; of our *European* Fruits in great quantities, Apricocks and Peaches growing Standard Trees, ripening their Fruit without the help of a Wall, as also Bonano's, Plantains, Oranges, Lemons, Citrons, &c. Common to the hotter parts of the World; so that this Island seems to be fit for producing the Fruits of both Hot and Cold Countries: of the first in its Valleys or lower Land, and of the last in the tops of the Mountains, where 'tis much Colder, and where many large Chesnut *Of the Name, Air, Inhabitants, &c. of Madera.*

D

A Voyage to JAMAICA.

nut Trees grow. The greatest part of the Island is one very high Hill, reckoned four Miles from bottom to top, which for the most part is clouded, the descent of which being very steep, makes the small Rivers very rapid, and their Cliffs or Banks very perpendicularly high. This Island is very fruitful, having formerly furnished great quantities of Sugar which was here planted, and was at first excellent, and yet what they have here is extraordinary but very little; the Reason of which is, that there are so many Sugar Plantations in the *West-Indies*, 'tis not worth their while to make it, although being once refin'd or clayed 'tis very White, and one Pound of it will do as much as a Pound and a half of any other; so that although they make some which is dearer, yet they find so much more Profit in manuring their Vines, that they scarce make what is sufficient for their own Spending in their Families and Sweet-meats, but buy that of their own Plantations in *Brasil* for that purpose. The greatest part of this Island is at present planted with Vines, the Soil being very proper, for it is rocky and steep; they keep their Vines very low with Pruning, in that agreeing with the Culture of the Vines in *France*, as also in that these Wines grow on the same Soil with those most esteemed there, as the *Hermitage* Wines, which grow on the rocky steep Hills on the sides of the *Rhosne*. The Grapes are of three sorts, the White, Red and great Muscadine, or *Malvasia*; of which three the first are most plentiful, for out of the White is made the greatest quantity of Wine, which is made Red by the addition of some Tinto, or very Red Wine made out of the Red Grapes, which gives it a deeper Tincture than that of *Champagne*, and helps it to preserve it self much better. It is sufficiently known that White-Wines, generally speaking, perish very soon, and that Red ones are much easier preserved, the deeper their Tincture be; so in *France* they suffer the Husks or Skin, and Juice of the Grapes to lie longer or shorter time in the *Cuve* together after bruising, according to the Stipticity or Tincture they desire, or which is all one, the time they would have their Wines to keep. The Virgin Wine, or that made of the Juice running of the Husks immediately without standing or pressure, is soon ready to drink, fine, and very soon perish'd, the Husk impregnating the Wine with something equivalent to Hops in Beer. The same likewise happens in Oil Olive; for it is to be observed, that that sort of Oil void of all manner of Tast and Smell, call'd Virgin Oil, which runs off the Olives without pressing, will without the addition of Salt, in two Months turn rancid, whereas that which has by strong pressing and standing been impregnated with some small parts from the Rind, or Stone of the Olive and Kernel, is able to keep for a very long time without any addition. The *Malvasia* or Wine made of the Muscadine Grape, does not keep, but Pricks very soon, and so is made in very small quantities. The great quantity of Wine here made, is that of the White mixt with a little Tinto, which has one very particular and odd Property, that the more 'tis expos'd to the Sun-beams and heat the better it is, and instead of putting it in a cool Cellar they expose it to the Sun. It seems to those unaccustomed to it to have a very unpleasant Tast, though something like Sherry, to which Wine it comes near in Strength and other Properties. It is Exported in vast quantities to all the *West-Indies* Plantations, and now of late to the *East*; no sort of Wine agreeing with those hot Places like this.

Of the Wines of Madera.

They have here some Corn of their own growth, about as much as may maintain them four Months of the year, but most comes to them from *Dantzick*, *Ireland*, *New-England*, &c. in Exchange for their Wines to be carried to the hot *East* and *West-Indies*, and some few Sweet-meats, as Marmelade of Quinces, Citron-Pills, &c. which they here make up with *Brasil* Sugar, or that of their own Island. The Sea round this Island is very deep, (as it is in most places where the Land is high) within a Mile of the Shore 'tis Fifty Fathom Water,

and

A Voyage to JAMAICA.

and one quarter of a Mile further to Seaward 'tis Fifty more. In deep Water the colour of it is Blue, and in shallow Green. It seems to me that the difference of the Colour of Sea-water (without troubling our selves with many Opinions about it) comes either from the depth of it, which when very deep and diaphanous is of a deep Blue as the Sky when clear, or if shallow it takes its Colour from the Colour of what lies at the bottom. And that it is so, appears by *Purchas*, p. 1131. where 'tis taken notice that Water appears Red, Green, or Dark in the Sea, according to the bottom, and that the *Red-Sea* is called so from Red Coral, or Coral-stone lying at bottom, making the Water, which is to be seen into Twenty Fathoms, look Red, or White, if White Sand is at bottom; or Green, if Green Oozy, *Id.* p. 1147.

The Sea hereabout is very well provided with *Albacores*, or *Thynni*, whose Description follows.

This Fish was Five Foot long from the end of the Chaps to that of the Tail, the Body was of the make and shape of a Mackarel, being roundish or torose, covered all over with small Scales, White in some places, and Darker colour'd in others, there was a Line run along each side. The coverings of the Gills of each side were made of two large and broad Bones covered with a shining Skin, the Jaws were about Six Inches long, having a single row of short strong sharp Teeth in them, and were pointed. The Eyes were large, and the Gills very numerous, behind which were a small pair of Fins. *Post anum* was a Foot long Fin, about Three Inches broad at bottom, and Tapering to the end. It had another on its Back answering that on the Belly, and from these were small *Pinnulæ* at every Two Inches distance to the forked Tail, which was like a New Moon falcated, before which on the Line of the two sides was a membranous thick horny Substance, made up of the Fishes Skin, stood out about three quarters of an Inch where it was highest, something like a Fin. It was about Three Foot Circumference a little beyond the Head, where it was thickest. The Eye was about an Inch and a half Diameter. The Figure of this Fish is here added, *Tab.* 1. *Fig.* 1. taken from a dried Fish, where every thing was perfect save the first Fin on the Back, which I suppose was accidentally rub'd off.

Albacores described.

It is frequently taken by Sailers with Fisgigs or White Cloath, made like Flying-Fish, and put to a Hook and Line for a Bait; The Flesh is coloured, and Tasts as the *Tunny* of the *Mediterranean*, from whence I am apt to believe it the same Fish. It is to be found not only about *Spain*, and in the way to the *West-Indies*; but in the South-Seas about *Guayaquil*, and between *Japan* and *New-Spain* every where.

This is called *Tannyes* of *Oviedo sum.* p. 214. *Alsicares* of *Terry*, p. 9. *Albocores* of *Mandelslo*, p. 196. *Dolphin* or *Tunin* of *Marten*, *Oreynus Rondeler*, p. 249. *Thunnus Gesner*. 1158. *Aldrovand*. p. 307. *Muss. Jrammerd. Rail. Hist.* p. 176. *Tab. M. 1. Cortt. Thynni Species ejusd. app.* p. 15. & 24. *Tab.* 9. *No.* 1. where the Figure seems not good. *Thynnus Bellon.* p. 106. *Salvian.* p. 124. *An palamite of Oviedo Sum.* p. 211 ? *Guarapucu Brasiliensibus, an Cavala Lusitanis, nostratibus Coningsvisch. Maregr.* p. 178 ? *Pisc. Ed.* 1658. p. 59 ? *vel an Corvata pixima ejusd.* p. 150 ? *Ed.* 1650. p. 51 ? *Tons of Escarbot Nova Francia*, p. 35. *du Ravenean de Lussan* p. 171. *An Albacorett. Pisc. Ed.* 1658. p. 73 ? *Toni di Fernan Colon vita di Christof.* f. 29. *An Ox-Eye of Anonymus Portugal. ap. Purchas*, p. 1313 ? *vel. Toninas Ejusd. ib.* p. 1314 ? *Tunnies of Francis Gualle Purchas*, 806. *Albacavas Ejusd.* p. 446. *Hakl. of Smith New England*, p. 227. of *Galvano Purchas*, in 42°. *North Lat. South Seas*, p. 1685. *Ton ou tosard de Conche*, p. 138. *An tonino Ejusd.* p. 142 ? *Ulasso a Tuny Fish of Daddeley*. p. 576. *Albacore* of *Ligon*. p. 6. *Aubeville*. p. 30. *An a Spanish Macquerel of Ligon ? Albachores Pyrard. de Laval.* p. 6. 137.

It has hard Fat, Flesh of a sharp Taft, opening the *Hæmorrhoids* either by its Acrimony, or because it breeds Melancholy Blood. *Rond.*

They are taken by the way (to the *West-Indies*) playing about the Ships, by Spears thrown at them. *Oviedo Sum.*

They (at *Maldives*) are taken with White-Line, where they Boil thele with Dolphins and Bonetos in Sea-water, and dry them after by Fire on Hurdles, which makes them keep a long time for Traffick. *La Val. p. 138.*

Dolphins, Boneto's, and other Fishes loving very deep Water, are also found here. This Sea has so great a Surff, that there is no Landing at the Town of *Funchal* without taking the advantage of coming in with a Wave, and being pull'd on dry Land with it, from whence you are again to be lanch'd to go on Board. The Ebbing and Flowing of the Sea is not here considerable, if I remember right. There are some few Towns in it, the Principal of which is *Funchal* or *Fontal* in the South-East part of the Island. The Town has about Ten thousand Inhabitants, whereof One hundred are for the Governors Guard, paid by the King. The Governor of this Island is a *Portuguese* sent from *Lisbon* hither, and lives in the Caftle of *Funchal*: he Commands on Shore, and cannot come off, having about Twelve hundred Crowns *per ann.* Salary from the King, besides what he can get by Trading. Here is a very fine Cathedral Church, and about Eight hundred Friars. They have here a large Hofpital, and in it a private Corner for those who are *incognito* to be treated for the Pox, a Diftemper very common in this Place. The Town of *Funchal* is well provided with good Water, and commanded by a Citadel, whither they retire in time of Danger. It has a Ciftern hew'd out of the Rock, to receive Rain-water, which maintains many People, and is very good. They cannot Hang any here, but only Banifh to the Cape *de Verd* Iflands. Confidering that this Ifland had not been very antiently Inhabited, being but difcover'd in the Fourteenth Century, and that Common Fame relates all the Inhabitants hereof to be Criminals banifh'd hither, I expected to have found a great deal of Barbarity and Rudeness here, and nothing almoft elfe; but on going afhore I was very much difappointed, for I have not feen any where more accomplifhed Gentlemen than here, having all the Civility one could defire; but moft of them whether bred to Letters or not, are fent for their Breeding to *Portugal*. The Scholars, whether Phyficians, Divines or Lawyers, are bred up at *Salamanca*, and thence return in fome time for their own Ifland to live. I met with a very Ingenious Phyfician here, who fpoke good Latin, and underftood his Profeffion very well. Their Manners are much the fame with thofe of the *Portuguefe*. Their Women never ftir abroad but to Mafs, and appear not in their Houfes to Company. They are very much ferv'd by *Negros*, and their Women come out of Bed the firft Week after lying In. They carry every thing on a Log drawn by Oxen, the Country being fo fteep and rocky, and the ways narrow, that no other Carriage can go. Every Tradefman wears his fhort Doublet, and for the moft part black Cloak, under it a long big hilted Dagger, with a fharp Knife in his Pocket. No Man here dares go in the Street after 'tis dark, left any who has a grudge at him fhould fhoot him, or left he fhould be taken in the dark for another Man. I was told half a Piece of Eight to a *Negro* would purchafe any Man's Life. Their Bread is good, and they Eat much of it, as alfo of Poor Jack. They are fo biggotted to their own Cuftoms, that the Soldiers before this Governor's time wore Cloaks, but neither himfelf wears them, nor will he fuffer his Guards to put them on. The King has about the tenth part of all Merchants Goods Exported or Imported into this Ifland, befides fomething to be paid the Friars. They have here the Inquifition, and are very ftrict even on Merchants themfelves: they compell'd the *French* Proteftants to change their Religion. A few

A Voyage to JAMAICA.

few Leagues West lies Three or Four small Islands or Rocks, called the Desarts, which are Till'd by Eight or Nine Men sent from *Madera*, One of them commanding the rest. The Commander some Years before carrying over his Wife and Daughters, he was kill'd, and they abus'd by the Murderers. They make here some red earthen Ware which is very thin and brittle, cooling Water, or whatever is put in them; the Red colour coming likely from the Iron lying in the Clay. I saw several pieces of Ore here which seem'd to be very rich in that Metal.

People here having a great Opinion of the Skill of *English* Physicians, they pray'd the Consul I might come ashore and see those who were Sick, among others they desir'd my directions in the following Cases.

A Clergy-Man of about Thirty Five or Forty Years of Age, had some while before been shot at by a mistake in the Night, by some who took him for one they ow'd some Prejudice to; the Gun was loaded with small Shot, which lighted about his Temples, he told me from that time he began to lose the sight of one Eye, till it was a while after entirely gone, and a Cataract (which I saw) grown in it; the other Eye, (which is usual, when one is hurt) decaying so much that he could scarce see any thing with it. I told him, I thought this might have been remedied if at first they had taken out the small Shot, which in all likelyhood had weakned the Eye, made a small breach or (lying near it) Compression of the Optick Nerve, by that means hindering the passage of the Images of Objects from the Eye into the Brain; but that now there seem'd to be no Remedies left, but to let some small quantity of Blood. Purge with *Pil. Coch.* and use *Millepedes* in great quantities for a good time, with a White Vitriol Water, outwardly dropt on the Eye, to eat away the Films as much as might be. *Observations of the Diseases of Madera.*

Many here had been very severely Clap'd, several of which (by means of an ill manag'd or protracted Cure) had been troubled with Caruncles and Carnosities, and now had great pain in making Water, and a *Stillicidium Urinæ*, so that they were forc'd to have dry Napkins applied to keep them from being wetted and Excoriation. These last Symptoms were common to those likewise who had their Carnosities eaten out of the *Urethra*, and cur'd; but violent pains in making Water remaining, (for which most part of the famous Physicians of *Europe* had been consulted without success.) I advised them to foment outwardly, to use inwardly large quantities of Decoctions of Mallows, Marsh-mallows, and other such Emollient and Anodyne things, to Emulsions and other temperers of heated sharp Urine, and to Gum Arabick powdered and taken in very large quantities.

I was desir'd by the Abbess of the Nunnery of the Order of *Santa Clara* to come to that Monastery, and give my Opinion concerning some under her care, who were Sick. I was brought first into a Room where one lay abed, and the others retiring she shew'd me a small Tumour upon the *Os Pubis*, it had a small blackish spot on the Head, and was very hard, she found very little pain in it at most times, but at others when she used to be purged by her *Catamenia*, she had most intolerable trouble there, and a great *Menstruum Fluxus* out of that Speck, with very huge pain. There was no Livid Veins, nor any sign of a confirm'd Cancer, but concluding it to be very much inclining that way, I desired them in the first place to abstain very religiously from all outward Applications, especially hot smelling Plaisters, which they told me had done her very much prejudice, I advised her next to a Steel Course, and some opening Medicines, as *decoct. Amar. Milleped.* easie Purges, &c. as the way to open Obstructions, sweeten the Blood, and send it the right way, but what success they had I know not.

F.   Although

14   *A Voyage to* JAMAICA.

Although this Climate be very hot, some of these were troubled with true Consumptions, for which I ordered them some easie Opiates, and other Medicines. I have observed the same Disease about *Montpelier*, amongst the Inhabitants of that place, though the Air be esteemed a Remedy for it.

But the greatest part of the Patients I had in this place were troubled with *Chlorosis*, a great many of them from their Single, Melancholy, and Sedentary Life, want of due and proper Exercise, &c. Falling into several kinds of several kinds of Diseases, bringing along with them variety of Symptoms, according to the different parts of the Body on which they fell, for their Blood wanting those Evacuations Nature design'd them, it is easie to imagine to what ill Circumstances in some time they may be brought. I generally for this ordered them first bleeding, thereby to avoid the danger there might be from too much Blood, it generally abounding in Persons thus Diseased, and then after some Vomits, or bitter easie Purgers, prescribed a Steel Course, with Exercise. Thus having seen most of those Sick in this place, I went away after having received a very handsome and neat Treat of Fruits, Sweet-Meats, &c. both the Preserves and Furniture of the Room being of the Nuns own Work, than which I never yet saw any thing of their kinds so pretty.

As for the Birds of this Island, those I saw were;

Of the Birds of Madera.   *Phasianus*, or the Pheasant.

Perdix Rufa, Aldrov. The Red-legg'd Partridge, which is in most parts of *Africa*, and some about *Montpelier*, and in *Italy*.

Merula Vulgaris, or the Common Black-Bird, all three to be met with in *Europe*.

Passer Canariensis, or the Canary-Bird of *Boile* of Air, 178. *Paxaros de Canaria de Lopez de Gomara, cap.* 224. Canary-Birds of *Galvanos, Port. Chron. Purchas.* 1673. They are here in the Fields in Flocks, sitting on the ripe-headed Thistles and Plants, feeding on their Seeds, and making a noise not unlike, but much pleasanter than our Sparrows. These Birds are brought into *Europe* in great quantites, and sold for Singing-Birds to be kept in Cages, but those of these Islands are so much accustomed to the heat of the Climate, that they do not thrive so well as such which breed and are sent into *England* for that purpose from *Switzerland*.

Wild Peacocks and Pigeons were here caught in abundance with Perches at first, *Cadam.* 105. And Peacocks are to be found now wild in some parts of the Continent of *Africa*.

The Plants I gathered, or saw in the Fields, were these.

Of the Plants of Madera.   *Oleastri Species ut quidam putant, ut alii Zizyphus alba. Gesn. hort. Germ. fol.* 269. *Olea Sylvestris folio molli incano*, C. B. *Pin. p.* 472. *Zizyphus Cappadocia quibusdam, olea Bohemica*, J. B. *Oliva Bohemica sive Eleagnos, Math. ed. Bauh. p.* 174. *Lugd. p.* 111.

Lapathum pulchrum Bononiense sinuatum, J. B. Fidle Dock.

Jasminum tertium seu humilius magno flore, C. B. *p.* 398. *Catalonicum. Park. Parad.*

Arum maximum Ægyptiacum, quod vulgo Colocasia, C. B. *Pin. p.* 195. This is here planted by River sides in great quantities for the Roots sake, which is eaten, and very much esteemed, the leaves being good for nothing but to wrap up things in.

Arundo Donax sive Cypria Ded. p. 602. *Arundo domestica Matth.* Δόναξ *Donax sativa nostra, Adv. Lob. & Pen. p.* 27. *Lugd. p.* 999. *Arundo secunda sativa seu Donax Dioscoridis & Theophrast*, C. B. *Pin. p.* 17. *Arundo Quarta aquatica quæ Donax vocatur, Lon. fol.* 173. *Arundo vel harundo magna textoribus*

A Voyage to JAMAICA.   15

textoribus experta Gesn. Hort. Ger fol. 248. The great *Spanish* or *Cyprus* Reed and Cane.

Ruta quarta seu ruta sylvestris minor, C. B. *Pin.* p. 336. *Ruta sylvestris per omnia similis Hortensi sed undiquaque minor*, Gesn. *Hort. Germ. fol.* 277. *Ruta sylvestris Peno & Lobelii*, Lugd. p. 973. *Ruta sylvestris tenuifolia*, C. B. *Not. Math.* p. 541. *Ruta sylvestris Lob. Ruta sylvestris minim.* Dod. p. 120. *Sylvestris montana ruta*, Bod. à Stapel not. in Theoph. p. 798. *Ruta sylvestris. Trag.* p. 69.

Hypericon minus. Dod. p. 75. *Hypericon exiguum*, Trag. p. 74. *Hypericum quintum seu minus supinum, vel supinum glabrum*, C. B. *Pin.* p. 279. *Supinum glabrum*, Ger. *Hypericum minimum Septentrionalium*. Bod. *à Stapel not. in Theophrast.* p. 1050. The least trailing St. *John's* Wort.

Muscus marinus plumiformis ramulis & foliis dentissimis capillaceis. Cat. pl. Jam. p. 6. *Tab.* 2. *Fig.* 1. This from a broad base sticking to Stones, or other Solids at the bottom of the Sea, rises to be about three Inches high, being divided into several Branches, and they into Twigs, which were subdivided into smaller Branches, set with long round short Leaves, no bigger than Hairs, coming out of opposite sides of the middle Rib or Stalk, of a glue or dark yellowish colour, which did not crackle under the Teeth: they look just like Feathers, and were more thick branch'd and set with Twigs than any of the *Abies-Marina-Belgica* kind I ever saw.

I found this thrown up by the Waves on the Shore of the Island of *Madera*, near the Town of *Funchal*.

Lenticula palustris sexta vel Ægyptiaca, sive stratiotes aquatica foliis subo majore latioribus, C. B. *pin.* p. 362. For the synonimous names of this Plant, as of the others here mentioned, I refer the Reader to my Catalogue of *Jamaica* Plants, p. 13. to which he may subjoin *Lenticula palustris ex insula Jamaicæ, sedi arborescentis foliis subrotundo molli.* Plukenet. *Alm.* p. 401.

I found this Plant either in the Island of *Madera* or *Barbades* floating on *Tab.* 2. *Fig.* 2. the Water, having several Capillary brown Fibres for its Roots, and appearing Nerves on the upper sides of the Leaves, which because, it seems to differ very little from that of *Alpinus*, this being not *Hirsute*, I take to be the same, and his differing from that of *Vestingius* but in little, I think them not to be two Plants.

It is used for the same Diseases as Plantain, either outwardly or inwardly in Juice or the Powder to a Drachm.

Because there is no account of the Seeds of this, or whether it has any or no, I think this a more proper name for it than that of *Stratiotes*.

Hemionitis Asari folio, Cat. pl. *Jam.* p. 14. *Filix hemionitis dicta Maderensis, pediculis splendentibus nigris, crenatis foliis Asari rotundioribus crenarum serpentis oblongo quadratis ob semina adnascentia per ambitum circumcirca restrictis.* Plukenet. *Alm.* p. 155. *Tab.* 287. *Fig.* 5.

The Root of this most elegant Plant was made up of many brown Fibrils, which towards the Surface of the Earth were covered with a *Ferrugineous* Down, the Stalks were many from the same Root, blackish, round and shining, about seven Inches high, on the top of which was a round Leaf, exactly like that of *Asarum*, about two Inches Diameter, having Veins running from the top of the Foot-Stalk as from a common Centre through the Leaf, which was of the consistence of *Hemionitis* or *Lingua Cervina*. Round the edges on the under-side lay the Seed in a Welt, being *Ferrugineous*, as of other Ferns, and making the Leaf appear as if it were indented.

It was brought from *Madera* to Dr. *William Sherard*, by one sent to that Island in search of Plants for Sir *Author Rawdon*, and by him given me.

Lonchitis aspera Maranthæ, J. B. *Rait Hist.* p. 139.

Adiantum

Adiantum ramosum majus, foliis seu pinnulis tenuibus longis profunde laciniatis obtusis, Cat. pl. Jam. p. 22. An Filix ramosa Canariensis rutæ murariæ pinnulis angustis, altius incisis, mediæ costæ alternatim alligatis. Plukenet. Almag. p. 156. Tab. 291. Fig. 2?

Tab. 2. Fig. 3. This rises to be a Foot and a half, or two Foot high, having a reddish pale brown Stalk, cornered in the inside, and round on the other, at Nine Inches or a Foots distance from the Ground branch'd; those Branches undermost, or next the Root, being the largest about a Foot long having their Twigs, on which stand the *Pinnulæ* or Leaves alternatively, they being long, thin, pale green coloured, and divided into long blunt narrow Sections or incisures, by several very deep *Laciniæ*.

It grew in the Island of *Madera*, about half a Mile beyond the Town of *Funchal*, by a Road side going towards the Mountain.

Gramen paniceum spica simplici lævi. Raii Hist.
Gramen dactylon Siculum multiplici panicula spicis ab eodem exortu geminis. Raii Hist. p. 1271. Plukenet. Tab. 92. Fig. 1.
Gramen tremulum maximum. C. B. Raii Hist. p. 1274.
Gramen miliaceum angustifolium altum locustis minimis, Cat. pl. Jam. p. 35. An Gramen montanum panicula miliacea sparsa. C. B. prod. p. 17? pin. p. 8? Theat. bot. p. 131?

Tab. 2. Fig. 4. This had a round small hard green Stalk or *Culmus*, frequently jointed, at each joint, having three or four Inches long narrow grassie Leaves, and rising to be four or five Foot high, the Panicle was about six Inches long. The little Twigs or Strings going out of the upper part of the *Culmus*, and to which the *Locustæ* were fastened were about two inches long, taking their beginning from the same part of the Stalk, standing round about like so many Rays from the Centre, at about an Inch distance more or less from one another after the manner of Oats. The *Locustæ* were not Scaly, but standing singly by one another, being many and small, having within clay-coloured *Glumæ* or Chaff, one shining roundish small Seed, like that of Millet. This same Plant which grew of *Italian* Seed in *Oxford* Garden, being given by Dr. *Sherard*, who found it in *Italy*, to Mr. *Bobart*, was much larger by Culture, than the same Wild.

Gramen avenaceum, panicula minus sparsa, cujus singula grana, tres aristas longissimas habent, Cat. pl. Jam. p. 35.

Tab. 2. Fig. 5. and 6. This Grass had a Panicle of about six Inches long, not very sparse, when ripe of a reddish yellow colour, the Spikes were placed alternatively at long Intervals, and had set on them by small Foot-stalks several very long Grains, each of which had on their uppermost ends three very long *Aristæ*, by which it may be sufficiently distinguish'd; the *Glumæ* were of the same colour with the Panicle, and not awned: the Spikes were not many in number.

Urtica, caule lignoso, foliis tenuioribus atrovirentibus, Cat. pl. Jam. p. 38. An urtica urens ramosa Lusitanica Comm. Cat. Amst. p. 369.

This had an upright corner'd woody Stem, solid, and having a Fungous Pith, being cover'd with a smooth reddish brown Bark, rising two or three Foot high, having Joints and Branches set opposite to one another, on which stand likewise opposite to one another at the Joints the Leaves on three quarters of an Inch long, Foot-stalks. They are very thick set with burning small Prickles, being Inch long, and three quarters broad at round base where broadest, from whence they decrease to their ends, being very much cut in, on the edges thin, and of a dark green colour.

This (which comes so very near the *Urtica urens minor*, C. B. that I doubt if it differ any other ways than in its Stature and Duration from it) found on the *Madera* Island, near the Town of *Funchal*.

Persicaria

Persicaria procumbens longissima, angustifolia, non maculosa, spica longiore, laxiori & graciliori. Cat. pl. Jam. p. 48.

The Root of this Plant has several Protuberancies here and there, as also great numbers of redish Brown Strings or Filaments scattered up and down in the muddy ground. The Stalks are spread round, trailing on the Surface of the Earth for about four Foot in length, they are round, redish, smooth, jointed at every Inches interval, having a swelling at every joint, and near the top one Leaf, exactly like that of the ordinary *Hydropiper*, only much narrower and longer; the Flowers stand on Foot-stalks, *ex alis sel.* and on the tops of the Branches, like those of the ordinary Arsmart, only they are not so closely put together, but more lax and slender, and to them follows in a green Husk a small shining black Seed angular, and having two prickly ends, very like to the *Persicaria pusilla repens. Ger. emac.* only the Stalks are much longer. *Tab.3.Fig.1.*

It grows in the Island of *Madera*, in a Rivers Bank, half a Mile beyond the Town of *Funchal* towards the Mountain, and in *Jamaica* on the moist muddy low Banks of the *Rio Cobre, &c.* very Copiously.

Blitum vulgare minus surrectum. Munt. pl. cult. p. 291. *An Blitum Virginianum Polyspermon erectum viride D. Sherard. Plukenet. Alm.* p. 68?

The other *Synonyma* of this Plant, and of the rest hereafter mention'd may be seen in my Catalogue of *Jamaica* Plants.

I found it in the *Madera* Island near the Town of *Funchal*, and it differed in nothing from the ordinary wild small white Blite, only it seemed to be more erect. *Tab.3.Fig.2.*

Psyllium majus erectum, C.B. J.B. *Raii Hist.* p. 881.
Convolvulus althea foliis Clus. rar. pl. hist. lib. 4. p. 49.

I found this plentifully in *Madera* Island near the Town of *Funchal.*

It is good to cure Wounds, *Clus.*

Salvia major, folio glauco, serrato. Cat. pl. Jam. p. 64.

This hath square whitish or *glaucous* Stalks, rising two or three Foot high, having two Leaves standing opposite to one another; on Inch Foot-stalks, being two Inches long, and one broad near the Base where broadest, being cut in very deep on the edges, of a dirty green colour on the upperside, and very white underneath, having one middle, and several transverse Ribs. *Tab.3.Fig.3.*

It grew near *Funchal* in the Island *Madera*, where I gather'd it without Flowers or Seed, so that I am not able to determine its Family; perhaps it may be a *Marrubium nigrum,* or of some other kind.

Horminum luteum glutinosum, C.B. *Raii Hist.* p. 547. *Colus Jovis Ger.* p. 769.
Origanum spica latioribus. Cat. pl. Jam. p. 65. *An origanum Madersense nostrati simile odoratius capitulis albicantibus Pluken. Alm.* p. 278?

I found this wild in *Madera* Island; it has very broad Spikes in which it seems chiefly to differ from *Origanum vulgare.*

Hedera terrestris. Cæsalp. p. 453.

I found this in the Island *Madera* near the Town of *Funchal.*

They use to Boil it in their Flesh Broths in *Germany, Cord.*

Trifolium bituminosum seu trifolium cæruleum aut violaceum bitumen redolens. Moris. Hist. pl. part. 2. p. 136.

I found it in the Island *Madera.* The Seed from *Italy,* in *Germany* produces one with the smell and tast, but the Seed of the *German* Sown has neither tast nor smell, *C. B.*

Fumaria quinta seu lutea, C.B. *pin.* p. 143.
Genistella tinctoria Ger. p. 1316.
Scorpioides bupleuri folio, C.B. *Raii* p. 931.
Cicer sativum, C.B. *Raii Hist.* p. 917.

A Voyage to JAMAICA.

Tithymalus perennis & procerior lini folio acuto. Cat. pl. Jam. p. 82. Tithymalus dendroides Linariæ foliis ex insula Canarina. Pluken. Alm. p. 369. Phyt. Tab. 319. *fig.* 5. *An Tithymalus linariæ folio lunato flore Moris. hort. Bles. p.* 343 ? *An Tithymalus Tingitanus linariæ foliis lunato flore. Herm. cat. pl. p.* 600 ? *Tithymalus Tingitanus elatior lunato flore linariæ foliis creberrimis Plukenet. Alm. p.* 372 ? This seem'd to differ in nothing from the *Tithymalus annuus lini folio acuto Magnol. in Botan. Monsp.* but in this that the Stalks were higher and woody.

Plantago quinquenervia cum globulis albis pilosis, J. B. *tom.* 3. *lib.* 31. *p.* 504. *Caryophyllus barbatus sylvestris annuus latifolius multis capsulis simul junctis donatus. Morison. Hist. pl. part.* 2. *p.* 568.

Lychnis hirsuta quarta, seu sylvestris lanuginosa minor, C. B. *pin. p.* 306.

Cistus folio oblongo, integro, glabro, subtus albido, vasculis trigonis. Cat. pl. Jam. p. 86. *Hypericum seu Androsæmum magnum Canariense ramosum copiose floribus fruticosum. Pluken. Alm. p.*189. *Phyt. Tab.* 302. *Fig.* 1. *An Hypericum seu Androsæmum Canariense non fætens capitulis brevioribus filamentis donatis, D. Bobart. ejusd. ib* ?

Tab. 3. Fig. 1. This Shrub was five or six Foot high, having a solid Stem covered with a light brown reddish smooth Bark, and towards its top being divided into many Branches going out opposite the one to the other, having likewise Leaves set on them one against another, some being larger than others. The largest are about an Inch long, half as broad in the middle where broadest, smooth, whole, of a pale green colour above, and white underneath, with one middle Rib, and some transverse Nerves, going from it to the sides of the Leaf, appearing on its under side. It has no Foot-stalk but out of the *Alæ* of the Leaves towards the top rises many brown Stalks supporting Flowers which are whitish with many *Stamina*, surrounded by a *Pentaphyllous Calix*, after which come Heads of the same colour, as big as a small Pea, being roundish, tho acuminated at top, made up of three *Loculaments* or Cells, having each on his top an *Apex*. In each of these Heads lies great quantity of small oblong Ash coloured Seed. The Head bruised smells very sweet.

It grew beyond the Town of *Funchal* towards the Mountain, on each side of the Road, in the *Madera* Island.

Geranium Altheæ folio, C. B. *Raii Hist. p.*1055.

Apocynum fruticosum, folio oblongo, acuminato, floribus racemosis. Cat. pl. Jam. p. 89.

Tab. 4. Fig. 1. This had woody Stalks round, and of the bigness of a Hens Quill, covered with a reddish brown Bark, the Wood being solid and white, having Leaves going out at about an Inch distance, always opposite to one another; they stand on half a quarter of an Inch Foot-stalks, are two Inches long, and about three quarters of an Inch broad, near the middle towards the Base where broadest, and whence they decrease, ending in a point which is not very sharp: there is one middle Rib, and several transverse ones running through the Leaf, which is undivided, smooth, of a yellowish pleasant green colour. *Ex alis foliorum,* toward the tops come three or four Inch long *petioli,* which are branched, and sustain several very small Flowers.

I cannot exactly remember the place where I gathered it.

Trifolium acetosum corniculatum luteum minus repens & etiam procumbens. Morif. Hist. pl. p. 183.

It takes out spots of Linnen, *Cam.*

Fœniculum vulgare. Ger. emac. p. 1032.

I found this in the *Madera* Island very plentifully.

Bupleuron primum sive folio rigido, C. B. *pin. p.* 278.

I found this in some of the Islands going to *Jamaica,* but where I do not remember.

'Tis a Sallet Herb, *Cæsalp.*

Bupleurum

Bupleurum tertium minimum. Col. min. cogn. stirp. p. 85, & 247.
I found this likewise in some of the Islands, but which I remember not.
Heliotropium majus. Gesn. hort. Germ. f. 261. *An Heliotropium Siculum majus flore amplo odorato. Boccon de Plant. Sic. p.* 90 ?
I found a Plant, something higher than the ordinary *Heliotropium majus* is, in *Madera* Island, but I take it notwithstanding to be the same, only it varied in Stature from the Soil, being in every thing else the same.
Solanum vanum seu fruticosum bacciferum, C. B. *pin. p.* 166.
Asparagus maritimus crassiore folio, C. B. pin p 490.
Clusius seems to make this a distinct Plant from the *prat. & marit.* saying they were differing though in the same place.
Hieracium stellatum J. B. *tom.* 2 *lib.* 24. *p.* 1014.
Hieracium fruticosum foliis tenuissime coronopi modo divisis. Cat. pl. Jam. p. 123. *An Hieracium fruticosum foliis angustissimis non descriptum. Hort. Lugd. Bat. Raii Hist. p.* 239 ? *Hieracium fruticosum angustissimo incano folio. Herm. cat. pl. p.* 316 ?

From one single, three or four Inches long, crooked, Root, rises a woody, solid, crooked, round, light brown Stalk, three Foot high, having several small Branches towards the top, and now and then tufts of Leaves, some bigger, others smaller, but all of them divided or laciniated very minutely, almost into Hairs like the Leaves of *Coronopus Ruellii* or *Sophia Chirurgorum.* The Flowers are severals at top, standing within a *Calix* made up of a great many small, long, and narrow Leaves, which are reflected when the Seed ripens, leaving many small black *Pappous* Seeds to be carried away with the Wind. *Tab.* 5. *Fig.* 1, 2.

It grew on the stony Hills to the Eastward of the Town of *Funchal* in the Island of *Madera.*

Alypum, sive Herba terribilis procerior, cortice cinereo scabro, folio acuminato longiore. Cat. pl. Jam. p. 124.

This rose much higher than the *Herba terribilis Narbonensium,* having a hard white Wood with a large Pith, a *Scabrous* or unequal light brown or grey Bark, the Branches towards their ends were very thick set with Leaves without any order: they were two Inches long, and a third part of an Inch broad in the middle where broadest, being narrow at the beginning, increasing to the middle, and ending in a point, equal at the edges, with one middle Rib, and several transverse ones of a yellowish green colour. Towards the tops of the twigs *ex alis fol.* come the Flowers, being several Heads round or Spherical, made up of many very small blue Flowers, with their *Stamina* set round very close together in the same Head, to which follows a very small grey *Pappous* Seed, all over downy. *Tab.* 5. *Fig.* 3.

Helichrysum secundum seu Helichryso Sylvestri flore oblongo similis, C. B. *pin. p.* 265. *prod. p.* 123. *Jacea Stœchadis citrinæ foliis prælongis paucis capitulo minore subrotundo aspero, Pluk. Alm. p.* 193.
It is good in Decoctions for the Colick. *Clus.*
Gnaphalium ad Stœchadem citrinam accedens, J. B. *tom.* 3. *lib.* 26. *p.* 160.
I found this both ramose and not ramose.
Chrysanthemum aquaticum Cannabinum folio tripartito divisi. Herm. cat. pl. p. 146. *An Eupatorium aquaticum Virginianum, Park. p.* 596 ? *Chrysanthemum Virginianum bidens cannabinum Pluken. Alm. p.* 100 ?
Erica folio coridis sexta, seu major scoparia foliis deciduis, C. B. pin *p.* 485.
Genista non spinosa prima, seu singulosa & scoparia, C. B. *pin. p.* 395.
Common-Broom.

The Flowers are eat in Sallets, although two Ounces of the Seed Decocted are a Vomit *Mes.* but not more than Radishes, &c. *Lob.* The water of the Flowers, or half a Dram of the Seed beaten, are good against the Stone, *Lem. Myrtus*

Myrtus septima, seu sylvestris foliis acutissimis, C. B. *pin. p.* 469.

I found this very plentifully growing wild in the Hedges by the way-sides in the Island of *Madera*. This is used for Currying Leather, as *Rhus* or *Lentisk, Cæsalp.* The ripe Berries are used for Sauce, *Math.* Before Pepper was found, as *Pliny* tells us, the Fruit of this was made use of in its place.

Lycium folio oblongo, serrato acuminato spinis minoribus armatum. Cat. pl. Jam. p. 171. This seemed to differ very little from the common *Lycium*, only the Leaves were longer, serrated, and pointed, and the Prickles were not so large.

Tab.5.Fig.4.

Palma prunifera foliis yuccæ, fructu in racemis congestis crassi formi, duro, cinereo, pisi magnitudine, cujus lachryma sanguis draconis est dicta. Comm. cat. Amst. p. 260. *An* Dragon-Tree *of Dampier, cap.* 16 ?

I found this in the Island of *Madera* in the Hedges very plentifully though not very large. It is found in the Island *Soccotora, Borneo, Canaries, Madagascar,* and (*Alnise de cadamosto ep. Ramn. pr. vol. p.* 105) at *Porto Santo*, where they cut the Trees at the Feet, and next Year find the Gum, which they Defecate in Water by Boiling and Purging. The Fruit is Yellow and Ripe in *March*, and good to Eat.

The Tree is pierced near the bottom, and so yields the Gum. The Fruit Cools and Alters, and is proper in Fevers. *Cinaber du Dioscorid. Thevet.*

It is adulterated with *Rubrica* and *Colophony. Cæsalp.*

Lobels Leaf is the *Spatha* in all likelihood. *Lugd.*

The Gum is used by Goldsmiths for a Foile and Enamel, and by Glassiers for colouring Glass, *Park.*

It is used to strengthen the Gums and Teeth, in bloody Excretions, Fluxes, &c. *Jonst.*

Opuntia maxima, foliis majoribus crassioribus & atrovirentibus, spinis minoribus & paucioribus obsitis. Cat. pl. Jam. p. 195. *An ficus Indica seu opuntia maxima, folio spinoso latissimo & longissimo. Herm. cat. pl. p.* 243 ?

This *Indian* Fig was in every part exactly the same with the Common, only each Leaf was broader, thicker, of a darker green colour, and not so prickly, having a very few white, short Prickles; and sometimes only one coming out at a hole very like that kind on which comes the Cochineel, only it is not quite so free of Prickles as that.

It grows in a Gully near the Town of *Funchal* in *Madera*, and in the *Canaries*.

On *Sunday* 23. Having taken Wines and some fresh Provisions on Board, we weighed Anchor and set Sail, we having little Wind ; two days after we saw the Body of the Island, being about Twenty five Leagues, or Seventy five Miles from us, and then we first took Dolphins with Fisgigs, or sharp arrow-headed or bearded Irons, fitted with Poles of about Ten Foot long, Lead for the more convenient striking them, and a Rope or Line tied to them to hold the Fisgig, which is shot at them by the strength of the Hand when they come within reach of those waiting for them, usually on some of the Yard-Arms, Beak-Head, or Poop; in which fishing the great matter seems to be to allow for the refraction of the Water. They were laid in wait for not only so, but likewise with Lines and Hooks, which were hung out baited with Rags in the shape of flying Fish, and so adjusted as to hang sometimes to touch the Water, at others not, according to the Waves, thereby imitating the Flying-Fish, which the Dolphins pursue with great greediness. Dolphins are reckoned the swiftest Swimmers that are, their Bodies being contrived for that purpose there is as much pleasure in seeing them pursue the Flying-Fish, as in Hunting or Hawking, the Flying-Fish getting out of the Water, where the Dolphins cannot

A Voyage to JAMAICA. 21

not far pursue them. They are likewise invited to the Waters Surface by throwing any thing upon it, they being voracious. They love the Company of a Ship, because of what scraps are now and then thrown over Board, or the Barnacles growing to the Ships sides, insomuch that I have been assured by some who have sailed in *Guinea* Ships, that they have had the same Sholes of Dolphins follow them for many hundreds of Leagues between *Guinea* and *Bar-* *Of Dolphins.* *badoes*; and Sir *Richard Hawkins*, in his Observations takes notice, that in some Voyages they had followed his Ships a Thousand Leagues, although they had strokes at them, and mark'd them several times with their Irons, by which marks they knew them to be the same Dolphins. *Battel apud Purchas*, likewise tells us, that a Shole of Dolphins follow'd their Ship Thirty Days from *S. Tome* to *Brasile*. And hence, it may be, it was this *Fish was thought by some to be the Dolphin of the Ancients, and to be enamoured of Men*. One thing very remarkable in this Fish, is the various Colours it puts on before it dies, being usually Yellow, with ranges or rows of small Blue, and round spots, which very strangely change, and afford that pleasure to the Eyes, that I confess I never saw any thing of this Nature so surprizing; but after all it comes to a very light Blue Colour when Dead, which it keeps ever after. In this property it is like the Chameleon, and this appearance seems very much to depend on the strength or motions of the Spirits, and Fluids, unto or from the Skin, by which its Surface is altered so as to make those several *Phænomena*. *Martens* tells us when Maccarel are alive they also give fine Colours. In the Dolphin we took, there was a Flying-Fish in its Belly, which was in part digested. The Fish its self when made ready was dry, though pretty good Victuals, and well tasted. The nearer the Head the more 'tis prised, altho I am apt to think, that if this Fish, so much commended by Sailers, were a Shore in a Market, where other Fish were to be had, it would not be counted so great a delicacy. There is a description of it in Mr. *Willughby*'s Book of Fishes, and a Figure, *Tab. O. 2.* which was taken when that Fish was dying, with the various spots and colours on it : whereas that of *Piso* under the name of *Guaracapema* and *Jo. de Laet*, was taken when the Fish was quite dead. This is the *Dorade* of *Lery*, and most of the *French* Voyagers, and the Dolphin of our *English*, and *Dutch* Seamen; and Sir *Richard Hawkins* calls it the *Dorado* of the *Spaniards*. Although *Piso* says that it leaps out of the Water after the manner of Porpesses; yet I never could observe that, unless very seldom when having pursu'd Flying-Fishes to the Surface of the Water, they give a small leap to take them in the Air. It may very well be the *Hippurus* of *Joand Jonst.*, it having all the marks of his imperfect description, though his Figure be not good.

It is the Dolphin of Drake *Hakl. Part. 3. p. 732.* Of *Cliffo ib. p. 750.* Of *Escarbot Nova Francia, p. 35.* Of *Hudson Purchas, p. 588.* Of *Leron, p. 4. 6.* Of *John Davis Purchas, p. 132.* Of *Battell. ib. p. 970.* Of *Jo. dos Sanctos, ib. p. 1546. Dorade de Richef.* 186. *de Dutertre*, 212. *Giltheads of Oviedo sum, p. 234. Hippurus Rondeletii Gesn. Willughb. p. 213. Tab. O. 2. Guarcapema Brasiliensibus, Marcgr. ed.* 1648. *p.* 160. *Guaracapema, Pison. p.* 48. *ed.* 1658. *Pesci indorati di Col. f.* 32. A delicately coloured Fish pleasant to look on, of *Ra-welfe, cap. 2. Hippurus Rondeletii, p.* 255. *Dorada Warakepemne, Laet. p.* 571. *Dorados of Anonym. Port. Purchas, p.* 1313. *An Aureus Piscis, Fernandez, p.* 872. *Dorade de Roveneau de Lussan. p.* 171. *De Lery, de Cauche, p.* 141 *Abbeville, p.* 36. An Dauphins. *Ejusd. ib.* : *Dorado* of the *Spaniards* of Sir *Rich. Hawkins, p.* 42. *Dorado* of *Mandelslo, p.* 196. Where it seems to be Confounded with a Dolphin. *Duurade de Pyrard. de la Val. p.* 137. A Dolphin of *Hawkins Hakl. p.* 520. Of *Boyle* of *Air*, p. 179. Of *Terry*, p. 10.

G   This

This Fish is found in most parts of the Hot *East* and *West-Indies*, in the South-Sea, and at the *Maldives*.

A Sparrow Hawk blown one hundred Leagues off Land.
When we were about one hundred Leagues off the Island *Madera*, we had a small Sparrow-Hawk had been blown either by Storm, or otherwise from Land, he lighted on the Ship, and was so faint and feeble that he droop'd his Wings and look'd pitifully, he sate on one of the Ropes, and would not stir till we were just going to take hold of him, and then he would remove to some other place, or go to some other of the Ships, where he could be more secure.

The place of meeting the Trade-Wind.
We had the Winds variable when we were in 25°. North Latitude; whereas in other Voyages other people have met with the Trade-Wind or Breeze in upwards of Thirty Degrees; but from *Friday* the 28. Day of *October* to *November* the 4. we had South-Westerly Winds, quite contrary to the Trade-Wind or Breeze, which may be supposed to come from the Suns being gone to the South this time of the Year, and probably if we had been in this Latitude, when the Sun had been in the Northern Signs of the *Zodiack*, we should have had the Easterly Winds, as Seamen meet them, sometimes much to the North, without considering the probable cause; but wondering at the variety produc'd by it.

The Fourth and Fifth we had a great Calm in about 24°. 30' North Latitude, with it a very great Sea, tumbling us from side to side of the Ship.

The Tropic-Bird.
The 5th. of *November* we saw the Tropic-Bird, or *Avis Tropicorum*, flying very high round the Ship, they are very easily known by two long Feathers in their Tails, the Icon and description in Mr. *Willughby's* Book of Birds is accurate enough, but to me it seems to be rather of the Gull, than Duck kind. They are common every where between the Tropicks, and rarely seen any where else, whence they have their name.

They are ordinarily met with first in the Voyage to the *West-Indies*, Three hundred and fifty Leagues off of *Dominica*, or *Defeada*, towards *Spain*, though in the third Voyage we made thither, we met with one in the mid-way between *Spain* and the *Canaries*, which every one wondered to see so near *Spain*. *Oviedo.* I suppose this accident might have happened when the Sun was North of the *Equinoctial*, and towards the *Tropick of Cancer*.

The Feathers in the Tail are made use of as Ornamental by the Savages in their Hair and Nostrils. *Du Tertre.*

This Bird is taken notice of by the following Writers.

Bird with long Tail, of *Fenton* or *Ward*, *Hakl. part.* 3. p. 767. *Fetu en cul on l' oiseau de Tropic, de Du Tertre,* p 276. *Rabo di Giunco Col. f.* 29. 32. *Rabo de Junco, de Oviedo, Hist. General, lib.* 14. *cap.* 1. *Coda di Giunco Ejusdem ap. Ramnus, fol.* 161. *Aves aliæ rabos de Juncos dictæ cauda Juncea, Aldrov. Ornith. Tom.* 3. p. 544. The Tropic-Bird of *Willughby*, *Angl.* p. 231. *Tab.* 76. Of *Smith* Summer Isles, p. 171. *Une espece d' Aigrette de Rochef.* p. 165. White Birds having in their Tails but two long Feathers of *Jo. Davis, Purchas,* 133.

Of Sharks.
In the Calms, all over these Seas, 'tis ordinary to have Sharks come about the Ships, we had some often came to ours: several of the Seamen immediately, on their appearance, took great Fishing-Hooks, with Iron Chains of a pretty length, fastened to a long and strong Line or Rope, and baited them with a piece of Salt Beef, or even Red Cloth, throwing it into the Sea in their sight. They come for the most part immediately and swallow it, the Seamen giving them Line to run where they please, wearying and wasting themselves. Then the Seamen pull them near the Ships side, and throw a Rope, with a Loop or a Noose, into the Sea, put it about their Tails, and so pull them out of the Water on Board, which never could be done, (without this help,) by the Hook or Line, because of the great weight, and efforts made

A Voyage to JAMAICA.

made by the Fish. When 'tis got on Board it makes mighty strokes with its Fins and Head, but mostly with his Tail, having therein a very great strength; which Mariners well-knowing immediately cut off, and then the Arteries by the *Spina dorsi* spurt out vast quantities of blood, till the Shark is weakned and Dead. This Fish is very well known to Natural Historians by the name of *Canis Carcharias & Lamia Rond. & aliorum Gesneri*; it has this particular to it, with some others of its own Tribe, that the Mouth is in its under part so that it must turn the Belly upwards to Prey, and were it not for that time that it is in turning, in which the pursued Fishes escape, there were nothing could avoid it, for 'tis very quick in Swimming, and has a vast strength, with the largest Swallow of any Fish, and is very devouring. It has several Ducts on the Head, fill'd with a sort of Gelly, from which, being press'd by the water, issues an unctuous, viscid, slippery, and mucilaginous Matter, very proper to make the Fish very glib to sail the readier through the Water. Most Fishes have something Analogous to this. He had Three Rows of moveable Teeth, and behind each Ear in a Cavity an almost fluid Body, which when in its natural bigness might be almost round, of the bigness of a Six Pence, and as thick as a Crown-piece; this being put into Paper, and dried a little, grows hard, and if touched falls into an extremely fine Powder, commended very much by all Seamen in the Stone, and difficult Labour, as a very great Remedy, and by them very carefully taken out and preserved: this is what is mentioned by some Natural Historians under the name of *Lapis tuberonum*, the *Spanish* name for a Shark being *Tiburon*. I once on opening one of the Female Sharks found the Eggs in the Ovary perfectly round as big as the top of ones Thumb, and at another time the *Fœtus* or young ones in their Coats, lodged in the *Uterus*, after the manner of our Viviparous Creatures, for upon cutting the Coats the small live Fishes came out, being able to Frisk and Swim up and down in the Salt-Water.

They are commonly about Ships in their way to the *East* and *West-Indies*, and about the Island of *Jamaica*, and are no bad Meat, though laced because thought to Prey on Men; they are of a prodigious bigness in the *Mediterranean*.

Those of *Nice* told *Gyllius* one of Four thousand Pound weight had been taken wherein was a Solid Man, and those of *Marseilles* took one had a *loricatum hominem* in him. Arrows of Savages are sharpen'd with the Teeth of these, being sharp and venomous, so as seldom to be cured, *Laet*.

One bit off a Seaman by the Leg to the Thigh. *Linschot*.

Massiliæ & Niceæ aliquando captæ sunt lamiæ in quarum ventriculo homo loricatus inventus, *Rond*. Wherefore he is of Opinion that this was the Fish, in which *Jonas* was for three Days and three Nights, called a Whale, the want of an *Aspera Arteria*, which (*in ordinary Whales is necessary for breathing*) straitens the *Gula*, makes the *Gula* in this Fish so wide that a Man may be swallowed entire. The Teeth which are said to be Serpents Tongues, are set in Silver, and hang'd about Childrens Necks to help their Teething, and likewise for Tooth-Picks.

They cut off with their double sets of Teeth an Arm or Leg, as with a Rasor, and eat Turtle, which Men feed on when taken out of their Bellies, *Col. 213*.

The Skin is rough drawn from Tail to Head, from one to five Fathom long. It bites great pieces from Whales, as if dug out with Shovels, and Devours all the Fat of Whales under Water. Train Oil is made of their Livers, and Men eat their Flesh hang'd up; they sell them in *Spain*. They eat many Men in washing, *Martens*. The Stones in the Head of this Fish are good for those who cannot make water, and for pain in the Liver, *Xim*. *Rochefort* calls these Stones their *Cervelle*, and he as well as *Du Tertre* assure us that

their

A Voyage to JAMAICA.

their Liver affords great quantity of Lamp-Oil, and that they are of Eighteen or Twenty Foot long about the *Caribes*.

Several of them set about one Seal to take him, encompassing and falling on him with their joint strength, *Oviedo*.

They go faster than a Ship will Sail, as much as a Man can out-run a Boy of Four Years old. They turn round the Ship, and go from side to side, and from Head to Stern. They follow a Ship Two hundred Leagues and may further. Are cut in pieces, dried in the Ropes, and eat in two or three Days, or fresh with Garlick Sauce. They are not good for Passengers. The Males have two *Penes*, in one eight Foot long, it was as long as from the Cubit to the middle Fingers end: the Female but one *Vulva*. They bring Thirty five Young ones at a Birth alive, *Oviedo*.

Divers in Pearl-Fishing are hurt by Sharks, *Laet. p. 668*.

They are taken by *Riverss* or *Remora* by the *Indians*, *Col. 112*.

They are in a Salt-Lake near the Sea in *Hispaniola*, *Oviedo. 32*.

They enter the Rivers, and are no less perillous than great Lizards or Crocodiles, devouring Mankind and Horses, *Oviedo sum*.

They are taken once or twice with the same Bait, *Terry*.

Thomas Smith, a Boy, swimming about the Ship by *Swrat*, had most of the outside of his Thigh bitten away, of which he died with bleeding, *Purchas Downton, p. 505*.

They are fished for Oil. One was pulled into the Sea by a Shark in the Night, *Knivet*.

They destroy Men at one Bite within a Salt Lake in the Country, *Martyr*.

They are mangled and made a Prey to others of the same Kind. *Ligon*.

I was told at *Montpelier* there was one there taken in the *Mediterranean* (if I remember right) forty Foot long. A Man bathing himself by *Port-Royal* in *Jamaica*, one swam by him and bit off great part of his Breast and Arm of which he presently died by the flux of Blood.

This is the *Tuburon* or *Hays* of *Linscloten*. *Tuberones, Laet. p. 571, 586, 669*. Shark of *Terry, p. 9*. Of *Hortop. Hakl. p. 487*. Of *William Finch*, *p. 416*. Of *Nicol. Purchas, p. 1257*. Of *Ligon, p. 6*. Of *Smith*, Summer-Isles, *p. 172*. *New-England, p. 227*. *Hay* of *Martens*. Dog-Fish, called by the *Poringals Tuburones*, Of *Knivet*, *Purchas*. *Canis Carcharias, seu Lamia Rondeletii & aliorum, Raii, p. 47. Tab. B. 7*. Where seems to be a bad Figure from *Gesner*. Sharks of *Anonymus Portugal* of *Brasile*, *Purchas, lib. 7. cap. 1. p. 1314*. Who tells that they Kill Men, and that they find in them Breeches, &c. when taken: as also that *Indians* use their Teeth for Heads of Arrows, being poisonous, *ib. Requiem à Abbeville, p. 30*. *Tiburon* of *Oviedo sum, p. 211. 212. Lamia Rondeletii, p. 390. Tiburo Ejusdem, p. 489*. where, in his account from *Lopez de Gomara*, he confounds the Crocodiles and Sharks. *An Maraxus, Ejusd. ib. ? Lamia* of *Bavolse* in the *Mediterranean*. *Kirms, Cetus*, the Whale of *Jonas, Math. cap. 12. 40. Tiburoni di Fernan, Colon. f. 122*. and 211. *Canis Carcharias, Bellon. p. 58. Lamia, Ejusd. p 98. Tiburonus Martyr. Decad. p. 251, 252. Tiburo Fernandez, p. 87. Carcharias Piscis, Gyllii, p. 567. cap. 99. El Tiburon de Hernandez à Ximenes, edit. p. 182. Requiem de Rochef. 191. Du tertre, 202. de Denis, Tom. 2. 272*. Sharks or *Tiburons* of Sir *John Hawkins. Hakl. Part. 3. p. 516. Tiburones de Oviedo, lib. 13. cap. 6. & cap. fol. 33*. A Shark called by the *Portuguese Tuberonet*, and the *Dutch Hayes*, of *Mandelslo, p. 197. & 219*. Shark of Sir *Rich. Hawkins, p. 42*. Of *Fenton* or *Ward, Hakl. p. 767. Boyle* of *Air, p. 179. Touberans de Feynes, p. 206. Lamie de Molinet, p. 203. An Peimoner, Pyrard. de la Val*, making wading from one *Maldive* Island to another difficult, *p. 72 ?*

We

We were now in very hot weather, which show'd its self on every one, not only by Sweating, but by their breaking out all over into little Whales, Pimples, or Pustles, (which is ordinary in other places of the same Latitudes.*) This besides its unseemliness in discolouring the Skin, which was very Red, was very troublesome by itching, and this last symptom was chiefly about the Back Bone, though sometimes these reddish Pustles covered the whole Body, and at other times they were as it were all gathered into one small Carbuncle, very uneasie and painful, and such an one I my self had, on the side of my Hand. I did not at all doubt but that these Eruptions were the effect of the Sun Beams, which throwing into our blood some fiery parts, put it into a brisker motion, whereby it was purg'd of those *Heterogeneous* and unaccustom'd Particles it had from the warm Sun, and perhaps by that fermentation was likewise clear'd of some other parts might be hurtful to it, and therefore instead of prescribing a Remedy for its Cure, I told those who importun'd me, that I thought this Distemper was the greatest advantage they could have, and that this was the effect of the change of Climate, and a proper seasoning, and what might secure them from future Sickness by purging the Blood from hot and sharp parts, and rather than check it, wish'd them to help the expulsion with a little *Flos Sulphuris* or any other innocent Diaphoretick. If their Humour was to be complied with, Bleeding first, and then Purging were infallible Remedies, which by cooling the Blood, diminishing its quantity, and at the same time making an irritation on the Glands in the Guts, causes the Humour to come that way, so in some measure by artifice supplying the natural Evacuation by the Skin. But because the other was Natural, and this Artificial, and not so certain nor safe, I was constantly for the first. Here were grounds to admire the contrivance of our Blood, which, on some occasions, so soon as any thing destructive to the Constitution of it, comes into it, immediately by an Intestine Commotion endeavours to thrust it forth, and is not only freed from the new Guest, but sometimes likewise what may have lain lurking therein (occasioning small disturbances without breaking out into any violent Disease) for a great while. And from hence it comes that most part of Medicines when they are duly administred in such cases are, not only sent out of the Body themselves, but likewise great quantities of Morbific Matter, as may appear very plainly in Salivation, where not only the Mercury, but likewise all the Humours causing those dreadful Pains, Ulcers and Disorders, are spit out together with it by the help of the Bloods Fermentation. If it be here objected that the Sun Beams are too spiritual, and cannot be thought to be so poysonous: I answer, that it is certain the Sun Beams when concentred will do by a Burning Glass most, if not every thing, can be done by a Fire, which every Chymist knows will not only add to Lead, and some other Bodies, those Particles that will weigh in a Balance, and that very considerably; but likewise will make such a change in Mercury barely Precipitated by it, as to make it almost as great a Poison as is commonly known.

This alteration of the Climate was discovered likewise by the very great and sudden Putrefaction of Urine, which in some few Hours would stink intollerably, and all other Fermenting Liquors would Sour immediately: Flesh, and all other Aliments would Corrupt and Spoil likewise in a little time: Tallow Candles would scarce be able to stand upright, and Butter would be of the same Consistence as if half melted over a Fire in *England*.

It is a commonly received Opinion by some Ingenious Men that Lice dye on change of the Winds from being variable to be constant, or passing the Equator; and that to the South of the Tropick of *Cancer* are none to be found, but this notion is certainly false; for although I think the great Sweatings, and little Apparel of the Inhabitants and Travellers in the *Torrid Zone*, occa-

*Great heat in Monotapa occasions Pushes to come out, Dos Sanct. Purchas, p. 1556.

Of the changes in the Blood from the alteration of the Climate.

Of other signs of the alteration of the Climate.

Of Lice in the Torrid Zone.

occasion less disturbance from this sort of Vermine, yet I am certain both *Indians*, *Moors*, and *Europeans*, who live there are subject to them, though they be not in so great plenty as in more Northerly Countries where the Inhabitants Sweat less, and go better Cloathed, in the Plies of which Apparel these Creatures find good shelter. I cannot on this occasion pass by a Matter of Curiosity relating to the *Plica Polonica*, a strange Disease frequent in *Poland*, which comes from the intangling many Locks of the Hair, and has very odd Symptoms attend it, one of which is, that on cutting of it off, it endangers the Person who had it. There are several Opinions about it, but many believe it to be only the effect of Laziness, and not Combing the Hair; I am very apt to believe this, because Dr. *Connor* gave me some of the *Plica* he cut off a Person in *Poland*, in which was an innumerable quantity of Lice and Nits lodged amongst, and at the bottom of the Hair. 'Tis very odd that when these Locks are cut off they should grow sickly, but it may be this way easily answered, that carrying off so much humor, or being a Nest for so many Lice, they do something like a Pea in an Issue, discharge the Blood, which when it is cut off is no longer done. That Lice constantly suck the Blood is certain from Persons Fluxed whose Blood is no sooner impregnated with Mercury than all these kind of Vermine swell, drop off, and dye.

When we came near to the Tropick we were call'd upon for our Tropick-Money, that is to say, we who had never before crossed that Line to the South, must now give either so much Money, as by the usage of Seamen we shall be tax'd, to make them Drink, or be duck'd thrice into the Sea from the Yard Arm, we chose rather the first, and so were free.

The sixth Day was taken a Fish which was thought to be a *Barracoda*, but was not so, it was taken with a Flying-Fish Bait, such as are made use of for *Boneto's*. I called it *Serpens Marinus, compressus, lividus*.

This Fish was three Foot seven Inches long, an Inch and a half broad near the Head, where broadest, having a long Head, sharp, or ending in a point, with the Mandibles Prominent, in which were many Teeth in each of the Jaws, very sharp and threatning, the Under Jaw was longer than the Upper, and ended in a *Callous* Substance, the Tongue was bony, and shap'd like an Arrow Head, the Gills very red: its Eyes were round of an Inch Diameter: it had two Fins at the Gills, one long one all over the Back, and one from the *Anus* way'd towards the Tail, which was forked, it was all over smooth without Scales, of a Livid colour, and its flesh was full of Bones.

Tab.1, Fig.2.

Its *Æsophagus*, if any, was very short, its Ventricle fill'd with small Fish like Anchovies. The Guts had one or two Circumvolutions. The *Cæcum* was very long, extending its self to the *Anus*, and fill'd with the Contents of the Stomach, its *Diaphragme Membranous*: its Liver large, with a Gall-Bladder, containing in it watery Gall. There were under the Guts two long Bodies went to the *Anus*, which I take to be the Kidnies: It had in it two sorts of Worms, the one round and Cristalline, the other long, rowling themselves spirally like a Snail.

This Fish was taken about the Tropick of *Cancer*, with a Bait, such as is used for Dolphins and *Boneto's*, viz. a flying-Fish, or Hook dressed like it.

It was a great dispute among the Seamen whether it should be eaten or not, most People saying its Livid colour was a sufficient Argument against it; but this is no manner of Reason, for Fishes much Swarthier are eaten in many places. One who had been a Privateer, and in the South-Seas, where sometimes they had hard Fare, and met with new kinds of Fishes, assured us, that in such Disputes they usually tasted the Heart, which if sweet they thought a good Argument for the use of them in Victuals, if not, they were usually rejected as poisonous, how far true my Relator must answer, for my part I believe it as little as the former: but there is a very good Reason to

object

A Voyage to JAMAICA.   27

object against the eating of a new Fish in the *West-Indies*, for there are certainly some of them, which if eaten, prove poisonous at some seasons of the Year, at least many People have told me so.

If this be the Fish spoke of by *Laet, p. 27.* where he calls it *Piscis congro forma similis venenatus*; he says 'twas venemous both to *Dutch* and *English*.

After two or three days Calm, being carried West with rouling Seas, on the Sixth we had about the Tropick a small North-Easterly Gale, which began low, rising by degrees till it came to a pretty strong Wind, and was fixed for a while between North and East, being concluded by the Seamen to be the Trade-Wind, which blows not always from the same Point, but generally *Of the* on the North-side of the Equinoctial within this Tropick: in this Sea it blows *Trade-Wind*. between North and East, though at several times of the Year 'tis usually more Southerly or Northerly according to the Suns place in the *Zodiack*. Neither does this Trade-Wind keep the same Point, but varies every hour, though those two Points are usually the utmost bounds: neither is the Trade-Wind constantly of the same vehemence, but sometimes blows very hard, though rarely Stormy; and at other times very easie, though seldom Calm. There is very seldom Rain with this Wind, though when it comes the Drops are thick, and it is violent. The Weather here was generally clear, though sometimes Hazy, especially about the *Horizon*, the *Zenith* being seldom overclouded.

We had several days without any remarkable matter happening, save that flying-Fishes were here very frequent: the Description and Figure of which is common in most Natural Historians, therefore I shall not say more, only that it is a kind of Herring with very large Fins, with which it can fly some time *Of Flying-* in the Air when pursued by Dolphins, *Bonito's, &c.* and that it is taken notice *Fishes.* of by Voyagers, and Natural Historians as follows, to me it seems to deserve the name of *Harengus Alatus*.

Flying-Fishes of *Cocks. Purchas*, 398. Of *Best. ib. p. 466. Purchas,* 37. *Pesci Rondini di Col. f.* 32. *Hirundo Salviani, p.* 185. *Volatiles Pisces, Laet. p.* 572. *Bokery*, a Flying-Fish of *Duddeley*, 576. *Hirundo, Bellon, p.* 193. *Mugil Alatus Rondelet. p.* 267. *Fernand. p.* 87. Flying-Fishes of *Clusse, Hakl, pars.* 3. *p.* 750. Of *Ligon. p.* 4. Of *Terry, p.* 11. *Hirundo Salviani, quoad Iconem. Willughby, Tab. P.* 4. *Poissons Volans de Cauche, p.* 140. *D. Lambert. p.* 42. *Albeville, p.* 30, 31. *Peces que bolan Lop. de Gom. cap.* 91. *Pirabebe* 2. *Pison. ed.* 1658. *p.* 61. Flying-Fishes of an *Anonymus*, Portugal in *Brasile, Purchas, lib.* 7. *cap.* 1. *p.* 1314. Of *Mandestlo, p.* 196. and 211. Of *Hawkins Hakl. p.* 520. Sir *Rich. Hawkins, p.* 42. Of *Drake, p.* 732. *Hakl. p.* 3. Of *John Davis, Purchas,* 132. Of *Layfield, ib.* 1197. *Poissons Volans de Rochef. p.* 183. *de Du Tertre,* 212. *de Feynes, p.* 205. *Du Molinet. p.* 205. *Du Pyrard. de la Val. p.* 6. *Linschot. Descr. Pescados Boladores, de Oviedo, lib.* 13. *cap.* 2. and *cap.* 5. *Volatori ejusd. sum, p.* 132. who is of Opinion they are not the *Golondrinos* of the *Spanish* Seas, at least he saw them not in those Seas; they fly Two hundred Paces, and when their Fins are dry fall down. They fly sometimes on one, sometimes on the other side; and are pursued and taken by *Gilheads*, and out of the Water by Sea Mews and Cormorants, *id. sum.*

They came on Board our Ship every where between the Tropicks, flying out of the Water and lighting thereon by accident. They are very good Victuals, tasting like a Fresh-Herring. They are common in most parts of the *East* and *West-Indies*; in *Japan*, and the Isles *Ladrones*, where they are eaten. They are sometimes more in one place of the Sea than another, for *Oviedo* who cross'd the Seas many times to the *West-Indies*, tells us that the Seas are like Provinces, some are Fertile in Fishes, others not, according to the Winds.

They

They fly till their Wings are dry says *Hawkins*, being pursued by *Giltheads*, otherwise called *Bonito's*, which they take with them, or White Cloath made into their shape, as a Bait, *id*.

They leap into our Boats, whence they cannot get, their Wings being dried, *Drake*.

They are near the Line, as well South as North, and beyond the *Cape of Good Hope*, *Pyrard. de La Val*.

Their Wings are given them by Nature to escape. *Abbeville*.

These Flying-Fishes bring after them another kind of Fish which preys on them, it is commonly called *Boneto's*, but I shall call it, *Scombrus major Torosus*, being of the same kind with a Mackarel, though it has been taken notice of by the following Writers, under other names, as

Bonito of *Drake*, *Hakl*. part. 3. p. 732. Of *Cliffe*, *ib*. p. 750. Of *Ligon*, p. 5. *Bonite de Cauche*, p. 141. *Abbeville*, p. 30. *An toni precioli*, *Col. f.* 74? who observed them in the Sea about *Espanola* in plenty. *La Bonite de Rochef*. p. 187. where is a bad figure, *de Du Tertre*, 214. where is a good figure, *Du Pyrard de la Val*, p. 6. 137. *De Ravenau. De Lussan*. p. *Bonito's de Francisco Gualle apud Linschot*, p 124. 2. P. *Purchas*, lib. 4. cap. 13. p. 806. *Hakl.* p. 406. Of *Mandelso*, p. 196. Of Sir *Richard Hawkins*, p. 42. *An Pelamys Bellonii, Gesner? Raii*, 186. *Tab. M*. 2 ? *An Alba Coretta*, *pis.* p. 73. ed. 1658 ? *Giltheads*, otherwise called *Bonito's* of *Hawkins*, p. 520. *Turbots* of *Oviedo*, *sum*, p. 211. 214. *Boniti de Laet*. lib. 10. cap. 11. p. 415. *in Peru*, 430. *Bonitoes* of *Terry*, p. 11. Of *Jo. Davis. Purchas*, 132.

This appear'd in every thing like a Mackarel in colour of Skin, smoothness, great and small Fins, Tail, &c. only it was much thicker in proportion to the length of its Body than the Mackarel, being One Foot and a half long;

Tab. 1. Fig. 3. it weighed Ten Pounds, and had in its Belly many of these young Fishes described before under the name of *Serpens Marinus*, &c. and young Flying-Fishes, which were most of them consum'd to the Bone, their Intestines be-

Of Boneto's. ing fill'd with blackish scaly Matter, and the Bones of these Fishes. The Lines on the side of the Figure of this Fish are not in the Fish, but mistaken by the Designer or Graver.

They are taken between the Tropicks, and in more Northerly Latitudes, when the Sun is in the Northern Signs, as the Dolphins are, either with Phisgigs being struck from the Yard-Arms, or by Hooks and Lines baited with Cloath in the shape of Flying-Fishes, and are reckoned very Savoury, they tasting like a Mackarel exactly. They are also taken at the *Maldives* in the South-Sea.

They were found in the South Seas between *New-Spain* and *China*, by *Francisco Gualle*; whence he Argues a Current and Strait there, these Fishes using such places as well as *Albacoras*, which (by the way) is Translated *Tunnies* in *Purchas*, lib. 4. cap. 13. p. 805.

They pursue Flying-Fishes, and were galled with Phisgigs following the Ship Five hundred Leagues. *Hawkins*.

They are less valued in *Peru*, because thought to breed Fevers and other Distempers, *Laet*.

The *Remora*, as it is commonly called, is also frequent here; it is described Of Remo- and figured in most Natural Historians, and is called *Reversus vel Inversus*,
ra's. *Laet.* p. 6. *Reverst, Col. f.* 112. *Iperuquiba & Piraquiba Brasiliensibus, Lusitanis Piexepogader & piexe piolho Nestratibus Suyger. Marcgr.* p. 180. *Raii,* p. 119. *Ap, Tab.* 9. *Fig.* 2. *Un remora de Molinet*, p. 205. Sucking-Fish of *Terry,* p 9. Fishes called *Guajcanus* or *Reversus*, taking other Fishes. Of *Martyr. Decad. Poissons semblables au pinaru de Cauche*, p. 143. *Echenoen seu Remora. Imperat.* p. 684. *Aldrovand. de piscibus*, lib. 3. cap. 11. p. 336. The Sucking-Fish.

The

The *Indians* take Fish with these hang'd at a Rope, the *Remora* apply themselves to other Fishes, and take them as *Tiburoni, &c. Col.* 112. They used to Fish with them, putting them out, and holding them in a Line, they wou'd fix on *Manati*, forcing them to dry ground, *Aldrov. lib.* 3. *Rondel. lib.* 15.

The Mariners do not observe any manner of Current in this great Ocean, but such as follows the Wind.

On the Sixteenth of this Month in the Lat. of 14°. 40´. North, and being about Five hundred Leagues distant from *Barbados*, which was near due East of us, in ordinary Weather one of the Sailors that was on the Forecastle took up a large live Grashopper, and brought it me, which thinking very strange, as being a very great way from Land, I immediately enquired as nicely into, as possibly I could, and was assured by him that gave it me, that it came not thither from Land with them, or by any accident, but fell down from the Rigging of the Ship; which he supposed might perhaps stop its course. A Seaman, on this occasion, averred to me that he saw some of the same fly through the Rigging of the Ship: and some other Seamen in the same Fleet we were in, made the same observation. When I came to *Barbados* I met there Sir *John Narborough*, who was then going to the Plate Wreck near *Hispaniola*, and thinking this very odd, and him a very experienc'd and observing Person in such matters: I asked him if ever he had seen the like, he replyed very often, and that it was very ordinary to observe them fly many Leagues from Land at Sea. The Description of this I then took as follows, and called it *Locusta maxima, cinereo-purpurea, Maculis Brunis.*

A *Of a Locust observed at Sea.*

This Locust from the Head to the end of the Wings was two Inches and a half long, its Body was two Inches in length, in the *Abdomen* were seven *Incisures*, it had two *Antennæ* each half an Inch long, a large Purple and Brown Head, with two lenticular Eyes each Prominent, three pair or six Legs, taking their Origin from the *Thorax*, the hindermost pair being thick at the Thighs and prickly, two Inches long, more than twice as long as those before, those in the middle longer than the foremost, the Wings membranaceous, of an Ash, inclining to Red or Purple Colour, with many brown spots on them: it had three Incisures on its Back, which was guarded, as it were, with armour.

*Tab.*1.*Fig.*1.

This same accident did *Vanderhagen* take notice of in his Voyage, as appears by the following Note, though he does not mention the distance from Land.

In reditu è S. Helenæ multa rufæ subalbida locustæ vischantur, aquæ innatantes; quarum & quædam in naves ipsas, advolitabant. Vanderhag. Excerpt. Cluf. Cur. post.

They are in great numbers (about *Senega*) cover the Ground and obscure the Air every third or fourth year, destroying all. *Cadamosto.*

They are salted and eat by the Æthiopians. *Escarbot nova Francia, p.* 210.

They consume, a Famine follows, and then a Plague. *Schnirdel.*

They destroy the Ground not only for the time, but burn Trees for two Years after; so that People (in *Ethiopia*) are forc'd to sell themselves and Children for Sustenance. *Jo. dos Sanctos.*

This Locust is the same with those eaten in *Barbary*, they dry them in Ovens to preserve them, then either eat them alone, or pounded and mixed with Milk. Captain *Dampier* has told me they taste like Shrimps, and they are without question fed on by the Inhabitants of many places of the World. That they are the Qails mentioned, Numb. 11. 31. seems to be plainly proved by *Ludolfus*, who in his Appendix to his Commentaries on his *Ethiopick* History has expresly at large discoursed very learnedly on this matter. Some years since many Locusts very like these came into *Wales*, where they lived a small time and dyed, I think through the inclemency of the Air; but by their course I remember

A Voyage to JAMAICA.

member it was likely they had come Originally from *Barbary*. Those who have a mind to consult Authors about them may see, among others these who discourse of them under the following Names, *viz.*

Locuste Rosse and *Gjalle di Cadamosto*, f. 17. and 109. Grashoppers of *Escarbot. Nova Francia*, 210. Of *Giros*. *Purchas*, 1425. Of *Jo. Dos Sanctos*, *ib.* 1554. Locusts of *Schnirdel. Purchas*, p. 1359. *Ludolf. cap.* 10. *lib.* 1. *vers.* 16. *Comm.* 1. 96. p. 168. *Numb.* 11. 31. *Selav, Coturnix* or Quails. *Juan de Barros, lib.* 1. *Decad.* 2. p. 16.

When we came into 13°. 10'. Lat. we went due West for *Barbados*, which is the way not to miss it, because Sailers being sure of the Latitude by Observation they keep in it least they should run by the Island, which being very low Land, and so consequently not visible far off at Sea, may be easily over-run by unwary Seamen. We saw here several Tropick-Birds, and Men of War Birds, the last of which is mentioned by the following Writers.

Caripira de Laet. lib. 15. *cap.* 13. p. 575. *Caripira* forked Tails, bringing news of Ships, *Anonymus Port. Purchas*, p. 1317. *Cuda inforcata*, that is the forked Tail. *Ovied.sum. p.* 202. An Sea-Mews or Cormorants, *Ejusd. ib.* p. 214 ? Birds which the *Portuguese* call *Garayes*, or *Rabos forcados*, with Tails like a Taylor's Shears, of *Mandelslo*, p. 196. *Rabi horcados todos Negros, de Oviedo, lib.* 14. *cap.* 1. *Rabi forcati. Ej. ap. Ramnus*, p. 161. *Fregattes de Raveneau de Lussan.* p. 116. and 218. who observed them at *Villia* and *Guatulco*, a Rock in the South-Seas. *Fregattes entierrement Noirs, Rochef*, p. 164. *Rabi-horcado Hernandez.* Birds attending the rising of Flying-Fishes of *Ligon*, p. 4. *Rabo-forcado*, and *Rabi-horcado, Cluf. exot.* p. 107. *Oyseaux faisans la Chasse aux Poissons. d'Abbeville*, p. 52. *Fourcades, Ejusd.* p. 53 ? *aves furcata Laet.* p. 601. *Fregattes du Tertre*, p. 269. *Ciseaux de Coufurier de Cauche*, p. 133. *Aves dicta Rabos forcados, cauda bifurcata, Aldrov. Ornith. tom.* 3. p. 544. Sea-Fowl chasing the Flying-Fish of *Hawkins, Hakl.* p. 520. A Man of War of *Ligon.* p. 61. Of *Jo. Davis Purchas*, 132.

Of Men of War-Birds.
This Bird seems very large, bigger than a Kite, and Black ; they fly like Kites very high, and often appear immoveable over the water, to wait for, and catch small Fish appearing on its Surface ; they are sharp winged, and their Tail is forked. When Flying-Fishes are persecuted under water by Dolphins, *Bonitos, &c.* They rise and fly for some space in the Air, and are often devoured by these Birds in that time.

We saw them first when we came near *Barbados*. The Sailers guess themselves not many days, or about Two hundred Leagues off the Islands when they spy them first, and it is wonder'd at how they can direct their Course to the Land at Nights, being so far distant, it seems no very strange matter, because they are very high in the Air, and can see Land much farther then those on the Deck or Top-Mast of a Ship. The Reason of their flying so high may be to have a greater Field before them, for Prey, because they may go where they see the Dolphins follow or hunt the Flying-Fishes.

They are commonly thought in the *West-Indies* to foretel the coming in of Ships, for when they see a Man of War-Bird come into their Ports, they reckon Ships will soon follow, and 'tis very often true, for they love to Fish in not very rough Weather, so that when it blows hard at Sea they come into the Ports and Bays to Fish, where the Wind is broken off by the Land, and the same Wind blowing them in, brings in the Shipping after them.

There are more of these in the firm Land of *America* than in the Isles. The *Indians* of *Cueva* say the *Axungia*, or Fat of them is very good for taking out *Cicatrices* and marks, and for the withering of Arms or Legs, and other Diseases. One of these Birds at *Panama* coming to take *Sardinas* that were a curing in the Sun a *Negro* broke his Wing with a Stick he had in his Hand ; the Body after it was clear of its Feathers was little bigger than a Pigeon. The

Wings

A Voyage to JAMAICA. 31

Wings being extended, no Man, though several tried, could reach with his Arms stretched out within four Inches of the tips of them, *Oviedo*.

Their Grease is a Soveraign Remedy for the Sciatica, and all cold Gouts, *Du Tertre*.

It follows the *Alcatraz* (or Pelecan) in the Air to catch its Dung for Food, *Col*. In the first Voyage to the *West-Indies* the *Spaniards* followed the flight of Birds as the *Portuguese* did in their Discoveries, *Idem*.

He persecutes the *Alcatraz* (or Pelican) for his Prey till he lets it fall, he catching it before it gets down, *Oviedo*, *sum. p.* 202.

It is very good Meat *Caucbe*.

The *Indians* love its Feathers, which they use with their Arrows, observing they last longer than other kinds. Their Grease takes out Scars, *Xim. Laet*.

We had also Boobies, a kind of Bird so called, as well as Noddies, came frequently here on Board the Ships; they are so called by Seamen, because they do not stir from you, but suffer themselves to be catch'd by the Hand, or light on Seamens Arms, being unaccustom'd to Men. The Booby is nearest to the Soland Goose of any Bird I know, therefore I shall call it *Of Birds called Boobies and Noddies.*

Anseri Bassano congener avis, cinereo-albus.

These are Grey and White of colour, large as a *Muscovy* Duck, White above and Grey below, the four Toes join'd by a Web as in the Figure, they fly over the Water as a Kite over the Land, and watch the Fish to take them so soon as they appear on the Surface of the Water: It has a long roundish Bill of a yellowish colour. We first met them when we approached the Island of *Barbados*, and had them afterwards in great numbers all along amongst the *Caribe* Isles: they are very numerous on *Redondo* a small Island not far from *Nieves*, and there they breed, it not being inhabited or resorted to, so that they are not disturbed with mankind. I shall next describe the Bird called a Noddy, and call it *Of the Booby, Tab.6.Fig.1.*

Hirundo marina, minor, capite albo.

The Noddy Bird was Eleven Inches long from the end of the Bill to that of the Tail, and Twenty six Inches from Wing to Wing extended; the Bill was streight, black, roundish, an Inch and a half long, having two large Apertures for the Nostrils, the Tail was Four Inches long, the top of the Head was White, all the rest of a dirty brown reddish colour, the Legs and Feet were Two Inches long: it had Four Toes, Three before, was Web footed, and of a dark Brown colour. *Tab.6.Fig.2.*

They feed on small Fish, and go out a great way to Sea, where when they meet Ships they pitch themselves without any fear of the Men, even sometimes on the Hands of the Sailers, if held out for them to Pearch on. They are mentioned by the following Writers.

Noddy of *Jo. Davis Purchas*, 132. who observed it at *Noronha*. *An* Birds on the Ship Galleries at *J. de Martin Vaz*. *Purchas, Wilsen, p.* 486 ? *Uccelli fimili al Gargino, Col. f.* 43 ? Noddy in *Greenland* or *Spitzberg*, *Purchas*, 472 ? Of *Pool, ib. p.*707 ? *Passere Sempie*, that is simple Sparrows of *Oviedo sum, p.* 203. *An Mallemucks* of *Spitzberg* ?

It is somewhat less than Sea-Mews, has Feet like a Malard, stands on the Water, lights on Ships, has a black Head, and its Shoulders Russet, and is not good to eat, *Ovied*. 203.

We took here a Shark which in his Stomach had remaining the Feathers of one of these Birds called a Noddy, the Flesh of which was dissolved into a kind of *Mucus* and Blood. It is not unlikely that this Bird which Preys on small Fish on the top of the Water, was on falling down to catch such Prey taken by this Shark. It is also likely that the Feathers, of this or other Birds, when taken, may serve to make the devouring Fish endure Hunger longer than otherwise it could. It being observed that Birds of Prey will endure Hunger longes

A Voyage to JAMAICA.

longer than others, and indeed there was Reason it should be so, they being not very certain to find their Food at certain Hours. I remember once to have found in an Eagle Shot in *Essex* and sent me by Mr. *Barret*, that the Hair of Hares and Rabbits were the only Contents of his Stomach, which probably might keep him from being very sensible of his wants.

We had near, but before we came in sight of *Barbados*, a sort of Sea-Snail with Barnacles sticking to it, floating in the Sea, there were several of them, and their Apertures were filled with a froth all standing in Bubbles. The Snail I described thus.

Of a Sea-Snail, Tab.1.Fig.4.
It was more flat or compressed than most of the *Cochleæ Marinæ* consisting only of Circumvolutions round the *Columella*, or *Axis* of the Shell. It had some visible *Oblique striæ* on the Circumvolutions, was brittle and thinner than any Marine Shell I ever saw; it was of a very fine Violet or Purple colour. These Shells floated on the Surface of the Sea, and had many *Bullæ* of a viscid froth came out of them such as is raised from Sope and Water. They were more compressed than that of the Streights of *Magellan*, or *Mediterranean*, as also smoother. Wherefore I have chose to call it *Cochlea Marina è cæruleo purpurascens, compressa, lævis, tribus volutis constans*. It is the *Cochlea Marina Vicesima tertia cæruleo-purpurascens* of Dr. *Lister Hist. Conchyl. Tab.* 572.

Of Barnacles.
There grew to this Shell on every side that particular kind of *Plurivalved* Shell-Fish made of several Shells called *Concha Anatifera*, figured in Dr. *Listers Historia Conchyl. Tab.* 439. and 440. which sticks to, and Breeds on any thing floating in the Sea, by a hollow Neck somewhat resembling a Wind-Pipe. It looks somewhat like a Cockle, and has in it some *Cirrhi* which has been taken for the budding Feathers in the Wings of young Barnacles or *Brent* Geese which were supposed to Breed out of Trees. These Birds used to come yearly to *Scotland*, and other Countries in great numbers from the North in Winter, and go away in the Spring. They used in Northern Countries to have drift Wood come from the North with these Shell-Fish sticking to them, and never observing the Barnacles Breed as other Fowles, thought they bred so, till the *Dutch* in their attempts for a North-East Passage found these Barnacles sitting on Eggs as other Geese. I shall not say any thing further, but refer the Reader to the Authors following, where he will find them treated of.

Arbor Anatifera prima seu Arbor ex cujus ligni putredine vermes, & ex his anates viventes & volantes generantur, C. B. pin. p. 513. *Arbres des Isles Hebrides, les troncs ou bois desquelles cheuz dans la mer, & pourris par l'eau Marine, se muent & changent dans quelqut temps en vers, puis en oyes ou canes vivantes*, De Duret. p. 287. *Britannicæ Conchæ anatiferæ Gallis Macreuses. An Sapinettes Nortmannorum eodem*, Lob. Obs. p. 655. *Britannicæ Conchæ anatifera*, Ej. Icon. p. 259. Park. p. 1306. Ger. emaculat. p. 1587. *Arbore delle anitre Durant. in Fig. Telline pedate di Imperat.* p. 683 ? *Ex surculis arborum concha anatiferæ*, Lugd. p. 1398. *Arbores conchifera vel anatifera falso dictæ ut tellina ac Balani*, J. B. Tom. 3. lib. 39. p. 818. *Arbores conchiferæ vel anatiferæ dictæ ut tellinæ & alia.* Chabr. p. 580. *Arbor admiranda Vicesima quinta in Orcadibus & Hebridibus insulis anatifera*, Jonst. Dendr. 471. *Concha anatifera Calceolar*, p. 25. *Olakis & concha anatifera Aldrov.* Ornith. Tom. 3. p. 174. and 548. Barnacles of *Hudson* or *Marcolino*, Purchas, p. 615.

The Twenty fifth of this Month at Noon we arrived at *Bridge-Town* in *Barbados*, and anchored in Ten Fathom Water, all the Ships and Forts saluting his Grace. The Island of *Barbados* had its name, or *Barbata* as *Martyr*, from a kind of Fig-Trees which are frequent in it, of which I shall give a Description hereafter; they have Filaments or Threads come out of their Tops and hang down in handfuls or Sheafs, and make the Tree look as if it was bearded: at least this was told

me

A Voyage to JAMAICA.

me there to be the reason of this Name given to it by the *Europeans*. It lies in about 13 *degr.* 10 *min.* Northern Latitude, and is about 36°. 55'. West of *Madera*. The Rains when we were here came on very violently, it having been dry and parch'd for many Months. They pour'd down very fast, insomuch that being at *Spikes*, a Town distant some Miles from *Bridgetown* whither we were going, it was thought we should not be able to get thither, because the Gullies or Brooks were believed not fordable, however we ventur'd and got safe; but the Freshes or Rains were such, that two of the Dukes Servants who hired a Boat, at an excessive rate, to carry them to *Bridgetown* by Water, were, by the violence of the Water running off the Island, carried to Sea, and neither they, the Boat, or Boat-men ever heard of after. It was desired by His Grace the Duke of *Albemarle*, that the Governor, Colonel (since Sir *Edwyn*) *Steed*, would please to take care to inform himself from the Neighbouring Islands, if by Wind or Currents they or their Vessel had been heard of among them; but neither during our stay, nor afterwards, had we any tidings of them: so that it was concluded they were lost either by being swallow'd up of the Sea, starv'd for want of Provisions, or thrown on some of the Neighbouring Islands, where they might suffer by the rage of the exasperated *Indians*.

The Island, (which is Twenty eight Miles long, and Fourteen broad,*) is not very high, but yet has several raised and depressed Grounds in it, which are generally Fertile. This lowness of the whole Island gives them more regular Breezes or Winds, so that they Grind their Sugar-Canes with Wind-Mills, and not Cattle, as they do in *Jamaica*, where the Land being higher stops the regular motion of the Winds or Breezes*. It is likewise from this low situation that it has at present, and has had so great a fruitfulness, though now it be fallen much off from what it was, through the great labouring and perpetual working of it out, so that they are now forc'd to dung extremely what before was of it self too Rank. The Duke of *Albemarle* having a Patent for all the Royal Mines in the *West-India* Plantations belonging to *England*, made great enquiry after Minerals, but received information of none, save only of an Hill where was a shining Substance which look'd very fine, and was lodged in Earth; some of this Earth was afterwards sent to *Jamaica*, but proved to be nothing but white or silver colour'd *Marcasite*, which on trial held no Metal, or so little as not to be worth while to look after. These *Marcasites* are very common in most places of the World yet discovered, and impose on People ignorant of these matters; Sir *Martin Forbisher*, a great Man, when he went to discover a North-West Passage, brought home a Ships lading of this from North *America*. In *Trinidad* are *Marcasites**. And I have some of this same Mineral Substance that I had from the Streights of *Magellan*, not to be known from those found in *England*. This *Marcasite* was discovered by the falling off of some run-away Grounds, as they are called, from the side of a Hill in which it was seen. These run-away Grounds come from great Rains, after which a parcel of Ground, as the side of a Hill with whatever is on it, falls off from the other part, and carries whatever was growing on it along with it and remains on another Mans Plantation, whose Property it becomes.

There is towards that part of this Island called *Scotland*, some Pits, out of which are taken what is in *England*, and other places, called *Barbados*-Tarr: it is of two sorts, one liquid which swims on the top of the Water, and is of the consistence of Common Tart, smells strong as *Petroleum*, and in every thing resembles it; the other is more solid, and seems to be a better sort of Pit-Coal. In several places of *America* these sorts of *Bitument* are found, and have several Names; the most common Name is *Mountjack*, by which 'tis known very well amongst the Privateers.

* *Ligon.* p. 26.

* *Barbados wants Night or Land Winds, because it has no Hills*, Boyl, *of Air*, p. 85.

* *Of the Island of Barbados.*

* *Duddeley,* p. 571.

Of Barbados-Tarr.

K   'Tis

A Voyage to JAMAICA.

'Tis called *Bitumen Nigrum*, by *Fragosas*, p. 92. who tells us, that 'tis found in *Cuba* of the consistence of Pitch, and used in cold Distempers, Hysterick Fits, and for Pitching Vessels. Pitch to trim a Ship of *Masham*. *Hakl*. p. 3. p. 695. Unctuous substance like Tarr and *Mountjack*, *Ligon*. p. 101. Stone-Pitch plentiful in *Guiana*, *Rauleigh*, 631. *Hakl*. Who says it will not melt with the Sun. Pitch melting not with the Sun, *Harcourt*, *Purchas*, 1287. *Fons aquâ piceâ*, softer than Tree-Pitch, fit for Ships, of *Martyr*. *Minero de pasta como pez, con lo qual, rebuelta con azeyte o sebo brean los Navios y empegan qualquiera cosa, Ovied. cap. 51. Pix quædam fossilis in Insula Trinidad. Lact*, p. 602. l. 17. c. 17. Who says 'tis easily softened by the Suns heat, and not fit for Ships. Pitch-Fountains near *Anna*, of *Furer*. *Purchas*, 1412. *Bituminous* Waters, and Soil in *Sustana*, makes People short lived, *Cartwright*, *Purchas*, 1435. Fountain of Pitch and Tarr, *ib.* 1694. There is without doubt great vertue in these *Petrolcums*, they are very penetrating, ease Pains, Aches, &c. There is distilled from them an Oil which is more piercing than the thing itself; and which I have been assured was a good Remedy in the Gout.

They at *Barbados* want Wood very much, both for all manner of uses in Building, and for Fewel. For Building the Inhabitants go to *Santa Lucia*, an Island within sight of this, to *Tobago*, where it is plenty, and the other Neighbouring Islands. Their Fewel is *Guinea* Corn-stalks, Cane-Trash, that is the dried mark remaining after expression of the Juice, &c. This Island is very strong, by the Inhabitants (who have been Ten thousand Foot, One thousand Horse, and Fifty thousand Souls, besides *Negros*, *Ligon*. p. 43. and 100. But now are about half that number,) all living near one another, well Disciplin'd, and in good order. It has many Rocks to Windward, in that part called *Scotland*, which defend the Coast on that side, and to Leeward there are Shelves, and few Harbors, and there are Batteries at every place where a *Canoa* can Land, to hinder any Enemies approach. The Principal Town, called *Bridgetown*, is large, and has Batteries and Guns mounted to defend its Road or Harbor: so that in the late *Dutch* War *De Ruyter*, their Admiral, in vain attempted it with a very strong Fleet. From this place goes about Three hundred Sail of Ships yearly in Trade to *England*, *New-England*, *New-York*, *Jamaica*, &c. Their Horses come from *Barbuda*. They having little Pasture-Ground, they have few Cattle or Sheep, those they have are fed on Cane-Tops, *Guinea*-Corn, or *Scotch*-Grass. Their Mutton, which is fed on Sugar-Canes, is very Fat, White, and Sweet. The Duke of *Albemarle*, who had a Patent to be Chief Governor and Inspector of all the *American* Plantations or Islands where he came, took a review of the Forces of this Place in three several Divisions, and was very well pleased with the great Reception and Entertainment he had here from Sir *Edwyn Steed* the then Governor. For my own part I lik'd so well the Dessert after Dinners, which consisted of Shaddocks, Guavas, Pines, Mangrove-Grapes, and other unknown Fruits in *Europe*, that I thought all my Fatigues well bestowed when I came to have such a pleasant prospect. I was told a Goose here at sometimes was worth Twenty Shillings. *Spanish* Money goes here Current, and all over these parts.

I saw here the Wild Goose of *New-England*, or Geese White and Gray of *Escarbot Nova-Francia*, p. 50. Geese of *Hudson*. *Purchas*, p. 602. Of *Saris*, *ib.* 372. Wild Geese of *Copland*, p. 466. who observed the same at the Cape of *Good Hope*. Of *Payton* who saw them in an Isle of *Ethiopia* 33°. 30'. Lat. 487.

I saw also the *New-England* Deer in a small Enclosure near the Church, which seemed the same in every thing with our Fallow-Deer.

Turner (ap. *Purchas*, p. 1265.) found Hogs, Pigeons, and Parots there.

A Voyage to JAMAICA.

The Springs here near the Shore, were overflowed by the Sea and brackish, which gave many the Flux; so when Besiegers (at *Ormus*) came to drink brackish Water they dyed of Fluxes, *Monox, Purchas,* 1798. and six hundred of the Earl of *Cumberland's* Fleet dyed of the Bloody-Flux at *Porto Rico, Purchas,* 1149. *Layfield,* 1167. which likely came from the same Cause.

Plants I observed at *Barbados*, which I did not take notice of in the other *Caribbe* Islands, were,

Filix non ramosa minor, caule nigro, surculis raris, pinnulis angustis dentatis raris brevibus acutis subtus niveis. Cat. pl. Jam. p. 20. Where may be seen its other Synonymous Names, and reference made to the Authors where it is already graved.

This has a solid Root, small, and having several Scales towards the top, covered with a Ferrugineous Moss, and many Filaments and hairy Fibers below, whereby to draw its Nourishment; it is of a dark Brown colour, from whence rise many Leaves, having blackish red shining Stalks, for the most part Triangular, rising a Foot high, at about Eight Inches from the Root having Twigs at about half an Inch distance, sometimes opposite to one another, sometimes alternatively, each Twig being about one Inch long, and very narrow; the *Pinnula* are short, very narrow, sharp, and leave a defect between each other, on the upper side, being of a dark green colour, and below having a White Meal all over it, making it extremely pleasant to look on.

It grew out of the Rocks in the shady Inland parts of *Jamaica*, and in *Barbados*.

Filix non ramosa minor, caule nigro, surculis raris, pinnulis latis dentatis subtus niveis. Cat. pl. Jam. p. 21.

This had Roots and Stalks exactly like the preceding, and was for Magnitude the same, the chief difference was in the Leaves or *Pinnula*, which were rounder and broader, otherwise the same, for they were covered over with a White *Farina* as it. Perhaps this may be only a variety of the former. *Tab. 7. Fig. 1.*

I found it in *Barbados* and *Nevis*.

Cyperus maximus paniculâ, sparsâ, foliaceâ. Cat. pl. Jam. p. 35.

This was in the Stalk Triangular, filled with a Pith like Rushes, and exactly the same with other *Cyperus* Grasses, only it was much larger in every part, and in lieu of a Panicle, its Head was made up of scaly Spikes Sparse, at the top of this was only somewhat smaller Leaves than were at bottom, that is, there were first several larger leaves standing under the Panicle, from the middle of which rose many Triangular, small, and shorter Foot-stalks, which *Tab. 8.* at top, instead of Spikes, had nothing discernable but small Leaves, from the middle of which went other smaller *Petioli*, on whose top were very small and reddish Leaves, especially on their under parts, something, or very like the Description and Icon of *Papyrus ex Ægypto Salmasio missa. Bod. à Stapel. Not. in Theoph. p.* 432.

I gathered it in *Barbados*, in standing Water, a little out of *Bridgetown*. Whether this be the same with other *Cyperi*, before their Spikes come out, or in its Panicle be like that of *Papyrus Niloticus* of *J. B.* which if *Cæsalpinus's* Description be true it imitates, I cannot tell, but am sure I never in all the Plants I chanc'd to see of it, saw any other Spikes than these Leaves.

Gom. cap. 163. Says that in *Peru* they used Barks or little Boats made of Rushes or Straw, which I suppose may have been a sort of this, which was used for that purpose by the *Egyptians*. And *Blasco Nunnez*, after being taken, complaining of it, 'twas answered that it was the only way of

Boats

Boats in that Country. Bull-rushes are now used by some for floating Bodies.

Cyperus maximus paniculâ minus sparsâ ferrugineâ capitulis compactis crassioribus. Cat. pl. Jam. p. 35.

Tab. 9.

This has a great many rough three-cornered grassy cutting Leaves, about three Foot long, and near an Inch broad at Root where broadest, and whence they decrease, ending in a point, and all of them make a large tuft round the Surface of the Earth where it grows. From among the middle of them rise three-cornered Stalks as big as ones Finger, solid, and filled with a rushy fungous Pith, about three Foot high, on the top of the Stalks under the Spikes, stand two Foot long Leaves, tho there are many others shorter, being of like make with the Leaves at bottom. Above these stand several Spikes on several three-cornered *Petioli* of diverse lengths, as in others of this kind, the longest being of Six Inches, and some of them being branched into Two or Three other *Petioli*, all of which have some Ferrugineous scaly, round, pointed *Apices*, or smaller Spikes, made up of Ferrugineous Scales lying on one another, sticking close to the tops of the *Petioli* which make the Panicle. The Seed is brownish, red, shining and Triangular.

It grew near *Bridgetown* in *Barbados*.

Gramen cyperoides paniculâ conglomeratâ è plurimis spicis cinereis constante. Cat. pl. Jam. p. 36.

Tab. 10. Fig. 1.

This had a roundish Tuberous Root which had many Fibers as those of a Leek, it was not Odoriferous but covered over with several dry Skins, and from thence rose several Four or Five Inches long, narrow Leaves of a grass green colour, like others of this kind; the Stalk was slender three-cornered, about a Foot and a half high, on the top of which stood several small *Spikes* clustered together without any Foot-stalks into one Head, each of them being made up of many Gray Scales between which lay roundish edged Seeds of a reddish colour; under this Head or Panicle stood some short Leaves, as in others of this kind.

It grew in *Barbados*, but where I do not remember.

Ricinus Americanus tenuiter diviso folio Breyn. cent. 1. p. 116. Cat. pl. Jam. p. 40. *Manſanilla de las avellanas para purgar de Oviedo,* lib. 10. cap. 4. *Ben Magnum Medicorum vulgo Wormmus,* p. 187. *Avellana Purgatrix, Consant.* p. 1. *An Castanea purgatrix,* Muſ. Moſe. p. 254?

Spanish Physick Nuts. I found these in Mr. *Draxe's* Plantation in *Barbados*, where they were planted in a Garden; they agreed as to every thing with *Breynius's* Description and Figure; they had Flowers of a fine Scarlet colour.

They were put out of use by *Mechoacan*, Frag. When *Hispaniola* was first discovered, the *Indians* used this much for Purging, then the *Spaniards* through necessity used the same, not without hazard of Life. They purge Phlegm and Choler violently up and down, their vehemence is taken off with roasting. They are good for the Colick, and they Purge being put into Glysters. Hot in the third, and Dry in the second Degree, Doſ. à ʒſ. ad ʒj. Tosted. *Oviedo* says one was killed in *Spain* with half of one; and yet Nine did not work in *Espanola*: they kill'd several at first, his own Children were almost dead, had they not been vomited with Oil.

The Fruit is an easier Purger than the Common-Physick-Nut, the Flower dryed and powdered, and given to half a Crown weight purges Hydropick-Water plentifully, *Du Tertre*.

Convolvulus exoticus annuus foliis myriophylli millefolii aquatici flore sanguineo. Mor. hist. p. 20. Cat. pl. Jam. p. 58.

I found

I found this in *Barbados* plentifully, though I cannot say that 'twas Wild, but for its beauty planted in Gardens.

The Inhabitants ascribe the Vertues of a Philtro to this Plant. The Juice of the Leaves makes an Errhine which Cures the Head-ach, *H. M.*

Tencroides filiculosam foliis laurinis, floribus galeatis & labiatis. Cat. pl. *Jam.* p. 64.

The Branches of this had a woody hard Stalk filled with a fungous Pith and were about the bigness of ones Little Finger, round, and jointed. At every half Inch, are Leaves set opposite to one another, each whereof has small or no Footstalks, being about Four Inches long, and One and a half broad in the middle where broadest, beginning narrow, encreasing to the middle, and thence decreasing 'till they end in a point, being smooth and equal on the edges. The tops of the Branches and Twigs are Six or Eight Inches long Spikes of Flowers set at small distances opposite to one another round the Stalk, after the manner of some of the Verticillated Plants, each Flower having a small Foot-stalk, being both Galeated and Labiated, of a whitish yellow colour, the *Galea* being Bifid, and the Lip Trifid, and having two *Stamina* with *Apices* standing out of the open Flower, to each of which succeeds an Inch-long Seed-Vessel brown, roundish, small next the Foot-stalk, and swelling towards the point, being distinguish'd in the middle by a Partition into two Cells, which contain some large Brown Seeds. *Tab.* 10. *Fig.* 2.

It grew in the Island of *Barbados*, where I gathered it, if I rightly remember, but this Description being mostly taken from a dryed Plant is not so exact as it ought to be, though it is plainly of the same kind with *Valli-upu-dali. H. Malab.* and I could not bring it so near to any of the *European* kinds as the *Teucria.*

Phaseolus utriusque Indiæ lobis villosis pungentibus minor. Herm. par. Bat. prod. p. 364. Cat. pl. *Jam.* p. 69. *Phaseolus Brasilianus siliqua deurente lanugine obsita ricini fructu hort. Reg. Par.* p. 140. The stinging Bean *Ger. emac.* p. 1215. *An phaseolus orientalis pruritum excitans hirsutie siliquarum fructu nigro splendente. Pluken. Alm.* p. 292?

It grows plentifully in *Barbados.*

The Root boiled and given provokes Urine, with Oil boiled it Cures the Gout and *Erysipelas*, with the Root *Cocinil* it is good in Purulent Urine; the Leaves beaten and applied are good for Ulcers. The Beans eaten are provokers to Venery. *H. M.*

Twelve Pods of this infus'd in two Quarts of Bear: and half a Pint of the infusion given every morning to drink, is a certain Remedy for the Dropsie. *Ray. Hist. pl.* p. 887. This Remedy has been tryed in the *East-Indies* by Mr. *Buckley*, and found successful there.

Lysimachia lutea non pappose, erecta, foliis glabris fructu caryophylloide. Cat. pl. *Jam.* p. 85.

This is in every thing the same with the *Camaraibeya Marcgr.* only somewhat lesser; the Leaves are not hairy but smooth, else the same with it in every thing besides. *Tab.* 11. *Fig.* 1.

It grew in *Barbados* in Watery places.

Malva, vel alcea fruticosa, ribesii foliis, seminibus asperis. Cat. pl. *Jam.* p. 96.

The Twigs of this Plant had Leaves set on them alternatively, having about half an Inch Foot-stalks, they were almost round, of about an Inch and a half Diameter, deeply laciniated, or cut in on the Edges into three parts, each whereof were indented and sinuated about the Edges; they were rough on the upper side, and of a durty green colour, and pale or whitish below, being somewhat like those of a Curran Bush, or *Ribes. Ex alis foliorum* come the Flowers in a *Calix* having almost no Foot-stalk encompassed by several *Tab.* 11. *Fig.* 2.

L *Foliola*

Foliola to which follows several large rough Lappaceous or Echinated Seeds, having many little Prickles on their outsides, being in shape and disposition like to the other Mallows.

I found it in the Island of *Barbados*.

Solanum bacciferum fruticosum, stipitibus & foliis majoribus, spinis ferocioribus armatis. Cat. pl. Jam. p. 108.

Tab. 11. Fig. 3.

This is in every thing like the *Solanum fruticosum bacciferum spinosum flore caruleo. Cat. pl. Jam.* p. 108. only larger, the Prickles are very red, sharp, thicker set, and much stronger, being somewhat like those of the Wild Rose, and not only set on the Stems, but likewise on the backsides of the Leaves along their middle Rib, being very like to it in other things: the Leaves are much larger, and the Fruit is the same, there being many flat whitish Seeds in a Pulp, as in others of this kind.

I had it in *Barbados* and the *Caribes*.

Solanum pomiferum tomentosum, fructu pyriformi inverso. Cat. pl. Jam. p. 108.

Tab. 12. Fig. 1.

This is in every thing like the *Jurepeba* of *Piso*, only the Fruit is as large as an ordinary Pear, of a yellow colour, turbinated, and exactly of the same shape, from a large round beginning growing smaller to the top, which is like a Nipple, the whole Fruit having the Figure of a Pear inverted.

It grows plentifully about *Bridgetown* in *Barbados*.

The Root (if this be *Juabeba*) is bitter, and opening especially clearing the Urinary passages, but because 'tis too bitter, it may be help'd with the *American* Liquorice. *Piso*.

Planta de qua moxa colligitur, forte artemisia vulgaris. J. B. *tom.* 3. *lib.* 26. p. 184. *Cat. pl. Jam.* p. 127.

I was told by an Ingenious Planter at *Barbados*, whose name I have forgot, but who took much pains about *Exotic* Plants, that he had cultivated an Herb from the *East-Indies*, which he conceived to be the Plant which there yielded them their *Moxa*; it seemed to me in every thing to be the same with the *Artemisia vulgaris*, J. B. For having had the favour of the aforesaid Gentleman to send to his Plantation for it, the Messenger brought rolled up in a piece of Paper, a Branch of this Plant, as it seemed, broken from the main Stem. It was about a Foot long, the Stalk round, reddish, solid, having a great Pith, and Leaves placed alternatively without any order, very like the Leaves of our ordinary Mugwort, they being laciniated after the same way, of a dark green colour above, and covered over beneath with a woolly *Tomentum* making them white. This Wool being gathered is perhaps what they call *Moxa*, the Leaves were in handling somewhat Odoriferous. Towards the tops, from the Bosoms of those small Leaves came without any Foot-stalks, some small roundish, striated, whitish, woolly Heads, which I take to have been the Flowers budding out, but because I did not see them, cannot affirm positively that 'tis the same, although if I remember right, the aforesaid Ingenious Person told me he knew no difference. This *Artemisia* is taken notice of by *Clusius* to have grown in all those parts of *Europe* he travell'd, and I saw it though not Wild, yet thrive very well in the *West-Indies* Gardens, and I see no Reason why we may not allow it to grow Wild in the *East-Indies* and to be that Plant from whence *Moxa* is gathered. The *Moxa* or Cotton may be either what's Natural on the back sides of its Leaves, or rais'd on it by Insects as the *Bedeguar* on the Wild-Rose.

Melanomma & melanoxylon arbor laurifolia nucifera gemmis nigricantibus Americana Plukenet. Phyt. tab. 205. *fig.* 3. *Cat. pl. Jam.* p. 135.

I found it in the Island of *Barbados*.

Exoedoxylum seu lignum odoratum Americanum folio amplo subrotundo profunde venoso fructu glandiformi beretini instar nullo calice donato. Plukenet. phytogr. tab. 176. *fig.* 1. *Cat. pl. Jam.* p. 136.

This

This Tree was large, and had Twigs covered with a brown Bark, under which was a hard Wood, the Leaves were placed at its ends without any order, standing on strong Inch long Foot-stalks. They themselves were roundish, about Six Inches long, and Four broad in the middle where broadest, being shining, and smooth on the upper side, but uneven on the under, occasioned by several Nerves running variously through it, making Furrows on the upper side.

I found it in the Island of *Barbados*.

Laurifolia venenata, folio leviter serrato, oblongo, obtuso, copiosum lac præbens. Cat. pl. *Jam.* p. 136.

The Trunk of this Tree was of about Two or Three Foot Diameter, rising Forty or Fifty Foot high, and yielding Milk in all its parts very plentifully as the Tree commonly called Milk-Wood in *Jamaica*, to which it was very like in all its parts, excepting that the Leaves of this had more transverse parallel Veins than it, they were not so much pointed, but more Round or Oval, and serrated very easily on the Edges.

I found it in the Roads every where in the Island of *Barbados*.

Fellers of Wood take care of the Milk of this Tree coming on their Faces or Eyes, it making them Blind for a Month: two Horses quarrelling in a Wood, this Juice coming into their Eyes Blinded them. *Ligon.* Their Boards when dried the Poison evaporates, and then they are made into Sugar-Pots, *id.* Its Shade is thought hurtful.

Periclymenum rectum, salvia folio rugoso minore subrotundo. Cat. pl. *Jam.* p. 164.

This was about the heighth of Garden-Sage, having towards the top rough four square, hoary, Stalks, on which stood Leaves opposite to one another, on a quarter of an Inch rough Foot-stalks, they were almost round, of about three quarters of an Inch Diameter, having one middle Rib, sending several Transverse Fibres through the Leaf, which is corrugated on its Surface, like the Leaves of Sage, and indented about the Edges, *Ex alis foliorum* rises several Foot-stalks about Two Inches long, sustaining a small Head made up of several pale small Flowers, whose *Oræ* are divided into several Sections, and to which, although I did not see the Fruit, yet I question not but the same kind followed, as to the others of the same Family to be described hereafter.

It grew in the Island of *Barbados*.

Christophoriana Americana Malabathri foliis acuminatis nervosis dentata. Plukenet. tab. 159. fig. 1. Cat. pl. *Jam.* p. 164.

I found it in the Island of *Barbados*, whence I brought some dried Samples of it.

Arbor Americana Malabathri subrotundis foliis, subtus lanugino ferruginea villosis. Plukenet. tab. 249. fig. 3. Cat. pl. *Jam.* p. 164.

This seemed to be in Flowers, Leaves, *&c.* in all things the same with the others of the same Family, only it was lesser in the Leaves than most of them, they were somewhat rusty on the back side, and corrugated above, in which it was differing from all the others. The Leaves stand opposite one to the other, and they were of a dark green colour.

I found it in the Island of *Barbados*.

Grossularia fructu non spinosa, Malabathri foliis oblongis, floribus herbaceis racemosis, fructu nigro. Aninga peri. Pis. (Ed. 1648.) p. 116. (Ed. 1658.) p. 218. Jonst. Dendr. 269. Cat. pl. *Jam.* p. 165. *An Arbuscula Jamaicensis Malabathri angustioribus foliis leviter crenatis superna facie per siccitatem nigris, subtus anicum argenteis & prælevore splendentibus.* Plukenet. Phyt. tab. 265. fig. 1. Almag. p. 30.?

This Shrub rises sometimes Ten Foot high, though very often 'tis about Five or Six only, its Trunk is small, branch'd, having a smooth whitish Bark: its Leaves stand on half an Inch Foot-stalks opposite to one another; are Six
Inches

Inches long, one and a half broad in the middle where broadest: of a very dark green colour, having three Ribs running through the Leaf from the Foot-stalks end, with transverse ones after the manner of the others of this kind, or the *Folium* us'd in the Shops: the tops of the Branches are Bunches of small white Flowers, to which succeed so many small, black, round, smooth crowned Berries, having in a purplish Pulp, several very small black Seeds.

It grows in a Gully near the Town of St. *Jago de la Vega,* in most Gullies in *Jamaica,* and in all the *Caribes.*

Piso, in his Description of this Plant, says, first, That the Leaves are *Lanuginosa,* and then *Veluti Lanuginosa,* which shews his slight way of describing. He says likewise that the Leaves powdered, or Juice of them, Cure fresh and inveterate Ulcers by the first intention, if they be put into them very deep.

Grossularia fructu non spinosa, Malabathri foliis, subtus niveis, fructu racemoso, in umbella modum disposito. Cat. pl. Jam. p. 165. *Sambucus Barbadensibus dicta foliis subincanis. Plukenet. tab.* 221. *fig.* 6.

This appeared in every thing to be the same with the former, only the Leaves were white underneath, and extremely pretty; the Fruit stands in an Umbel at top, after the manner of Elder Berries, they making a more Horizontal Surface than the preceeding, whose Berries are not so numerous, and some plac'd higher, others lower.

I found it in *Barbados* or *Nieves,* which of them I remember not.

Lycium forte, foliis subrotundis integris, spinis & foliis ex adverso sitis. Cat. pl. Jam, p. 171.

Tab. 11. Fig. 4.

This had a white Wood which was hard and solid, a Bark redish-gray coloured, and somewhat *Scabrous,* or rough on the Twigs, on which were set Leaves opposite to one another, of about three quarters of an Inch Diameter, being almost round, only somewhat pointed towards both ends, without any Foot-stalks, and having two reddish, long, sharp, and strong prickles, rising *ex eorum alis.* They are of a dark green colour above, and smooth.

It grew in *Barbados,* where I gathered it.

Arbor mali persicæ foliis angustis, oblongis, acuminatis, ex adverso sitis. Cat. pl. Jam. p. 108.

Tab. 13. Fig. 1.

This Tree was one of the largest size, growing very high with a great Trunk, and having the Leaves standing on the Twigs, which are somewhat prickly, at about an Inch distance, always by pairs opposite to one another, without any Foot-stalks, each being about Four Inches long, and one broad near the beginning where broadest, and whence they decrease, ending in a point being equal on the edges, and green, with one middle, and some transverse Ribs. Because of the imperfect description I have of this, I cannot place it better.

I gathered it in *Barbados.*

Prunifera vel nucifera seu nuci prunifera arbor Americana præcelsa angustis laurifoliis late virentibus mastichen odoratam fundens. Plukenet. tab. 217. *fig.* 5. *Cat. pl. Jam. p.* 108.

This is one of the largest Trees, and highest of the Island of *Barbados,* where it grows every where, and is in use for all sorts of Buildings. The Twigs were brown and smooth, having Leaves with very short, if any Foot-stalks, being themselves about three Inches long, and about an Inch broad in the middle where broadest, and whence it decreases to both extremes, being of a very curious green colour, smooth and shining, somewhat like to Bay-Leaves, having one middle, and several transverse Nerves running very curiously through the Leaf, which is hard and not succulent. The Fruit was a turbinated small Plum of the bigness and shape of a Hazel-Nut, having under

a Mem-

A Voyage to JAMAICA. 41

a Membrane a thin Pulp, covering a very large and smooth Stone, which is hard, and includes a white Kernel.

It grew every where in the Island of *Barbados*.

Malus arantia, fructu rotundo maximo pallescente humanum caput excedente. Cat. pl. Jam. p. 112. *Venew Sinensium Martin.* Atl. Sinens. Lusitan. *Jambos. Raii hist. plant.* p. 1793. *Malus Arantia utriusque India fructu omnium maximo & suavissimo; Belgis orientalibus l'ompelmus, Virginiensibus nostraticus (ab inventoris nomine qui ex Ind. orient. ad oras Americanas primo transtulit) Shaddocks audiunt.* Pluken. Almag. p. 239.

This Tree is in every thing like an Orange-Tree, only larger, the Leaf has *Tab.* 12. a small Leaf before the other larger, as has the ordinary Orange. The Fruit *Fig.* 2, 3. is round as big as a Mans Head. The Rind is yellow and smooth, not thick, and the Pulp is very Aromatick, besides it has a sweetish four Tast. There is a variety or another sort of this with the Pulp and Rind of an Orange colour.

They are planted in *Jamaica*, and thrive extremely well, though I must confess, I think, that as in *Jamaica* their *China* Oranges are better than those in *Barbados*; so in *Barbados* their Shaddocks surpass those of *Jamaica* in goodness.

The Seed of this was first brought to *Barbados* by one Captain *Shaddock*, Commander of an *East-India* Ship, who touch'd at that Island in his Passage to *England*, and left its Seed there.

After Ten Days stay at *Barbados* we set Sail, and came the next Morning, *December* 6. in sight of *Santa Lucia*. This is inhabited by a small number of Of Santa People from *Barbados*, (within sight of which it lies) who keep it on the ac- Lucia. count of its Wood, which it has in plenty, and they at *Barbados* very much want. It has been disputed by the *French* whether the *English* were Proprietors of it, or they; but I was told that being in the Possession of the *English* at the time of the Signing the Treaty of Neutrality with *France* in 1687. it should remain quietly to them hereafter. I have heard that it abounds with great variety of Serpents.

The same day we had sight of *Martinico*, by some *Matalina*, *Matinino* or Of Marti- *Martinino*, an Island belonging to the *French*: called by *Columbus Matinine*, nico. distant Ten Leagues from *Dominica*, Col. 195. In 14°. North-Latitude, and 322. Longit. *Philips. Hakl.* 477. It was the first Plantation the *French* had in the *Caribes*, and if I was not misinform'd, the Mother of their other Plantations: the Inhabitants are reckoned Twelve hundred.

We came the Seventh in sight of *Dominica*, which is an Island belonging Of Domi- to the *Caribe Indians*, who are at present Inhabitants of it. It was discovered nica. in *Columbus's* Second Voyage to the *West-Indies*, after Twenty Days Sail of Seven hundred and fifty, or Eight hundred Leagues from *Gomera*, and in Fifteeen Days from the *Canaries* by *Laudoniere*. It was so called because discovered on a *Sunday*†. It has two Hot Baths in it, and used to afford Refresh- † Col. 93. ment to the *English* Sailing that way*. *Sir Anthony

Afterwards we came in sight of *Guadalupe*, which is an Island inhabited by Shirley, 599. the *French*. It had its Name from *S. Maria di Guadalupe*, and was discovered Purchas. *November* 4. by *Christopher Columbus* in his Second Voyage. Of Guada- lupe.

On *December* 8. we came within Thirteen Leagues of *Monserrat*, so called from its heighth †, and discovered in *Columbus's* Second Voyage: it is not very ‡ Col. f. 97. large. This Island is very well furnished with good Water. It has a River, Of Monser- a small Town, and many Sugar-Works. It has about Two thousand Whites, rat. (mostly *Irish*,) on it, and is Subject to the *English*. The Captain-General of the *Leeward Caribe* Islands, who generally Resides at *Nieves*, has always the Command of it, and a Deputy-Governor there. They are furnished here with some Money for Traffick.

M Between

42 A Voyage to JAMAICA.

Of Redondo.
† *Col. 97.*

Between *Monserrat* and *Nieves* lies a very small Island called *Redondo* or *Rotonda*, discovered by *Columbus* in his Second Voyage, who gave it the Name of *Santa Maria Rotonda*, from its Figure †. It consists of one Rock very Perpendicular and high, looking like a Pyramid, and as if there were nothing but Rock; but I was inform'd by those who have been upon it, that there is on its top an Acre or two of very good Ground, that it has a very good Landing Place, and a Well of very good fresh Water. It has also great store of *Iguanas* of a blackish colour. Many Boobies, and other Birds that come hither to lay their Eggs at proper Seasons.

On *Friday* the 9th. of *December* we came to Anchor in *Nevis* Road, in Seven Fathom Water.

* *Harcourt.*
Smith.
† *Laet. 26.*
Of Nieves.

Nieves, sometimes *Mevis* or *Meves**, was inhabited in 1628 †. It Consists of one Mountain of about Four Miles heighth to the top, whence is an easie descent to all parts of the Island; but steepest towards the Town where is the Road. They have neither Springs nor Rivers, but have what Water they make use of from Cisterns receiving the Rain-Water. The Ground is cleared almost to the top of the Hill, where yet remains some Wood, and where are Run-away *Negros* that harbour themselves in it. There are about Two thousand Inhabitants here, who being gathered together for the Duke of *Albemarle* to Review, I found more Swarthy, or of a yellowish sickly look, than any of the Inhabitants of these Islands. The Town or Road is well fortified with Batteries, and a Fort. They have little Money, but Buy and Pay with Sugars which are blackish. Their Horses, which are small, as well as many of their Provisions, come from *Barbuda*, an Island not far distant where Cattle are bred. I went to the top of the Hill to gather Plants, and though it had not did not Rain at bottom; yet I was taken there in so great Showers that I was wet unto the Skin. There is here an hot Spring affording a constantly running Rivulet of Water, made use of for all purposes as common Water. I found here the following Plants.

Filix arborea ramosa, caudice non diviso, pinnulis angustis raris obtusis integris. Cat. pl. *Jam. p. 22.*

As to the Trunk, and the manner of its growth (as well as I can remember) this was the same with the Tree Fern of *Jamaica*, in the whole Face of the Tree and Leaves, resembling a young Tamarind-Tree. The Foot-stalks and middle Ribs of this Trees Leaf were not prickly, but their Stalks smooth, large, and of a reddish green colour: the Branches and Twigs rose out of it alternatively, and the *Pinnulæ* were blunt, even, not dented about the Edges, and were about half an Inch long, being join'd at bottom to the Twig, but having a defect between them because of their narrowness. The *Pinnulæ* were of a dark green colour, and pretty thick.

This I gathered in the Mountain which makes the Island of *Nieves*, towards the top, in a Gully where it grew very plentifully.

Gramen dactylon bicorne tomentosum maximum, spicis numerosissimis. Cat. pl. *Jam. p. 33.*

Tab. 14.

This has many strong Thongs, or large white Filaments, which are Two or Three Inches long tapering, and taking firm hold of the Earth. The Stalk rises Three, Four, or Five Foot high, and has at bottom many Leaves enclosing it, and one another, after the manner of some of the *Cyperus*-Grasses; each of them being Two Foot long. with a sharp Back, being harsh, narrow, and something like those of the *Cyperus*-Grasses. The Stalk is streight, round, pale green, as big as a Goose-Quill, made up of Four or Five Joints, the spaces between them being covered with a Leaf, from the uppermost of which towards the top go Twenty, Twenty four, or a great many *Petioli* or Foot-stalks jointed, and from whose Joints go other smaller *Petioli*, making in all, both a little under, and at the top itself, a vast number of *Panniclæs*, each of

which

which is divided into Two Spikes standing like Horns, after the manner of *Gramen ischæmon bicorne*. Each of them has very much long, soft, very white Down, or *Tomentum* much finer than Cotton, and sometimes one may perceive among them some few Coarser Hairs which I take to be *Arista*.

This varies in being sometimes of a reddish or purplish colour, which I suppose proceeds only from the Age of the Plant.

It grew under the top of the Hill or Mountain making the Island of *Nieves*, and near the *Angels*, on the other side of the River, as well as in several other places of the Island of *Jamaica*.

The *Indians* told *Piso* that the Roots of this beaten, and given with any convenient Liquor, was a proper Remedy to expel Poison.

Gramen avenaceum, panicula minus sparsa, glumis alba sericea lanugine obductis. Cat. pl. Jam. p. 35.

The uppermost Joint of this Grass had a Six Inches long grassie Leaf, which by its under part enclosed the *Culmus* beneath, this Joint was about a Foot long, by which one may guess that the whole Grass was very high; the Panicle was at top, being about Four Inches long, not very sparse, made up of many *Petioli*, or Spikes standing round, or taking their original alternatively without any order from the top of the *Culmus*: each of these *Petioli* had fastened to them by very small and short Foot-stalks, several reddish, oblong, pointed Grains or Seeds, something of the shape of Oats, lying between two *Glumæ* or chaffy Membranes, which were on the out side covered over with a pretty long silken *Lanugo*, *Tomentum*, or Cotton, which distinguishes it sufficiently from others near akin to it. Tab. 14. Fig. 2.

I found it in *Madera*, or one of the *Caribes*, and if I remember right it was in the Island *Nieves*.

Urtica racemosa, fruticosa, angustifolia, fructu tricocco. Cat. pl. Jam. p. 38.

This has a cornered woody Stalk covered with brown at bottom, but at top green Bark, having many fierce Hairs on them, and Leaves coming out alternatively, first on one side, then on another. They have half an Inch Foot-stalks, with many strong Hairs, the Leaves themselves being Three Inches long, and not an Inch broad near the round Base where broadest, from thence they decrease to their tops, where they end in a point, being deeply serrated on their Edges, and having little Hair on them. Towards the tops of the Branches come *Ex alis foliorum* an Inch and a half long Strings, to which stick the Fruit, which at first is very small and *tricoccous*, growing larger and rough, with long Hairs on its out-side, each of the three round sides containing one large striated roundish Seed. Tab. 16. Fig. 1.

I found it in one of the *Caribe* Islands, and cannot positively affirm it to be stinging, but believe it to be that of *Piso*.

Piper sarmentosum folio minori, latiori, & tenuiori, atroviridi. Cat. pl. Jam. p. 45.

This is in every thing like what in *Jamaica* is called *Spanish Elder*, only the Leaves are thinner, broader (especially at Base,) of a darker green colour, having their Ribs less apparent than it, and being smooth, otherwise as to manner of growing, &c. 'Tis exactly the same.

I found it in one of the *Caribes*.

Amaranthoides fruticosum, foliis longis, angustis, subtus niveis. Cat. pl. Jam. p. 48. *An Polygonum erectum lignosum rorismarini foliis Virginianum.* D. Banister, Pluk. Alm. p. 302?

The Stalk of this Plant was streight, woody, covered with a smooth, reddish brown Bark; at every Inch or two having Leaves greater and lesser, about 1, 2, or 3, the largest having a very short Foot-stalk, being about two Inches long, and three quarters of an Inch broad in the middle where broadest, from the Foot-stalk increasing to the middle, and thence decreasing and ending in a point, being smooth, dark green on the upper side, and very Tab. 7. Fig. 3.

white

44   *A Voyage to* JAMAICA.

white underneath. *Ex aliis foliorum* ſtands without any Foot-ſtalk a round conglomerated Head for Flowers, made up of many white dry Membranes laid very cloſe *Squammatim* one by another like the others of this kind.

I found it in *Madera* Iſland, or one of the *Cariles*.

Aparines folio anomala vaſculo ſeminali rotundo multa ſemina minutiſſima continente. Cat. pl. Jam. p. 50.

Tab. 7.
Fig. 4.

The Branches of this were woody, covered with a ſmooth *Bark* about a Foot and a half long, cornered, and having Leaves plac'd at the Joints two always oppoſite the one to the other, being about an Inch and a half long, and about an eighth part of an Inch broad at Baſe where broadeſt, and whence it decreaſes, ending in a point, being ſmooth, equal on the Edges and Carinated; *Ex aliis foliorum* comes a roundiſh ſmall Body, a little prickly or hairy at top, which augments 'till it is round as big as a Pepper Corn, as it were crowned at top, and which contains within it one Cell, which is full of ſmall Seed like that of *Henbane*, ſticking to a Body which is in its Center.

I found it in *Barbados*, or one of the *Caribes*.

Colocaſia hederacea ſterilis latifolia. Plumier, p. 37. fig. 51. lit. a. & fig. 52. Cat. pl. Jam. p. 63.

This I obſerved in the Woods of the Iſland of *Nieves*.

On taſting the end of the Stalk of this Plant Father *Plumier* found his Mouth ſo inflamed that he could not ſpeak for two Hours, but was forced to keep his Mouth open, his Tongue hanging out. *Oxycrat* took away the Inflammation, but the *Acrimony* of the Juice had ſo burnt his Tongue and the Roof of his Mouth, that he could not taſt any thing in Ten Days.

Pulegium longiſſimis latiſſimiſque foliis. Cat. pl. Jam. p. 64.

Tab. 7.
Fig. 5.

This had fourſquare hollow Stalks, having Joints at Two or Three Inches diſtance, at which ſtand the Leaves oppoſite to one another, being about an Inch and an half long, and three quarters of an Inch broad, ſmooth, equal on the Edges, being broadeſt in the middle, and pointed at both ends, having one middle Rib, and ſeveral lateral apparent ones going to the Edges, without a Foot-ſtalk, but having a Membrane ſurrounding the joint where 'tis ſet on, encloſing the Stalk and ſeveral Hairs or Threads, or ſoft Prickles, ſome longer, ſome ſhorter, as well as Branches, having ſmaller Leaves. Towards the top come at the Joints ſeveral *Verticilli*, being Heads pretty thick ſet round the Stalk, made up of ſeveral Flowers, *Apices* or Seeds, under which are generally Two Leaves as the others below, only much ſmaller.

It grew in one of the *Caribes*, but which I do not remember, neither were my Obſervations about it very exact when I gather'd it, moſt being taken from the dry'd Plant, ſo that I am not certain if it Smells, or be a *Pulegium*. Mr. *Petiver* had it both from *Guinea* and *Barbados*.

Legumen trifolium ſub terra fructum edens. Raii hiſt. pl. p. 919. Cat. pl. Jam. p. 73.

I found this in ſome of the *Caribe* Iſlands, but where I remember not.

Althæa ſpicata betonicæ folio villoſo, ſpica breviori & laxiori. Cat. pl. Jam. p. 97.

Tab. 14.
Fig. 4.

This has ſeveral woody, round, reddiſh Branches, hollow, and having Leaves ſet on alternatively ſtanding, on half an Inch Foot-ſtalks, being of about an Inch long, and three quarters of an Inch broad near the round Baſe where broadeſt, and whence they decreaſe, ending in a Point, being Serrated about the Edges, extremely hairy, hirſute, or woolly, of a yellowiſh green colour, *ex alis fol.* come ſmall Branches, on which, and on the ends of the Twigs, come in *Pentaphyllous, villoſe calices,* the Flowers, Spike faſhion one above another, after the manner of *Althæa Americana pumila flore luteo Spicata Breynii*, to which follows ſeveral Seeds, about five in number, of the ſhape of thoſe of Mallows, and ſet round after the ſame manner.

I met

A Voyage to JAMAICA.

I met with this Plant in one of the *Caribe* Islands.
Colutea affinis fruticosa Pimpinella folio, siliquis falcatis bovinorum cornuum in modum dispositis. Cat. pl. p. 142.

This Branch was covered with a long *Ferrugineous* Wool, making it rough, under which was a hard Wood. It was frequently divided into Twigs which were set pretty thick with Leaves at about an Inch distance asunder, each of them being about an Inch long, and half as broad near the *Tab.* 16. Base where broadest, and from whence they decrease towards their ends to a *Fig.* 2. Point, being very deeply cut in on the Edges, and smooth, standing on *Tab.* 14. Foot-stalks. *Ex alis foliorum* come the Pods. They are always two, stand-*Fig.* 4. ing like Bulls Horns, being hairy, about an Inch long, crooked, pointed, round, and having within them several small Pease.

I found it in one of the *Caribes*.
Laurocerasi foliis siliquosa fruticosa. Cat. pl. Jam. p. 153.

The Twigs of this had under a thin green Bark a soft whitish Wood and large Pith, the Leaves stood at the Joints opposite to one another, and sometimes alternatively, having one third of an Inch long Foot-stalks, they were four Inches long, and about two broad near the middle, towards the farther end where broadest, being narrow at both extremes, having one mid-*Tab.* 16. dle, and some crooked transverse Ribs being equal, succulent, and like to the *Fig.* 3. Leaves of *Laurocerasus*. It had at top a short crooked pointed Pod, with two eminent Nerves on its Valves, parallel to the large one on its Back.

In grew in the *Caribes*, but where I remember not.
Arbor mali Persica Mameya dicta foliis subrotundis, acuminatis, ex adverso sitis. Cat. pl. Jam. p. 180.

This Tree had Twigs cover'd with a smooth reddish Bark, and Leaves coming out of the Joints by Pairs, set opposite to one another at about two *Tab.*7.*Fig.*2. Inches distance, having very small or no Foot-stalks, being about two Inches long, and one broad, near the round Base where broadest, and whence it decreased, ending in a Point, being smooth even on the edges, and having Veins running through its Leaf, very regularly after the manner of *Mammee*.

I found it in one of the *Caribes*.

The Bath here is taken notice of by some Travellers, as *Harcourt* and *Of the bat* Smith. The first says that it cures the Leprosie: is good in Coughs, it curing *Bath at* the Author, who drank and bath'd. It also remedies burning with Gun-powder, *Nieves*, and swell'd Legs. *Harcourt. Purchas.* 1282. The second tells us that it cur'd their Men in two or three Days, who were tormented with a burning Swelling, as scalding from the Dew of Trees, *Smiths obss. p.* 57.

That hot natural Waters are sometimes when cold little else than common Water, and used for it, is not only apparent from this, but likewise *Purchas*, who informs us that Seamen furnish their Ships at *Tidore* with Water hot at issuing out, but cold when it has stood. *Purchas*, 44. And that at *Dehage* is a Stream of hot Water, which put into a Vessel becomes cool and healthsome, and is so little differing from fair Water as to be drunk, and serve other uses to the Inhabitants instead and in want of it. *Elkington. Purchas lib.* 4. 523. Which, by the way, may let us see that 'tis not improbable that hot Waters owe their warmth to something without, and not mix'd in them.

The Miners being sent out in this Island found here a Mineral out of which Alum could be made. So *Hawks ap. Hakl.* 3. *p.* 469. tells us 'tis natural to *New-Spain*. And 'tis without question to be found in several places of *America*, though in most of them it will not defray the Cost of Making and Carriage to *Europe*, where is its great demand for the use of Dyers.

The Captain General of these Islands, which was Sir *Nathanael Johnson*, resided during our being here at *Antego*. This Island is not far distant from *Of Antego,* Nieves, and is thought to be on many accounts preferable to it. It was dis-

N cover'd

A Voyage to JAMAICA.

cover'd in *Colons* Second Voyage, and called by him *Santa Maria del Antigua*, 97. There were reckon'd Two thousand White Inhabitants on it at the time of our being here. It was told me when I was at *Nieves* that it was very difficult of Access; and dangerous for Ships, not well acquainted with the Shoals about it, to land there.

Of Barbuda. — *Barbuda* is depending on these Islands: it is small, mostly Pasture, and breeds great store of Cattle of all sorts, with which Colonel *Codrington* (to whom it belongs) keeps several Vessels that are always Trading with the other *Caribe* Islands. It has also Provisions more than sufficient for the Consumption of its Inhabitants. The Proprietor keeps a large Family for its Defence.

We came from *Nieves* Road on the Eleventh of *December*, in five Hours time to the old Road in St. *Christophers*, which is on its South side, and fortified with a strong Fort belonging to the *English*. This Island has no Harbour or Road on its North side. It has a ridge of Hills runs through its middle, lying East and West, as does the Island. There are deep Gullies, Rivulets, (or Torrents with steep Banks) on each side of these Hills. This Island has the best Water of any of the *Caribes* and the Inhabitants look whiter, less sallow, and are of finer Complexions than any of the Dwellers on the other Islands. It was inhabited (at the time of my being here) by *French* and *English*; the *English* being in the middle, and the *French* at both extremes. The Governor Colonel *Hill* Treated his Grace the Duke of *Albemarle*; and the *French* Governor hearing of his coming ashore sent him a Complement by an Officer. We sail'd along its South side, and came in sight of a great Hill which is called the Sulphur Mountain. It is bare, and I was told had great quantity of Brimstone in it. This Island was planted by Captain *Warner* with Fifteen People in 1623. *Smiths Obss.* p. 51. they living on Tortle. *ib.* There were in this Island strong Hedges made of that sort of *Acacia* described by *Aldinus*, and that sort of Flower called White Lilly in *Jamaica*, in abundance.

Of Statia. — We past in sight of St. *Eustache* commonly called *Statia*, a small round Island, some Leagues West, or to Leeward of St. *Christophers*. It belongs to the *Dutch*, who inhabit it, though it has had several Masters in time of War.

Of Saba. — After passing this Island we came in sight of *Saba*, which likewise is inhabited by the *Dutch*, though in War it has been several times attempted and taken from them by other Nations.

Of Santa Cruz. — The next Island was *Santa Cruz, Santa Croce*, call'd *Açay*, of *Martyr. Decad.* and *Ajay* of *Oviedo.* The *Caribes*, its Inhabitants, eat Men, and some of them were sent into Spain. *Oviea.* This is larger than most of these *Caribe* Islands: and has been formerly inhabited by the *Spaniards*, but now is by a small number of *French*. It lies East and West, and has a ridge of Hills runs through its middle.

Of Porto Rico. — On *Tusday* the Thirteenth of *December*, we came in sight of *Porto Rico*, by the *Indians* called *Burichena, Martyr. Decad.* and *Boriquen* by *Oviedo*. This Island commonly called by Sailers St. *Johns*, or St. *Juan de Puerto Rico*, lies West of this last. It is large and well inhabited by *Spaniards*, who have a Governor here. It has a ridge of Hills running through its middle East and West likewise. Sir *Richard Greenvile* who landed and fortified on this Island, p. 151. left it because of the *Muskites* stinging them there, *ib.*

Of Mona. Of Hispaniola. — The next Isle we sailed by was *Mona* a small Island East of *Hispaniola*, after which we came to Sail along the side of this last mention'd Island. It is very long, and has a ridge of high Hills cover'd with Wood, goes through it East and West, as it lies it self. It has lying off of it *Alta bela*, famous for Tortles, where are a great many Eggs laid by them in the Sand, which are there hatch'd. To the South, and near the West-end is another Island called *Isla de Vac.* — *Isla de Vacas, Isle de Vaches*, or corruptly by the *English* Isle of *Ash*: at this place

A Voyage to JAMAICA.

place the *English* used to fish for Tortoise, &c. But now it is inhabited by the *French*, they pretend to hinder them.

The *French* inhabit one half of *Hispaniola* or St. *Domingo*, and are possess'd of that part of it which lies to the North, the *Spaniards* that to the South. It had according to *Casas* three Millions of Inhabitants, whereof not two hundred remain'd in his time. It was called *Quirqueia* and *Haiti* by the Natives, the first from its Vastness, the last from its Roughness. *Martyr*. It was called likewise *Cipango* from its Gold Mountains, and by the *Spaniards* first *Isabella* from the Queen, then from *Hispania*, *Hispaniola*, and was counted Four hundred and fifty Miles long, and Three hundred broad by *Martyr*. And One hundred and fifty Leagues long, and Eighty broad by *Oviedo*. It is in 18 or 20°. Lat. *Id*. There was One Million two hundred thousand Inhabitants on *Hispaniola* at first, who being not used to Labour, were kill'd most of them by it. *Martyr*.

Of Hispaniola.

Near *Jamaica*, (which is Seventeen Leagues West from *Hispaniola*,) we met the *Faulcon* Frigat sent from that Island to wait for us; in a little time we came in sight of *Morant*, the most westwardly Point of *Jamaica*; and on the Nineteenth of *December* came into *Port-Royal* Harbour.

A

THE
Natural History
OF
JAMAICA.

BOOK I.

OF THE
Plants of *JAMAICA*.

CHAP. I.
Of Submarine Plants.

Hose growing in the Seas about *Jamaica*, may be divided into such as are of a stony Substance, as Corals, under which are comprehended Pores, stony Mushromes, &c. those of a woody, or horny and tough Substance, with a coralline Incrustation; and those of an herbaceous or soft consistence, as the *Fuci*; but they have this common, that all of them smell very fishy, or strong of the Sea.

They are likewise saltish to the Taft, and for the most part, upon drying, there sticks to them a White Substance, chiefly made up of Sea-Salt; and from this Concretion, or rather Salt, (without the addition of any *Alkali*) sticking to the Leaves of the Delisk (the humid parts of the Salt-water being evaporated) it is that that Plant is made delightful to the *Irish* Palats.

Most of these Plants vary very much in their Colours, not only those Sea-Shrubs, which have incrustations over them, when in their Maturity or State, being sometimes Purple, or White on the same Branch; but likewise, being thrown on Shore by the Waves, dashing against Stones, &c. they lose soon, most, or all of their Crust, and change their Face extremely. The same cause makes all Corals here vary very much, being extremely White as Snow, if they have lain on the Shore, exposed to the injuries of the Air, and dashing of the Sea, for some time.

But

But the most common variety of these Submarine Plants comes from the *Conferva Marina*, and Froth of the Sea, which makes them, by being cast on them, and dried with them, of a very Green, or dirty Yellow Colour.

Several of this Tribe, and more than of others, have been in *Europe*, they being to be met with at the first landing of Seafaring Men, and not growing in the inland parts, whither Sailers seldom go; add to this, that they are for the most part from their salter, harder, and drier Consistence, easilier kept from Corruption, and by their Beauty more tempting than other Vegetables; so that the Cabinets of the Curious, Shops of Drugsters, Surgeons, Barbers, and Apothecaries have abounded with greater Varieties of this, than any other kind.

The most part of Corals themselves having striated Lines, and many of them starry Pores, which are put by most Authors for the Characteristicks of Pores, as distinct from Corals ; I shall not give those Corals I met with in *Jamaica*, the names of Pores which seems to belong to most, if not all of them to be found here; but only call them *Corallia porosa*, to distinguish them from the more solid.

I shall put the *Corallines* with the *Frutices Marini*, or Sea-Shrubs, which are made up of a horny, tough Matter, with a Coralline incrustation, because most of them have a Nerve or Thread running through them, which although the last herein described seem to be quite destitute of, yet because of its not being erect, and smallness, it shall be reckoned by me a Coralline.

For the name *Spongia Spuria*, I make use of it here to signifie any fibrous, reticulated spongy Body, growing in the Sea, and thrown up by the Waves, which notwithstanding it be very like in many things to a true Spunge, it is yet harder ; so that 'tis not easily pressed, nor has much elasticity, neither doth it suck up Water as that does, which are the marks of true Sponges. These Bodies would by some be called *Alcyonia*.

The way of Propagation of this whole Tribe, which lies much from our Sight or Observation, has been very obscure, it seems to be different in the several Kinds of them, as to the harder or stony Plants, there is some account of their Propagation, in giving a Description of the *Astroites*, or star Stone : there is likewise somewhat of the growth of Corals, in the account of some of those I found growing; and as to the *Fuci*, or those of a softer or herbaceous Consistence, their Seed has been discovered, (and shewed me first) by the Industry of the Ingenious Herbarist, Mr. *Samuel Doody*, who found on many of this Kind solid Tubercles, or risings in some Seasons, wherein were lodged several round Seeds, as big as Mustard-Seed, which, when Ripe, the outward Membrane of the Tubercle breaking, leaves the Seed to float up and down with the Waves. This Seed coming near Stones, or any solid Foundation, by means of a Mucilage it carries with it, sticks to them, and shoots forth *Ligula* with Branches, and in time comes to its Perfection and due Magnitude.

Tab. 17. Fig. 1.
Tab. 18. Fig. 1.
Tab. 19.

1. *Corallium asperum candicans adulterinum. Cat. pl. Jam. p.1.* J. B. *l.39. c. 33. Raii hist. p 62. Gypsum Coralloides Boet. de Boot. p.321. Mus. Swam. p. 19. Corallium fossile exalbidum, pennam gallinaceam crassum, ramulis aliàs pluribus, sapius binis aliàs ampliatis.* Luid. *lith. Brit. p. 6. No. 92, 93. Tab. 3. No. 92. An Corallii albi species minima duas uncias lata, è Scopulo gypseo enata. Mus. Swam. p. 17 ?*

One Kind, or Variety of this, *Tab.* 17. *Fig.* 1. was broad at Base, about two Inches, and about one quarter of an Inch thick, three or four Inches high, whitish, smoother than any of the other Corals I met withal here, though a little rough, and having some few Pores, solid, and white within when broken, continuing

continuing broad to the end, which were round tips like those of Harts-horn, half an Inch long, standing out from the margin of the broader part every way, which was mostly flat, though sometimes branched; these points or tips, which are sometimes crooked, growing larger, united, and having their Interstices between them filled up, make the broadest part of this Coral, it appearing to be nothing else than the tips join'd together, and having holes in those places where the space between them is larger, and not so soon fill'd as the others, something resembling those round Cakes of Bread made by the *Jews* at *Easter*. And of this, as to manner of growth, there is very great variety, the chief mark of it being from its colour and solidity; the most remarkable varieties being in the under parts of it being broad or roundish, the Interstices filled up or not, and its round Branches growing parted from one another every way, or being extended one way.

It is to be found on most of the Shoals about the Island, and very plentifully on the Northern Coast of *Hispaniola*, where the Silver Ship was wreck'd, much of this kind having been fished up with it.

There are many varieties of this growing on Sea Shrubs, of various shapes and figures, but all agreeing in being more stony and solid than the other Kinds, although it be porous within, and rough on the outside. *Tab.* 18. *Fig.* 1.

One sort of it I have growing all round, a common Glass Bottle, this Bottle was by chance or Earthquake, thrown into the Sea in *Port-Royal* Harbour, and thence taken up by a Diver, whom the Reverend Mr *Scambler* sent to the bottom of the Sea to take up Money and Goods he had lost in the great Earthquake. It is figured. *Tab.* 19. This shews the Seed of Coral to be liquid. I think the kind of *Fossil* Coral found in *Oxfordshire*, and mentioned by Mr. *Luid. Lithophilae Brit. p.* 6. *No.* 92. & 93. seems to be this. It is there figured, *No.* 92. *Tab.* 3.

This is pretty well figured by *Gesner*, and *J. B.*

II. *Corallium Album pumilum nostras. Cat. nost. p.* 1. *Raii hist. p.* 62. *Tab.* 18. *Tab.* 18. *Fig.* 2. *Fig.* 2.

I could not find any great difference between the Coral found on the West Coast of *England*, and this in *Jamaica*, and very little between it and one I had given me by Mr. *George Handisyd*, who brought it from the Streights of *Magellan*, with many other Natural Curiosities: this last was a little higher, more branched, less crooked, and slenderer: That I gathered in *Jamaica*, was not over an Inch high with many very crooked Branches, smooth and solid: it grew on Rocks, Shells, &c. in the bottom of the Sea, as also in roundish lumps or masses, whence it is sometimes cast on Shore. It is used in the *London* Shops for White Coral; and *Gerard* describes this *English* one, giving the figure of the White Coral, taken from *Lobelius*'s *Icones*.

This sort which grows in *Jamaica*, is found about *Falmouth* in *England*, not growing (as I have heard) but thrown thither by the Storms, broken to pieces.

III. *Corallium album porosum maximum muricatum. Cat. pl. Jam. p.* 1. *Tab.* 18. *Tab.* 18. *Fig.* 3. *An* White Coral of *Davis. Purchas, lib.* 3. *cap.* 1. *sect.* 5. *p.* 135. & *Fig.* 3. *lib.* 1. *cap.* 6. *sect.* 2. *p.* 448. & 449? *Corallium album Linschot, cap.* 92.? *vel de Bry,* 3. *Part. Ind. or. cap.* 1. *p.* 7? *Corallium album non fistulosum medulla intus radiata. Mus. Swammerd. p.* 17?

This from a stony and yellowish root or base, somewhat like in substance to Freestone, sends up several Stems as large or larger than ones Thumb, a Foot and a half high, of a yellowish White Colour, having several crooked

Branches

Branches join'd or anastomos'd frequently one with another, tapering towards the top, and ending milk White like the tip of a Horn, this end under water, was always White, as well as soft, and contain'd a milky viscid Juice, smelling very strong of the Sea. The outward Surface of this Coral, was, from what we may call the Root, upwards, rayed, striated, or waved by many Lines on its Surface, appearing very well to the Microscope, between which were long vacuities, or empty spaces, and all along there were a great many asperities, muricated prickles, or small eminencies, likened to the Leaves of Lavender-Cotton by *Clusius* in his description of the lesser sort, hollow on the upper side, and convex below; so that they are fit to hold Water, or any Substance for nourishment of the Plant, and within these small Tubercles are many Lines, going from the Center round to the Circumference, as in several other of the Corals, and this starry radiated cavity grows lesser, going into the middle of the Branch, which has altogether from bottom to top, in lieu of a Pith, such a radiated Cavity in its middle, on the sides of which is a stony Substance amongst which lies many Pores, interspers'd without any order. In time these under Stems and Brances have their Pores fill'd, and their Tubercles levell'd, with a stony matter filing them up, and making them look quite of another Face and Consistence, and of a yellowish or greenish Colour without, which is altogether solid, without visible Pores, and stony; and it is observable that if any tops of Coral have been broken accidentally, they are as they ly prostrate on the tops of others, consolidated and united very firmly together, and with those under them, as it were by some sort of Cement.

This I have found frequently in the *Caribes*, and in *Jamaica*, in shoal Waters, or rocky Banks, where they are produced in very great abundance, and look, when one is over them in a Boat, with their white tips, just like Stags Horns, and if taken up they smell exceedingly of the Sea.

In *Columbus*'s second Voyage, with their Anchor points, in St. *Martin*'s Road, they took up Coral, which gave them good hopes of other things: that being thought a rich Commodity, which I am apt to believe must be meant of this, as also that which *Knivet* mentions to grow in *Brasile* by Cape *Frie*, this being the most common Coral in the *West-Indies*.

Having seen some pieces of Coral, in Mr. *Charlton*'s curious Collection, which came from the *East-Indies*, that as to Substance, Surface, Colour and Magnitude, agreed with this, I am apt to believe, that his from the *East-Indies*, is the same with what I found in the *West*, therefore doubt whether this be that mentioned by *Davis* to grow about *Madagascar*, and *Baixos das Chagas*, and by the same Author, as well as *Lancaster* and *Hawkins*, to be found on the Shoals about *Sumatra*, and by *Payton* at *Priaman* and *Tecoo*; *Linschoten* near *Mozambique*, and by Hatch near *Jacatra* or *Batavia*, in the first of which places 'tis observ'd to cut the Cables of Ships, when they ride out of the Rivers Channel, where is Oze; and at the others to cause or grow on dangerous Shoals, agreeable to what *Hughes* observes, that Boats are lost on it at *Jamaica*. I am likewise of opinion that this is what *Pyrard de la Val* tells us, grows about the *Maldives*, hindering the Inhabitants wading from Isle to Isle, and which they beat with small Stones, and boil with their Cocos Water making their Honey and Sugar (called *Jagra*) form, (or Kern) using it as temper for the Sugar, of which more hereafter.

There are great varieties in this, brought both from the *East* and *West-Indies*, one of which having lain expos'd to the Waves, being dash'd or tumbled by them, and so freed from part of their outward Coat, is what I suppose *C. B.* calls *Corallii albi varietas aspera & striata*.

IV. *Coral-*

The Natural History of Jamaica. 53

IV. *Corallium porosum, album, minus, muricatum. Cat. pl. Jam. p. 2. Tab. 17.* Tab. 17. *Fig. 2. & Tab. 18. Fig. 4. Substantia lapidea frutic̄em Corallii albi repræsen-* Fig. 2. *tans minutissimis punctis & porositatibus scatens, Mich. Rup. Bess. Gaz. Corallii* Tab. 18. *albi species, Clusio dicta planta saxea abrotonoides, Musf. Swammerd. p. 18. Co-* Fig. 4. *rallii albi species gypsam materiam referens atque adeo Corallium gypsoides dicendum cum base gypsea ac magna. Ejusd. p. 19. An perus marinus Coralleides seu pseudo Corallina. Velsch. hecatost. p. 26?*

This, in every thing, is like the former, only 'tis not over six Inches high, the Stalks or Branches, are no larger than a Man's little Finger, but more numerous, and thicker together, and the under part, or what, for analogies sake, may be term'd, the Roots, or rather old Stalks petrified, are larger and longer.

It is to be found with the former very plentifully, off of *Pelican Point* near *Port Royal*; and if it be the same with what is mention'd by an Anonymous *Portugal* of *Brasile* in *Parcheu*, to grow in *Brasile* copiously, is there made into Lime. The great Naturalist Mr. *James Petiver*, my very good Friend shewed me this lately sent him from the *East-Indies*.

I suspect the above described Coral, clear'd of its muricated prickles by the Waves, &c. to be *Corallium octavum sive album stellatum minus. C. B. pin.* though, if the figure of it be true (being hollow) it cannot be the same.

V. *Corallium, porosum, album, latissimum, muricatum. Cat. pl. Jam. p. 2. Tab.* Tab. 17. *Tab. 17. Fig. 3. Corallii alii species maxima, ramis latissimis ac compressis, duo-* Fig. 3. *rum ferme pedum altitudine, latitudine trium; si digitis pulsetur sonum metallicum edens. Musf. Swammerd. p. 19.*

This looked rather like a flat Stone than Coral. This piece I describe was about half Inch thick, at bottom eight Inches broad, expanding or extending its self by degrees, being two Foot long, and broadest at top, where it was two Foot broad, and divided as it were into several *Laciniæ*, lying now and then over one another, which being struck, give a metallic sound. The Surface of this, towards its top, is very rough with small muricated hollow Tubercles, and now and then some small tips, which are also muricated and White at top; the inward Substance is more solid, ponderous and white than that of the before described Corals. A great many of these Stones, coming, as it were, from the same Root, inclose one another somewhat after the manner of the *Squama* of *Bulbous* Roots, or rather the *Petala* of Roses, only they stand sparse, at distance one from another, especially their broad ends, so that under water they look like the Leaves of a Book expanded.

It grows in very many shapes, the *Laciniæ* being sometimes hollow something like a Tunnel, and at other times rounder, and the edges divided into variously shap'd; but for the most part very broad, flat, parts or Segments.

It is to be met with off of *Pelican Point* by *Port Royal* Harbour.

VI. *Corallion minimum capillaceum. Cat. pl. Jam. p. 2. Tab. 20. Fig. 1.* Tab. 20.

This seems the same with the Corallines for substance, smell, colour, &c. Fig. 1. only is not near so large, neither is it discernibly jointed, it is not much bigger than a Hair, and seems to have no thread within, but smells of the Sea, crackles under the Teeth, stands more upright, as Coral, and not bending, because of the want of joints, in which it differs from the Corallines, being otherwise very like the *Corallina minima capillacea*, mentioned hereafter. It is figured *Tab. 20. Fig. 1.* both as it appears to the naked Eye, Tab. 20. and by a Microscope. Fig. 1.

It was thrown upon the Sea Banks near the Palisadoes by *Port Royal*.

P VII. *Lapidis*

VII. *Lapidis astroitidis sive stellaris tertium genus.* Boet. de Boodt. l. 2. c. 146. fig. B. Cat. p. 2. *Astroites undulatus, Musf. Swammerd.* p. 6. *Astroites undulatus major concavus, capitium reprsentans, duorum pedum circumferentiam habens, Ejusd. ibid. An hydatites sive Cymatites. Velscb, hecatost. p. 60. Tab. 21 ?* Brain Stones.

This is frequently much bigger than ones Head, roundish at top, or a Hemisphere, having many undulated Furrows on its top, imitating somewhat the *gyri* or *anfractus* of a Man's Brain, from whence they are commonly call'd Brain Stones; these Furrows have Lines go from their middle to each side, they are White, and make as it were high Banks in the Stone, which are sometimes waved like the Sea; from whence the Name *Astroites undulatus.* There is some variety in this as to the bigness of the *Sulci.*

They grow in all the Seas about this Island, and vary very much as to bigness and manner of growth, having very often within them small White Shell-Fish. This is mentioned to be found by my good Friend Dr. *Charles Leigh,* in his Natural History of *Lancashire, &c.* Tab. 2. Fig. 1.

VIII. *Lapidis astroitidis sive stellaris primum genus.* Boet. de Boodt. lib. 2. cap. 146. Cat. p. 2. *Pietra Stellaria di Georgio Transfer. Lapillus in India ? Capite draconis erutus. Marsil. Pacin. de vit. caelitus comparanda, lib. 3. cap. 15. Lapis Stellaris, Card. de rer. varietate, lib. 7. Draconites sive dracontias. Agricol. lib. 6. de Nat. fossil. Astroites distinctissime Stellas amulans, Musf. Swammerd. p. 6. Astroites Gesner. de Fig. lap. p. 35. Stellatus lapis Aldrovand. Musf. Metall. p. 872. Fig. p. 877, 878, 879. Astroites* or starry Stones of Dr *Plott's* Natural History of *Oxfordshire,* p. 87. Tab. 2. Fig. 6, 7, 8. *& p.* 130. Tab. 8. Fig. 2. *Millepora Imperati* p. 720. *sive Astroites ligneus elegantior stellulis totam massam permeantibus Luid. Lith. Brit.* p. 9. No. 160. Tab. 2. *Lithostrotion sive hasaltes minimus striatus & stellatus, Ejusd.* Tab. 23. *An Astroita congener radularia cretacea, Ejusd.* p. 10. No. 176. Tab. 2 ? & Tab. 3. No. 98 ? *Asterias Cardioides Major. Velsch. hecatost.* p. 60. *Asterias stellis majusculis. Asterias stellulis minoribus. Ejusd. Cometites. Ejusdem. An Rhodites, Ejusdem ? An Asteria confuse Stellas reprasentans, Ejusd. ibid ?* Star-Stone, Tab. 21.

This is for Colour, Substance, *&c.* exactly the same with the former, only instead of undulated *Sulci* or Furrows, are only small holes, pipes, or spots, in whose cavities are *Stria;* or partitions coming from the same Center to the Circumference.

This is as frequent in all these Seas as the former, and is found in *Europe* in several places at Land, petrified and not petrified as Sea-Shells. Mr. *Charlton* gave me a piece of it capable of polish, petrified and transparent like an Agat, only the *Stria* were White. These Stones are most certainly bred from a Seed; for in the places where they are most to be found in shallow Sea-water, I have seen what we may call their Seed or Spawn very frequently, *viz.* a mucilaginous, crystalline, clear Body, of the same shape, and with the same spots on their Surface, only no *Strie* going from the Center of the holes, which when taken up in the heat, does out of the Water corrugate and contract its self into narrower dimensions, turn *Opaque* and of an Ash, or pale Yellow Colour, in which notwithstanding may plainly be seen the rudiments of the Stone.

These *Astroites* Stones are taken up from the bottom of the Sea by Divers, and brought on Shore not only to burn with Timber to make Lime; but likewise to build, and that not only ordinary Houses, but Forts to resist the battery of Cannon.

This

The Natural History of Jamaica. 55

This Stone moves if put into Vinegar. The Powder of it, (according to some Natural Historians) to four Grains, is good for the Pestilent Contagion and Worms, against which last it is so effectual, that worn it hinders the breeding of them if it touches the naked Body, and keeps off Tremblings, Apoplexies, and other sudden Diseases. It brings Victory to those who carry it, and helps Diseases of the Liver and Lungs, and cleanses the Blood: hang'd up in a Room it hinders the coming thither of Spiders, or other venemous Animals. The whiter, with the Stars well formed, are your best, and are set in Rings worth more or less, according to the fancy of the Buyer.

This Stone being, before its Original was known, brought by Empyrics to *Florence*, was pretended to come from the East, and that there it was generated in the Head of a Dragon, having Stars on it; it was supposed to receive great Vertues from the *Cauda Draconis*, or Constellation of that name, famous amongst Astrologers, and therefore was held to be very good for many Diseases, as well to preserve from them, as to Cure them, in so much that one *Transfer* in his Bills at *Vienna*, besides the Vertues abovesaid, gave out, that worn it was able to preserve from the Gout, comforted the Sight, generated Hair, hinder'd Miscarriage, expell'd the After Birth, kept away Fear and fearful Dreams, clear'd the Understanding, and comforted the Melancholy, helpt the Falling-Sickness, and kept from being hurt by any wild fierce Beast, &c. Which all of them deserve no Credit, being built on Fancy, or desire of Gain: only the Powder of it, as of Coralline, may reasonably be supposed to Cure Worms.

This kind of Coral is very frequently found in Quarries, and near the Surface of the Earth in several places of *Europe* and *England*. There are many varieties of it to be found growing under Water, and cast upon the Shores of the Seas adjoining to *Jamaica*, and the *Caribe* Islands. There is no difference between those to be found in *Jamaica* and *England*, excepting what may be easily accounted for, from that Fossil in *England*, its having lain long in the Earth, whereby some of its Substance is sometimes lost, and at other times its pores and cavities are filled with earth, sparry, or flinty Matter. This Stone I have some pieces of by me, that were given me by Mr. *Beaumont*, which are as hard, diaphanous, and capable of Polish as any Flint or Agat. (*Tab.* 11. *Fig.* 1, 2, 3.) shews the Original of this Astroites, when beginning to grow and sprout up from the bottom of the Sea. 'Tis first a Jelly like the white of an Egg or faln Star, then grows more *Opaque*, and if drove ashore, or taken up in this state, comes to corrugate and dry up into these forms. (*Fig.* 4, 5, 6, 7, 8.) shews varieties, and different views of the Stone or Coral its self, when come to its full ripeness, state and maturity. (*Fig.* 9.) shews also one of these Corals or Stones taken up at *Faringdon* in *Barkshire*. This Stone has lost all the Rays in the starry Pipes or Cavities, save one which yet remains intire, but it retains some part of them at the Circumference, which makes the Rays or *Striæ* in the sides of those cavities. It is also more ponderous from the under part, which is filled with stony and earthy Matter. This is to be found in the place abovementioned in great plenty. (*Fig.* 10, 11.) shews the same Stone turned into a Flint, which yet retains its Figure as to be easily seen, looking on its top to have the Stars, and long-ways to have the *Striæ* or Rays this Coral has. This polishes as Agat, and is as hard and diaphanous as it.

This Stone or Coral has likewise been found in *Oxfordshire* by Dr. *Plott*, and by *Aldrovandus*, I think, in *Germany* and *Italy*, and *Bellonius* says they build with them in some places of *France*. In *Lancashire* by Dr. *Leigh*, as appears by his History of that Country, *Tab.* 1. *Fig.* 5. and by Mr. *Luid* in *Gloucestershire*, as appears by his *Lithophylacium Brittanicum*.

It is very strange that so much of this, which is only now naturally to be found growing in hot Climates, should be met with in such plenty dug up in other places of *Europe* and *England*, and that under the Surface of the arable Earth, in such plenty and manner, as if it had sometimes in former Ages naturally grown there, which must have then been not only warmer, but covered with Sea-water.

IX. *Lapidis astroitidis sive stellaris secundum genus. Bort. de Boodt. lib. 2. cap. 140. Cat. p. 2. Lapis sigstein à Germanis dictus, multas in corpore suo veluti rosas, à natura pictas ostendens. Gesn. de lap. Fig. p. 38. Astoria rosam referens. Muf. Swammerd. p. 6.* Sea-Rose.

It is found on the Shoals with the other Kinds of *Astroites*.

X. *Fungus lapideus major undulatus. Cat. p. 2. Tab. 18. Fig. 5.*
The upper part of this *Fungus Lapideus*, or Coral, was an Hemisphere, or rather half of an Oval Figure, almost as big as ones Fist, having on each side of it large strait cavities able to receive ones Finger, just like the *Lapis Astroites Undulatus*, made up of the *Striæ*, after the manner of those Stones, only larger, underneath it was likewise striated, after the manner of the other *Fungi lapidei*, each of the *Striæ* coming from the same *Petiolus* or small Footstalk as from a Center, and running to the Circumference.

I found it on the Shore with the others of this Kind.

XI. *Fungus lapideus minor. Cat. p. 2. C.B. pin. 375.*
This is frequently found on the Shores of this Island.

XII. *Frutex marinus Elegantissimus Clusii. Cat. p. 3. Besl. Bester fascic. Frutex marinus major. Ejusd. ibid. Frutex marinus elegantissimus Clusii purpurascens, flavescens, & cortice suo denudatus. Muf. Swammerd. p. 21. An Corallina reti-formis, purpurea, ramosa nervis tenuissimis. Plukenet. Almag. p. 118?*

I can add very very little to *Clusius* his Description but only this, that sometimes from the middle or lateral Branches come smaller fans, so that 'tis not always plain, but has several other Branches or small fans, rising out of both sides of it, which is taken notice of in one belonging to the *Royal Society*, by Dr. *Grew*.

It frequently grows on the Rocks at the bottom of the Sea, in all the hot *West-Indies*, and sometimes on Shells, which together with them, not being a firm Foundation, are thrown on all the Shores of this Island.

This, as I think, being clear'd by the Waves, Stones, and Rains, of its outward Coralline Crust, which is sometimes Purple, and sometimes White, makes quite a different Figure, and so gave occasion to *Clusius* to make it a different Plant, and from him *Caspar Bauhine*.

'Tis used to Fan the richer sort, to keep them cool while they eat or Sleep. *Hughes*: or to keep away Gnats, Muscato's, or Merrywings.

XIII. *Corallina fruticosa humilior & crassior ramis quaquaverfum expansis superficie tuberculata. Cat. p. 3. An quercus Marina Theophr. Baf. Best. fasc: Corallina cinerea tuberculis incrustatæ ramis flexilibus. Muf. Swammerd. p. 22 a*

The heighth of this Coralline Shrub was about six Inches from a solid Foundation, about the bigness of ones little Finger, being branched every way to tapering ends like Corals. The Crust was generally whitish, friable, having its Surface rough, and was of the thickness of a Crown Piece; in the middle of which was a horny substance, blackish and tough, as in other of these Coralline Shrubs, the *Icon* of the *Plantæ Marinæ Resedæ facie*,
Clus.

Cluf. agrees pretty well to this. I have sometimes thought that this may perhaps be the following Coralline not grown to its due magnitude.

It grew in the Sea every where about this Island, whence it was thrown up by the Waves with other Recrements of the Sea.

XIV. *Corallina fruticosa elatior, ramis quaquaversum expansis teretibus. Cat. p. 3. Tab. 22. Fig. 1, 2, 3. An Corallina pulcherrima, Corteniana, seu arbuscula marina Tab 22. coralloides, ramosissima foliis teretibus, abrotani fœmina instar verrucosis. Plukenet. Fig. 1, 2, 3. Almag. p. 118 ?* Sea Feather.

This has a two or three Inch long diameter'd broad woody Pedestal, spread on any Stone or stable Body in the bottom of the Sea, from whence rise Stalks about two, three, or four Foot high, tough, woody, as large as ones Finger, round and divided into several small Twigs spread on every hand, of about three or four Inches long. All of them are incrustated with a Coralline Matter, of a yellowish, Purple, or whitish Colour, having some small holes, or asperities in its Surface, and very often *balani*, or the *corallium album candicans adulterinum* J. B. sticking to them. I suspect the Trunk and Branches of this, cleared of their Crust, may sometimes pass for black Coral, making some of those kind of Plants called by Mr. *Thornefort* in his *El. Bot. Lithophyta.*

It grows on the Rocks at the bottom of the Sea, and from thence the whole, or some side Branches, by the tumbling of the Sea are cast ashore, where by the Waves 'tis either wholly clear'd of its incrustation looking like a Shrub, without Leaves, or partly clear'd, looking yellowish, and tasting very salt.

Tab. 22. Fig. 1. Shews this whole Plant contracted, or less than the natural bigness. *Fig. 2.* A Branch of it to the natural bigness, and *Fig. 3.* a small Twig of that Branch.

XV. *Corallina fruticosa, ramulis & cauliculis compressi, quaquaversum expansis, purpureis elegantissimis. Cat. p. 4. Tab. 22. Fig. 4.*

This, which appeared to be only part of the Plant, had its inward Stalks and Branches about a Foot high, being roundish, blackish, lignose, tough, frequently branch'd, and covered with two *Lamina*, or Plates of a Coralline substance, and most elegant Purple Colour, though in some places 'tis whitish, of a saltish tast, and Sea or Fishy smell, as others of this Kind; the Plates or incrustation of this are about the thickness of a Sixpence, and from the breadth of half an Inch it has at bottom, decrease towards their tops, and look somewhat like to the Lead wherewith Glasiers join their Pannels of Glass one to another in making their Windows.

I found it thrown upon one of the Cayos off of *Port Royal.*

XVI. *Corallina opuntioides, ramulis densioribus, & foliis magis sinuatis atque corrugatis. Cap. p. 4. Tab. 20. Fig. 2. An Corallina scutellaris alba, rosarii instar perforata, Plukenet. Almag. p. 182 an Corallina & opuntia Jamaicensis, cum orbiculis plurimis quasi pendulis per siccitatem nigris, Ejusd. ibid?*

This seem'd to be in every thing the same with the *Scutellaria sive opuntia marina*, J.B. one Leaf growing out of another, after the manner of *Opuntia*, only the Branches were in this more numerous, and the Leaves on their Convex or upper part more corrugated, and on their under part more concave or sinuated, whereby it really differs from that of *J. B.* The Leaves were tied together by a Thread made of many Filaments collected together, somewhat like Thread made of Flax, and the Leaves were harder and more stony, breaking between the Teeth, and tasting Salt, smelling likewise of the

Q   Sea.

Sea. It is sometimes green or blackish by *Conferva* froth, or other substances sticking to it.

This Coralline was very plentifully thrown on the Sea Banks every where upon this Island.

Tab. 20. *Fig.* 2. Shews it both in its natural bigness and magnifi'd.

XVII. *Corallina major, nervo crassiori fuciformi, internodia breviora nectente. Cat. p.* 4. *Tab.* 20. *Fig.* 3. *Corallina fistulosa Jamaicensis, candida cum internodiis brevissimis & quasi filo trajectis nostratibus,* White Bead Bandstring *dicta, Plukenet, Almag. p.* 118.

This was very like our *English* or Shop Coralline, only larger, and the joints much shorter, appearing something like old fashion'd Bandstrings, it was longer likewise than it, and not so brittle, saltish and smelling of the Sea. It has a String as big as a coarse Thread in its Center, of the same substance with the *fuci*, being of a yellowish Colour, roundish, sometimes plain, sometimes curled. Properly speaking, these Corallines are rather incrustations by a Coralline Matter, than any thing else, this being a White Coralline, roundish, with short Zones, like small Beads strung on a nerve.

It is thrown on the Banks of the Sea by the Palisadoes near *Port-Royal*.
Tab. 20. *Fig.* 3. Shews it in its natural bigness, and magnifi'd.

XVIII. *Corallina nervo tenuiori, fragiliorique internodia longiora nectente. Cat. p.* 4. *Tab.* 20. *Fig.* 4.

This has a great many Stalks and Branches coming from the same spongy Root, sticking to the Stones in the bottom of the Sea, they are of the bigness of Shop Coralline, and about two Inches long, spread on every hand, and made up of several long joints, being white, polite, brittle, saltish, and smelling of the Sea, different from our Shop Coralline, in that the joints are much longer, and that it has within it a brittle Thread, on which the joints are strung like Beads.

It grows with the former.

In *Tab.* 20. *Fig.* 4. It is figured in its due bigness, and magnifi'd.

XIX. *Corallina minima capillacea. Cat. p.* 4. *Tab.* 20. *Fig.* 5. *An Corallina fossilis capillaris, Lnid. lithophyl. Brit. p.* 7 ?

This was the least of all the Corallines I here met with, it was about two Inches long, having Branches and Twigs in very great plenty, being not ordinarily jointed to appearance, for the most part smooth, very white, no bigger than the Hair of ones Head, if so big, smelling strong of the Sea, and crackling under ones Teeth, as the others of this Kind. Sometimes this is jointed.

It was thrown up with the other Corallines on several Banks of the Sea round this Island.

Tab. 20. *Fig.* 5. It is represented in its due bigness, and magnifi'd.

XX. *Fucus marinus vesiculas habens membranis extantibus alatas. Cat. p.* 4. *Tab.* 20 *Fig.* 6.

This has a dark coloured, tough, roundish, crooked Stem, about nine Inches high, having many crooked Twigs very thick set, with Bladders full of Air. The Bladders themselves are roundish, or rather Triangular, having an extant *foliaceous* membrane at top incircling it, and three other extant *ale* underneath, making it look Triangular, or something of the shape of a Funnel, being pyramidal, and of a dark brown Colour, or blackish when dry'd like Glew, smelling strong of the Sea, and tasting salt like other *Fuci*.

Besides

Besides it has several round small Protuberancies over its Surface, supposed to be the Seed.

It grows on the Rocks, covered with the Sea, on all the Coasts of this Island.

XXI. *Lenticula marina serratis foliis. Cat. p. 4. Lob. olf. p. 653. Sargassa de Bry Ind. or. part. 5. p. 40. Zargosso Musf. Swammerd. p. 23. Acinara, agresto marino, terza specie. Imper. p. 645.*

This is usually about a Foot long, having tough, small, dark brown, or blackish Stalks, on which come several Inch and half, or two Inches long Leaves, being not over the eighth part of an Inch broad in the middle where broadest, serrated about the edges, being of a dark brown Colour like other the *sui*, it has many round air Bladders, coming out from the Stalk on small Footstalks, like in magnitude, shape, &c. to Lentils, which give it the name.

It grew on the Rocks about this Island, whence it is thrown upon the Shore, and carried with the Currents through the Gulf of *Florida*, all along the Coast of North-*America*, in great abundance, where I gathered it, and with it took up several small Crabs which were alive, and which shall be described hereafter, in their proper place.

Fernan Colon, in the life of his Father *Christopher*, *f.* 29. tells us that in the year 1492. in the first Voyage for *America*, that they were very much frighted in meeting with this, which was so thick as to retard their sailing, judging by it that they were not far from Rocks or Land, especially when they found a live Crab (*Gambaro vivo*, ill translated in *Purchas* a Grashopper) among it. *f.* 74. he farther says that they met with this in the Bay of *Samana* in *Hispaniola* near the Land in Shallow Water, from whence they supposed what they met with before, had been when ripe separated and carried into the Sea. And *f.* 75. he relates that in their first Voyage they found of it two hundred and sixty three Leagues West of *Ferro*, whence in their return they conjectur'd, while they sail'd amongst it, that they were not so near that Island as one hundred and fifty Leagues, though in their second Voyage. *ibid. f.* 93. they found none of it four hundred Leagues West of *Gomera*. He says also, *f.* 110. that it was to be found between *Cuba*, *Jamaica*, and the West end of *Hispaniola*.

Oviedo's account of it is, that 'tis found in several places, and quantities according to the Seasons, Winds, and Currents, even sometimes half way to the *Indies*.

Thevet and *de Lery* mention, that in their Voyage from *Brasile* for the space of fifteen days, they met with so much of this under the Tropic of *Cancer*, that they were forc'd to cut their way through it, though on their being apprehensive of Shoals, and sounding with their Lead, they had no ground by fifty Fathoms Line. The like says *Acosta*, adding that as they clear'd the bundles of it from the Ship more rose (as they supposed) from the bottom of the Sea. *Linschoten* relates much the same, and that he found it off the Cape *Verd* Islands, forty Leagues from Land, from 20°. to 24°. of Latitude. *Welsh* in his Voyage observ'd it off of *Guinea*, from 30°. to 38°. North Latitude. *Cliffs* under the Tropic of *Cancer* for one hundred Leagues. *Payton* from 22°. 3´. North Latitude to 30°. *Pyrard de la Val* from 21°. to 30°. *Finton*, that they judged it to be driven from the Cape *Verd* Islands, by the East Winds, of which opinion was *Mandelslo*, who saw it from 20°. to 24°. Lat. one hundred and fifty Leagues off of *Africa*, though he says some were of opinion it came from the *West-Indies*. From all which it is very likely that it may grow as well at Cape *Verd* Islands, as in the *American* Ocean and Isles, and that it is carried to Sea, and floats in it

by

by means of the Winds and Currents in several places, and is no certain sign of the place where a Ship is, as is thought by many People.

The Uses of it are mentioned in *Acosta*, to be eaten greedily by Goats, and to be pickled with Salt and Vinegar, and so used instead of Sampier when it is wanted, and that a Seaman much troubled with Sand and gross Humors, eating of it raw and boiled found so much benefit by it, that he carried it with him to Land, to use there for the same purposes.

XXII. *Lenticula marina foliis latis brevibus serratis. Cat. p. 5.*

This is exactly the same with the former, only the Leaves are not over half so long, and twice as broad, being about three quarters of an Inch long, and about a third of an Inch broad, serrated, and having Bladders like it, being very thin and transparent.

It is frequently cast on the Shores of this Island.

XXIII. *Lenticula marina foliis latis brevissimisque. Cat. p. 5.*

This is the same with the two preceding, only the Leaves are more numerous, being very thick set, shorter, and not altogether so broad as the immediately preceding, neither is it (for the most part) serrated, being of a very dark brown Colour, and having smaller round Bladders.

It grows very plentifully on the Rocks covered with the Salt-water, whence it is sometimes cast by the Waves on the Shore.

Tab. 20. Fig. 7.

XXIV. *Fucus feniculaceus, seu coralloides lenta feniculacea, cauliculis longioribus gracilioribus & densioribus. Cat. p. 5. Tab. 20. Fig. 7.*

The Root of this is broad, sticking to, or incrustating Stones in the bottom of the Sea, in which it grows. The Stalks are several, rising from its Base, two or three Inches long, being divided into several Branches, which are round, smooth, and something like Fennel, branched into Twigs, of a strong Sea smell, and of a purple, whitish, or yellowish Colour, like the *Coralloides lenta feniculacea* of *J.B.* only the Stalks are slenderer, more numerous, and not so much branch'd or curled.

I found it on the greater Stones under Water, near the Bridge at *Passage* Fort.

Tab. 20. Fig. 7. Shews it in its natural bigness and magnifi'd.

Tab. 20. Fig. 8.

XXV. *Fucus feniculaceus, seu coralloides lenta feniculacea minor. Cat. p. 5. Tab. 20. Fig. 8.*

This fixes it self very firmly to the Rocks and Stones covered with the Sea-water, by several Filaments, from whence rises many roundish, crooked, very small, pellucid, pale yellow Stalks, branched into smaller, or almost capillary Twigs, being in all not much over an Inch long, tough, not crackling under the Teeth, and having sometimes a white Crust over it, though for the most part none; smelling strong of the Sea, and looking exactly like the *Coralloides lenta feniculacea* of *J.B.* only in every thing smaller.

It grew on the Rocks under the Sea-water about this Island, whence it was very often cast upon the Shore.

Tab. 20. Fig. 8. Shews it in its natural bigness, and magnifi'd.

XXVI. *Fucus feniculaceus minimus. Cat. p. 5.*

This was about an Inch in Circumference, being made up of many small Stalks of a purple Colour, round like those of Fennel, and observing the like manner of division in the leaves as it, for which reason I put it here, and not among the *Musci*, though it be so small as to require a Microscope to view its parts.

It

It was thrown up amongst the other Recrements of the Sea, along the Shores of this Island.

XXVII. *Fucus minimus denticulatus triangularis.* Cat p. 5. Tab. 20. Fig. 9.
From a broad Base or beginning rises up a Stalk, woody, tough, of a yellowish purple Colour, almost round, about two or three Inches high divided into many Branches, which are for the most part crusted over with a White Coralline substance, and have many deep incisures, small Teeth, or harmless Prickles, in three rows along its edges, being for the most part Triangular, Smelling very much of the Sea, and not crackling under the Teeth.
It is frequently cast up on all the Shores of this Island.
In *Tab.* 20. *Fig.* 9. It is represented in its due bigness, and magnifi'd.

XXVIII. *Fucus marinus coralloides minor fungosus albidus teres segmentis in summitate planis.* Cat. p. 5. Tab. 20. Fig. 10. *An corallina fistulosa flexilis seu corallina geniculata mollis Americana segmentis latis & compressis.* Pluken. Tab. 168. Fig. 4 ? Almag. p. 118 ?
From the same beginning rise two or three round Stalks, about an Inch and a half high, being bigger than those of Shop Coralline, white, fungous, not crackling under the Teeth, divided into several Branches and Twigs on every hand, they being at top flat or plain, and very thin, of a white Colour, and differing from those underneath, which are roundish, all of them smelling of the Sea.
It is thrown up by the Waves on the Banks of the Sea in several places.

XXIX. *Alga angustifolia vitrariorum.* C. B. pin. p. 364. Cat. p. 5.
I found this sort of Grass with its Roots growing in the oazy Grounds in the bottom of the Sea, off of Point *Pelican* near *Port Royal*; and I am apt to believe that growing there the same with ours in *England*, to be different, viz. much broader than that about *Montpelier* and *Venice*, describ'd by *J.B.* and both of them from a third sort of it I found here, which was much narrower than any of these Kinds, and is frequently thrown up on the Shores of the Island with the other Excrements of the Sea, which therefore may be call'd

XXX. *Alga marina graminea angustissimo folio.*
Cushions stuft with this are thought, by some, good for hydropick and gouty People. *Cæsalp.*
It is used to stuff Beds, to wrap Glasses in, and to make Hay. *Lugd.*
Mats made of it, or otherwise used in bedding it destroys Buggs, which is by its smell. *Lacuna.*

XXXI. *Alga Juncea five juncus marinus radice alba geniculata.* Cat. p. 5. Tab. 22. Fig. 5. Manati Grass. *Corallina aut potius alga nodosa, vitrariorum æmula segmentis tubulosis,* the Manittee Grass, *Jamaicensibus dicta.* Pluken. Almag. p. 119.
This seemed to be a sort of Rush growing at the bottom of the Sea. It had a jointed round Root, whitish without, and sending out at joints several hairy, white Filaments to draw its Nourishment. From thence sprang several Foot long, round, green Rushes, not so big as a Hens Quill, when wet being round and hard, but when dry shrivelled and black.
'Twas very plentifully floating along the Coast of the Island by Point *Pedro*, and is thought, with the foregoing *Alga*, to be the Food of the Fish *Manati*.

R   XXXII. *Fucus*

XXXII. *Fucus maritimus gallo pavonis pennas referens.* C. B. *prod.* p. 155. *Cat.* p. 5.

This grows sticking to the Stones in the bottom of the Sea, whence it is thrown on Shore in several places about *Port Royal*; and of it there is a variety, with thicker and whiter Leaves, which is nothing but an incrustation of a Coralline white Matter over it.

XXXIII. *Alga latifolia prima sive muscus marinus lactucæ folio.* C. B. *pin.* p. 364. *Cat.* p. 5. *Lichen Marinus Platyphyllos.* Plukenet. *Alm.* p. 226. *Fucus marinus lactuca folio. Boh. hist. Ox. part.* 3. p. 645. *Fucus lactucæ folio. Tournefort el. bot.* p. 443. *Inst.* p. 568. Oyster green.

This is commonly thrown up on all the Shores of this Island.

It is adstringent and drying, hindering all Fluxions, as Gout, &c. *Casalp. Matth.*

The *Bryon Thalassion*, of *Theophrastus* and *Pliny*, is called in *Northumberland* Slanke, which in Lent the Poor People Seeth, and that with Leeks and Onions : they put it in a Pot, and smore it as they call it, and then it looketh black, and then put they Onions to it, and eat it. *Turner.*

XXXIV. *Pila marina subrotunda, compressa, mollis. Cat.* p. 6. *Tab.* 23. *Fig.* 1.

This was very soft, roundish, compress'd, of an Inch and an half Diameter, white, woolly, and made up of innumerable short white Filaments, interwoven one within another, sticking to a *Fucus faniculaceus.* It was something like the *Alga pomum Monspeliensium,* J. B. only not so round, being compressed, and more soft. Perhaps it may be a Ball voided out of the Stomach of some Fish.

I found it on the Shore in several places, where it had been thrown up by the Waves of the Sea.

XXXV. *Spongia globosa Imperat.* p 385. C. B. *pin.* p. 568. J. B. *tom.* 3. *lib.* 39. p. 816. *Cat.* p. 6.

This in its inward texture was exactly like a Sponge, when prest yielding, and when not prest, by its elasticity, gaining again its former Dimensions, being roundish oblong, not so big as ones Fist, and having many round holes or cavities, as large as a Goose Quill, in its body, and on its top, where were several extant blunt ends standing up, making its Surface uneven, being somewhat harder than the ordinary Sponge, and not so fit for use: smelling very much of the Sea, like others of this Kind.

It is frequently thrown upon the Sea-Banks by the Waves.

XXXVI. *Spongia dura seu spuria major, alba, fistulosa, fibris crassioribus. Cat.* p. 6. *Tab.* 23. *Fig.* 2.

This was five Inches long, near as big as ones Arm or Wrist, being hollow within, something flat and compress'd being made up of Filaments, much like the former, only they are much stronger more lignose or woody, and of a paler Colour, with larger holes in it, but whither it be a variety of that, or a differing Plant, I cannot tell, but am inclinable to think that 'tis quite differing.

It was thrown up by the Sea Waves on its Shores in several places.

XXXVII. *Spongia dura seu spuria maxima ramosa fistulosa. Cat.* p. 6. *Tab.* 24. *Fig.* 1. *An Srongia Americana capitata & dignitata.* Plum. Tournef. *Inst.* p. 576 ? *An rarissimum spongia genus.* Raii hist. app. p. 1850 ? *Spongia novi orbis textura*

textura laxa cincinnata, cylindri cavi figura, ligniculâ circumnascenti. Plukenet. Alm. p. 356. Phyt. tab. 322. fig. 3?

From a Foot Diameter'd, large, broad, roundish Root, spread over the Corals or Stones, in the bottom of the Sea, rise many Foot and a half long, hollow, round, tubulous, blackish Bodies or Pipes, each whereof is about an Inch Diameter, the hollow or cavity in its middle being half an Inch over. The other or fibrous part round it, is made up of many strong Filaments or Fibers, interwoven one within another, having sometimes large vacuities between. These Fibers are of a yellowish Colour, and have a mucilaginous blackish coloured fishy Matter lying between them, of which when clear'd, their reticulated contexture looks very pleasantly. The hollow in the middle has usually in it that kind of louse or crustaceous Animal, sticking to the Gills of Fish in these Seas, and between them very often, the small *Stella marina minor radiis echinatis* of which more hereafter.

This is frequent on the Shoals off of Point *Pelican* near the Coral Rocks, growing in somewhat deeper Water, &c. from whence pieces of it are frequently cast ashore.

There is great variety in this, being very often much larger, and sometimes smaller than that here described. Very often parts, either of the Root or Branches, are driven ashore, and there, according to the time they have lain, put on several Faces and Colours, by the washing away of its mucilaginous Body. They sometimes are branched, and at other times appear in various shapes like Gloves or Harts Horns, from whence *Clusius* gives the same several names, and from him *Caspar Bauhine* was led into the same error.

XXXVIII. *Spongia dura seu sparia ramosa altissima.* Cat. p. 7. Tab. 23. Fig. 3. *An Spongia Americana longissima funiculo similis Plum. Tournefort. Inst. p. 576?*

This appeared to be much the same with the former, only it was smaller, higher, more branched, harder and not hollow, ending in a point, the outward side having very often in some places on it a stony or woolly yellowish Matter.

It is to be found amongst the Coral Rocks with the former.

XXXIX. *Spongia dura seu spuria, superficie apicibus acutis exstantibus aspera, intus cavernosa.* Cat. p. 7. Tab. 23. Fig. 4.

This came very near the texture of Sponges, being harder then they, sometimes as large as ones Fist, for the most part flat, broad, and compressed, having within it several cavities, larger than to be able to receive ones Thumb, roundish, and of several Figures, besides innumerable smaller round holes, through every part of it, the Colour of it was whiter than that of Sponge, and on its Surface were several sharp pointed asperities larger than Pins Heads standing on it, of the same Colour and Substance as its self, and not hollow.

It is frequently cast up on the Shore, with other Excrements of the Sea.

XL. *Spongia minor & mollior medulla panis similis, fibris tenuissimis.* Cat. p. 7. Tab. 23. Fig. 5.

This in its texture came near that of a Sponge, only it was much finer and lighter, being roundish and branched, not so big as ones Finger, being smaller in some places, and larger in others, where were Tubercles, with round

round holes or cavities in them, the colour was like that of a Sponge, and the Fibers much finer than those of any of this kind.

It was cast up on the Banks of the Sea by the Waves, sometimes, though not very frequently.

XLI. *Pila marina velut ex ampullis constans ad Spumæ similitudinem.* C. B. *pin.* 368. *Cat. p.* 7. *Vesicaria marina in pilam conglomerata ex ostrearum testis. Plukenet. Alm. p.* 385.

This is frequently cast upon the Banks of the Sea by the Waves, beyond the Palisadoes near *Port Royal*, &c.

XLII. *Vesicaria marina non ramosa, ex ampullis majoribus, paucioribus, sinuatis, auriculæ instar constans, cujus superficies, favi instar notantur. Cat. p.* 7. *Tab.* 24. *Fig.* 2.

There were only two or three of these Bladders joined together, they were sinuous here and there, with cavities, and look'd something like an Ear; and were marked on the Surfaces like Honey Combs, each of them being as big as twenty of the ordinary Bladders of *Vesicaria marina non ramosa*.

It was cast upon the Shore with the other.

XLIII. *Vesicaria marina non ramosa, è vesiculis infundibuli forma, membrana undulata extante coronatis constans. Cat. p.* 7. *Tab.* 24 *Fig.* 3.

There was an oblong String or Ligament, which seemed to fasten this to some solid Body at the bottom of the Sea, from it arose a great many Pyramidal Bladders, like to so many Funnels, beginning narrow, and increasing from the point at bottom for about three quarters of an Inch to the top, where was an undulated Crown or ledge surrounding the top. These Bladders were dark Yellow, like Glew or Parchment, some of them had inclosed a small *buccinum*, and others a hole, out of which I suppose it had crept, such Bladders being often found empty. The whole was very elegant, and doubtless the *ova* of some *buccinum*.

It was thrown up with the former about the month of *December*, on the great Sea side, beyond *Port Royal* near the *Palisadoes*.

These three last recited Substances, are, without question, nothing else but the Membranes, or Shells of the Eggs or Spawn of Fishes, though they have been reckoned Plants by the greatest number of Naturalists.

CHAP.

The Natural History of Jamaica. 65

CHAP. II.

Of Mushromes, Mosses, &c.

THE number of Mushromes I observed in *Jamaica* were very few, two of the ground kind, one whereof was edible, the other poisonous, and as many of those of Trees, one exactly resembling *Jews* Ears, growing after the same manner on the Trunks of Trees, and the other coming out of the ends of the Roots of them: my observations on these proceed no farther than what easily appeared of them without any very strict Scrutiny. I took the three first to be *Europeans*, and I am apt to think the fourth so likewise.

As to Mosses, I think following the ordinary division, they had best be divided into those on Trees, and those on the Ground, each of which may again be subdivided into those with broad Leaves, called *Lichens*, and those with narrow long Leaves. I think it past doubt, that most, if not all, of them do propagate themselves by Seed, it being plain to the Eye that many of them have it, though so small, as in a calm day to go away like Smoak, which doubtless, with Wind, may be carried very far, and planted on the tops of Houses. This may be the reason why many of this kind are not only common to *Jamaica* and *Europe*, but even some of them are to be found in *Peru*, and the Streights of *Magellan*, as by Specimens brought thence upon the Jesuits Bark, and by Mr *George Handisyd* does plainly appear. This kind of Plant growing on Walls is with some reason supposed to be that which is translated Hysop in our Bibles, where *Solomon* is said to * have spoken of Trees, from the Cedar of *Lebanon*, to the Hysop on the Wall.   * 1 *Kings* 4. 33.

As to the *Lenticula aquatica & equiseta*, I leave them here where I find them, till future Observation shall discover a more proper place.

I. *Fungi Math. p. 776, &c. Cat. p. 8.*
They grow in the *Savannas* after Rain.

II. *Fungi albi venenati viscidi*, J. B. tom. 3. lib 40. p. 816. Cat. p. 8.
They grew with the other *esculent* one, whereby several people have kill'd themselves with them in this Island, mistaking one for the other.
They cause the Hiccough, ulcerate the Guts, make pale the Body, stop the Urine, bring Cold, stop the Arteries, bring Tremblings and Death: besides all hot *Alexipharmacks*; Nature has one excellent one, the Juice of the Herb *Nhambu* and *Jaborandi*, which are Diuretick and Sweat, and fortifie the Native Heat, if, presently after the Poison, it be given them in Wine. *Piso.*

III. *Fungus noxius primus, vel membranaceus auriculam referens, sive Sambucinus*, C. B. Cat. p. 8. *Agaricus auriculæ forma, Tournef. el. bot. p.* 441. *Inst. p.* 562.
This is to be found on several Trees of this Island. It agreed in every thing with that of *Europe* so exactly, that I observed no difference, but took it to be the same.

IV. *Fungus ramosus minor, corrugatus, ex albido luteus, è radicibus arborum proveniens.* Cat. p. 8. *An coralloides ramosa, Nigra, compressa, apicibus albidis Tournefort Inst. p.* 565?

This *Fungus* was about an Inch long, coming out from the ends of the small Roots of Trees, it begins very narrow, growing in breadth to its end, where it is flat, ramose, or deeply cut, and jagged and curled, or corrugated, of a yellowish white colour.

I found it in a Gully, where it grew out of the ends of the Twigs of the Roots of the greater Trees, appearing out of the Earth, after being wetted with Rain.

V. *Muscus arboreus ramosus,* J. B. *tom.* 3. *lib.* 37. *p.* 764. *Cat. p.* 9. *Lichen cinereus latifolius ramosus, Tournefort. el. bot. p.* 438. *Inst. p.* 550. *Muscofungus arboreus capitulis rostratis. Bob. hist. Ox. part.* 3. *p.* 634.

It is common on all Trees in *Jamaica,* especially when decaying.

It is in *England* eaten by wild Mice.

It is used by Perfumers. *Imp.*

VI. *Muscus arboreus* 3. *seu enm orbiculis,* C. B. *pin. p.* 361. *Cat. p* 9. *Muscofungus arborum angustior peltatus & scutellatus. Bob. hist. Ox. part.* 3. *p.* 634. *An Muscofungus arborum capillaceus sentis amplis per ambitum pilis radiatis Ejus. p.* 635? *Lichen cinereus vulgaris capillacio folio minor. Tournef. el. bot. p.* 438. *Inst. p.* 550.

I found this Moss on the Twigs and Branches of old Trees, in several places of this Island, especially towards the North parts. C. B. made this the same with the precedent. *J. B.*

VII. *Muscus tenuis & capillaceus cinerei coloris è ramis ilicis dependens, Clus. rar. plant. hist. p.* 23. *Cat. p.* 9. *An Muscus arboreus aurantiacus staminibus tenuissimis ex insulis Fortunatis? Plukenet. Phyt. Tab.* 309. *Fig.* 1. *Alm. p.* 254. *Muscus Cinereus è ramis arborum dependens, Canariensis, ex Staminibus Crassioribus geniculatis, in tenuissima & longissima fila ramulosus. Ejusd. ibid? Idem colore viridi. Ejusd. ibid?* Long Moss with which the Trees of the Island *Plata* were much overgrown. Of *Dampier Cap.* 6. *Muscus albus & incanus è ramis abietis dependens. Bromel. p.* 68. *Muscofungus arborum nodosus sive geniculatus. Bob. hist. Ox. part.*3. *p.*634. *An Muscofungus arboreus Canariensis ex staminibus Crassioribus geniculatis in tenuissima & longissima fila ramulosus Ejusd. ib. p.* 635? *Muscofungus arboreus vulgaris comosus Cinereus. Ejusd. ib. Muscofungus arboreus Canariensis capillaceus aurentiacus. Ejusd. ib?*

I found this hanging down from the Branches of old Trees in the North side of this Island.

It is not the *Luscus, &c. p.* 2. *Am. Fig.* 39. of *Theod. de Bry,* for that, in my opinion, is the *Viscum Caryophilloides tenuissimum è ramulis arborum Musci in modum dependens foliis pruina instar candicantibus, &c.* To be described hereafter.

This is made use of by Perfumers for their Powders. *Cæsalp.*

The Vertues ascribed to this are many, as that it is Astringent, stops Bleeding, and with Oil of Roses Cures the Headach, comforts the Stomach, restores lost Appetite, sweetens the Breath; the infusion in Wine causes Sleep, stops Fluxes and Vomiting, *&c.*

The colours and largeness, of some of the varieties of this, are taken notice of by *C. B. p.* 361. *Unus tenuior, Crassior alter, brevior, alius prolixior; omnes candescentes, pauci rutilantes, quandoque nigri,* where 'tis plain some of the Kinds abovementioned are taken as varieties by him.

I have received of this from Mr. *Charleton,* who had it from the *East-Indies,* by the name of *Moxa,* and in it were to be seen Heads or Cups, such as are supposed to contain the Seeds of this and other Kinds.

VIII. *Mus-*

The Natural History of Jamaica.

VIII. *Muscus arboreus septimus sive pulmonarius*, C. B. pin. p. 361. Cat. p. 10. *Muscofungus arboreus Platyphyllus ramosus è viridi fuscus*. Bob. part. 3. hist. Ox. p. 634.

It grows in the Woods on the large Trees.

It is good for most Diseases of the Lungs, given in Powder either to Man or Beast. *Lon.*

It is Cold, Dry, Adstringent, and stops all Fluxes in Women. *Math. Dod.*

IX. *Muscus crustæ aut lichenis modo arboribus adnascens cinereus*. Raii hist. p. 116. Cat. p. 10. *Muscofungus Lichenoides minor Cinereus vulgatissimus*. Bob. hist. Ox. part. 3. p. 634. *Muscus aridus crustatus*. Park. p. 1313.

It grows on old decaying Twigs, or Trees in the Woods, very frequently.

X. *Muscus crustæ modo arboribus adnascens flavus*, Raii hist. p. 116. Cat. p. 10. *Muscofungus Lichenoides minor vulgatissimus flavus*. Bob. hist. Ox. part. 3. p. 634. *Muscus Crustaceus bracteolatus flavus*. D. Pet.

It grows on the old decaying Twigs of Trees in the Woods, very frequently.

XI. *Muscus arboreus minor cavus corniculatus albidus*. Cat. p. 10. *Muscofungus montanus Corniculatus*. Bob. hist. Ox. part. 3. p. 632.

This was near an Inch high, white, hollow, tapering, and very often divided into two Branches, something like a Stags Horn, having at the bottom many small very deeply divided Leaves, somewhat like the *Muscus crustæ aut lichenis modo arboribus adnascens cinereus, Raii*, sticking to the sides of fallen Timber.

I found it in a Wood between Mountain River, and Colonel *Copes* Plantation, growing out of a rotting Tree, lying cross the Path.

I have found one much like this in *England*, in *New-Hall* Park, growing out of Timber, only much smaller, and more dark in Colour. *C. B.* confounds this with *filix sax. atilis Trag.* J. B.

XII. *Lenticula palustris, Trag.* p. 689. Cat. p. 10. Ducks-Meat.

'Tis very common on all standing Waters in *Jamaica.* And in *Java*, as *Bontius* tells us.

It is accounted Cold and Moist in the second Degree, and reckoned good in all outward hot Swellings, or Diseases of the Skin, it stops all Fluxes of Blood from Cold, and is good against all Inflammations and Ruptures. *Dorst.* The Water is good against inflam'd Livers, *Lon. viz.* Cloaths being dip'd in it and applied. It is good in the Gout applied. In Inundations it takes Root, and comes to be not unlike *sisymbrium aquaticum*. *Dal.*

It is eaten greedily by Ducks and Hens, mix'd with Bran. *J. B.*

XIII. *Lenticula palustris quarta sive quadrifolia*, C. B. pin. p. 362. Cat. p. 11. This is common in all clayie, shallow, standing Waters in *Jamaica.*

XIV. *Fucus sive alga Capillaceo folio* 1. *vel alga viridis Capillaceo folio*, C. B. pin. p. 364. Cat. p. 11.

It is to be found in all standing Waters, or places where Rivers run slow.

Pliny, and many Natural Historians, say that 'tis a Sovereign and perfect Remedy against broken Bones.

XV. *Mus-*

XV. *Muscus terrestris repens quartus, sive Muscus filicinus major*, C. B. pin. p. 360. Cat. p. 12. Tab. 25. Fig. 2. aaaaa.
It is to be found in the shady Woods of *Jamaica*.

XVI. *Muscus terrestris repens septimus, sive muscus denticulatus major*, C. B. pin. p. 360. Cat. p. 12. *Muscus Americanus denticulatus minor. Tournef. Inst.* p. 556.
This grows very plentifully on the shady Hills near the Banks of the *Rio Cobre*, &c. Compar'd with that of *Europe*, I could not see any difference.

XVII. *Muscus terrestris latioribus foliis major seu vulgaris. Raii, hist.* p. 122. Cat. p. 12. *Muscus squammosus major sive vulgaris. Tournef. Inst.* p. 553.
'Tis sometimes to be met with near the Tree Roots in the Woods.

XVIII. *Muscus terrestris repens minor, ramulis circa extremitates conglomeratis foliis capillaceis.* Cat, p. 12. Tab. 25. Fig. 1.
This appear'd to be in every thing the same with the *Muscus terrestris vulgaris minor adianti aurei capitulis, Raii.* Only the Stalks were much stronger and and larger, having no Branches or Twigs 'till towards their ends, where were very many close set together, the Twigs being reddish, and the Leaves exactly like those of that of *England*, only somewhat more narrow.
It grew in the inland shady Woods by the Roots of the Trees.

XIX. *Muscus terrestris repens major, ramulis circa extremitates conglomeratis, foliolis multis & minimis capillaceis, caulem occultantibus.* Cat. p. 12. Tab. 25. Fig. 2.
This has many three or four Inches long, black, strong Branches, or Strings larger than Threads, creeping on the Surface of the ground, on whose ends come many Twigs set close together, or conglomerated, of about half an Inch long, which are set with almost indiscernible pale green Leaves, very thick, so that no part of the Stalk is visible, the upper Leaves covering the under. Although I never saw this headed, yet I believe it has Heads, therefore put it here.
It grew on Mount *Diablo*, and other woody *Mediterranean* places.
There is one in *England* much like this, only the Twigs not so much conglomerated and larger, as are the Leaves, which are pellucid. It is to be met with in barren Grounds.

XX. *Muscus terrestris minor repens, cujus ramuli foliis multis & minimis seriatim quadrato ordine dispositis cinguntur.* Cat, p. 12. Tab. 25. Fig. 3.
This is very like the former only in all its parts less, the Twigs are not so thick set together, but more sparse in every part of the creeping Stalk, the Leaves are smaller than any of this Kind, and covering the Stalk as the precedent, but so orderly as that there remains a Furrow between their *Series*, making them striated.
It grows with the former. This differs from the *Muscus rupestris Virginianus, &c. Ban. Cat. M.S. Pluken. Alm.* p. 248. Dr. *Plukenet* in his *Mantissa*, p. 248. doubts whither it be the same with this or not.

XXI. *Adiantum seu polytrichum aureum medium, capitulo proprio pediculo insidente.* Cat. p. 12. Tab. 25. Fig. 4.

This

This is in Stalks, Leaves, Heighth, &c. exactly like the *Adiantum seu Polytrichum aureum medium*, Raii. But instead of having its Flower or Head at top of its Stalk, it has a reddish Foot Stalk, as long as its Stalk, rising from the Root without any Leaves, on which grow the Heads which I suppose are the same with the others of this Kind, for the Heads were fallen off when I gathered the Plant.

I found it in the mountainous inland Woods of this Island, as Mount Diablo, &c.

XXII. *Adiantum aureum medium ramosum. Tab.* 25. *Fig.* 5.
This had a Stalk about five Inches long, of the same colour, thickness, and substance with those of this Kind, which towards the top was branched into eight or nine Branches, about three quarters of an Inch long, each of which had capillary Leaves, and Heads just like the others of this Kind.

I had it from Mr. *Geo. Handisyd*, who brought it from the *Magellan* Streights. It was figured here through inadvertency.

XXIII. *Polytrichum aureum tertium seu minus*, C. B. *pin. p.* 716. *Cat. p.* 12. Little Goldilocks, or golden Maidenhair.

The Decoction of this Plant Cures baldness, Expectorates tough Humours, is Diuretick, and good for whatever Maidenhair is proper, discussing *Struma*, &c. *Fuchs*.

It is used in magical Arts for a Philtre, &c. *Trag.* &c.

It expels wonderfully the Stone, and is used against Witchcraft, &c. *Schmenckfeld*.

The Nail of the Toe of the right Foot being anointed with the Juice of this Plant, is said to put away the Cararact of the left Eye, and *Vice versa*, H. M. *This seems to have been an abuse put on the Authors of this Book, who did not discover it, perhaps more such inadvertencies may be therein.*

XXIV. *Muscus saxatilis vel lichen* 1. *petræus latifolius sive Hepatica fontana*, C. B. *pin. p.* 362. *Cat. p.* 13. Common Ground Liverwort.

'Tis very common on the shady moist Brinks of all Rivers in this Island.

It is good against the Inflammations of the Liver, hot and sharp Agues, and Tertians coming from *Clusier. Ger.*

And against all Obstructions of Liver and Spleen. *Dorst.*

The Powder of the Plant with Sugar, is used for all the abovemention'd Diseases. *Lon.*

This is the Lichen of *Dioscorides, Galen* and the *Greeks*, and whatever Vertues are attributed to this *Hepatica* by *Serapio* and others, are said to belong by *Dioscorides* to Lichen, as applied by way of Cataplasm to stop the Hemorrhoids, Inflammations, and Ring-Worm, and outwardly used with Honey to Cure the Jaundice, or ill colour of the Skin; therefore it cannot be called *Hepatica* according to the Opinions of the Antients. *Fuchs.*

It Cures the Jaundice, if the Decoction be taken to the quantity of a Quart; it Purges gently adust and tough Humours, and being many days repeated, I have seen many cured by it of the *Maligna Scabies & Ulcera excedentia*, but the Decoction must be fresh every day with Whey. *Cesalp.*

It stops Blood, *Dioscorid*. and is good against Burns. *Turn*. The Powder of it taken with Sugar remedies the Diseases of the Liver and Lungs. *Trag.*

XXV. *Equisetum* 13. *sive foetidum sub aqua repens.* C. B *pin. p.* 16. *Cat. p.* 13. Stinking Water Horse Tail.

This grows in most Rivers through the Island.

It is used for cleansing Houshold Goods. *J. B.*

XXVI. *Equisetum majus aquaticum.* J. B. *tom* 3. *lib.* 36. *p.* 729. *Cat. p.* 215. This was found in the Marshes of *Jamaica* by *James Harlow*, who brought it from thence. Dr. *Sherard* communicated it to me.

CHAP. III.

Of Ferns or Capillary Plants.

THE Tribe of Ferns in *Jamaica* are very numerous, and strange in their manner of growth. Many of them are Scandent, more than one white on the Back; some have Trunks like other Herbs or Trees, and some perfect Stalks. But whatever they are, they shall by me be ranged as those of *Europe* are by Mr. *Ray*, according to the division of their Leaves: and whereas it is commonly held that no Fern, has, properly speaking, a *Caulis* or Stalk, yet it will hereafter plainly appear they have. To avoid a great deal of trouble in the disposition of them, I shall consider the Stalks of those climbing Ferns only as Roots. The divisions of the Leaf its self arising from such Stalk shall determine its place without having regard to this Stalk, which ought otherwise to be first taken notice of in the division. Without doubt all hereafter named Ferns are such: only there are six I call *Phyllitidi Scandenti Assinis*, which I am not sure are Ferns, and therefore not having seen their Seeds, I give them this, till a better place is found. The name *Trichomanes* shall signifie with me the second divided Fern with broad small Leaves, and the word *Lonchitis* those with longer, though they agree not in every particular with the *Europeans*. The general divisions are, first, those with undivided Leaves. Secondly, Those with once divided into *Pinnæ* only, then into those with the middle Rib, a Twig or *Surculus*, and on it *Pinnulæ*, which I call *Filices non Ramosæ:* and the third is the ramose Kind. Many of these Ferns are indented about the edge when young, and afterwards are plain with a ferrugineous Welt on the outside, and sometimes they are sinuated; many other Varieties there are, appearing really differing Plants, which may be easily found out by any, who considers their several descriptions.

It is no great wonder that in so great a number of new and strange Ferns as are mentioned in my Catalogue, and described hereafter, I was put to it to find words, to describe some of them. I think the terms I have made use of are very plain, though Dr. *Plukenet* is pleased in his *Mantissa p.* 83. to find fault with one of them, *viz. Surculus,* and to recommend in its place *Ala.* I continue to think that Word more proper than the other, for 'tis taken by the best, and even purest Writers, for the *Germen Annotinum,* or yearly Sprout or Twig, to which are contiguous the Leaves of Trees. Now if any Word can be fitter to signifie the last division of the Stalk of a Fern, to which the *Pinnulæ* or leaves are fasten'd, I leave any body to judge. The Word *Ala* is used by the best Herbarists in many, and very uncertain significations, and even when it is applied to the wings or Leaves of Ferns, is sometimes taken for the *Pinnæ,* or Leaves themselves, sometimes for what I call the Twig or *Surculus,* and sometimes for the whole *Ramulus* or Branch, so that it would have been very confounding, and not have explained sufficiently my meaning. These Words are so often used here, and the Cuts of the Ferns, either figured elsewhere, or in this

Book

Book, make my acceptance of them so plain, that there can be no doubt about them. As to the Etymology of the Word *Surculus*, its Definition from *Festus, &c.* I take them not to be material Arguments, and could prove my Sense of it to agree very well with what he and the best Authors say, were it needful to insist on this matter.

All these sorts of Ferns, no question, may be made use of for the same purposes, as those of the same Kinds in *Europe*.

I. *Lunaria elatior Adianti albi folio duplici spica. Cat. p.* 14. *Osmunda Americana filicula foliis Tournefort. El. bot. p* 437. *Osmunda filicula folio major. Ejusd. Inst. p.* 547. *An filix saxatilis ruta muraria foliis Americana seu adianthum album folio filicis ex insula Jamaicensi. Plukenet. Alm. p.* 150?

This is sometimes about a Foot, but mostly six or seven Inches high, having a very slender green Stalk, at coming first out of the Earth, being of a dark colour. At about four Inches from the Ground, out of one side of the Stalk goes one Branch, to which are alternatively set on Twigs, which have several broad irregularly figur'd roundish *Pinnulæ*, sometimes deep cut, at other times a little indented on the edges, being of a pale green colour, like to *adiantum album*, and having many Furrows appearing radiated. Out of the *Ala* or bosom of this Branch rise two round, small, green, two Inches long Stalks, towards the tops of which are several small Bunches of first green, afterwards ferrugineous Dust, like to that of *Lunaria, Osmunda Regalis,* or the other Ferns. The Root is like that of *Polypodium*, and is covered with a blackish Hair, having several *Fibrils* like the Roots of other Ferns.

It grew on a Rock by the Banks of *Rio-Cobre*, below the Tow., on the same side of the River.

This Plant is perfectly differing in all its parts from the *Filix non ramosa pinnulis crebris obtusis crenatis. Cat. p.* 21. As one may easily see by comparing their Descriptions and Figures, so that one would wonder how Dr. *Plukenet* came to doubt whither they were the same in his *Mantissa, p.* 78.

II. *Lunaria elatior matricaria folio spica duplici Cat. p.* 14. *Tab.* 25. *Fig.* 6. *Lunaria racemosa cicutaria foliis Jamaicana. Plukenet. Alm. p.* 208.

This is much higher than the former, rising a Foot or more from the Ground. In its higher Spikes, which are double, it exactly agrees with the former, only the Leaves or *Pinnulæ*, are longer, narrower, not quite cut in to the middle Rib or Twig, and of a paler green colour, something in their Divisions, like the Leaves of *Matricaria*.

It grew with the former.

III. *Phyllitis non sinuata minor apice folii radices agente. Cat. p.* 14. *Tab.* 26. *Fig.* 1. *Filicifolia phyllitis parva saxatilis Virginiana per summitates foliorum radicosa breviore & latiore folio, Plukenet. Alm. p.* 154.

This has a small, scaly black Root, with many long Fibers drawing its Nourishment, of a dark brown colour. The Leaves are many, rising from the same Root, of a different magnitude, having no Foot-Stalks, the largest being two Inches and a half long, and about half an Inch broad near the middle where broadest, increasing from the Root thither, and thence decreasing, growing very narrow, and ending in a point. This point bows down its self to the Ground, strikes Fibers, takes Root, and sends out rounder Leaves, in time growing longer, and with their ends taking Root, and so propagating its self. The Seed lies in round spots on the back of the Leaf of each side of the middle Rib.

It

It grew in a thick, very high, and shady Wood, at the bottom of Mount *Diablo*, beyond the *Maggoty Savanna*.

It differs from Mr. *Banister*'s *Phyllitis saxatilis Virginiana per summitates foliorum prolifera*, Which is sinuated at Base, and has a Foot-Stalk.

IV. *Phyllitis arboribus innascens, folio non sinuato tenuiori rotundis pulverulentis maculis aversa parte punctato. Cat. p. 14. Lingua cervina longis angustis & undulatis foliis. Tournef. Inst. p. 545.*

This has a long knobbed Root, having fastened to it several Scales, or remains of Leaves dropt of, and a great many reddish brown Fibers, interwoven one within another, having a three Inches long green Foot-Stalk. The Leaf is nine Inches long, two broad in the middle, where broadest, not falcated or sinuated at beginning like *European* Harts-Tongue, but very narrow, and ending in a sharp, or sometimes blunt point, being very green, thin, and shining, and having on its backside several round ferrugineous spots in which lie the Seeds.

It grows mostly on the Trunes of great, chiefly old, Trees, as Misseltoe, and is to be found on such Trees in the Path going to sixteen Mile Walk, and in shady places of the Hills in *Liguanee* by *Hope* River. Sometimes when old Trees fall down, this will then grow on the Ground.

V. *Phyllitis non sinuata foliorum limbis leviter serratis. Cat. p. 14.*

This, which seemed to be a small starv'd Plant, had a Root made up of many brown Fibres, sending up some eight or nine Leaves without any Foot-Stalks, being about three Inches long, three quarters of an Inch broad, near the further end where broadest, of a yellowish green colour, and smooth, dentrated about the Edges, being narrow at beginning, they increase to near the end, and then decrease to a blunt point.

It grows in the great shady Woods, in the inland parts of the Island.

VI. *Hemionitis peregrina Cluf. rar. plant. hist. lib. 6. p. 214. Cat. p. 14.*

I had this given me by. Dr. *William Sherard*, who had it of one, who gathered it in *Jamaica* or *Madera*.

VII. *Filix Hemionitis dicta Maderensis, hedera arborea aliquatenus amula, seu foliorum basi auriculis linis utrinque donato. Pluken. Alm. p. 155. Phyt. Tab. 287. Fig. 4. Hemionitis Jamaicensis hederaceo folio, lineis seminiferis tenuissimis in dorso notato. Boh. hist. Ox. part. 3. p. 560. An Hemionitis Lusitanica elegantior Tournef. el. Bot. p. 436 ? Inst. p. 546 ?*

This was brought me with the former, and was gathered by *James Harlow* in *Madera*, if I rightly remember. I question much whether this be really differing from the former.

Though Dr. *Plukenet* tacitly confesses in his *Mantissa, p. 82.* that he had made two of this Plant in his *Almag. p. 155.* yet any body who compares his Figure of this Plant, *Tab. 287. Fig. 4.* and mine of the *Hemionitis peregrina foliorum segmentis sinuatis, &c. Tab. 16. Fig. 2.* figured and described hereafter, will find them vastly differing.

VIII. *Hemionitis folio hirsuto & magis dissecto seu ranunculi folio. Cat. p. 14. Filix hemionitis dicta sanicula foliis villosa. Plukenet. Alm. p. 155. Phyt. Tab. 291. Fig. 4.*

This had a great many fibrous, and threeady black Roots, sending up several smaller, and two six Inches long shining black cornered Stalks, being covered over with a Ferrugineous Hair or Moss, the Leaf stands at top, being divided into three parts or Segments, very deeply cut in, even almost to the

the Center, the two undermost Sections having Ears, or *appendicula*, making the Leaf appear divided into five Sections. The division in the middle is the larger, being an Inch and a half long, and about half as broad in the middle where broadest, being easily dented on each side, rough, of a yellowish green colour, ending each of them in a point, and having one Purple middle Rib coming through them from the Foot-Stalk, as from a common Center. From these middle Ribs go several transverse Fibers, on which is a great deal of Ferrugineous Moss, which is the Seed. The whole Leaf is very like that of the *ranunculus pratensis repens hirsutus*. C. B.

It grew in the shady Banks of a gully in a Wood, between the Town Savanna, and Two-Mile-Wood.

IX. *Hemionitis foliis atrovirentibus maxime dissectis seu filix Geranii Robertiani folio.* Cat. p. 15. *Filix hemionitis Americana petrosolini foliis profunde laciniatis* Pluknet. *Phyt. Tab.* 286. *Fig.* 5. *Almag.* p. 155. *Filix hemionitis Jamaicensis foliis Geranii Rupertiani quodammodo æmulis.* [*Ejusd.* Ibid. *Adiantum monophyllum Americanum foliis profunde laciniatis ad oras pulverulentum.* Bob. hist. Ox. part. 3. p. 592.

This as to Root, Stalk, &c. is the same with the former, only the Leaf is smooth, of a dark green colour above, being very much more dissected or laciniated with Ferrugineous Lines all along the Margins of the back sides of the Leaves, the Dissections appearing like the *geranium Robertianum* Leaves as to manner of incisures, &c.

These two Plants may be reduced to the second Division of Ferns more properly, but because of their great affinity with the *hemionitis peregrina*, I have put them here.

It grew with the former.

How Dr. *Plukenet* came to make two of this one Plant, as he does in his *Alm.* p. 155. and *Mantissa* p. 82. he best can tell.

X. *Hemionitis peregrina foliorum segmentis sinuatis longioribus, & magis acuminatis seu hedera folio anguloso.* Cat. p. 15. Tab. 26. Fig. 2.

This had several long fibrous Roots uniting in a knob, which sends up three or four Stalks, with Leaves very like the *Hemionitis Peregrina Cluf.* only it was not so broad nor thick in the Consistence of the Leaf, and had not such Auricles as it, and the Segments into which it was divided, which were three, were longer, more sharp pointed, and sinuated about the Edges, which makes me of opinion that it is a differing Plant.

It was brought from *Jamaica*, and given me by Dr. *Sherard*.

XI. *Phyllitis minor scandens foliis angustis.* Cat. p. 15. *Filix sarmentosa bifrons, seu, Dryopteris scandens Jamaicensis inter filicem & lycopodium media vel Filicis & lycopodii Compos.* Plukenet Tab. 290. Fig. 3. Alm. p. 156. *Lingua cervina scandens caulibus squammosis.* Tournefort. el. Bot. p. 435. Inst. p. 545.

This has a compressed Stalk, not so big as a Hens Quill, covered over very thick with very many ferrugineous Hairs or Moss, like that of the other Ferns. This sometimes turns Gray, mounting up straight forty Foot high, sticking close to the Bark of any Tree, by many hairy small fibrils, of a reddish Brown colour. On each side of this Stalk, which is sometimes branched, goes a Leaf or *Pinna* at about half an Inches distance, being plac'd alternatively, each of these Leaves is about four Inches long, and three quarters of an Inch broad in the middle where broadest, being narrow at beginning, and ending in a point, smooth, of a light Green colour, and having one middle Rib, on each side of which is a row of ferrugineous Spots, in which lies the Seed.

It grew on the Truncs of the Trees in going up Mount *Diablo*, towards the North side.

I cannot imagin what Dr. *Plukenet* means by his Title to this Plant, why he should call it in his *Alm. p.* 156. *Inter filicem & lycopodium media*, and correct it in his *Mantiss. p. 84.* calling it *Filicis & lycopodii Compos.* Neither can any Body tell but himself, unless he has taken the top of the Scandent Stalk of this Plant for a Head like that of *Lycopodium*.

XII. *Phyllitis scandens minima musci facie foliis membranaceis subrotundis. Cat. p.* 15. *Tab.* 27. *Fig.* 1. *Filix hemionitis lichenoides Americana fungi auricularis Casalpini æmula radice reptatrice Plukenet. Almag. p.* 185. *Fig.* 3.

This had a flat black Stalk covered with a Hair like other Mosses, applying its self to the Rocks, Stones or Trees, and rising seven or eight Foot high, putting out at more or less distance, small roundish membranaceous yellowish green Leaves, like those of the *Polytrichum aureum minus foliis subrotundis.* They grow sometimes longer, having incisures on their edges; the Plant looks somewhat like a Moss, but if narrowly view'd seems nevertheless properly to belong to this place.

It grew on the moist Truncs of Trees, by *Rio d' Oro, Orange* River, and *Archers* ridge, between sixteen Mile Walk and Saint *Maries.*

XIII. *Phyllitidi scandenti affinis major, folio crasso subrotundo. Cat. p.* 15. *Tab.* 28. *Fig.* 1.

This has a cornered compress'd striated green Stalk, having Leaves coming out on each side of the Stalk, alternatively at an Inch distance, under each of which is a broad fungous or spongy gray Clavicle, sticking very firmly to the Truncs of Trees, or sides of Rocks, by that means mounting sometimes to thirty Foot high Each. Leaf has no Foot-Stalk, is about an Inch and a quarter long, and an Inch broad near the round Base where broadest, being almost round, somewhat shap'd like a Heart, smooth, thick, juicy, of a pale whitish green colour, almost like those of *Orpin.*

It grows on the Truncs of the taller Trees, or moist inland Rocks, in the Woods, in most places of this Island.

XIV. *Phyllitidi scandenti affinis minor folio crasso oblongiori. Cat. p.* 15. *Tab.* 28. *Fig.* 2.

This is exactly the same with the former, only the Leaves are longer and narrower, they being an Inch and half long, and about three quarters of an Inch broad, at round Base, where broadest, and a little more frequent than the former. Of this there are, as to bigness and manner of growth, some small varieties.

It is to be found on the Truncs of Trees on Mount *Diablo, Archers* ridge, *Orange* River Banks, and other the *Mediteranean* Woods in this Island.

XV. *Phyllitidi scandenti affinis minima, folio crasso oblongiori. Cat. p.* 15. *Tab.* 28. *Fig.* 3.

This, as to Stalks, Clavicles, manner of growth, &c. is the same with the precedent, only these parts are in every thing lesser, the Leaf not being over three quarters of an Inch long, and one quarter of an Inch broad at round Base where broadest. The Leaves are set at much lesser distance.

It grows with the former.

XVI. *Phyllitidi scandenti affinis minor, foliis subrotundis, acuminatis, ex adverso sitis. Cat. p. 15. Tab. 28. Fig. 4.*

This creeps up the Truncs of Trees after the same manner as the preceding, by a cornered white Stalk, putting forth at about half an Inches distance at the Joints, some small *Capreoli*, or rather *Cirrhi*, like those of Ivy, which take hold of the Barks of the Trees they Climb. At the same places come out the Leaves opposite to one another, having Foot-Stalks about the eighth part of an Inch long Foot-Stalks, being almost round, about three quarters of an Inch long. They are half an Inch broad near the round Base where broadest, whence they decrease, ending in a sharp point, being smooth, equal about the edges, and of a light brown colour. What Flower or Seed it makes, I know not; but by its likeness to the preceding, I bring it hither.

It grew on the Truncs of the Trees near the Banks of *Rio d' Oro*, coming from St *Maries* to sixteen Mile Walk.

XVII. *Phyllitidi scandenti affinis minor graminifolia, folio oblongo acuminato, foliorum pediculis alis extantibus auctis. Cat. p. 15. Tab. 27. Fig. 2. Phyllitis minor scandens salicinis foliis acuminatis viridibus. Bobart. hist. Ox. part. 3. p. 558:*

This has a like Stalk with *Cirrhi*, as the former, only a little larger, the Leaves come at Joints singly, not two opposite to one another, having about half an Inch long Foot-Stalks, with extant striated membranaceous *Alæ* or *Appendices*, inclosing the Stalk almost round, and looking like the small or first Leaf of an Orange Tree, the Leaf its self is an Inch and half long, and near half an Inch broad at round Base where broadest and whence it decreases to the point having one middle Rib, and several transverse ones running to the edges being smooth, thin, shining, grassy, somewhat like the Leaves of *Phyllitis*, and of a pale green colour.

It grew on the Trees with the former.

XVIII. *Phyllitidi scandenti affinis major graminifolia, folio oblongo acuminato, foliorum pediculis alis extantibus auctis. Cat. p. 15. Tab. 27. Fig. 3.*

This is in every thing the same only larger in all its parts, the Foot-Stalks being two Inches long, the Leaves grassie, four Inches long, and half as broad.

It grew with the former.

XIX. *Polypodium altissimum. Cat. p. 15. Polypodium Jamaicense majus & elatius, alis longioribus, punctis aureis aversa parte notatis, Bob. hist. Ox. part. 3. p. 563. An Polypodium majus Africanum pediculis foliosis lobisque planis. ejusd. ib. p. 564? An filix Polypodium dicta minima Virginiana platyneuros Plukenet. Alm. p. 153. Phyt. Tab 289. Fig. 2?*

This *Polypodium* (which had Roots with Tubercles, Fibers, and reddish Moss or Hair, as the ordinary *Polypodium*) rose to be three Foot and a half high, having a reddish green, large, smooth, strong and long Foot-Stalk. The *Pinnæ* were set on by pairs to the middle Rib, opposite to one another, at about an Inches distance, join'd at Base by a Membrane running along the middle Rib, making it alated as with others of this Kind. The *Pinnæ* each of them are six Inches long, and about three quarters of an Inch broad in the middle where broadest, beginning narrow, increasing to the middle, and ending in a point, having one middle Nerve, on each side of which, on the backside of the Leaf, is a row of Ferrugineous Spots, being the Seeds. The Leaves are of a yellowish green colour, and smooth.

The

The *Pinnæ* of this, when young, are so broad as to have their edges lie over one another, which makes it seem a different Plant.

It grew in the Inland mountainous parts of this Island.

XX. *Lonchitis palustris maxima. Cat. p. 15 Filix Americana maxima aurea non ramosa alis integris alternis planis, Bob. hist. Ox. part. 3. p. 571. Lingua cervina ramosa aurea. Plum. Tournef. Inst. p. 546.*

A great many Leaves rise from the same knobbed large Root, to four or five, or even nine or ten Foot high, having a greenish cornered irregularly shap'd middle Rib, as large as ones little Finger. To this at every Inch, or Inch and a halfs distance are set the *Pinnæ*, from the bottom to the top of the Stalk alternatively from the opposite sides of it. Each *Pinna* a Foot long, and three Inches broad in the middle where broadest, being narrow at beginning, and ending obtusely. It is green, smooth, and directly of the shape, colour, and consistence of Harts-Tongue, having one middle Rib eminent on the backside of the *Pinna*, which is sometimes all, and sometimes half covered over with a rusty colour'd Moss, in which lies its Seed. The Leaf or Branch has at its end, or that of the middle Rib, one single *Pinna* at top.

It grows in the Marshes near *Black River* Bridge, going to *Old Harbour*, and by the *Salt River* near *Passage Fort*.

It is used instead of Thatch to cover Houses.

It is also used to stop Dysenteries, and the violent motions of the *Iliac* Passion, by boiling the Root, and drinking the Decoction.

The Decoction of the Root is excellent in Obstructions of the Spleen, Quartans, Scurvey and Melancholy, especially if *Sarsa* and *China* be added. A Salt made of the Leaves by Chimistry is an excellent Remedy against Ulcers, and carious Bones of the Toes and legs, being very drying. *Bont.*

The Root is used by the *Malabars* and *Javans* in bitings of poisonous Beasts. *Bont.*

It is Temperate, and opens Obstructions of the Liver. *Bontius.*

If this be what *Flacourt* mentions to grow in *Madagascar*, he says it is there made into Hats, Garlands or Crowns, which they wear on their Heads, because of its good Smell. And that it is Cordial, and good for Liver and Spleen, as our Maiden-Hair.

XXI. *Lonchitis Asplenii facie pinnulis variis, viz. subrotundis, & ex utroque latere auriculatis. Cat. p. 16. Tab. 29. & 30. Fig. 1. Adianthum seu filix trichomanoides Jamaicensis; pinnulis auriculatis dentatis, ad basin amplioribus radiculas ex nutante apice ad Terram demittens. Plukenet. Almag. p. 9. Tab. 252. Fig. 4. Filicifolia lonchitidis facie Jamaicensis, ad basin uniuscujusque pinnæ, binis auriculis obtusis donata ambitu æquali & aversa parte ferrugineis punctis duplice ordine notata. Ejusd. Alm. p. 152. Phyt. Tab. 286. Fig. 2. Filix Jamaicensis simpliciter pinnatis asplenii foliis aversa parte duplici serie punctorum notatis Ejusd. Alm. ibid. Phyt. Tab. 290. Fig. 1. An filix minor Africana lonchitidis folio pinnulis auriculatis planis. Ejusd. Phyt. Tab. 89. Fig. 7? Polypodium Americanum medium foliis variis pinnulis obtusis binis macularum ordinibus insignitis. Bob. hist. Ox. part. 3. p. 563. Lonchitis Jamaicensis elatior pediculis molli lanugine pubescentibus alis brevioribus utrinque auriculatis rarius dispositis, Ejusd. ib. p. 568. Lonchitis minor Jamaicensis non dentata alis angulis crebris atrovirentibus, utrinque auriculatis. Ejusd. ibid. 569. Polypodium incisuris asplenii. Tournefort. El. Bot. p. 432. Inst. p. 541.*

The

The Face of this Plant, and difference of the Leaves or *Pinnulæ* make it very difficult to assign it a right place, for almost every several middle Rib or Stalk has several differing Kinds of *Pinnulæ*. It sometimes is about a Foot and a half long, has a pale green Stalk, which is somewhat hoary, and at bottom has roundish *Pinnulæ*, resembling the Leaves of *Nummularia*, and those are placed at some distance from one another. See *Tab.* 29. & *Tab.* 30. *Fig.* 1. *lit. a a a a a a.* On other Stalks at greater or lesser distances come *Pinnulæ* or Leaves that are Oblong, and something auriculated on both upper and under side, and then above them to the point are rounder Leaves. On other Twigs the Leaves are joined close to one another, after the manner of those of *Asplenium*. All of these sorts of *Pinnulæ* have Seed on the back parts, lying in Ferrugineous Spots along the middle Rib, as others of this Kind. See *Tab.* 29. & *Tab.* 30. *Fig.* 1. *lit. b b b b b.* Sometimes again the Leaves will be Oblong and auriculated of each side, both upper and under, and disjoined without any cohesion up to the top. At other times they will be auriculated, ?disjoined, and towards top grow weak, trailing and touching the ground take Root, and propagate the Plant, so that I have not seen in any Plant so great sporting of Nature as in this. Another variety is in the Leaves, which are serrated, or, as it were, made up of *Pinnulæ*, after the manner of the *Chamæfilix*, which take Root when they touch the ground, as the preceding variety doth, and this is so very odd, that it would impose almost on any one to make them believe it a differing Plant. See *Tab.* 29. *lit. c c c.* all the *Pinnulæ* of this Fern are shortest at bottom, and longest in the middle. The Roots were many black *Ligulæ* united, making a strong Foundation for the Plant.

It was brought from *Jamaica* by Sir *Arthur Rawdon*'s Gardener, and communicated to me by Dr. *Sherard*.

It appears by *Tab.* 29. *Fig.* 1. *lit. bb, &c.* that it has two sorts of Leaves, those of *Asplenium*, and auriculated on both sides, so that I cannot but wonder why Dr. *Plukenet* should think that it cannot be the same with the *Filix non ramosa scolopendrioides* of *Plumier*, because that is not auriculated. *Vid. Mant. p.* 79.

XXII. *Lonchitis altissima, pinnulis utrinque, seu ex utroque latere auriculatis.* Cat. p. 16. *Tab.* 31. *Lonchitis glabra minor Plum.* p. 19. *Fig.* 28. *An Filix, seu lonchitis aspera Jamaicensis elatior pinnis longioribus utrinque auriculatis, subtus punctatis, & obtusioribus denticellis spinosa. Plukenet, Almag. p.* 152? *Lonchitis Americana glabra alis latioribus & brevioribus crebrius dispositis. Bob. hist. Ox. part.* 3. *p.* 568.

That part of this *Lonchitis* which came to my Hands was about three Foot long, having a very strong middle Rib or Stalk, furrowed forward, and round backwards, reddish and shining, being set at every three quarters of an Inch with *Pinnulæ* or Leaves about an Inch long, and half as broad, being whole without incisures to near their tops. Each *Pinnula* had two Auricles, one of each side, as well under as above, at the Base of the *Pinnula*, that above being most discernible. They have two rows of ferrugineous Spots, one on each side of the middle Rib.

This was brought from *Jamaica* by Sir *Arthur Rawdon*'s Gardener, and communicated to me by Dr. *Sherard*.

XXIII. *Lonchitis altissima, pinnulis variis non laciniatis.* Cat. p. 16. *Tab.* 32.

This *Lonchitis* was about five Foot high, having about a Foot and a half long, roundish, dark brown, or blackish Foot-Stalk; the *Pinnæ* were about two Inches long, about three quarters of an Inch broad at Base, where they parted from the middle Rib, from whence they decreas'd to the end, which was roundish and blunt, not at all laciniated, having one middle Rib on each side of which were many ferrugineous round Spots or Seeds: there was about half an Inches empty space between the parts of the *Pinna*, which were plac'd alternatively on each side of the middle Rib.

I found it in the inland mountainous parts of this Island on Mount *Diablo*.

It appears plain that this is not the same with the last save one, which Dr. *Plukenet* in his *Mantissa*, p. 79. thinks may be the same with this.

XXIV. *Lonchitis major, pinnis latioribus, leviter denticulatis, superiore latere auriculatis.* Cat. p. 16. *An lonchitis foliis superius incisis major. Plum. Tournef. Inst. p. 539 ?*

This rises a Foot and a half high, the Stalk being blackish at its coming out of the Earth, having *Pinnæ* set alternatively at about a third part of an Inches distance from one another, by very short Foot-Stalks: the *Pinnæ* set in the middle are largest, being an Inch and a quarter long, and about half an Inch broad at the Base, where broadest, from whence they decrease, ending in a point, being serrated, or cut in on the edges, and having on the uppermost edge of each *Pinna*, an Ear or *Appendicula*, and on the backside several vermiculated ferrugineous Lines, in which is the Seed, on each side of the eminent middle Rib.

It grew in the inland woody parts of this Island.

It is very evident that this is perfectly distinct from the *Lonchitis major, pinnis angustioribus leviter denticulatis superiore latere auriculatis* to be described presently, as appears by their Descriptions and Figures, contrary to what is affirmed by Dr. *Plukenet* in his *Mantissa*, p. 79.

XXV. *Lonchitis minor pinnis latioribus leviter denticulatis, superiore latere auriculatis.* Cat. p. 16. Tab. 33. Fig. 1. *Filix minor non ramosa Jamaicensis alis uncialibus, acuminatis latioribus.* Bob. hist. Oxon. part. 3. p. 572. *An lonchitis foliis superius incisis Minor, Tournefort. Inst. p. 539 ?*

This was exactly the same with the precedent, only not half so large, so that it may be thought no variety, but a distinct Species. The Seed lay on its Back after the same manner, and the *Pinnæ* were thinner and narrower.

It grew on the Rocks near the Banks of *Rio d' Oro*, near sixteen Mile Walk.

Dr. *Plukenet* questions in his *Mantissa*, p. 79. if this be the same with his *Almag. Filix seu lonchitis aspera Jamaicensis pinnula lata, brevi, angulosa*, p. 152. By the name I should not have taken it to be the same.

XXVI. *Lonchitis major pinnis angustioribus leviter denticulatis superiore latere auriculatis*, Cat. p. 16. Tab. 33. Fig. 2. *An Filicula lonchitidis folio auriculata & dentata, Plukenet. Mant.* p. 77 ? *Vel an Filix mas non ramosa Marilandica, pinnulis angustis rarioribus, profunde dentatis, superna parte auriculatis, Ejusd. ibid.* p. 78 ? *Filix minor non ramosa Jamaicensis, alis uncialibus angustioribus.* Bob. hist. Ox. part. 3. p. 572.

This has a Root having many long capillary Fibers, being black, scaly, and covered with a ferrugineous Mois. The Stalk is greener than the former, having *Pinnæ*, set opposite or alternatively, they are in every thing exactly like them, only very narrow, being not half so broad, but in Auricles, Nerves, Spots, &c. exactly the same.

It grew with the other, in the more inland woody parts of the Island.

XXVII. *Lon-*

The Natural History of Jamaica. 79

XXVII. *Lonchitis major pinnis longis angustissimisque. Cat. p.* 16. *Tab.* 34. *Lonchitis minor Jamaicensis pinnulis angustis longis ad basin auriculis rotundis donatis, Bob. hist. Ox. part.* 3. *p.* 568.

This has a large Root, solid, and of a whitish green colour within, towards the top covered over with a ferrugineous Hair, and sending up several Leaves above a Foot high, having greenish, white, and smooth Foot-Stalks, about five Inches long. The *Pinnæ* are set on to the middle Rib, sometimes opposite, (chiefly below,) and sometimes alternatively having about the sixth part of an Inch distance between the pairs, each of them being about two Inches long, and a sixth part of an Inch broad at beginning where broadest, and whence they decrease, ending in a point, being not at all serrated, very smooth, even, and of a grassy green colour. The round auricles mention'd by Mr. *Bobart* are very small.

It grew out of the fissures of the Rocks in the shady Road, going to sixteen Mile Walk.

It seems to me impossible that this should be what Dr. *Plukenet* thinks, *Mantiss. p.* 81. *viz. Filix Jamaicensis Jacea majoris Gerardi aut potius centaurei majoris lutei Parkinson. æmula falcatis foliis integris & margine æquali. Almag. p.* 154.

XXVIII. *Polypodium minus pinnulis raris subtus cinereis. Cat. p.* 16. *Filix polypodium dicta minima Jamaicensis foliis aversa parte ferrugineo pulvere aspersis ritu circumquaque respersis. Plukenet. Alm p.*153. *Phyt. Tab.*289. *Fig.*1. *Polypodium minus Virginianum foliis brevioribus subtus argenteis. Bobart. hist. Oxon. part.* 3. *p.* 563. *Tab. an.* 2. *Plant. Capill.*

This has a round, blackish, hard Root like other *Polypodies* only smaller, covered in some places with a ferrugineous hairy Moss, and creeping for a good length. From thence go many long fibrils into the Earth, of a reddish brown colour. At about half an Inches distance, this Root sends out Leaves, which have gray, or light brown coloured Foot-Stalks two Inches long, the whole Leaf and Foot-Stalk being about four Inches high. The *Pinnæ* are not so frequent as in the precedent, but more rare, set almost opposite one to the other, about three quarters of an Inch long, and an eighth part broad, not laciniated, nor sinuated, thick, on the upper side of a yellowish green colour, on the under gray or whitish, with some rows of ferrugineous Spots on their edges, being the Seed.

It grew on the Rocks, in a Wood on the side of a Mountain, near Mr. *Elletson*'s Plantation in *Liguanee*.

XXIX. *Lonchitis minor pinnulis angustis crebris tenuibus atrovirentibus. Cat. p.*16. *Filix seu Lonchitis Jamaicensis polypodii folio pediculis nigris. Plukenet Almag. p.* 152.

This has a firm knobbed Root, as big as the top of ones little Finger, covered towards the top with scaly brown Moss, and having very many dark brown Fibres going deep into the Earth. It sends up Leaves from six Inches to two Foot long, their Foot-Stalks being brown, an Inch and a half long, the *Pinnæ* are set on to the Stalk, or middle Rib, not opposite to one another, but alternatively, being cut into the middle Rib. These are about three quarters of an Inch long, and not over a Twelfth part of an Inch broad at beginning where broadest thin, and of a dark green colour, each having one middle Rib, on each side of which lies its Seed in redish Spots : sometimes this is two Foot high, and the *Pinnæ* almost opposite though very rarely.

It grew on the woody Mountains near Mr. *Elletson*'s Plantation in *Liguanee*.

XXX. *Lon-*

XXX. *Lonchitidi affinis arbor anomala folio, alato è pinnarum crenis fructifero.* Cat. p. 16. *Hippoglosso forte Cognata Surinamensis; foliis oleandri serratis, in crenarum extremo flosculos perminutos sanguineos gerentibus vel forte hemionitidi affinis.* Breyn. pr. 2. p. 57. *An Hemionitidi affinis Americana epiphyllanthos folio simpliciter pinnato, hyppoglossi æmulo, radice reptatrice lignosa ad foliorum crenas florida angustiori & longiori folio ramosa caulescens.* Plukenet phyt. Tab. 247. Fig. 4?

This Tree riseth to about eight Foot high, having Stems not much thicker than ones Thumb, whose Wood is white and solid, with small Pith, and Bark is smooth, of a light gray colour, a little striated, and something like that of Fir, to whose *Cyma* or springing *Gemma* the top of the Tree is exactly like. The Leaves came out near the tops of the Branches without any order, are winged four Inches long, the *Pinna* being set on to a middle Rib, at about half an Inches distance, alternatively. They are about seven or nine in number, having an odd one at the end of the middle Rib, each of them is about an Inch and an half long, and three quarters of an Inch broad near the middle where broadest, shining something like the Leaves of *Laurus Alexandrina*, being of a dark green colour, smooth, having one middle Nerve, and several lateral ones. The principal of these Nerves end in some little small notches at the Margin, in which come first a russet or ferrugineous Moss, and then out of the middle of that a very small russet colour'd Flower on a small Foot-Stalk, after which follows sticking close to the Margin of the Leaf, the Fruit, which is round, no bigger than a small Pins Head. It is made up of a great many, almost round, dark brown or ferrugineous Seeds, set round in a fungous Body, as may be easily seen by a Microscope. The Roots are for the most part long Threads and Filaments, running into the crannies of the Rocks, seeking Nourishment to the Plant.

There is a variety of this with much broader Leaves.

It grows among the woody Hills on the Honey-Comb Rocks, near Mr. *Batchelor's* House, &c. very plentifully.

I think this a properer place for this Plant, than to be put in another, as Dr. *Plukenet* would do in his *Mantiss.* p. 81.

XXXI. *Trichomanes majus pinnis sinuatis subtus niveis.* Cat. p. 17. Tab. 35. Fig. 1. *Adianthum seu trichomanes maximum Americanum subtus argenteum pinnulis productioribus serratis summo caule involutis ex insula Jamaicensi.* Plukenet. Alm. p. 9. *An Trichomanes argenteum ad oras nigrum.* Plum. Inst. p. 540?

This Plant has several long Filaments of a dark brown colour, coming from a solid Oblong small Root, covered over at the top with a blackish hairy Moss like other the Ferns. From thence rise many Leaves about a Foot long, having reddish brown, roundish, shining Foot Stalks, and middle Ribs, on which the *Pinnæ* are set sometimes opposite to one another, sometimes alternatively, rarely towards the bottom, but near the top as thick as they can stand by one another, each *Pinna* being about half an Inch long, and a quarter broad at Base, where broadest, and for the most part ear'd, from whence it diminishes, ending in a point. They are of an irregular Figure, notch'd about the edges with some sinuations, and small incisures, being green above, and very white beneath, having on their Margin or edges, a ferrugineous Line or Moss, containing its Seed.

It grew on the side of a woody Hill, amongst the Rocks, near Mr. *Elletson*'s Plantation in *Liguanee*.

The odness of the Title made me doubt if this was Dr. *Plukenet*'s *Adianthum* abovementioned, but he himself is positive in it. *Mantiss.* p. 9. therefore he must explain how it comes to be *pinnulis productioribus serratis summo caule involutis.*

XXXII. *Tri-*

XXXII. *Trichomanes majus totum album pinnis aculeatis Trapezii figura. Cat.p.17. Tab. 36. Fig.1.& 2. Filicula maritima ex insulis Caribais, seu adianthum maritimum segmentis longioribus, angustis, auriculatis & crenatis, pediculo atro nitente Pluk. Almag. p.152. Phyt. Tab. 286. Fig.1. Filix seu lonchitis aspera Jamaicensis, nostrati similis, foliis tamen longioribus, à latere tantum superiori auriculatis, angutis per ambitum, & creberrimis denticellis spinosum. Ejusdem Alm. p.152. Lonchitis pinnulis angustioribus leviter denticulatis superiore latere auriculatis D. Sloane. Bob. hist. Ox. part. 3. p. 566. quoad descr. Lonchitis Jamaicensis glabra pinnulis alternis profundius dentatis, superiore latere auriculatis. Ejusd. ibid. An filix seu lonchitis aspera Jamaicensis pinnula lata brevi angulosa. Plukenet. Alm. p. 152? An lonchitis auriculata & serrata. Plum. fig. 29?*

This has a great many long Filaments and Fibers for its Roots, of a dark brown colour, having towards the top, where the Root is round and solid, ferrugineous Hairs. From the Roots rise the Leaves, which are from about six Inches Inches to a Foot and an half in length, having a green Stalk or middle Rib, to which are set alternatively green *Pinna*, as thick as they can stand by one another from the very Root, each of which is much larger than of the ordinary *Trichomanes*, and of an irregular Figure like to a *Trapezium* having very small prickles at their corners, and a scarce discernible, middle Rib, on each side of which is a row of Seeds in small ferrugineous Points or Spots. This Plant, sometimes has Incisures appear very plain on its edges, keeping the same Figure and Magnitude in is *Pinnula*. At other times the *Pinnula* are very long, taper to the end, and have round Incisures, as appears by *Tab. 36.* where these several varieties are graved.

It grew out of the Crannies or Fissures of the drier Rocks on the Road-side going to sixteen Mile Walk.

XXXIII. *Trichomanes majus, nigrum pinnis leviter dentatis Trapezii figura. Cat. p. 17. Tab. 35. Fig. 2. Lonchitis Americana minor, pinnulis alternis obtusioribus, modice dentatis & superiore acie pulverulentis. Bob. hist. Ox. part.3. p.567. An Adianthum sive filix trichomanoides pumila pinnulis auriculatis, ad petiolum angustis, per ambitum minutissime denticulatis. Plukenet. Alm. p. 9 ? Phyt. Tab. 251. Fig. 4? vel Filix minor Jamaicensis alis subrotundis ad pediculum angustis minutissime dentatis, Bob. hist. Ox. part 3. p.573?*

This rises, from the same sort of Root as the precedent, to about six Inches long, the Foot-Stalk being black and shining sometimes hairy, and three Inches long, as is also the middle Rib, to which are joind *Pinnulæ* set alternatively very close together, of the shape of a *Trapezium*, lesser than the former, of a darker colour, and without prickles. It has some very small jags on the edges, which on the uppermost side (when it has attained its due Magnitude) turns into a ferrugineous Welt, being the Seed, and then it does not appear serrated.

It grew out of the Rocks on the Banks of *Rio d' Oro*, and other the rocky, inland, woody parts of the Island.

XXXIV. *Trichomanes majus nigrum, pinnis Trapezii figura latissimis tenuibus. Cat. p. 17. Adianthum Jamaicense, lonchitidis amplioribus foliis non ramosum pediculo atro nitente. Plukenet. Alm. p. 11. Phyt. Tab. 252. Fig. 5.*

This had a long creeping Root exactly like that of *Polypodium*, which has many Fibrils for drawing its Nourishment. From thence rose up several black shining polish'd Stalks about a Foot high, to which, towards the top, were fastned three or four pair of *Pinnulæ* or Leaves, which were very large, and of the Figure of a *Trapezium*, and withal very tender and thin, of a light green colour, like to the *Adiantum foliis Coriandri*. C. B. They measured

Y two

two Inches the longest way, were serrated when young, but when old have a rusty coloured Welt round the edges, and are sinuated, so that it would in differing Ages impose on people, making them believe it were a differing Plant.

It was brought from *Jamaica* by Sir *Arthur Rawdon*'s Gardener, and communicated from him to me by Dr. *Sherard*.

XXXV. *Trichomanes foliolis longioribus eleganter superna præsertim parte laciniatis.* Cat. p. 17. Tab. 36. Fig. 3. *An trichomanes major foliis longis auriculatis.* Banist. Cat. Raii. hist. p. 192. ? *Adianthum seu trichomanes Bermudense maximum serratis foliis auriculatum.* Pluknet. Almagest. p. 9 ? Phyt. Tab. 125. Fig. 1 ?

The Root of this has many capillary brown Fibers, from whence rise a great many Stalks, black, shining, and small, about six Inches high, being very thick set, even almost from the Ground, at about an eighth part of an Inches distance, with *Pinnæ*, sometimes opposite, sometimes alternatively, not half an Inch long, nor an eighth part broad, very much cut in on the edges, especially the upper side, having on the backside of the Leaf, some large ferruginous Spots, in which are the Seeds.

It grew out of the Fissures of the Rocks, of each side on the *Rio d' Oro*, near Mr. *Philpot*'s Plantation in sixteen Mile Walk.

XXXVI. *Trichomanes, foliolis dentatis, superiore latere auriculatis.* Cat. p. 17. *Adianthum seu filix trichomanoides Jamaicensis pinnulis auriculatis ad basin strictioribus & rarius dentatis major.* Pluken. Alm. p. 9. Phyt. Tab. 253. Fig. 5. *Chamæfilix Jamaicensis trichomanoides alis oblongis ad basin strictioribus & auritis.* Bob. hist. Ox. part. 3. p. 573. *Adianthum Americanum alis amplioribus ex angusto principio sese dilatantibus & dentatis.* Ejusd. ibid. p. 591. *An Adiantum Bermudense minus pinnulis obtusis & dentatis, lonchitidis in modum auriculatis.* Ejusd. ibid. vel *Adianthum sive Trichomanes Bermudense maxmum serratis foliis auriculatum.* Pluken. Almag. p. 9? Phyt. Tab. 125. Fig. 1 ? *Trichomanes Americanum latifolium dentatum.* Tournefort. El. Bot. p. 431. Inst. p. 540.

This has a great many reddish brown hairy Fibrills, or Filaments for Roots, sending up Leaves an Inch or Inch and half long, being altogether green: there are four or five, or twice that number of pairs of roundish, green, quarter of an Inch Diameter'd *Pinnæ*, sometimes set opposite, sometimes alternatively to the green middle Rib. Sometimes they are indented, or cut in on the edges, with *Auricles* or small *Appendices* upwards, at other times not indented, having one odd one at the end larger than the rest. On the backsides of these *Pinnulæ* are wormy, long, ferrugineous Spots.

The differing Figures of this Plant in several States or Ages have given occasion for some mistakes about it, it having several Faces in several Seasons.

It grows out of the Fissures of the moister Rocks near the Banks of *Rio d' Oro*, towards the North side of this Island, as well as in *Barbadoes* and *Bermudas*.

XXXVII. *Filix maxima in pinnas tantum divisa oblongas latasque non crenatas.* Cat. p. 17. Tab. 37.

This rises to about four Foot high, having a light reddish brown coloured Stalk, which near the Root has some ferrugineous Moss or Hairs, and at about a Foot and a half high, begins to have *Pinnæ*. They go out alternatively at about three quarters of an Inches distance out of opposite sides of the middle Rib, which has an odd *Pinna* or Lobe at the end, closing the Stalk. Each of the *Pinnæ* is about six Inches long, and one and an half
broad

broad in the middle where broadest, from a very short, or almost no Foot-Stalk increasing to the middle, and thence decreasing to a point, being equal on the Margin, very smooth, having one middle Rib eminent on the under side, from whence go transverse Ribs to the sides of the Leaf, of a pale green colour in several things resembling the Leaves of Harts-Tongue.

It grew on the sides of Mount *Diablo*, in going to the North side of this Island.

The whole Face of this Plant differs from the *Lonchitis palustris maxima* before described, so that there is no manner of Reason to think them the same, as Dr. *Plukenet* does in his *Mantiss* p. 81.

XXXVIII. *Filix major in pinnas tantum divisa oblongas latasque non crenatas à basi rotundiore ad apicem se sensim angustantes.* Cat. p. 17. *Filix Jamaicensis pastinaca aquatica foliis alternis crassiuscule dentatis pediculo splendente nigro.* Plukenet. Alm. p. 152. Phyt. Tab. 285. Fig. 2. *Filix Americana maxima non ramosa, lobis integris acutioribus, & ad margines rotundius crenatis, maculis rotundis pulverulentis aspersis.* Bob. hist. Ox. part. 3. p. 571. *An Lingua cervina ramosa, nigris tuberculis pulverulenta.* Plum. Tournef. Inst. p. 546?

The Stalk of this Fern was about two Foot and a half high, it was cornered and only branched into *Pinnæ*, which agreed in most things with the precedent; only at roundish Base, where by a small Foot-Stalk they were set on to the middle Rib, they were broadest, *viz*. about three quarters of an Inch, and then decreased, ending in a point, being about three Inches long, they were not so regularly striated by parallel Lines as the former.

It was brought from *Jamaica* by *James Harlow*, Sir *Arthur Rawdon*'s Gardener, and by Dr. *Sherard* given to me.

This is not the *Adianthum Jamaicense lonchitidis amplioribus foliis non ramosum, &c.* Pluken. Phyt. Tab. 252. Fig. 5. which I believe is what I call p. 81. *Trichomanes majus nigrum pinnis trapezii figura, latissimis tenuibus*.

XXXIX. *Filix major scandens in pinnas tantum divisa oblongas latasque non crenatas.* Cat. p. 17. Tab. 38. *Filix Jamaicensis pinnatis fraxini foliis tenuissime crenatis apicibus mucronatis.* Plukenet. Almag. p. 153. Phyt. Tab. 286. Fig. 3. *Filix scandens Jamaicensis, pinnatis fraxini foliis* Bob. hist. Oxon. part. 3. p. 571. *Lingua cervina scandens, foliis laurinis serratis* Plum. Tournef. Inst. p. 545. *An Filix Jamaicensis pinnatis fraxini foliis tenuissime crenatis foliorum apicibus obtusis & subrotundis.* Plukenet. Phyt. Tab. 286? Fig. 4. *An Filix Jamaicensis alato pediculo fraxinella foliis crenatis radice radic e repente.* Ejusd. Almag. p. 154. Tab. 287. Fig. 3 ?

This had a crooked Stalk of a dark brown colour, cornered, sometimes smooth, at other times covered with a ferrugineous hairy Moss, as big as ones little Finger, sending out of each side many several Inch long hairy *capreoli*, sticking to the Barks of Trees, and rising by them to twenty or thirty Foot high, being greenish, made up of many white Filaments within. These Stalks or Roots send out Leaves at about two Inches distance a Foot and a half long, whose *Pinnæ* are set on to the middle Rib, alternatively at about half an Inches distance, being about three Inches long, and three quarters of an Inch broad, in the middle where broadest, being narrow at beginning, and end, having one middle, and several transverse Ribs, thin, smooth, and of a pale green colour, like the Leaves of Harts-Tongue.

There is a variety of this, whereof the Leaves are very superficially indented, and smaller; and likewise another which is rounder at end, being I suppose from the Age of the Plant.

The Juice mix'd with the Oil of *Sergelim*, (or *Sesamum*) Ginger, and long Pepper, and anointed on the Head, Cures the Cholerick Headach. *H. M.*

It grew on the Trees, on the South side of Mount *Diablo*, and other the Trunes of large Trees, in the mountainous, woody, inland parts of this Island.

XL. *Filix major scandens in pinnas tantum divisa, oblongas latissimas non crenatas. Cat. p. 18. Tab. 39.*

This had a Stalk with Fibrils, as other the scandent Ferns, by which it climbed the Trees as they, being like them in every thing. From this rose a pale green cornered middle Rib, like that of other Ferns, to about a Foot and a half high, towards the top of which were two or three pair of *Pinnæ*, with an odd one at the end, they were much broader than the precedent, being about five Inches long, and half as broad, having one middle Rib, and several lateral Fibers, being broadest in the middle, and decreasing both ways, ending in a point, and intire without any indentures. Sometimes the number of *Pinnæ* was only two, and sometimes three.

It was brought from *Jamaica* by Sir *Arthur Rawdon*'s Gardener, and communicated to me by Dr. *Sherard*.

Any Person who compares the Description of this Plant, and its Figure with Dr. *Plukenet*'s *Filicifolia lonchitidis facie Jamaicensis, &c. Alm. p.* 152. *Phyt. Tab.* 286. *Fig.* 2. will not doubt with him, *Mantiss. p.* 79. whether it be the same.

XLI. *Filix major in pinnas tantum divisa, oblongas, angustasque, non crenatas. Cat. p.* 18. *Tab.* 40. *Filix non ramosa alis longis salicinis alternis non crenatis. Bob. hist. Ox. part.* 3. *p.* 572.

This had sometimes a blackish, and at other times a light brown coloured Stalk, or middle Rib, rising to two Foot high, at about 9 Inches from the Ground, be set with *Pinna*, which were sometimes opposite one to another, and sometimes set alternatively at about three quarters of an Inches distance. They were about four Inches long, and not over half an Inch broad in the middle where broadest, equal on the edges, beginning narrow, and ending in a long narrow point, being of a dark green colour above, and underneath having an eminent middle Rib, and some transverse ones, wholly covered over with a ferrugineous Moss, in which is the Seed.

There is another variety of this, differing only in magnitude, being not over half so big, in every thing else the same.

They grow very plentifully on the Banks of the *Rio d' Oro*, near *Archers Ridge*, between sixteen Mile Walk, and St. *Mary's* in the North-side.

This is not the *Filix non ramosa Jamaicensis pinnatis foliis integris, &c.* of Dr. *Plukenet*'s *Almagest. p.* 154. figured by him *Tab.* 285. *Fig.* 1. *Phyt.* as he doubts in his *Mantiss. p.* 81. This may appear by this Description, and their Figures compared. What he means by saying I had *rectius* have made it this, when he is of another Opinion the Line before, is to me a Riddle.

XLII. *Filix minor plerumque trifida, pinnis oblongis latisque non crenatis. Cat. p.* 18. *Tab.* 41. *Fig.* 2. *Filix Jamaicensis amplissimo folio phyllitidis facie triphylla. Plukenet. Almag. p.* 154.

This has a solid roundish Root, covered over with a hairy ferrugineous Moss, and many long Filaments of a dark brown colour, drawing its Nourishment as other the smaller Ferns. The Stalk is of a light brown colour, striated, about nine Inches high, at the top divided for the most part into three Leaves or *Pinnæ*, that one in the middle or top of the Stalk being largest,

largest, about four Inches long, and one and an half broad in the middle where broadest, being narrow at beginning and end, smooth, having one middle, and some transverse Ribs of a dark green colour, something like Harts-Tongue.

It grew in the shady thick Woods on the South side of Mount *Diablo*.

How Dr. *Plukenet* comes to be angry with me for calling this *minor*, *Mantiss.* p. 81. or how he comes to call it *folio amplissimo* in his *Almagest*, he best knows.

XLII. *Hemionitidi affinis Filix major Trifida auriculata pinnis latissimis sinuatis. Cat.* p. 18. *Tab.* 42. *Filix seu hemionitis dicta Caribaeorum amplissimis foliis trifoliata. Plukenet. Almag.* p. 155. *Phyt. Tab.* 291. *Fig.* 3.

This has a Root four Inches long, made up of many round black Scales, lying on one another, with a great many black Fibers and Filaments, drawing its Nourishment. The Stalk rises two Foot high, at coming out of the Earth it is covered with a ferrugineous Moss,] is smooth, of a light reddish brown colour, and divided towards the top into three broad Leaves or *Pinna*, two being set opposite one to another, and one being at top, which is the largest, being ten Inches long, and four broad, near the Base, where broadest, variously sinuated, or cut in on the edges, making small irregular Segments on the Margin, of a yellowish green colour, and thin, having one middle Rib, and several transverse ones, along which are rows of ferrugineous Spots, where lies the Seed. The undermost pair of *Pinna* have Ears, or *Appendicula*, taking their beginning from the Base of each *Pinna*, and being like it in every thing, only smaller.

It grew on a woody shady Hill, near the Banks of the *Rio Cobre*, by the Orange Walk in the Crescent Plantation. If this be what *Piso* means, he says 'tis very opening, purging of Choler, cutting and aperitive above the *European Polypodium*.

XLIII. *Filix major in pinnas tantum divisa, raras, latiores, oblongas, striatas, ex adverso sitas, & non crenatas. Cat.* p. 18. *Tab.* 41. *Fig.* 1.

This, from a black knobbed tuberous Root, rose to about two Foot high, by a cornered light brown Stalk, which when about nine Inches high, had *Pinna* set on it, always opposite to one another, at about an Inches distance. Each of them were four Inches long, and more than an Inch broad in the middle where broadest, being narrow at beginning and end, having one middle, and several transverse parallel appearing Ribs, or *Striae*, of a dark brown colour, equal at the edges or Margin, of a fresh green colour, and smooth.

It grew on the sides of Mount *Diablo*, in the middle of the Island.

This is not the *Adianthum Jamaicense lonchitidis falcatis foliis, &c. Plukenet. Alm.* p. 11. *Phyt. Tab.* 253. *Fig.* 1. This of Dr. *Plukenet* I conjecture to be the *Adianthum nigrum majus non ramosum, pinnulis & surculis rarioribus crassis & trapezii in modum figuratis* described by me hereafter, which F. *Plumier* has figured *Tab.* 47. under the Title of *Adianthum nigrum ramosum pulverulentum & falcatum*.

XLIV. *Filix major in pinnas tantum divisa, raras, oblongas latasque crenatas. Cat.* p. 18. *Filix Jamaicensis foliis semel subdivisis, pinnulis obtusioribus costae ad nascentibus sorbi aucupariae foliis quodammodo referentibus. Plukenet. Alm.* p. 155. *Phytogr. Tab.* 291. *Fig.* 1. *Filix Americana major non ramosa, alis integris amplis profundius crenatis, ad quemlibet nervum obliquum binis macularum ordinibus notatis Bob. hist.* Ox. part. 3. p. 571. *Filix Jamaicensis hirsuta sorbi aucupariae foliis quodammodo referens. Ejusd. ibid.* p. 576. *Filix minor sive Chamaefilix Jamaicensis*

Z

tensis in pinnas tantum divisa ala longiore costam claudente. Ejusdem, ibid. p. 577. Lingua cervina non ramosa rotundius crenata. Tournef. Inst. p. 545.

This Fern rises to about three Foot high, having a shining light brown, or gray coloured Stalk, almost square, to each side of which, at about nine Inches from the ground, and an Inches distance from one another are placed alternatively, the *Pinnæ*. They are about five Inches long, and about an Inch broad in the middle where broadest, very regularly cut in on the margins, so that they appear like *Pinnula* join'd at the edges. The colour of the *Pinna* is on the upper side dark green, on the under side are many eminent middle transverse Ribs, along which, on each side are rows of ferrugineous Spots, containing the Seed. The Root is black and large, with many large black Strings or *Ligula* going from it into the Earth.

It grew on the Banks of *Rio d' Oro*, near sixteen Mile Walk.

A young Plant of this is described by Mr. *Bobart. p. 577.* One a little larger, *p. 576.*

XLV. *Filix minor in pinnas tantum divisa, crebras, plerumque ex adverso sitas, oblongas, latasque, crenatas. Cat. p. 18. Tab. 43. Fig. 1.*

This is in every thing the same with the former, only not over half so high, the *Pinna* are more frequent, set for the most part opposite to one another, though sometimes alternatively, and are five Inches long, not being over three quarters of an Inch broad at Base where broadest, and whence they decrease to the point, being regularly cut in on the edges, and having on their backside a row of ferrugineous spots on each side of the eminent middle Rib.

It grows on Mount *Diablo*, near *Archers Ridge*, and other inland woody parts of this Island.

XLVI. *Filix minor in pinnas tantum divisa, raras, oblongas, latasque crenatas. Cat. p. 18. Tab. 43. Fig. 2.*

This is in every thing the same with the former, only, although as high, yet 'tis in every thing lesser, the *Pinna* a little more frequent, shorter and narrower by much, than that immediately preceding, being not over half an Inch broad at Base where broadest, ending in a point. And in this, which seems to be quite different from the former, there are some varieties.

It grew with the former.

XLVII. *Filix minor, in pinnas tantum divisa, crebras, non crenatas, inferiore latere auriculatas & rotundis pulverulentis areolis aversa parte notatas. Cat. p. 18. Tab. 44. Fig. 1. Lonchitis minor, alis crebris non crenatis, inferiore latere auriculatis. Bob. Mor. hist. Ox. part. 3. p. 567.*

This Fern rises about a Foot and a half high, its Stalk at coming out of the Ground, being covered with a hairy ferrugineous Moss, having *Pinnæ* set on to the middle Rib, at about four Inches from the Ground, and one third of an Inch from one another. For the most part they come alternatively out of each side of the middle Rib, being near three Inches long, and not over half an Inch broad at their auriculated Base where broadest, and whence they decrease to the point. They are smooth, of a dark green colour, having on their under side an eminent middle Rib, on each side of which are ferrugineous round spots, containing the Seeds.

There is a variety of this, having the *Pinnæ* broader, and coming out opposite to one another.

It grew on the sides of Mount *Diablo* very plentifully.

XLVIII. *Filix minor, in pinnas tantum divisa, crebras, non crenatas, auriculatas, & lineis pulverulentis aversa parte notatas.* Cat. p. 18. Tab. 44. Fig. 2.

This in magnitude, manner of growth, &c. seemed to be exactly the same with the former, only the *Pinnæ* in lieu of one Auricle on the under side, had two small ones, one on its under, and another on its upper side, and in lieu of spots on its backside, containing the Seed, there are two Lines running by the side of the middle Rib, containing the same.

It grew with the former.

No body can doubt whether this be the same with Dr. *Plukenet*'s *Filicifolia lonchitidis facie Jamaicensis, &c.* Alm. p. 152. Phyt. Tab. 286. Fig. 2. when this described here has very long *Pinnulæ*, and Lines on its Back, and the other has spots, &c. as appears by comparing their Descriptions and Figures.

XLIX. *Filix minor, russa lanugine tota obducta, in pinnas tantum divisa, raras, non crenatas, subrotundas.* Cat. p. 19. Tab. 45. Fig. 1. *An Filicula crispa lanugine hepatici coloris vestita.* Plukenet. Alm. p. 150? *An Filix hirsuta & lutescente pulvisculo sordescens.* Tournef. Inst. p. 537?

This was about a Foot and a half high, its Stalk was round, and all covered with a ferrugineous coloured Hair, the *Pinnæ* were placed about an Inches distance, each being fastned to the middle Rib, by a very small Foot-Stalk. Each of the *Pinnæ* were about an Inch long, half as broad at round Base where broadest, and whence they decreased to the end, which was round, they were all covered over with a rusty woolly Hair, sufficiently distinguishing them from any others of this Kind. The *Pinnæ* were likewise all covered over with Seed on the back of the Leaf.

Sir *Arthur Rawdon*'s Gardener brought it from *Jamaica*, and Dr. *Sherard* communicated it to me.

When Dr. *Plukenet* pleases to figure and describe the abovemention'd *Filicula* we shall find if it be the same.

L. *Filix minor, in pinnas tantum divisa, ex adverso sitas, oblongas, latasque non crenatas, infimis ad basin auriculatis.* Cat. p. 19. *Filix seu polypodium Jamaicense, hydrophylli Morini fere divisuris, summo folio raris denticulis profundius crenato.* Plukenet. Alm. p. 153. Phyt. Tab. 289. Fig. 3. *An filix seu polypodium Jamaicense latifolium pinnis infimis auritis.* Ejusd. Alm. p. 153?

This has a roundish small Root, with many round Thongs and Filaments, drawing its Nourishment. From this rise Leaves about a Foot high, having a cornered ferrugineous Stalk, to which about six Inches from the Ground, are set opposite to one another, the *Pinnæ*, being about four pair, with an odd one at end. They are about three Inches long, and more than one broad in the middle where broadest, from whence they decrease, ending in a point, being of a rusty green colour, having a middle Rib, and some transverse ones. The undermost pair of *Pinnæ* have an Ear or *Appendicula* at Base, shap'd in every thing like the *Pinnæ*, only not so large, and more blunt. The uppermost Leaf or *Pinna*, is somewhat sinuated or indented.

I found it in the mountainous and woody inland parts of this Island.

This is not the *Filix parva Virginiana, pinnulis productioribus, &c.* Plukenet. Phyt. Tab. 287. Fig. 2. Almagest. p. 152. as he seems to conjecture, in his Mantiss. p. 79. but that abovementioned of his Alm. p. 153. Phyt. Tab. 289. Fig. 3.

LI. *Phyllitis ramosa trifida. Cat. p. 19. Tab. 45. Fig. 2. Filix Jamaicensis non ramosa trifoliata angustifolia caule lævi. Bob. hist. Oxon. part. 3. p. 572.*

This, from a blackish tuberous Root, with many *Ligulæ*, sends up five or six Stalks about a Foot high, having a reddish cornered shining middle Rib, to which at about two Inches or less interval, are set opposite to one another, (after the manner of the *phyllitis multifida*) *Pinnulæ* or Leaves. Three *Pinnulæ* for the most part stand on the same very short common reddish Foot-Stalk, that *Pinnula* in the middle being the longest, *viz.* an Inch and an half long, and not over a quarter of an Inch broad in the middle where broadest, smooth, of a yellowish green colour, with reddish Ribs, and of the consistence of the Leaves of of *Phyllitis*, or *Hemionitis Multifida*, and being covered all over on their Backs with ferrugineous Powder, supposed to be the Seed.

This in every thing comes very near the *Lingua cervina foliis costæ innascentibus. Tournefort. El. Bot. p.* 431. *Tab.* 324 or *Hemionitis altera. Dal. Lugd.* 1218.

I gathered it in the woody inland parts of this Island.

By the Figure and Description of this, compared with the *Filicifolia phyllitis dicta, seu lingua cervina minor crispa, folio multifido ramosa Plukenet. Alm. p.* 154. *Phyt. Tab* 248. *Fig.* 2. one would wonder how this Author should affirm these Plants to be the same, as he does in his *Mantissa, p.* 81.

LII. *Phyllitidi multifidæ affinis filix scandens, in pinnas tantum divisa, oblongas, angustas non crenatas. Cat. p.* 19. *Tab.* 46. *Fig.* 1. *An Filix Scandens periploca foliis Jamaicensis. Plukenet. Alm. p.* 156 ?

This has a round Root the top of which is covered with a blackish Hair, having many strong Filaments drawing its Nourishment. From hence rises a round, smooth, small, shining, reddish brown Stalk, turning its self round any Tree it comes near, rising to twelve or fifteen Foot high, at every three or four Inches distance, putting forth Leaves, standing for the most part opposite one to another, on sometimes an Inch, or two Inches long round Stalks, divided usually into four *Pinnæ* or Leaves which are three Inches long, and about an Inch broad at the Base where broadest, whence they decrease to the end, being very green, and in consistence, smell, &c. like the Leaves of *Phyllitis*.

It grew plentifully on Mount *Diablo*, *Archers Ridge* in Mountain River Plantations, and in several other the mountainous, and inland woody parts of this Island.

LIII. *Phyllitidi multifidæ affinis, filix scandent in pinnas tantum divisa, oblongas, angustas laciniatas. Cat. p.* 19. *Lonchitis scandens, Brasiliana pinnulis eleganter laciniatis Tournefort. El. tot. p.*430. *Inst. p.* 519.

This, as to manner of growth, &c. is the same with the former, but differs in this, that the *Pinnæ* are broader, shorter, and much more laciniated, or cut into the edges, so that some of the Leaves are almost palmated, especially the under ones, those towards the top of the Stalks being like the former, only *Laciniated* on the edges, where on the Sections on the backside of the Leaf is a ferrugineous Moss covering the Segments, which is the Seed. And of this sort again there are infinite varieties of the Leaves.

This grew about the Trunks of Trees in a Wood near Captain *Heywood*'s House, in St. *Maries*, in the North side of the Island of *Jamaica*.

I think the Title I give this very plain, *viz.* that this Fern comes near to the *Phyllitis Multifida*, a Plant very well known in *Europe*.

LIV. *Filix*

LIV. *Filix non ramosa maxima, surculis raris, pinnis variis, inferioribus scil. oblongis acutis, superioribus verò absentis subrotundis. Cat. p. 19. Tab. 47. Filix sarmentosa, foliis variis, imis longioribus denticulatis, superioribus rotundis planis ex insula Jamaica. Pluken. Alm. p. 156. An Filix ad alas foliosa Tournef. Inst. p. 537?*

This great Fern rises to four Foot high, by a reddish brown Stalk, as big as ones Finger, having Twigs plac'd alternatively, at two or three Inches distance; about a Foot and a half long. They are beset on each side with *Pinnæ* alternatively at more or less distance, each of which is about four Inches long, and not an Inch broad near its beginning where broadest, and whence it decreases, ending in a point, being a little indented on the edges in the broadest part, and every where of a Grass-green colour, and having its Seeds lying in a ferrugineous Line or Welt along its edges: the tops of the lower Twigs, and the whole ones at top are broad and sinuared, or waved after the manner of *Asplenium*.

It grows about Mount *Diablo* in several Places.

What Dr. *Plukenet* means by *Sarmentosa* in his Title, I cannot guess.

LV. *Filix non ramosa major, surculis raris, pinnulis crebris, latis, minimis, brevibus, aculeatis. Cat. p. 19. tab. 48. An Filix non ramosa Jamaicensis pinnatis foliis integris serratis ad basin apophysi parva donatis subtus duplici ordine punctorum ferrugineorum agminatim insignitis. Plukenet. Alm. p. 154. Phyt. Tab. 285. Fig. 1? Filix non ramosa Jamaicensis elatior, alis crebris longissimis angustissimis dentatis. Bob. Mor. hist. Ox. part. 3. p. 575.*

This Fern rises above two Foot high, the Foot-Stalk at coming out of the Ground being swelled and covered with a ferrugineous Moss, the rest being of a greenish colour. It is about a Foot high before any Twigs are set on it, which go out alternatively at half an Inches distance, being more than six Inches long, and not over a quarter of an Inch broad, each of the *Pinnulæ* being not over the eighth part of an Inch long, and half as broad, sharp at the ends, and as thick set on to the Twig or middle Rib, as they can stand by one another, leaving no defect or empty space between them, and being of a yellowish green colour.

It grew by the Banks of *Rio Cobre*, in the Road going from Town to sixteen Mile Walk.

LVI. *Filix non ramosa major, surculis raris, pinnulis longis, angustis, rarissimis. Cat. p. 19. Tab. 49. Fig. 1.*

This rises to three Foot high, having Twigs going out of each side of the Stalk alternatively at about more than an Inches distance from one another, though sometimes 'tis not so much. Every Twig is about three Inches long, and three quarters of an Inch broad at Base where broadest, the *Pinnulæ* are about one third of an Inch long, and very narrow, having a large defect between them, making them appear very rare or thin set. On the backside of the Leaf they are mostly all covered over with ferrugineous Spots of Seed. Of this are some varieties.

It grew on the Banks of *Rio d' Oro*, between sixteen Mile Walk and St. *Maries*.

Dr. *Plukenet* doubts if this be his *Filix Jamaicensis Jaceæ majoris Gerard. aut potius centaurii maj. lutei Parkinson. æmula, falcatis foliis integris & margine æquali. Almag. p. 154.* By the Description and Figure of this it seems impossible to be it.

A a LVII. *Filix*

LVII. *Filix non ramosa major, surculis crebris, pinnulis longis, angustis.* Cat. p 19. Tab. 50. Fig. 1. & Tab. 51. *An filix palustris mas non ramosa pinnulis obtusioribus planis ex insula Bermudensi.* Plukenet. Phytogr. Tab. 243. Fig. 6 ? Almag. p. 151 ?

This rises to about two Foot high, having a greenish yellow middle Rib, or Stalk, to which at one third of an Inches distance are set the Twigs, sometimes opposite, at other times alternatively, being each about seven Inches long, and an Inch broad at the Base where broadest. The *Pinnulæ* are each half an Inch long, join'd to one another at their middle Rib, or original, and thence grow narrower to their ends, they are sharp, being of a yellowish green colour, and there is a defect between the ends of each of the *Pinnulæ*. Of this there are several varieties, those most remarkable being in the narrowness and breadth, sharpness or bluntness, and crookedness or streightness of the *Pinnulæ*, two of which varieties are figured Tab. 50. Fig. 1. & Tab. 51.

It grew by the inland woody and shady Rivers of this Island.

LVIII. *Filix non ramosa major, surculis raris brevioribus, pinnulis crebris latis brevibus non aculeatis.* Cat. p. 20. Tab. 48. Fig. 2. *Filix Indica major alis longis rotundius crenatis binis macularum ordinibus ad oras notatis.* Arana-panna. H. M. p. 12. Tab. 31. Bob. p. 3. hist. Ox. p. 575. *quoad. descr.*

This was the same with the preceding, only the *surculi* were not so long, and the *Pinnulæ* were not aculeated, but blunt and roundish, almost like those of *Asplenium.*

It was brought from *Jamaica* to Dr. *Sherard*, and by him communicated to me.

LIX. *Filix non ramosa major, surculis crebris, pinnulis brevioribus, angustis.* Cat. p. 20. Tab. 50. Fig. 2.

This is in every thing the same with the former, save one, only the Twigs are not so broad, being not over two thirds of an Inch broad, that is, the *Pinnulæ* are not over one third of an Inch long, and not quite so narrow at bottom, so that there is less void space between them, and of this Kind there are several varieties as to length of *surculi,* &c. The Seed lies in round ferrugineous Spots on each side of the middle Rib running through each *Pinnula.*

I found it on the rocky Banks of *Orange* River, and *Rio d' Oro,* between sixteen Mile Walk and St. *Maries.*

This is not the *Filix non ramosa Jamaicensis, pediculo albicante, alis marrubii aquatici fere divisuris,* &c. Pluken. as appears by its Figure and Description compared with those of his, Phyt. Tab. 290. Fig. 2.

LX. *Filix non ramosa minor, surculis crebris, pinnulis longis, latisque.* Cat. p. 20 Tab. 50. Fig. 3.

This has a light brown greenish shining Stalk, rising to about nine Inches high, or sometimes a Foot and a half, to each side of which, at about one third of an Inches distance are set Twigs about an Inch and an half long, and near an Inch broad at Base where broadest, the *Pinnulæ* being joined to near their ends, and being of a dark green colour, much resembling the precedent, only the *surculi* are broader, or which is all one, the *Pinnulæ* longer. The *Pinnulæ* are also broader, so that whereas in that going before there is a defect or empty space between the ends of the *Pinnulæ,* here it is none or very small, and the *Pinnulæ* lie over one another. The Root is knobbed, and has many Fibrils of a cinereous colour, sending up three or four Stalks from it, by which it is sufficiently distinguished from the *Filix minor palustris.*

Rali,

The Natural History of Jamaica. 91

Raii hist. p. 146. to which it comes very near. The Seed lies in round spots on the backsides of the Leaves.

It grew in the inland woody parts of this Island, about *Rio d' Oro* and *Orange* River Banks plentifully.

LXI. *Filix non ramosa minor, surculis crebris, pinnulis brevissimis, angustis. Cat. p.* 20. *Tab.* 52. *Fig.* 1. *Filix non ramosa Americana major, alis oppositis longis angustis profunde dentatis, Bob. hist. Ox. part.* 3. *p.* 575.

This is in every thing the same with the preceding, not being much over a Foot high, the Twigs three Inches long, and half an Inch broad. The *Pinnulæ* are a quarter of an Inch long, and join'd together almost to the end, being mark'd on the back with ferrugineous Spots, and are on the upper side of a yellowish green colour.

It grew on the Banks of *Rio Cobre* below the Town of St. *Jago de la Vega*, on the same side of the River.

LXII. *Filix non ramosa minima, surculis crebris, pinnulis brevissimis, angustis. Cat. p.* 20. *Tab.* 52. *Fig.* 2.

This has a great many black Filaments coming from a dark brown, roundish, small, solid Root, from whence rises a Stalk about six Inches high, having Twigs and *Pinnulæ* set just like the foregoing, only in every part much smaller, especially the *Pinnulæ*, which are join'd so far together, that the *surculus* seems rather Indented than to have *Pinnulæ*.

It grew with the former.

LXIII. *Filicula non ramosa minima, surculis crebris, pinnulis angustissimis, raris. Cat. p.* 20. *Tab.* 49. *Fig.* 2. *Filix Jamaicensis non ramosa, ex una radice cespitis instar contexta, pediculis numerosis, foliis auriculatis profunde sectis, Plukenet. Alm. p.* 150. *Phyt. Tab.* 283. *Lonchitis tenuifolia parva Jamaicensis, pinnulis oppositis profunde dentatis & utrinque auriculatis. Bob. hist. Ox. p.* 567.

This seem'd to be much the same with the former, only it was not over four or six Inches high, the Twigs were very short, set to the Stalk alternatively, and on them the *Pinnulæ*, which were not united to one another by the Margins, as those of the preceding, but cut in to the very Twig where there was any empty void space, or defect between them, occasion'd by their being so narrow, or rarely plac'd. The Seed comes in round Spots like that of *Adiantum Nigrum officin.* on the back sides of the *Pinnulæ*.

It grew in the inland parts of the Island by the shady Rivulets.

LXIV. *Filix non ramosa scandens, pinnulis latis, longis, crebris, obtusis & dentatis Cat. p.* 20. *Para-panna-Maravara. H. Mal. part.* 11. *p.* 31. *Tab.* 15.

This has a cornered dark green brownish Stalk, by the means of several fibrils and clavicles, taking hold of the Truncs of Trees, and rising by them twenty or thirty Foot high, putting out, at about nine Inches distance, Leaves about a Foot and an half long. It has a shining dirty green coloured Stalk, which from its middle upwards, at about an Inches distance alternatively, is beset with Twigs about four Inches long, and an Inch and an half broad, each of which has many *Pinnulæ* about three quarters of an Inch long, and not half so broad, indented about the edges, and of a lively shining dark green colour.

This is described by Mr. *Bobart* in the *hist. Ox. part.* 3. *p.* 578. under the Title of *Filix non ramosa major, surculis raris, brevioribus, pinnulis crebris latis brevibus non aculeatis. p.* 20. of my Catalogue.

It grew on the Truncs of the larger Trees in going up the sides of Mount *Diablo, Archers* Ridge, &c.

It

92   The Natural History of Jamaica.

It is given against Endemial Diseases, cures intermitting Feavers, stops Coughs, and opens the Belly. The Juice mix'd with the Blood of a Hen, cools the heat coming of Gun-Powder, or hot Oil. *H. M.*

LXV. *Filix non ramosa minima, caule nigro, surculis raris, pinnulis angustis, raris, brevibus, acutis, subtus niveis.* Cat. p. 20. Tab. 53. Fig. 1.
This was in every thing the same with the *Filix non ramosa minor, caule nigro surculis raris, pinnulis angustis, dentatis, raris, brevibus, acutis, subtus niveis.* Cat. Jam. p. 20. described above, p. 35. only not over half its bigness, so that 'tis not the same with that, nor consequently with Dr. *Plukenet*'s, *Adianthum Calomelanos,* &c. which I have in my Catalogue made the same with it.
It grew with the former in *Jamaica*.

LXVI. *Filix non ramosa major, caule nigro, surculis raris, pinnulis angustis, raris, longis, dentatis.* Cat. p. 20. Tab. 30. Fig. 2.
This agreed very exactly with the preceding Fern, only it was four times as large as it, and the *Pinna* were much longer, and indented on the edges, so that it might almost seem referable to the ramose Kind, the backsides of the Leaves or *Pinna* are not covered over with a white Meal as the precedent, but are gray or light rusty. This Meal is the Seed. The upper sides of the *Pinna* were of a very dark green colour.
It grew on the Rocks upon the Banks of *Orange* River, and *Rio d' Oro,* near Mr. *Philpot*'s Plantation in the North side.
It is easie to see this is not the *Adianthum Jamaicense lonchitidis falcatis foliis,* &c. *Plukenet. Alm.* p. 11. *Phyt.* tab. 253. fig. 1. as Dr. *Plukenet* questions in his *Mantissa* p. 6. That is another Plant before described.

LXVII. *Ruta muraria accedens Filicula non ramosa minima, pinnulis subrotundis profunde scissis.* Cat. p. 21. Tab. 52. Fig. 3.
This had a small solid black Root, covered with a black hairy Moss towards its top, and many dark brown Filaments, whereby it draws its nourishment, from whence rise nine or ten Leaves about three Inches high. The Stalks are of a dark green colour, and at an Inches distance from the Ground, are divided into several Twigs, set alternatively, those in the middle being largest, about three quarters of an Inch long, made up of *Pinnula* set alternatively, being very small, roundish, deeply cut in on the edges, of a pale green colour above, and underneath having very many ferrugineous Spots, in which lies the Seed.
It grew among the Rocks on the Banks of *Rio d' Oro,* between sixteen Mile Walk and St. *Maries*.

LXVIII. *Ruta muraria accedens Filix minor non ramosa, pinnulis subrotundis, profunde scissis.* Cat. p. 21. *Filix pinnulis cristatis Plumier.* p. 16. Fig. 25. *A. Filix minor Jamaicensis alis obtusis, pinnulis cristatis profunde scissis. Bob. hist. Ox. part.* 3. p. 581.
The top of the Root of this Fern sent up a great number of Leaves, about a Foot high, whose Stalks were of a dark green colour, cornered, and about six Inches from the Ground, divided into Twigs, set opposite to one another underneath, and alternatively above, at about three quarters of an Inches distance asunder. The Twigs at bottom, or nearest the Root, where largest, were about an Inch and an half long, having *Pinnula* or Leaves plac'd alternatively on them, being about nine pair, with an odd one at end. They had a small Foot-Stalk, were roundish, very deeply cut in on the edges, of a dark green colour above, and underneath covered very thick with large and many rusty Spots, in which lay its Seed.

It

It grew by the sides of *Rio d' Oro*, near Mr. *Philpot*'s Plantation, between sixteen Mile Walk and St. *Maries*.

LXIX. *Ruta muraria major, foliis variis, scil. oblongis integris, & subrotundis serratis. Cat. p.* 21. *Tab.* 52. *Fig.* 2.
This has a small solid Root, covered with some scaly rusty Hair, and having many long Strings and Filaments, whereby it draws its Nourishment. From hence rise many Leaves about six or nine Inches high, having pale green Stalks, divided towards the top into several Twigs, coming out for the most part opposite to one another, having set on them *Pinnula*, or Leaves, two, or three pair, with an odd one at the end, they being almost Round or Oval, much larger than those of Wall-Rue, and very orderly indented about the edges, of a pale green Colour. In the middle of these Leaves rise up others from the same Root, having higher Stalks, and the *Pinnæ* set after the same manner, only longer and narrower than the first Leaves. They are without incisures, and have on their backsides, by the Margin, a ferrugineous Welt or Line, in which lies the Seed.
It grew out of the Fissures of the Rocks, on the Banks of *Orange* River, and *Rio d' Oro*, in the middle of this Island.
This Plant is perfectly differing from *Adianthum foliis inferioribus Coriandri, &c. Plukenet. Alm. p.* 9. and all the Plants this Author conjectures to be the same with it, in his *Alm. & Mantissa, p.* 5.

LXX. *Ruta muraria maxima, foliis oblongis, crenatis. Cat. p.* 21. *Tab.* 46. *Fig.* 2.
This had a solid strong Root, covered with many large brown Scales, and having many capillary Fibrils, from whence rose many dark green Stalks about a Foot high, divided into many Twigs, set sometimes alternatively, and sometimes opposite to one another, at about half an Inch's distance one from another. On these were plac'd three or four pair of *Pinnula* or Leaves, with an odd one, being much lesser than the former, or those of *Ruta Muraria*, crenated or dented on the edges, being smaller, more oblong, and less round than the preceding.
It grew on the rocky Banks, of *Orange* River, and *Rio d' Oro*.

LXXI. *Filix non ramosa minor, pinnulis crebris, obtusis, crenatis. Cat. p.* 21. *Tab.* 54. *Fig.* 1.
This had a black oblong Root covered with many ends of the Foot-Stalks of Leaves dropt off, and having long Thongs and Fibrils deeply fix'd in the Ground. From hence rise green Stalks, blackish at the Root, about a Foot and a half high, divided into Twigs, set alternatively, on which the *Pinnula* are very thick plac'd, so as to leave no empty space between them, being large, broad, blunt, indented, and of a dark green colour.
It grew by the *Rio d' Oro*, between St. *Maries*, and the North-side.

LXXII. *Adiantum nigrum maximum, non ramosum, ptunis crebris, majoribus, crassis, & Trapezii in modum figuratis. Cat. p.* 21. *Tab.* 55. *Fig.* 1.
This has a black Root, covered with a ferrugineous hairy Moss, having many black Fibers running into the Ground. It rises two Foot high, by a strong black Triangular Stalk, covered with a hairy ferrugineous Moss, from whence, at about a Foot from the Ground, proceed, at half an Inch's distance, alternatively, the Twigs, which are thick set with *Pinnulæ* or Leaves, alternatively, each being an Inch long, and not over half so broad, of the figure of an irregular Lozenge, or *Trapezium*, being thicker, and of a darker

B b   colour

The Natural History of Jamaica.

colour than those of the *Adiantum foliis Coriandri*. The Seed lies in a rusty Line along the Margin of the *Pinna*. Of this there are some varieties as to largeness, &c.

This grew on a shady Gully's Banks, beyond *Troopers Quarters*, near the Town of St. *Jago de la Vega*, and in other inland Woods of this Island.

Piso commends this, and says, that 'twas commonly used in *Brasile* for Expectorating tough Phlegm, and for other the uses, are usually made of *Europæan* Maiden-Hair.

I take this to be *Conambai-miri five adianti species prior vel Avenea Lusitanis. Pis.* the Figure and Description agreeing with it as his second Kind of the same agrees with the *Filix non ramosa minor caule nigro, &c.* described above p. 35. of this Book.

LXXIII. *Adiantum nigrum majus non ramosum, pinnulis & surculis rarioribus, crassis & Trapezii in modum figuratis.* Cat. p. 21. *Adianthum Jamaicense lonchitidis falcatis foliis, ramosum pediculis splendentibus nigris.* Plukenet. Almag. p. 11. Phyt. Tab. 253. Fig. 1. *Adiantum Jamaicense lonchitidis falcatis foliis ramosum, pediculis splendentibus nigris,* Bob. hist. Ox. part. 3. p. 587. *An Adiantum nigrum pinnulis lonchitidis serratis minus.* Plum. p. 32. Tab. 48. *Lonchitis ramosa, pediculis nigris, pulverulenta.* Plum. Tournef. Inst. p. 539 ?

This as to heighth, stalk, &c. was the same with the foregoing, only the Twigs came out more rarely, *viz.* at about an Inch's distance, and were not so thick set with *Pinnulæ* or Leaves, so that there was a considerable empty space between them, but for shape, seed, thickness, &c. they were the same, though much smaller.

It grew in the inland Woods of this Island.

LXXIV. *Adiantum nigrum minus non ramosum, pinnulis majoribus crassis Trapezii in modum figuratis.* Cat. p. 21.

This had a small repent Root, having black Fibers to draw its Nourishment, from whence sprang many Leaves six Inches high, with Twigs and *Pinnulæ*, exactly like the first *Adiantum nigrum maximum, &c.* only the Margin of the *Pinnulæ* were very easily indented, they were not altogether so thick or dark coloured.

It grew by the way side on *Archers* Ridge, and in other the inland woody parts of this Island.

This which is figured by Dr. *Plukenet* Phyt. Tab. 125. Fig. 2. has only a Stalk, *surculi & pinnæ*, and is therefore, according to me, not ramose, notwithstanding what Dr. *Plukenet* says in his *Mantissa*, p. 5. I take it to differ from the immediately preceding, as may appear by comparing his own Figure and F. *Plumiers*, Tab. 47. This last I take to be another Fern of the Doctors own figuring Phyt. Tab. 253. Fig. 1. He likewise doubts I have made a third Plant of this one, *viz. Trichomanes majus, totum, album, pinnis aculeatis, trapezii figura,* which is his *Filicula maritima ex insulis Caribbais, &c.* Phyt. Tab. 286. Fig. 1. a quite different Plant, as is plain from what is said above.

LXXV. *Adiantum nigrum non ramosum majus, pinnulis majoribus tenuibus in Trapezii modum figuratis.* Cat. p. 21. Tab. 55. Fig. 2.

This rose to about a Foot and an half high, having a very polite, black shining, Stalk, with no ferrugineous Hair on it, the *Pinnulæ* being frequent, and exactly like those of the precedent Kind, not so much indented, but of a yellowish green colour, and very thin. On the upper Margin of them, in a ferrugineous Welt, lies the Seed. The Stalk was divided into two or three Twigs.

It grew with the former.

LXXXI. *Adian-*

LXXXI. *Adiantum nigrum non ramosum majus, surculis raris, pinnulis densis, crassis, minimis, cristatis, & Trapezii in modum figuratis.* Cat. p. 21.

This rose to about a Foot and an half high, the Stalk being black, very polite, shining, and set alternatively with Twigs, at more than an Inch's distance from one another. They are five Inches long, and very thick set with *Pinnulæ* alternatively, so that there is no defect or void space between them, each of them being of the figure of a *Trapezium*; thick, very small, of a dark green colour, with a rusty Line on its edge, wherein lies the Seed.

There is a variety of this, which by accident is sometimes branched. Another having only three *Surculi*, being, I suppose, young.

There is another variety, with more frequent Twigs, and the *Pinnulæ* not figur'd exactly like a *Trapezium*, but Semicircular, a little indented, and like that of *Adiantum fruticosum Brasilianum Cornuti*. This variety seems to be that described by *Breynius*.

I found this on the woody Mountains near Mr. *Elletson*'s Plantation in *Liguanee*, and on *Archers* Ridge, near sixteen Mile Walk, &c.

This is very much commended by *Piso* for opening of the Wind-Pipe, and all the Diseases of the Lungs, either in Decoctions or Syrups.

LXXXII. *Adiantum nigrum majus non ramosum, surculis è pediculi communis summitate, tanquam centro, prodeuntibus, & stella in modum radiatis.* Cat. p. 22. *Adiantum Jamaicense, pinnulis auriculatis ramosum, quinis ramulis ex eodem cauliculi puncto expansis, ornatum.* Pluknet. *Almag.* p. 11. *Phyt. Tab.* 253. *Fig.* 3. *Trichomanes Americanum radiatum.* Bob. *hist.* Ox. *part.* 3. p. 591. *Trichomanes Americanum radiatum.* Tournef. *el. bot.* p. 431. *Lonchitis radiata polytrici facie, Ejusd. Inst.* p. 539.

This had a Root with a great number of blackish brown, long Filaments, variously interwoven, from whence rose many round, black, very polite, and shining Stalks, about a Foot high. From the top of this, as from a common Centre, went nine Twigs about six Inches long, standing round at an equal distance from one another, beset with *Pinnulæ* which were of a dark green colour, set thick by one another, like in Figure and Consistence to the precedent, and having on its Margin ferrugineous Lines, in which was its Seed and on the upper side next the Stalk, an Auricle or *Appendicula* to each *Pinna*.

It grew in the Woods near Captain *Drax*'s Plantation in the North side of the Island by the Old Town of *Sevilla*, &c.

This is only divided into a Stalk, *surculi & pinna*, and according to my method, not *ramose*, so that it may be said to have *surculi*, and be radiated, though not *ramose*, notwithstanding what Dr. *Pluknet* says in his *Mantissa*, p. 5.

LXXXIII. *Filix arborea ramosa, spinosa, caudice non diviso, pinnulis latis, densis, brevibus, tenuibus, minutim dentatis*, Cat. p. 22. Tab. 56. *An Filix Jamaicensis prælongis & angustis alis filipendulæ accedentibus, pediculo senticoso rubente.* Pluknet. *Almag.* p. 156? *Filix Jamaicensis non ramosa, pediculo insco spinoso, sorbi aucupariæ pinnulis.* Bob. *hist.* Ox. *part.* 3. p. 578. *Filix arborescens, spinosa, pinnulis in summitate serratis.* Plum. *Tournefort. Inst.* p. 537.

This has a Trunc twenty Foot high, as big as ones Leg, (after the manner of Palm-Trees) undivided, and covered with the remaining ends of the Foot-Stalks, of the Leaves fallen off, which are dark brown, as big as ones Finger, two or three inches long, thick set with short and sharp prickles. At the top of the Trunc, stand round, about five or six Leaves, about six Foot long, having a purple Foot-Stalk, very thick beset with short, sharp prickles on its

back-

backside. At about a Foot distance from the Trunc, each Leaf is divided into Branches set opposite to one another, plac'd near the bottom, at about six Inches distance from each other. The Branches are a Foot long in the middle of the Leaf where longest. The Twigs come out of the Branches alternatively, being an Inch and an half long, and about two thirds of an Inch broad in the middle of the Branch where broadest, being made up of *Pinnulæ* about one third of an Inch long, and half as broad, blunt, easily indented about the edges, of a dark green colour above, pale green below, very thin, and so close set to one another that there is no defect or empty space between them.

It grew in a Gully between *Guanaboa*, and St. *Faiths*, as also on Mount *Diablo* in great abundance.

The *Specimen* which Mr. *Bobart* saw of this Fern, was only a side Branch, so that it is no wonder he calls it *non ramosa*.

From these Trees growing on the Mountains of *Hispaniola*, the *Spaniards* argued the fertility of that Soil, making Ferns grow to such a vast bigness, which in *Europe* were so inconsiderable, not considering that the Ferns in *Europe* and here, were quite different Kinds one from the other.

LXXXIV. *Adiantum nigrum, ramosum, maximum, foliis seu pinnulis, obtusis, varie sed pulcherrime sinuatis & dentatis.* Cat. p. 22. Tab. 57. Fig. 1. & 2. *Lonchitis altissima, globuligera, minor.* Plum. Tournef. Inst. p. 538.

This rises four or five Foot high, having a smooth, reddish brown, shining Stalk, as big as ones Finger, which is divided into Branches alternatively, going out of opposite sides of the Stalk, having Twigs, thick set with *Pinnulæ* or Leaves, after the manner of *Adiantum Nigrum Off. J. B.* These *Pinnulæ* are thick, blunt, variously sinuated, or deeply cut in on the edges, especially on their upper sides, and indented about their round ends, something like the *Filicula fontana major, sive adiantum album filicis folio, Pin.* of a dark green colour, and shining on the upper side, and below, having its Seed lying in round ferrugineous Spots, especially near the greater sinuations by the edges.

It grew on the sides of Mount *Diablo* very plentifully.

LXXXV. *Adiantum nigrum, ramosum, maximum, foliis seu pinnulis obtusis tenuibus, regulariter minutissime & pulcherrime sectis.* Cat. p. 22. Tab. 57. Fig. 3.

This had a Stalk of the same bigness, heighth, and colour, covered with a rusty dust. The Branches and Twigs were likewise the same, only the *Pinnulæ* or Leaves were thinner, deeplier, and more regularly cut in on the edges, of both sides, having no sinuations, being of a dark green shining colour above, underneath of a paler, where are smaller round ferrugineous Spots, in which lies the Seed.

It grew with the former, and on the Banks of *Rio d'Oro*, and *Orange* River going to St. *Maries*, in the North side.

LXXXVI. *Adiantum ramosum scandens, pinnulis seu foliis, oblongis, profunde laciniatis, pellucidis.* Cat. p. 22. Tab. 58. *An Filix scandens adiantho Narbonensi similis Jamaicensis.* Plukenet. Alm. p. 156 ? *Adiantum Jamaicense ramosum & pellucidum, pinnulis angustis crebrioribus ut plurimum peltatis, apice filamentoso è medio exeunte.* Bob. hist. Ox. part. 3. p. 589. *An Adiantum scandens, foliis tenuissime sectis & retusis.* Plum. Tournef. Inst. p. 543 ?

This has a Stalk not so big as a Goose-Quill, roundish, black, covered towards its top with a ferrugineous Moss, and having very many Filaments or Clavicles, by which it takes firm hold of the Barks of the Trees, and rises to fifteen or twenty Foot high, turning its self round. At every

Inches

Inches distance, it puts forth Leaves about a Foot long, having about two Inches of their Foot-Stalk naked. This Foot-Stalk afterwards divides its self into Branches, sometimes set opposite to one another, but mostly alternatively: the Branches have their Twigs, on which grow the *Pinnulæ* or Leaves, being long, deeply cut in on the edges, very thin, pellucid; of a yellowish green colour, having some dark opaque Ribs running through them, and a woolly Hair on them, and the Seed on the ends of their Segments in a little Cup.

It grows on the Trunes of the larger Trees on Mount *Diablo*, and *Archers Ridge*.

It is very plain that this is not the *Adianthum Petræum perpusillum Anglicum foliis bifidis vel trifidis Newtoni. Raii hist. p.* 141. nor any of those Plants concerning which Dr. *Plukenet* raises doubts in his *Mantissa, p.* 5.

LXXXVII. *Adiantum nigrum ramosum maximum, foliis seu pinnulis tenuibus, longis, acutis, spinosis. Cat. p.* 22. *Tab.* 57. *Fig.* 4. *Filix ramosa Jamaicensis cicutæ majoris foliis, sive adianti nigri vulgaris pinnulis amplioribus, Bob. hist. Ox. part.* 3. *p.* 584. *An Filix ramosissima cicutæ foliis. Tournef. Inst. p.* 537 ?

This rises three Foot high, having a reddish coloured smooth Stalk, divided at one Foot and a half from the Ground into several Branches, having their Twigs, and they their *Pinnulæ* or Leaves, after the manner of the *Adiantum Nigrum Officin.* J. B. only they are longer, thinner, sharper at point, having there a very little prickle, as well as others, much smaller. (so as to be scarce discernible,) along their Margin, where are no incisures or very small ones. They have on their back parts, many round, rusty Spots, after manner of the other Ferns.

It grew on Mount *Diablo* very plentifully.

This is sometimes of a pale green colour, with almost pellucid thin Leaves or *Pinnulæ*.

This cannot possibly be the *Filix Africana florida similis, &c. Plukenet. Alm. p.* 156. *Phyt. Tab.* 181. *Fig.* 5. as that Author doubts in his *Mantiss. p.* 83.

LXXXVIII. *Adiantum nigrum ramosum majus, foliis seu pinnulis tenuibus, longis, acutis, spinosis. Cat, p.* 23. *Filix Jamaicensis ramosa adianti nigri pinnulis angustioribus. Bob. hist. Ox. part.* 3. *p.* 584.

The Root of this was knobbed, and had many blackish Thongs run from it into the Earth, to receive its Nourishment. It differed only from the former in bigness, and so perhaps may be only a variety. Of this I have several Samples differing in bigness and colour.

They grew on Mount *Diablo* with the preceding.

LXXXIX. *Adiantum nigrum ramosum minus, pinnulis minoribus, tenuibus, obtusis, crenatis. Cat. p.* 23. *Tab.* 54. *Fig.* 2. *Filix Jamaicensis humilior acuta alarum pinnulis inferioribus, brevioribus. Bob. hist. Ox. part.* 3. *p.* 576.

This is about the ordinary bigness of the *Adiantum Nigrum Officin.* J. B. has a whitish or pale green Foot-Stalk, the *Pinnulæ* or small Leaves lesser, thinner, and not cut in on the edges, being not so sharp, and without those small prickles the precedent Kind has. It has a small creeping Root, like that of *Polypodium*, with hairy fibrils, by which it draws its Nourishment, and sometimes comes to be a Foot high.

I found it grow with the others.

XC. *Adian-*

XC. *Adiantum nigrum ramofum minus, ramulis furculis & pinnulis raris, minimis, fubrotundis. Cat. p. 23. Tab. 13. Fig. 2.*

This rifes to nine Inches high, having black Stalks at coming out of the Earth, covered with a rufty Mofs, having Twigs towards the top, coming out at near an Inches diftance from one another, fet oppofite for the moft part. On thefe come the *Pinnulæ*, not frequently, but rarely plac'd, being the fmalleft of any of this Kind, leaving a confiderable defect between each other, being varioufly finuated or cut in on the edges, fo that they appear divided into *Pinnula*, making it *ramofe*, and of a dark green colour.

It grew amongft fome Rocks below the Town of St. *Jago de la Vega*, near the River.

XCI. *Adiantum five capillus Veneris. J. B. Rail hift. p.* 147. *Cat. p.* 23.
This was brought from *Jamaica* by *James Harlow*, and given me by Dr. *Sherard*. It ought to have been among the Ferns that are not ramofe.

XCII. *Adiantum nigrum majus, ramofum, coriandri folio. Cat. p.*23. *Adiantum fruticofum coriandri folio Jamaicenfe, pediculis foliorum politiore nitore nigricantibus, forte adiantum fruticofum Brafilianum. Pluken. Almag. p. 10. Phyt. Tab.* 254. *Fig.* 1. *An Adiantum fruticofum Æthiopicum pinnulis amplis, fubrotundis, fuperne dentatis, media pinnarum parte petiolis infidentibus. Ejufd. Alm. p. 10. Phyt. Tab. 253. Fig. 2? Adiantum vulgari fimile & ramofiffimum. Plum. Tournef. p. 543?*

This rifes two or three Foot high, having a ftrong, black, very polite, and fhining Stalk, branch'd out at unequal intervals alternatively. Thefe Branches have Twigs fet alternatively with *Pinnulæ* or Leaves, in Foot-Stalks, fhape, colour, bignefs, thinnefs, feed, incifures, &c. agreeing exactly with the *Adiantum foliis coriandri*. C. B. only this is ramofe or branch'd, and is much larger and taller.

It grew on the fides of a fhady woody Gully, beyond *Troopers Quarters*, and in other great Woods of this Ifland, and in *Bermudas*.

XCIII. *Adiantum nigrum, ramofum, maximum, foliis majoribus Trapezii in modum figuratis. Cat. p.*23. *Tab.* 59.

This was much larger than the former, having a very black and polite Stalk, fhining, rifing three or four Foot high in Branches and Twigs exactly like it, the *Pinnulæ* were as to colour, thinnefs, feed, &c. exactly the fame, only they were of the figure of a *Trapezium*, and twice or thrice as large, and very little cut in on the edges, where on the two fides, making the Angle oppofite to the *Petiolus*, were the Seeds in a ferrugineous Welt.

It grew in the more inland large Woods of this Ifland.

This is defcribed by Mr. *Bobart. hift. Ox. part.* 3. *p.* 587. under the Title of *Adiantum ramofum foliis trapezii dentatis,* Plum.

XCIV. *Filix ramofa maxima fcandens, ramulis raris, pinnulis crebris, latis, brevibus, obtufis. Cat. p.* 23. *Tab.* 60. *Filix ramofa Malabarica, alis integris, alternis & acutis rotundius crenatis para-panna-mara-vara, H. M. Bob. part.* 3. *hift. Ox. p.* 583.

This had a Stalk as big as ones Thumb, applying its felf to the Trunes of Trees, and taking faft hold of their Barks, like Ivy, by means of many Fibrils and Clavicles it ftrikes therein, rifing by this means fifteen or twenty Foot high, and being covered over with a rufty coloured Mofs, as moft Ferns are. About five or fix Foot from the Ground, it has Leaves going

out

The Natural History of Jamaica. 99

out of opposite sides of the Stalk, being five Foot long, and branch'd a Foot from its beginning, the Foot-Stalk and Branches being of a gray colour, covered with a rusty Moss. The Branches come out alternatively, at two Inch's distance asunder, those next the Stalk or undermost, being the largest and longest. The Twigs have their *Pinnulæ*, which are large, whole, broad, obtuse, frequent, leaving scarce any empty or void space between them, being cut in to the very middle Rib, on the undermost Branches, but on the upper join'd to their ends almost, and of a shining green colour.

It grew in the inland Woods on the Roads side between *Guanaboa* and Colonel *Bourden*'s Plantation, on the side of Mount *Diablo*, and *Archers Ridge* very plentifully.

Whether this be really differing otherwise than in magnitude from the *Filix non ramosa scandens pinnulis*, &c. is to me doubtful, this being described for that by Mr. *Bobart*. 'Tis really ramose.

XCV. *Filix ramosa major, caule spinoso, foliis seu pinnulis rotundis, profunde laciniatis, seu cerefolii foliis.* Cat. p. 23. Tab. 61. *Filix ramosa Jamaicensis fumaria foliis, pediculis & rachi medio aculeatis* Plukenet. *Alm.* p. 156. *Eadem non spinosa lævis Ejusd. ibid.* Bob. hist. Ox. part. 3. p. 584.

This has a long Root like Polypody, towards the top covered with ferrugineous Hair or Moss, at the bottom of which are several Filaments or Threads of a dark brown colour. From this Root rise several Foot-Stalks cornered on one side, and round on the other, of the bigness of a Swans-Quill, of a gray colour. This Stalk is thick set with short sharp prickles, as well as the Branches, which go out opposite to one another, at six Inch's distance, the Stalk rising three or four Foot high. The Branches have their Twigs set alternatively. On which are the *Pinnulæ* or Leaves, being roundish, and very deeply cut in on the edges, after the manner of Chervil, to which it is like, being of a yellowish green colour, having the Seed in little Spots on the ends of the Segments of the back side of the Leaves.

It grew near the open Ground by *Rio d' Oro*, near Mr *Philpot*'s Plantation.

XCVI. *Filix ramosa major, pinnulis crebris, brevibus, latis, obtusis, subrotundis.* Cat p. 23. *Filix Jamaicensis seu polypodium Cicutaria latifoliæ fœtidissimæ foliis quodammodo convenient, pinnulis amplis, mucronatis, circa margines, serris latiusculis profunde sinuosis.* Plukenet. *Almag.* p. 153. *Phyt.* Tab. 289. Fig. 4. *Filix non ramosa major Jamaicensis, lobis longis quercinis polypodii divisura.* Bob. hist. Ox. part. 3. p. 574. *Forte, Filix arborescens caudice spinoso, ramosa, alis latis mucronatis, polypodii divisura, Ejusd. hist. Ox. part. 3. p. 583.*

The Root of this Fern is roundish, large, having a great many black fibers, and its top covered with a reddish Moss or Hair, as well as the Stalks at their beginning. They rise to be two or three Foot high, being light coloured green, having Branches rarely plac'd at two Inches distance, for the most part opposite to one another, tho sometimes alternatively. The lowermost Branches are the largest, being divided into two Inch long Twigs, like Oak Leaves, whose *Pinnulæ* or little Leaves, are at the undermost divisions longer than the others, being set on to the very Twig and indented; but on those above broad, short, obtuse, whole, and for the most part united almost to the very ends, so that I was very much in doubt whether I should not call this whole Twig an indented *Pinnula*, and reduce this to the not *ramose* Kind. It has very many ferrugineous Spots on the back-sides of the Leaves.

There

There is a variety of it in the broadness of the Twigs, and so consequently of the *Pinnula*.

It grew in the Woods by the *Crescent* Plantation, and in all the inland Woods of this Island.

This is not the *Filix Jamaicensis foliis semel subdivisis, pinnulis obtusioribus costæ adnascentibus, sorbi aucuparia quodammodo referentibus. Plukenet. Phyt. Tab. 291. Fig. 1.* as Dr. *Plukenet* conjectures in his *Mant. p.* 80. but is that figured by him, *Tab.* 289. *Fig.* 4. as is apparent to any body that compares their Figures and Descriptions.

XCVII. *Filix ramosa major, pinnulis longis, acutis, raris, falcatis. Cat. p.* 23. *Filix non ramosa Jamaicensis pediculo albicante alis marrubii aquatici fere divisuris quarum pinnulæ a tergo linea candidissima aspergine conflata & per ambitum ducta crenata sunt. Plukenet. Almag. p.* 153. *Phyt.* 290. *Fig.* 2.

This has a Stalk rising about three Foot high, as big as ones Finger, of a pale green colour, and smooth, being at about a Foot's distance from the Ground, divided into Branches set alternatively, about a Foot long. On which at about an Inches distance from one another, alternatively are plac'd the Twigs, being about nine Inches long, and made up of *Pinnulæ* about an Inch long, crooked or falcated, from their Base, where they are for a little united, and broadest, decreasing by degrees, ending in a point or *spinula*, being falcated or crooked, having a defect between each other, of a light green colour on the upper side, and underneath having a ferrugineous Welt below its edge, wherein lies its Seed. The Leaves are somewhat indented before the Seed makes a welt, which inclines me to believe this to be the same with *F. Plumiers Filix latifolia dentata & adlacinias molliter aculeata*.

It grew in the inland parts of this Island.

The Powder on the back side of the Leaf is commended by *Piso* in ill Ulcers.

XCVIII. *Filix ramosa major, hirsuta, ramulis raris, pinnulis asplenii, scil. crebris, latis, brevibus, subrotundis, non dentatis. Cat. p.* 23. *An Filix Jamaicensis ramosa, pediculis Muscosis, pinnulis rarioribus, dentatis. Plukenet. Almag. p.* 155 ?

This has many Stalks rising from the same Root, to be about two Foot high, being each of them as big as ones Finger, very hairy, and at about a Foot from the Ground, divided into Branches, the lowermost whereof are about nine Inches long, set almost opposite to one another. They have Twigs more than an Inch long, set alternatively, being made up of broad, short, for the most part whole, though sometimes, easily notch'd *Pinnulæ* or Leaves, which are roundish at their ends, often united for some space by their edges, of a pale green colour, being not only in its Stalks, Branches, and middle Ribs, but all over covered with a whitish, strong, short hair, distinguishing it sufficiently from all others akin to it.

It grew by the Banks of *Rio d' Oro*, near Mr. *Philpot*'s Plantation between sixteen Mile Walk, and St. *Maries* in the North side.

XCIX. *Filix ramosa minor, hirsuta, ramulis raris, brevibus, pinnulis subrotundis, folii apice radices agente. Cat. p.* 24. *Filix non ramosa, pediculo hirsuto coriandri foliis Americana, Plukenet. Almag. p.* 153. *Phyt. Tab.* 284. *Fig.* 5. *Filix minor Jamaicensis pediculis villosis, alis amplioribus oppositis quercinis. Bob. hist. Ox. part.* 3. *p.* 576. *An Filix villosa pinnulis quercinis. Tournef. Inst. p.* 537 ?

The Root of this was tuberous and knobby, having many two or three Inches long fibrils, and was covered with a hairy ferrugineous Moss, from whence rose three or four Stalks nine Inches, or a Foot high, being very slender,

slender, of a brownish red colour, and having much hairy Moss, of the same colour, on them. The *Ramuli* are somewhat like an Oak Leaf, rarely placed opposite to one another, at about an Inch's distance, being short, *viz.* not an Inch long. The *Pinnulæ* or Leaves are very few, broad, and roundish, somewhat rough or woolly, and have some ferrugineous round spots or Seed by which it is propagated. Besides this way of Propagation, it has another, which is, that the end of the Leaf leaning on the Ground, takes Root, and grows into another Plant, after the manner of the Stalks of the common *Rubus*.

It was brought from *Jamaica* by *James Harlow*, a Gardener sent thither by Sir *Arthur Rawdon*, and from him communicated to me by Dr. *Sherard*.

This is not the *Adianthum, seu filix trichomanoides Jamaicensis, pinnulis auriculatis dentatis, &c.* Plukenet. *Alm.* p. 9. *Phyt. Tab.* 253. *Fig.* 4. as that Author supposes in his *Mantissa*, p. 5. but his *Filix non ramosa pedicula hirsuto, &c. Phyt. Tab.* 284. *Fig.* 5. as may easily appear to any who will take the pains to compare their Descriptions and Figures.

C. *Filix ramosa major, ramulis raris, ex adverso sitis, pinnulis asplenii, scil. crebris, latis, brevibus, subrotundis, non dentatis.* Cat. p. 24. *Tab.* 62.

This Fern was about two Foot high, it had a brown Stalk, which at six Inches distance from the Ground, was divided into Branches, set opposite to one another both at bottom, and towards the top, at two Inches distance from each other. The undermost Branches, or those nearest the Root were longest, being about five Inches long, on which were plac'd about an inch long Twigs, made up mostly of four pair of *Pinnulæ*, which were united together for a little way, short, broad, whole, of a dark green colour, and almost round, so that each Twig look'd something like a short Leaf of *Asplenium*.

It grew in the Inland woody parts of this Island.

This is not the *Filix Jamaicensis foliis semel subdivisis, &c.* Plukenet. *Alm.* p. 153. *Phyt. Tab.* 291. *Fig.* 1. as that Author conjectures in his *Mantiss.* p. 80.

CI. *Filix fæmina seu ramosa major, pinnulis angustis, obtusis, non dentatis, impari surculum terminante longissima.* Cat. p. 24. *Filix fæmina ramosissima Jamaicensis, pinnulis alas claudente longissima.* Plukenet. *Almag.* p. 156.

This is very like the *Filix fæmina*, *Ger.* or the *ramosa major pinnulis obtusis, non dentatis*, C. B. having a reddish brown, smooth, shining Stalk, rising three or four Foot high, the Branches standing sometimes opposite, sometimes alternatively, on which are plac'd the Twigs, along which are set, after the same manner, the *Pinnæ* or Leaves. They are narrower, having a void space between them, and are more rarely plac'd than those of the *Filix fæmina*, *Ger.* being long, harsh, of a dark, or dirty green colour, at the end of each Twig having one odd *Pinnula*, twice as long as any of the side ones.

It grew in the inland *Savannas* of this Island.

I am apt to believe this to be what *Lery* means by *Feugiere*, this being so like our ordinary *Filix fæmina*, as to impose on most people, making them believe it the same, and he reckoned likewise that his *Fengiere* of *Brasile* was the same.

CII. *Filix fæmina seu ramosa major, pinnulis angustissimis rarissimisque.* Cat. p. 24. *Tab.* 63. *Filix fæmina ramosissima Jamaicensis pinnulis alas claudente longissima, pinnulis angustioribus.* Plukenet. *Alm.* p. 156.

This rises to about five Foot high, having a very strong Stalk, cornered, as big as ones Finger, of a black colour at bottom, and reddish green above, having Branches sometimes opposite, sometimes alternatively, on

D d   which

which come the Twigs, which are beset with *Pinnæ*, much narrower than any of this Kind, so that there is a very large defect or empty space between them, by which they may be easily known from any other of this Kind.

This grew in the inland parts of this Island in the *Savannas*.

CIII. *Filix fæmina seu ramosa major, dichotoma pinnulis lonchitidis, scil. longis, angustis, non dentatis.* Cat. p. 24. *Filix Jamaicensis dichotomos seu ramis bifidis, fæminæ nostratis pinnulis ramosissima.* Plukenet. Alm. p. 156.

This Fern rose to about seven or eight Foot high, having Stalks as big as ones Finger, being smooth, shining, roundish, of a reddish green colour, always divided into two Branches, standing opposite to one another, and they into two Twigs standing in the same manner, which are for the most part about three Inches long, and made up of many Inch-long *Pinnæ*, join'd at their bottoms to one another by a narrow membrane running along the Twig or middle Rib, and thence growing very narrow, they end bluntly, leaving a very considerable defect, or empty space between them, and being of a grass green colour on the upper side, and paler underneath. At every one of the larger divisions of the Stalk stand Twigs with *Pinnæ*, as in the tops of the Branches.

It grew in *Jamaica* on the *Moneque Savanna*, and in going down Mount *Diablo* thither.

It was observed in *Martinique* by F. *Plumier*. I find that it grows likewise in *China* by a Draught of it taken from the Life in that Country, and given me by Mr. *James Cuningham*.

CHAP. IV.

Of Herbs with grassie Leaves.

Herbs with grassie Leaves and less perfect or Stamineous Flowers which are Culmiferous, are divided into those with large Seeds, or *Corns*, and those with lesser Seeds called *Grasses*. There are very few Corns here, the *European* Kinds not ripening well: the others, as Rice, *Guinea* Corn of two sorts, and Maiz ripen very well, and give great increase, especially the two latter, but are the Food only of some few of the meaner sort of People and Cattle: *Cassada* Bread with Yams, and other Roots and Flower, coming from other parts where Wheat is plentiful, being the chief Subsistance of the Inhabitants.

I doubted very much whether I should find in the *American* Islands any Grasses, at least in Plains as our Fields in *Europe*, but I found many grassie Plains, and in them Kinds of Grasses analogous to those of *Europe*, and two which I could not find different from them. What the design of Nature was in their Production seems hard to discover, for in these Islands they had no large Fourfooted Beasts but one, till *Europeans* landed there, unless it be said that as Corn with greater Seeds are for Man's Nourishment, so these were appointed for the Food of Birds and Insects, which feed on them and their Ripe Seeds.

Grasses are well divided into those spiked or panicled, which are made up of many Spikes; of the first there are some few, whereof that *Panicum spica divulsa* seems to belong to the Panicled.

The

The Natural History of Jamaica. 103

The panicled contains under them the Reeds, which are large Grasses, the *Gramina Ischæma*, or *dactyla*, which are the most numerous and best feeding Grasses here, being that of which their Pastures are for the most part full. Of this there are two Kinds, one whose several Spikes making up the Panicle, take their beginning from the tops of the *Culmi*, as from their common Centre, which is common to all of this Kind, which have been known hitherto; but the other Kind which I call *Paniculâ longâ* is new in all its Species, none of them, or at least very few having been in *Europe*, or taken notice of before as such, their Spikes taking their beginning one over another at the upper end of the *Culmus*, and not just at the top, being somewhat like those *European* Grasses, called *Gramina panicea, spica divisa.*

Those Cyperus-Grasses which are very large, or have sparse Panicles, I call *Cyperi*, and those remaining, with triangular Stalks, *Gramina cyperoidea*, and between them and Rushes I have put two by the names of *Juncus Cyperoides*, because they seem to partake of both Kinds, having a tuberous sweet smelling Root, no Leaves, a Sheath like Rushes inclosing the under part of the Rush, and above some *foliola*, a Panicle or Spike at top, like those of the *Cyperi* or *Cyperus* Grasses, and this is also a new Kind.

The word Spike is here taken for a single Head, not branched into several Panicles.

1. *Oryza, Raii hist.* 1240. *Cat. p.* 24. *Ind. Or. part.* 6. *p.* 83. *Worm. Musf. p.* 150. *Nieuhof. p.* 86. *Mirand. Sin. & Europ. p.* 880. *Mus. Swammerd. p.* 13. *De Flaccourt. p.* 114. *De Marini. p.* 56. *De Feynes. p.* 107. & 141.

This Grain is sowed by some of the *Negro's* in their Gardens, and small Plantations in *Jamaica*, and thrives very well in those that are wet, but because of the difficulty there is in separating the Grain from the Husk, 'tis very much neglected, seeing the use of it may be supplied by other Grains, more easily cultivated and made fit for use with less Labour.

Rice is the commonest of all Grains, in most of the warm Countries and Islands in the *East-Indies*, from whence it has gone into some Countries and Islands of the same temperature in the *West-Indies* as may be more particularly gathered from the Writers mentioned in my Catalogue. But the Bay of *Bengale* is the place where most grows, and whence most of that used in *Goa, Malabar*, the *Moluccas* and *Sumatra* comes, so that if the Vessels miscarry from thence their Inhabitants suffer Famine.

It is sown in Marsh Land, that is very moist or overflowed with water, or steeped eight days in the River in *Paniers*, (according to *Cauche*) and the Earth is plowed or trodden with Oxen, that it appears Mud; if there be no Water in the Grounds where 'tis sown, they water it as *Albert* tells us they do, every forty hours, in *Egypt*, *Duart de Menses* about *Sofala*, and the Writers of *China* tell us they do there by artificial Channels. When it is reap'd they put it into Stacks, and then in most places beat it out of the Husk by Pestles and Mortars, and Winnow it, or clear it in a Hand-Wood-Mill (*Lembere tom.* 1. *p* 51.) or tread it out by Oxen in a large hard Floor by Buffaloes drove round so as they may tread on it all. *Dampier cap.* 15. *Mandeslslo* says, *p.* 166. that in *Japan* they keep it in the Ears, and beat it out as used, drying it over night in the Chimney Corner in bundles, and next morning beating it out clean in a Mortar.

It is in several Countries manag'd by a several way, sometimes if too thick in coming up 'tis planted thinner, and *Le Comte* says 'tis in *China* planted in Sheafs or Bundles, the better to resist the Winds

It is used for Food in most Countries where it grows, 'tis boiled in water, and so used as Bread, and is likewise mixed with Milk, Broath, &c. and made into many kinds of Messes.

There

There is also made of it a Drink or Wine, for which it is boiled, and then set a working, and from thence is distilled a vinous inflammable inebriating Spirit, called *Arack*, as also of it is made a Vinegar, as many Writers tell us.

To make Leven for this Rice Drink, in *Japan* they chaw Rice-Meal and Spit it into a Pot, *Mandelslo* p. 166. who likewise adds, that to make the Wine they add to the Rice some Honey or Sugar. *Id. p.* 156.

Rice, either in Substance or Decoction, is thought to be an extraordinary Adstringent or Binder.

The Meal of it strowed on the marks left by the Small-pox, helps them. *C. B.*

The Decoction is good against the Poison of Arsnick, Quick-Lime, or *Cantharides*. *Id.*

Riolan says the Husk of it is poisonous, and the Flower very ill smell'd. *J. B.*

II. *Milium Indicum arundinaceo caule granis flavescentibus Herm. Cat. p.* 425. *An Ampembe de Flacourt. p.* 118? *Mengrelie milium Texag. p.* 68? Guinea Corn of *Dampier. cap.* 3. *An Milium Indicum sacchariferum altissimum semine ferrugineo Breyn. prod.* 2. *p.* 72? *An Milium Indicum arundinaceo Caule semine fusco glumis splendentibus atris. Plukenet. Almag. p.* 250? *Milium Indicum, panicula sparsa erecta. Tournef. Inst. p.* 515?

This rises to eight or nine Foot high, has a hollow reddish coloured *Culmus*, or Stalk, jointed at every nine Inches distance, every joint having a Leaf by its Foot-Stalk inclosing the internodium to the next joint, being grasse, a Foot long, and Inch broad near the joint, whence it decreases, ending in a point, having a white middle Rib. Sometimes some smaller Spikes come out, *ex alis foliorum*, near the top; but that on the top is an oblong, roundish Head, seven Inches long, and three broad, near as big as ones Fist, having many small Branches, or Strings very close compacted together, on the Tops of which come in Follicles, yellow *Stamina*, as in others of this Kind, and to them follows in two brown Follicles, a round Seed of a whitish yellow colour, not so big as that sort of Barley call'd commonly Pearl Barley.

It is planted every where in *Jamaica* for Provision, yielding very great increase.

It is thought to Nourish little, and to be Adstringent as Rice.

It is dry, and is good in Dysenteries. Cakes are made of its Flour. In *Cercyra* it feeds Pigeons, and in *Sicilia* Fire-wood is made of its Stalks. *C. B.*

It is sown at a Foot distance, three or four Grains into a hole.

The Figures and Descriptions of *Sorgum* in most Authors, agreeing with this, was the occasion of my putting that in my Catalogue for a Synonimous name, although I am sure that one sort of the *Italian Sorgum*, (which has a white Seed that is flat, and the Panicle as it were compressed or flat, whereas that of this has a Panicle standing out on all sides,) is really different from this.

III. *Panicum Indicum spica longissima.* C. B. *Theat. Bot. p.*523. *pin. p.* 27. *Cat. p.* 26.

This differs not from the precedent, save in that the Head or Spike is above a Foot long, being largest at bottom, where it may be about three quarters of an Inch Diameter, tapering to the top. It has lesser Grains or Seeds than the former, many of them being set on the same common Foot-Stalk inclosed in *gluma*, and those set so close together, that it makes an even Surface, and appears like the common *Typha Palustris.*

It is to be met with in some *Negro*'s Plantations, though not so commonly as the former.

It came from *Guinea*, and Perroquets fed on it by the way. *Cluſ.*

Johannes Leo ſays that this ſort of Grain makes Bread, is uſed to be boiled in Milk, and to feed Birds.

IV. *Frumentum Indicum Mays dictum*, C. B. *Cat.p.26. Frumentum Turcicum Duran. p.68. & 112. Maïs Americanor. Contant. p.2. Bled de Turquie de Flacourt. p.127. Millet ou bled de Turquie de Biet. p.334. Gros mil, mais ou bled de Turquie de Rochef. Tabl. p. 48. Triticum Turcicum muſ. Swammerd. p. 13.* Corn and Maiz of *Dampier*, *An Mill de Maïs p. 80, 84, 101, & 136? Mays granis aureis, albicantibus, violaceis, ſpadiceis, nigricantibus, rubris. Idem ſpica albo-ſpadicea, rubro-ſpadicea, aurea & alba, albo punctis ſpadiceis notata, alba-violacea punctis ſpadiceis notata, albo lutea violaceis punctis & cæruleis notata, albo lutea rubris punctis notata, rubra nigra & ſpadicea, cærulea lutea violacea & alba. Tournef. Inſt. p. 531.*

This is every where planted, and gives ſeveral Crops, every year ripening three times, or in four months after planting.

It is of ſeveral ſorts, being the Grain is ſometimes yellow, dark red, or whitiſh, *&c.* which, becauſe I have ſeen ſeveral of them on the ſame Stalk, I take to be only varieties. The beſt enumeration of theſe varieties is in *Tabernæmontanus*'s Hiſtory of Plants, and Mr. *Tournefort*'s *Inſtitutiones*.

The *Indians* uſed to grind the Grain between Stones, and it was thought wholeſomer in *America* than wheaten Flour. It increaſes mightily, every Spike having many Seeds, though not ſo many as *Abbeville* tells us, who ſays that every Grain has in two months and an half, or three months; four, five, or six Stalks, every Stalk six or seven Spikes, and every Spike six, seven, or eight hundred Grains, and this three times a year.

This is the moſt common and natural Grain in the *Weſt-Indies*, and has been from thence communicated to other parts of the World; eſpecially the hotter parts of *Aſia, Africa* and *Europe*, though it is found in very Northern Countries of *America* naturally, and is able to endure great degrees of Heat and Cold, as may appear to any one who pleaſes to peruſe the ſeveral paſſages about it, mentioned in the Authors recited in my Catalogue of *Jamaica* Plants. The beſt account of its agriculture and uſe in the *Indies* is given by *Oviedo* in his *Coronica de las Indias, lib. 7. cap. 1.* to this purpoſe. They (the *Indians*) cut down and burn the Woods, (places where Graſs grows, not being ſo Fertile) whoſe Aſhes is as good as Dung. *Indians* go a pace aſunder, making holes in the Ground, and putting in four or five Seeds into every hole, covering it by the mould with their Feet, then going a pace forward they do the ſame. They ſteep it a day or two before, doing this after Rain, the ſharp Stick entring eaſily three or four Inches into the Earth. It is ripe in three or four months, or in *Nicaragua* in ſix weeks; but then 'tis ſmall, and not ſo good as that of four months, that being done by watering. They weed it, when it ripens Boys ſit on Trees and *Barbacoas* to preſerve it from Parrots and Birds. The Spikes are guarded from the Sun by Leaves, and are gathered when dry. Birds having Beaks like Parrots deſtroy it. In the Continent 'tis deſtroy'd by Deer, Swine, Cats, and Monkeys. It is harder in the Iſles to keep, becauſe of the wild Kine, Swine, Dogs, *&c.* from *Spain*. One Meaſure gives in Crop from ſix to one hundred fold. The *Indians* eat it roſted when young and tender, otherwiſe give it their Cattle. In the main Continent it is ground in a hollow Stone by a round one, as Painters do their Colours, with ſome water. The Paſt made into Balls is wrap'd in one of the Leaves, and Boil'd or Roſted, and eaten whilst hot. Many sorts of Cakes are made of it, the Bread will not keep past two or three days, growing musty, and spoiling the Teeth. A Drink is made of it, and its Flour corrects stinking water. All which he knew, having cultivated it twenty years.

The Juice of the Stalks or top affords a kind of Honey or Sugar, and they with the Leaves afford a sustenance for Cattle, and materials for *Indian* Baskets. The *Indians* made intoxicating Drinks of this at *Mexico*, and other parts, before the *Europeans* knew them. This Grain was transplanted from *Brasile* to St. *Thomas* by the *Portuguese*, and from thence to *Guinea*.

It is best preserv'd from Weavils in its Husk.

It is now used many ways, Rosted before it be quite ripe, Raw made into Meal, into Cakes, or Boil'd, made into Mault or otherwise for Drink.

Jo. dos Santos a *Portuguese* Writer tells us how about *Sofala* they make it into both Meat and Drink. But 'tis agreed upon that it affords very little Nourishment, and it hurts the Teeth, so that it is seldom now used but by Slaves, and as Food for Horses, Cattle, and Poultry, for upon it they thrive very much. Formerly *Hariot* tells us one Man's Labour in a day, in twenty five Yards Square of Ground fed a Man for a year.

V. *Gramen caninum maritimum spicatum quartum* C. B. *Cat.* p. 19. *Gramen caninum spicatum foliis brevibus maritimum*. *Bob. hist.* Ox. part. 3. p. 178. *An Gramen caninum maritimum spicatum foliis angustis longioribus. Ejusd. ib?*

I could not observe any difference between this Grass describ'd by *Caspar Bauhine*, and that here, it being only a little larger, which I take to be a variety from the Soil.

It grew every where by the Sea side, creeping very far, and covering large pieces of Ground.

The Vertues are the same with those of the *Gramen caninum*. *Park*.

VI. *Gramen spica brizæ singulari, locustis majoribus, villosis, purpurascentibus. Cat.* p. 30. *Tab.* 64. *Fig.* 1.

This has some small fibrous Roots from whence rises a frequently geniculated compress'd *Culmus* upwards of two Foot high. It has hard, yellowish green coloured, narrow, nine Inches long Leaves, up to the top of the Stalk, out of the *Alæ* of which go Branches, on the tops of which stands one compress'd Spike of about three quarters of an Inch long, made up of large, yellow Chaff, hairy at the end, about six or seven *Locustæ*, pretty large, being plac'd on each side of the Spike, something like those of the *Gramen spica brizæ majus* C. B. *prod.* only the *Glumæ* or *locustæ* are larger, hairy, and there is but one Spike on the top of each Branch, which as well as the rest of the Plant, is inclining to a purple or red colour.

It grows in the *Savannas,* especially those about Seven Plantations.

VII. *Gramen paniceum maximum, spica divisa, aristis armatum. Cat.* p. 30. *Panicum vulgare, spica multiplici, longis aristis Circumvallata. Tournef. El.* p 416. *Inst.* p. 515. Scotch Grass.

The Stalk or *Culmus* of this rises straight up about four or five Foot high, being sometimes branched, and having several protuberant Joints, the *Internodium*, or space between them, being six Inches in length: it is as thick as ones Finger, and is in part filled with a white spungy Substance, at every Joint is an arundinaceous Leaf, taking its beginning from the lower, and covering the *internodium* to the next Joint, and there stands out a grassie Leaf a Foot long, and an Inch broad at the beginning, whence it tapers to the end. The top or Spike is a Foot long, and is divided into several Spikes, about an Inch and an half long, each of which has a great many Seeds set to

the

the Spike very thick close to one another, without any Foot Stalk, lying in a rough, pale green Husk or Follicle, having a half Inch long Aune, or rough *Arista* at its end, making it look somewhat like to the Grain of Barley.

It is planted in moist Ground all over the Island for Provision for Cattle, but grows wild, as I was inform'd, at *Wagne* water, or *Agua alta* in the North side, and in that part of *Barbados* called *Scotland*, whence the name. After its being found very useful in *Barbados*, and had been there planted for some time, it was brought hither, and is now all over the Island in the moister Land by Rivers sides, planted after the manner of Sugar Canes, by burying the *Culmus* with a Joint, which strikes Root, and seldom misses to prosper, and to feed and fatten extremely Cattle of all sorts, as well Cows and Oxen for the Market, as Horses and Cattle for Teams and Riding. It is likely that this way of Agriculture might be useful in other places, if rightly managed.

VIII. *Gramen paniceum majus, spica simplici lævi, granis petiolis insidentibus.* Cat. p. 30. Tab. 64. Fig. 2.

This was very like the precedent, only seemed not quite so large, the Leaves were much shorter, being not over four Inches in length, from its beginning, where it was an Inch broad, tapering to the sharp point, being striated and grassie, of a blewish pale green colour. The Spike at top was about three Inches long, made up of many crooked, strong *Petioli*, of about an Inch long at the under part of the Spike, but not one quarter so much above, so that the Spike is pyramidal, every one of these *Petioli* grows larger at top, and sustains one Grain on its point, which is contained within two *Glumæ* striated, and like the foregoing, only the *Aristæ* are not so long, nor the *Gluma* rough, seeming only to differ from one of its Spikes, in having a long *Petiolus* to every Grain.

I found it in *Guanaboa*, with *Scotch* Grass, or the preceding.

IX. *Gramen paniceum spica simplici lævi.* Raii hist. p. 1261. Cat. p. 30. *Panicum vulgare spica simplici & molliori.* Tournef. El. p. 416. Inst. p. 515.

I could not find any difference between the Plant describ'd by Mr. *Ray* and this here, therefore conclude it to be the same.

It grew in the *Savanna*, between *Black River* Bridge, and the Town of St. *Jago de la Vega* in great plenty.

X. *Gramen paniceum minimum humi stratum, spicâ divisâ muticâ, foliis variegatis.* Cat. p. 30. Tab. 64. Fig. 3. *An gramen serpentarium Zeylanicum Breyn.* pr. 2. p. 54?

This has several thready Roots, which united send out very many two Inches long, broad Grass Leaves, spread on every hand, lying on the Surface of the Ground, and when young, being mark'd in several places with transverse Lines or *Fasciæ* of a brown colour, making it look very pleasant, which when the Plant grows old, or is dry, are obliterated. From these Leaves come many Stalks or *Culmi*, about three Inches long, consisting of so many reddish Joints, with a Leaf to every one of them, the tops of which are about an Inch long, and divided into several small Spikes, of a pale green colour, made up of many shining triangular Seeds, of a yellowish colour, inclosed in a pale green coloured Follicle or Chaff without Aunes, the Seed and Husk lying close to one another by the Stalk, after the manner of the other panic grasses.

It grew in the *Savannas* about the Town of St. *Jago de la Vega*.

XI. *Gra.*

The Natural History of Jamaica.

XI. *Gramen echinatum maximum spica rubra vel alba. Cat. p.* 30. *An Gramen tribuloides spicatum maximum Virginianum D. Doody Ejusd. Almag. p.* 177 ? *An Gramen marinum echinatum. D. Spragg. Raii hist. p.* 1928 ?

This Grass has several two or three Inch long thready Roots, sending out several Inch and an half long grassie Leaves, of a yellowish green colour, from the middle of which rise up several six Inches long Stalks or *culmi*, being jointed; the Joints are three quarters of an Inch distant one from the other, at which are now and then Branches which are crooked, having Leaves, and at the top an Inch and a half long Spike, of little Burs, or large roundish prickly Seeds, sometimes of a reddish, and sometimes of a green colour. The Prickles being long, strong, and sharp, standing on every side, having within them some oblong, large, flat whitish Seeds. Of this there are of various bignesses.

From the Roots go sometimes reddish, jointed Branches, on which at the Joints grow tufts of smaller Leaves, very thick set alternatively making this part of the Grass creeping, have a different Face from the other erect. I supposed this to have given occasion to Dr. *Plukenet* to mention and grave it twice, as I have taken notice in my Catalogue *p.* 30. So that for ought I know these Grasses may be the same, notwithstanding what is said by the Doctor in his *Mantiss. p.* 96.

This is not properly an aculeated Grass, because the Leaves are no *utriculus*, nor aculeated as that of *Italy*, but it has echinated Burs, as the *echinatum*, with several Seeds in them.

It grows in all Plantations at all times, when there are Rains, as well in *Jamaica* as the *Caribbees*.

It is very troublesome to Travellers on Foot, these small Burs or echinated Seeds, sticking close to their Garments, especially their Stockings, and pricking their Legs.

XII. *Gramen maritimum echinatum procumbens culmo longiori & spicis strigosioribus. Cat. p.* 30. *Tab.* 65. *Fig.* 1.

This had a fibrous Root, sending out many trailing, round, yellowish, crooked jointed *Culmi*, or Stalks, about a Foot and a half long, the Joints being an Inch and an half distant one from the other, at each of which is a Leaf inclosing the *internodium* to the next Joint, as with a Sheath, being two or three Inches long; likewise green colour'd, harsh, something like the *Cyperus* Grasses. At the top stands an Inch and an half long Spike, set round at a short interval from one another, with lesser small Burs, or roundish echinated Seeds, having on every side of them several strong, sharp Pricles, being first green, then of a Straw colour.

It grew on a small Island, called *Gun Cayos*, off of *Port-Royal* Harbour.

The largeness of the former seems to make it rather be the *Gramen tribuloides spicatum maximum Virginianum, D. Doody*, than this here described, notwithstanding what Dr. *Plukenet* says in his *Mantiss. p.* 96.

XIII. *Arundo sacchariferа, C.B. Cat. p.* 31. *Tab.*66. *de Bonton. p.*82. *de Biet. p.*336. *de Marini, p.*58. *de Feynes, p.* 160. *& 166. de Rochef. Tabl. p.* 58. *Canna, Saccharina de Nieuhof. p.* 89. *Azucar de Esquemeling, p.* 58. *Fare de Flacourt, p.* 120.

This has a jointed Root with many Fibers, as other Reeds, sending up an usually eight or nine Foot long jointed solid *Calamus, Culmus*, or Stalk as big as ones Thumb, or sometimes Wrist, according to the Ground in which it grows; the Joints are sometimes farther distant from, and sometimes nearer one another, generally about four Inches long; the outside of the *Calamus* is of a yellowish green colour, smooth, shining, and within is a white Fungous, or

rather

rather fibrous sweet juicy Pith. The Leaves, by their Foot-Stalks, or under parts, inclose the *Culmus* or *Internodia*, and are broad, of a lively yellowish green colour, striated, and like the others of this kind, or those of *Donax*. At the top of the Stalk comes the Panicle, which is about two or three Foot long, being branch'd from the bottom to the top, into many Spikes or Branches about a Foot long, each of which is subdivided into smaller Twigs, which are jointed, easily broken, having at every Joint alternatively the *Stamina* and Seed, which are very small, and a great quantity of Down, or *Tomentum*, after the manner of other the Canes, only in this the *Tomentum* sticks to the outside of the *Locustæ*, as at their Base, whereas in the others 'tis contain'd within them.

Sugar making is so commonly known, and its Refining, that I shall say nothing of it, save that Sugar is the Juice of the Cane boild into a Salt by the help of what the Sugar-Makers call Temper.

I tried to Boil the Sugar-Cane Juice, without any mixture, to Sugar, but it would not coagulate, keen, or granulate into the form of Sugar, because it wanted what they, in the making Sugar, call Temper, which is made of an Infusion of Wood-Ashes and Quick-Lime, and which must differ in quantity according to the Soil in which the Canes grow.

These Canes are planted in all the Lowlands of this Island, and never miss to thrive if placed where there is Rain, and the Soil rich and moist. They seldom thrive or are good if planted on Hills, or in those Valleys where Rain seldom falls. *Martyr* tells us that, when he wrote, the Sugar-Canes in *Hispaniola* thrive extremely, growing as big as ones Wrist, high as a Man, and putting twenty or thirty Stalks from the same Root, whereas those of *Valentia* had only five or six, so that in the year 1518. there were twenty eight Sugar-Works there.

XIV. *Arundo maxima folio dentato.* Cat. p. 32. *Roseaux de Bouton.* p. 32. ? The wild Cane.

This rises to fifteen Foot high, it has a Stem or *Culmus*, about the thickness of ones Arm; being hollow, hard, and having very frequent Joints, at every Joint having a Partition or *Diaphragme*, it is covered with a Clay coloured Skin, and remainders of the dry Leaves: it has at the Joints very long, narrow, small, dark green Leaves, like others of the Reeds, being very thick set with Indentures, or Prickles on its Margin, making it rough downwards. At some of the Joints, now and then, come out Branches two or three Foot long, be set with lesser Leaves than the former ; and sometimes there are Tufts of smaller, and narrower Leaves come out together at the top, making a large Bunch, and upon the top of it is a Joint as small as ones Finger, straight, Clay coloured, smooth, and full of Pith, holding a two Foot long chaffie or downy Panicle, (like other of the Reeds,) whereof all the Spikes look one way.

It grows on all sides of the *Rio Cobre*, and in Marshy Grounds.

The Cane, split, is made use of for Laths, and to make up the walls or sides of Houses with Mortar.

The tender tops of these Canes are cut into transverse slices pickled, and made use of as other Pickles, as the Bambo's in *East-India* are with *assa fœtida*, Salt, Vinegar, and Garlick Pickle.

Masegrave says it was made use of in *Brasile*, when made hollow, *viz.* clear'd of its *Diaphragmes*, to carry water for Travellers.

I am apt to believe *Masegrave* described this twice in his third Page, under the name of *Jataboca*, and immediately after under the name of *Uuba*. For this Plant, most part of the year, has no Panicle, and then appears as the *Jataboca*, having several Branches with Tufts of small Leaves, but no Panicle,

F f

Panicle, which it having another time of the year, may be his *Uba* or *Arundo Sagittaria*, but if his Descriptions or Figures were more exact, one could tell better what to say to them.

Indians make Arrows of them. *Benz.*

Du Tertre was mistaken in giving this the name of *Roseaux d' Espagne.* Savages, who are pox'd, use to rub themselves over with the Ashes of this Reed to cure their Disease.

The *Patagons* make their Arrows of Reeds, an Ell long, with Heads very artificially framed of Flint-Stone. *Hakl. p.* 3. *p.* 751. but I believe they differ from these made of this Reed.

XV. *Arundo alta gracilis, foliis è viridi cæruleis, lacustis minoribus. Cat. p.* 33. *Tab.* 67. The Trumpet Reed.

This puts forth Roots from every Joint, and sends up Stalks, or *Culmi*, they are round, hollow, jointed at every two Inches distance; of a Clay Colour, and about the bigness of ones little Finger. The Leaves come from the Joints. Their Foot-Stalks, or under part, covers the whole *Internodium*, and the Leaf rises at the upper Joint where 'tis near half an Inch broad, and tapers for more than a Foot in length, where it ends in a Point of a blewish green colour. The Stalks rise fourteen or fifteen Foot high, the top is a Panicle of a Foot long, branch'd out into many rough Spikes, being a Foot long, standing like those of other Reeds, and containing, in a downy Matter, within Chaff, the Seed, scarce discernible, plac'd rarely on it, here and there of a light brown colour, in every thing like those of the other Reeds.

They grow going to the *Laguna* above the Ferry, and in the *Laguna* near *Passage Fort* very Plentifully.

XVI. *Gramen dactylon bicorne tomentosum minus. Cat. p.* 33. *Tab.* 68. *Fig.* 2.

This has several long, strong, white, crooked Threads at bottom, to draw its Nourishment, from whence rise several harsh, narrow, nine Inch long Leaves, having sharp Backs like the Cyperus Grasses, and being reddish when dry. From the middle of these rise several *Culmi*, or Stalks having about three Joints, being a Foot and an half high, swell'd at each Joint, and having there Leaves swelling and covering the Stalk a little way, out of the *Ala* of which rises, as out of an *utriculus* (after the manner of *Gramen dactylon Siculum, &c. Rail*) a small *Pediculus,* or *Culmus,* whose Panicle is made up of two Spikes, standing on the top like a pair of Horns, as the other sorts of this *Gramen dactylon bicorne,* only they are shorter, being not over three quarters of an Inch long, having very much long, soft very white Hair, or *Tomentum,* much finer and softer than Cotton.

It grows on *Palmetto Savanna,* towards Sir *Francis Watsons* by seven Plantations, on Lime-Tree *Savanna* very plentifully, and on a *Savanna* by Mr. *Batchelor's* Plantation going towards *Black River,* though rarely.

XVII. *Gramen dactylon spicis brevibus crassis plerumque quatuor cruciformiter dispositis. Cat. p.* 33. *An Gramen ischæmon Malabaricum speciosum, longioribus, & mucronatis foliis Plukenet. Phyt. Tab.* 300. *Fig.* ? *Alm. p.* 175 ? *Mantiss. p.* 94.

It has a creeping Root, hoary Leaves, a Span long jointed Stalk, and at the top four Spikes, for the most part, each of which is thicker and shorter than any other of this Kind, being but half an Inch long, sometimes of a reddish, and sometimes of a white Colour, set cross ways, and in every thing agreeing with *Alpinus's* Cut and Description, so that I doubt not but that 'tis exactly the same.

It

It has sometimes five or six Spikes, or three, so that I doubt whether it may not be the same with *Vestingius*'s, *gr. Stellat. Ægypt*.

It grows every where by the Way sides, and in the *Savannas*.

The Root and whole Plant are boiled, and the Decoction used in difficulty of making Water, the Stone, Womens Obstructions, &c. *Bont*.

The Seeds are very much used by those troubled with the Stone in their Bladders or Kidnies. The decoction of the Roots is used by those who are taken with the Small-pox and Measles, or suffer Obstructions of the *Menses*. A Decoction of the Seeds, somewhat bruised, are good for the *Petechiæ*, and the whole Plant, especially the Roots, are useful in Wounds and Ulcers. The Root is cold and dry, and of subtle parts, and therefore its Decoction is used to promote Sweat. *Alp*.

XVIII. *Gramen dactylon elatius spicis plurimis tomentosis. Cat. p. 33. Tab. 65. Fig. 2. An Gramen dactylum Indicum spicis villosis subrubescentibus. Bob. hist. Ox. part. 3. p. 185?*

This has a very strong fibrous Root, broad Leaves of a pale yellowish green Colour, like those of Oats. The *Culmus* or Stalk is knotted, rising three Foot high, at the uppermost Joint it is divided sometimes into two tops, the one being in Flower, the other not. Several Spikes, *viz.* four, five or seven, come from the same top or Centre, all hanging downwards, each is four Inches long, and very hairy, downy or woolly. This stands above most other Grasses in the *Savannas*.

It grows in the *Savanna* by two Mile Wood, and most other Plains, very plentifully.

It is very certain that this is different from the *Gramen digitatum hirsutum. J. B.* which Dr. *Plukenet* thinks may be the same with it, in his *Mantiss. p. 95.*

XIX. *Gramen dactylon procumbens, crassum & viridius, culmo reclinato. Cat. p. 33.* Dutch Grass.

This Grass has a fibrous Root, from which spring several very green Leaves and Stalks, as from a Centre, both lying along on every side on the Surface of the Ground. The Stalk is one Foot long, the Spikes, at top, usually three or four very green, broad, and large, all coming from the top of the Stalk.

It grows by Highway sides in low Grounds in *Jamaica*, as well as *Barbados*.

It is esteemed the best fatning and feeding Grass for Cattle.

This bruised in the Mouth, or chaw'd, and put to a bleeding Wound, stops the *Hemorhage*. I saw once a Black stop a bleeding Artery with it, which Sympathetic Powder, and other Adstringent Medicines would not do.

XX. *Gramen dactylon spicis gracilioribus plerumque quatuor cruciformiter dispositis. Cat. p. 33. Tab. 68. Fig. 3.*

This has a deep fibrous Root, short and narrow Leaves, a jointed, crooked, slender, white Foot and an half long Stalk, bearing for the most part at top four white slender Spikes, standing cross ways, though sometimes they are three, six, or five in number. On them stand several Seeds contained in two ear'd Husks.

This is the most ordinary Grass in the *Savannas*. Its Stalks are there remaining dry most part of the year.

This is very different from the precedent Grass wich Dr. *Plukenet* conjectures in his *Mantissa. p. 94.* to be the same with it.

XXI. *Gra-*

XXI. *Gramen dactylon bicorne repens, foliis latis brevibus.* Cat. p. 33. *Gramen dactylum repens Indicum spica gemella Bob. hist. Ox. part.* 3. p. 185.

This Grass has a jointed Root, creeping, and striking fibers from the Joints of the Root, as well as Stalk, where it touches the Ground, like the *Caninum repens.* It has broad and short Leaves. The Stalk is a Foot long, at its end having two Spikes standing opposite to one another, which are made up of several small flat Seeds, lying *imbricatim* one on another the length of the Spikes, and of this there is a variety, the Spikes of some being much grosser than others.

It grows in moist low Grounds or Pastures in *Jamaica*, and most of the *Caribes*.

XXII. *Gramen dactylon bicorne spicis purpurascentibus majus.* Cat. p. 34. *Tab.* 65. *Fig.* 3.

This has a crooked repent Stem, the Grass broad and short, the Stalk fourteen Inches high, the Spikes, always two, standing not so horizontal, but more towards a perpendicular than the precedent, like a pair of Horns, with many purple or blackish *Stamina* on them.

It grows in Holes and Places, where water has stood in the *Savannas*.

This is not the *Gramen parvum Gangeticum, &c. Plukenet. Phyt. Tab.* 91. *Fig.* 6. as he conjectures in his *Mantissa*, p. 93. This appears by comparing the Figures of his and this.

XXIII. *Gramen dactylon bicorne spicis purpurascentibus minus.* Cat. p. 34. *Tab.* 68. *Fig.* 1.

It is in every thing like the immediately precedent, only in every respect smaller, and usually grows in the same places.

XXIV. *Gramen dactylon bicorne minimum aristis longis armatum.* Cat. p. 34. *Tab.* 69. *Fig.* 1.

This has several very small white fibrils for Roots, from which rise very many very small, narrow, capillary, pale green Leaves, about an Inch long. From the middle of these rise very small jointed round Stalks or *Culmi*, about two or three Inches high, having so many Joints, each Joint having a Leaf. At the top stands its Panicle, divided into two Spikes, like two Horns, three quarters of Inch long, having a few Seeds, each of which has two long *Aristæ* or Awns.

I found it in a small Wood near the Banks of *Rio-Cobre*, below the Town on the same side of the River.

XXV. *Gramen dactylon majus, pannicula longa, spicis plurimis nudis crassis.* Cat. p. 34. Tab. 69. Fig. 2.

This has several fibrous Roots from which rise many Leaves, inclosing the Stalk, and one another of each side with a hard sharp edge or back, being about a Foot long, very green, and something like the *Cyperus* Grasses. From the middle of these (being very many of these Leaves together, making a great Tuft) rises several three or four Foot long, solid Stalks or *Culmi*, as big as a Hens Quill, having so many Joints, and at every Joint a Leaf like the others below. The Panicle is a Foot long, towards and at the top divided into many small, three or four Inch long Spikes, not only at top but below it. Each of these has two rows of small, roundish, compress'd Grains, lying one way (the Back of the Spike being naked) *imbricatim* one over another, each of which contains within a pale green, or

reddish

The Natural History of Jamaica. 113

reddish Husk, or Follicle, a compress'd, roundish, pale, yellow, shining Seed.

It grows in the *Savannas* near Mr. *Batchelor's* House very plentifully.

XXVI. *Gramen dactylon, alopecuroides facie, pannicula longissima e spicis plurimis tomentosis constante.* Cat. p. 3. Tab. 70. Fig. 1.

This rises to about four Foot high by a jointed *Culmus*, whose Leaves are grassie, of about a Foot long, with a proportionable breadth. At top it has a Panicle appearing to be a Spike of about a Foot long, it is made up of many Spikes, some of them upwards of an Inch long, rising from all sides of the Stalk, or top of the *Culmus*, close by one another, having very small whitish *Locustæ* on each side with a great quantity of white, long, soft Down, or *Tomentum*, making it appear something like a Fox-Tail-Grass, if one look not very narrowly into its manner of growing.

It grew in Mrs *Guys* Plantation in the open Ground at *Guanaboa*, by her Plantain Walk.

Dr. *Plukenet* in his *Mantiss. p. 95.* questions if this be not the *Gramen digitatum hirsutum*, J. B. with how much Reason any body may see. I think I had some Reason from the Title of that of Dr. *Herman* to judge it to be that described by him, though he, who knows nothing of it, says I did it *inepte*.

XXVII. *Gramen dactylon pannicula longa, e spicis plurimis gracilioribus purpureis vel viridibus mollibus constante.* Cat. p. 34. Tab. 70. Fig. 2.

This has several fibrils for its Root, from whence rises a crooked Stalk or *Culmus*, about a Foot high, made up of three or four Joints, each having a three or four Inch long grassie Leaf, covering the *internodia* of the Stalk, which at about six inches from the Root is divided into many slender Spikes, making a six inches long Panicle. The Spikes stand out on every side of the *Culmus*, towards and at the top, at some small distance from one another, each of them being about two Inches long, very slender, soft, purple, or green, and made up of several naked Grains, or Chaff (*locustæ*) set to it by Tufts alternatively, first on one side, then on another.

It grew in Mr. *Batchelor's* Plantation near the red Hills.

This can not be the *Gramen Ischæmon Virginianum, numerosis spicis, &c.* *Pluken. Alm.* p. 175. which Dr. *Plukenet* conjectures may be the same, in his *Mantiss.* p. 94. this having neither black spots, nor hirsute or undulated Leaves.

XXVIII. *Gramen dactylon, pannicula longa, spicis plurimis gracilioribus & longis.* Cat. p. 34. Tab. 70. Fig. 3.

This has a fibrous Root, many Stalks a Foot and a half long. Its Spikes at top are many very small, or slender and long, the Panicle being divided into Spikes before it comes to the top of the Stalk.

It grows every where in the *Savannas*.

This is not the *Gramen Ischæmon Virginianum numerosis spicis, &c. Plukenet, Alm.* p. 175. having neither hirsuted, spotted, nor undulated Leaves, as he says it has, *vid. Mantiss. p. 94.*

XXIX. *Gramini tremulo affine, panniculatum elegans majus, spicis minoribus & longioribus.* Cat. p. 34. Tab. 71. Fig. 1.

This has a fibrous Root, from whence rises a round, pale, green, solid Stalk or *Culmus*, about a Foot and a half high, having Leaves nine Inches long at bottom, incompassing the Stalk. The Panicle is six inches long, the top of the *Culmus* being branch'd out into several Branches, on which are set

G g   several

several very small, long, compress'd Spikes, by small Stalks or Strings, after the manner of *Gramen tremulum*, only the *Petioli* are stronger, so as not to quake. Each of them are made up of very many small *Glums*, Scales, or Chaff set in a double row, being sometimes white, and sometimes purple.

I found it in the inland parts of the Island.

By the Figure of that Grass mentioned by Dr. *Plukenet Alm. p.* 176. and called *Gramen amoris India orientalis, panicula sparsa, &c.* figured *Phyt. Tab.* 190. *Fig.* 3. compared with the Description and Figure of this, 'tis plain this, and that mentioned by him are two Plants contrary to what he conjectures in his *Mantiss. p.* 95.

XXX. *Gramini tremulo affine, paniculatum elegans minimum. Cat. p.* 34. *Tab.* 71. *Fig.* 2. *An Gramen paniculatum ex oris Malabaricis panicula delicatiore* Plukenet. *Phyt. Tab.* 300. *Fig.* 2 ? *Gramen Jamaicense nostrati pratensi simile panicula compactiore. Ejusd. Alm. p.* 176. *Gramini pratensi minori simile Curassavicum panicula speciosa. D. Sheyard. ib*? *Gramen paniculis elegantissimis minimum.* Tournef. *Inst. p.* 322. ?

This has a great many white thready Roods, and many very small, narrow, pale green, a little rough Leaves, a *Culmus* or Stalk about three or four Inches long, having so many Joints, and at every Joint a Leaf, inclosing the Stalk. Near the top the Stalk is divided into several Foot-Stalks standing sparse on every hand, and sustaining several small, white, chaffie Spikes, made up of very small, white, scaly Chaff, compress'd, lying on one another in a double row, very elegantly, after the manner of *Gramen tremulum*, but having its *Petioli* so strong as not to quake.

It grows very plentifully in the *Savanna*, by the Town of St. *Jago de la Vega*.

This only seems lesser than the *Tsiampullu, H. M.* but differs from the *Gramen amoris alterum paniculis strigosioribus magisque Sparsis. Raii Cat. pl. exter.* not being the same with it, as Dr. *Plukenet* conjectures in his *Mantissa, p.* 95.

XXXI. *Gramen miliaceum, sylvaticum, maximum, semine albo. Cat. p.* 34. *Tab.* 71. *Fig.* 3.

This has a *Culmus* or Stalk, several Feet long, slender and weak, not able to support its self without being sustained by leaning on neighbouring Trees and Shrubs, amongst which it grows. It is hollow, jointed, and branched at the Joints, every Joint having a Leaf about three Inches long, and three quarters of an Inch broad in the middle where broadest being striated, and of a yellowish green colour. At the top of the Branches is a Panicle, made up of several Branches which have crooked *Petioli*, on the ends of which is a white, shining, roundish Seed, on one side flat, on the other roundish as big as a large Pins Head, lying in a purplish naked Husk, opening like the other Millets.

It grows in most of the Woods of this Island, and those of the *Caribes*.

By the Description and Figure of this, 'tis plainly different from the *Gramen Miliaceum latiori folio Maderaspatannm*, Plukenet. *Alm. p.* 176. *Phyt. Tab.* 189. *Fig.* 4. which is contrary to the conjecture of that Author in his *Mantiss. p.* 95.

XXXII. *Gramen miliaceum majus, panicula minus sparsa, locustis minimis. Cat. p.* 34. *Tab.* 72. *Fig.* 1.

The Panicle of this Grass was about six Inches long, made up of several Spikes, which lay so close to the *Culmus*, and were, especially at top, so short that

that the Panicle was not sparse, but set almost like a Spike, in so much that I took it for a Panic Grass. The Seeds were very small, oblong, compressed of a whitish colour, and shining, lying in two white Membranes or *Locusta*, which were covered with two green ones, both very small Foot-Stalks, as others of this Kind. The upper Joint of the *Culmus* was very long, and I believe the Grass very large, although I remember neither its Leaves nor Joints.

I do not remember the particular place of its growth, but think I found it in *Jamaica*.

Dr. *Plukenet* thinks that this Plant may be his *Gramen Miliaceum Americanum majus panicula minore*. *Alm.* p. 176. *Phyt. Tab.* 90. *Fig.* 7. or that this last recited Plant may be my next following Grass, but he is strangely mistaken, for there is no resemblance, as any one may see by their Figures and Descriptions. See his *Mantiss.* p. 95.

XXXIII. *Gramen miliaceum, panicula viridi, vel purpurea*. *Cat.* p. 34. *Tab.* 72. *Fig.* 2.

This Grass has a jointed Stalk a Foot long, and Leaves, the under part whereof encloses the Stalk, which is rough, nine Inches long, one broad near the Stalk, where broadest, and whence it decreases to the end, being, towards and at the top, rough, having a Panicle about three Inches long, made of several two or three Inch long purple, or green Spikes, standing sparse after the manner of millet on every hand. Each of these is made up of a great many roundish, naked, purple, or green *Locusta*, sticking to the Spike by a small *petiolus*, having a very small Grain or Seed, within a Follicle, like that of Millet.

It grows in clayie moist Grounds in several Plantations.

XXXIV. *Gramen miliaceum viride foliis latis brevibus, panicula capillacea, semine albo*. *Cat.* p. 35. *Tab.* 72. *Fig.* 3.

This has several Fibers for a Root, shooting forth a very frequently jointed, and sometimes branched *Culmus*, or Stalk, one Foot and a half high. Every Joint has one Leaf, the underpart of which covers part of the next *Internodium*, is rough, and of a pale green colour; the other is about an Inch or more long, and half as broad, of a very green colour, and hairy on the edges. The Panicle is at top three Inches long, made up of several Spikes or Branches, on which are set, by very small, long, *Petioli*, no bigger than Hairs: the Seed being very small, roundish, white, and lodged in green Chaff like other the Millets.

It grew in the Woods that were dry and shady.

XXXV. *Gramen pratense paniculâ & foliis angustissimis, spicis brevibus muticis locustis minimis*. *Cat.* p. 35. *Tab.* 73. *Fig.* 1.

This Grass has many small, thready, white, and capillary Roots, which being join'd together makes a great Tuft, and send forth a great many five Inches long, narrow, or almost round Leaves, being dry, and of a pale green colour. Amongst these comes up the Stalks, round, solid, hard, smooth, one Foot and a half high, of a clay colour, having small Leaves to nine Inches high, whence it is a very narrow Panicle, being divided into many three quarters of an Inch long Branches, sometimes black, and sometimes gray, having several small, oblong, reddish Seed, in a gray, or black, naked Husk, both Seed, and it, being so small, as scarce discernible to the naked Eye.

116   The Natural History of Jamaica.

It grows in most *Savannas*, particularly, in great abundance, towards *Black River* Bridge, beyond two Mile Wood, on the left Hand of the Road going thither.

It has very small nourishment, notwithstanding which Cattle eat it in dry and scarce times, when they grow very big in their Paunches with the great quantity of this Grass, not being satisfied with little.

XXXVI. *Gramen avenaceum sylvaticum, foliis latissimis, locustis longis non aristatis, glumis spadiceis.* Cat. p. 35. Tab. 73. Fig. 2. Wild Oats.

This Grass has a great many three or four Inches long Filaments, with lateral Hairs or Fibrils, by which it draws its Nourishment, which being united in a roundish Root, send forth several Leaves incompassing the Stalk, and one another by their under parts, or Foot-Stalks, which are striated, of a light brown colour, and about nine Inches long. The other part of the Leaf, leaving the Stalk, the higher it is, is the larger, those uppermost being six Inches long, and two broad, beginning narrow, by degrees growing larger, and ending in a point, being striated, thin, hard, rough and grassie, with a middle Rib, eminent on the back side. The Stalk is about a Foot and a half high, having below two very short Joints, is branched out about a Foot from the Ground into several Branches, whose Twigs have several half Inch long naked *Locusta*, sticking to them alternatively without *Petioli*, having a blackish Chaff or *Gluma*, in which is a long roul'd up Membrane, looking like Oats or Corn.

It grows every where in the inland high shady Woods.

It is thought to be the most nourishing and fatning Grass for Cattle in the whole Island, and is counted as good for that purpose as Oats.

XXXVII. *Cyperus longus odoratus, panicula sparsa, spicis strigosioribus viridibus.* Cat. p. 35. Tab. 84. Fig. 1. *An Acorus Brasiliensis aromaticus minor, Capicatinga, aliis Jacare catinga Pisonis.* Bob. hist. Ox. part. 3. p. 246?

This had a long, roundish, frequently jointed Root, reddish on its outside, and whitish within, very odoriferous, creeping under the Surface of the Ground, and making a large Turfe or Tuft, from whence rise up many Leaves triangular, carinated, with an eminent sharp cutting back, of a very dark green colour, larger, but otherwise in shape, &c. exactly the same with the others of this Kind. From among these Leaves rises several triangular solid, dark green, striated Stalks, two or three Foot high, having a rushy Pith, and at its top several smaller Leaves, but of the same shape with those at bottom, standing under the Panicle, which is very sparse, having, besides some shorter Spikes, a great many standing on Foot-Stalks, above small Leaves, some whereof are a Foot high; each of the Spikes being long, very small, roundish, of a pale green colour, made up of several green Scales, between which, and the Stalk, lies the Seed, which is oblong, and of a pale yellowish colour.

It grows by the Rivers sides in *Jamaica*, and most of the *Caribes*.

If it be *Piso's Capicatinga*, he tells us, it is not only given by its self, or mixt with other things to cut cold peccant humours, but is likewise used against Poysons, whence may be gathered that 'tis hot intensly, and of a thin consistence.

XXXVIII. *Cyperus rotundus, panicula sparsa, spicis strigosis ferrugineis.* Cat. p. 35. Tab. 74. Fig. 2.

This had a round tuberous Root, as big as a large Hasel Nut, having many fibers at its bottom, of about an Inch long. It is of a solid substance, within odoriferous, and aromatick to smell and tast. 'Tis covered over with

several

several red, dry Membranes, and has soft triangular graffie Leaves, about one Foot and an half long, like others of this Kind. At its Root, from the middle of these Leaves rises a Foot, or a Foot and an half high, solid, Stalk, triangular, and filled with a rushy, spungy Pith. At its top stand three or four Leaves, which are soft and graffie, as the others, of about six Inches long and shorter; above which come several sized *Petioli*, sustaining many long ferruginous Spikes, standing out sparse on every hand, each being long, round, slender, and containing between the Scales many oblong, whitish, cornered Seeds, making in all a very elegant Head.

It grew near *Bridge* Town in *Barbados*, and in the Marshes by the *Rio-Cobre*, above the Ferry, towards the fresh water *Laguna* plentifully.

However rude the Labour or Travel in Childbed of the *Savages* is, the Powder of this, of the weight of a Crown taken in White-Wine, makes them be speedily delivered. *Tertre. Rochef.*

XXXIX. *Cyperus, panicula maxime sparsa, ferruginea compressa, elegantissima.* Cat. p. 35. Tab. 75. Fig. 1. *An Cyperus Americanus panicula aurea maximâ.* Tournef. Inst. p. 527?

This had some few dark brown, or reddish Roots, sending up some two, or three Foot and an half long Leaves, inclosing the Stalk, and one another, below, very narrow, or almost round above, striated, and having a Pith like Rushes. From these Leaves rise a blunt three cornered Stalk, solid, not jointed, filled with a rushy Pith, two Foot and an half high. At the top stands two or three Leaves, (one whereof is a Foot long) under the Panicle, which is very sparse and elegant, made up of a great many Spikes, standing on the tops of three or four Inches long Foot-Stalks, some on none, and others, of all intermediate lengths, which send out round them other smaller *Petioli*, making it proliferous, each Spike being compress'd, broad, one third part of an Inch long, made up of two rows of ferrugineous Scales, one of each side, being plac'd one over another, in each of which lies a black three cornered Seed.

It grew in the Sea Marshes near the landing place at Mr. *De la Crees* in *Liguanee*.

The Description and Figure of this Cyperus, make it plainly different from the *Gramen cyperoides Maderaspatanum panicula magis sparsa & speciosa.* Plukenet. Alm. p. 179. Phyt. Tab. 192. Fig. 2. though Dr. *Plukenet* in his *Mantissa*, p. 97. thinks it may be the same.

XL. *Cyperus rotundus gramineus fere inodorus, panicula sparsa compressa viridi.* Cat. p. 35. Tab. 76. Fig. 1.

This has one round reddish Tuber, smelling not very much. 'Tis as large as a Pea, and white within. It is join'd to the other Roots by a small Fibril, which with other Fibrils united in an oblong Root, send out several five Inches long, blewish, green, graffie Leaves, inclosing the Stalk, and one another at bottom. In the middle of these Leaves comes a triangular four or five Inches long, not jointed, Stalk. The Top, or Panicle of this Grass is divided into several Inch long Foot-Stalks or *Petioli*, on every side whereof, as well as at the top of the Stalk, are small scaly, reddish green, one quarter of an Inch long Spikes, made up of two rows of green Scales, and all inclosed by three or four Inch long graffie Leaves, standing round the top of the Plant under the Panicle very much resembling the *Gramen pulchrum parvum panicula lata compressa.* J.B.

It grows in the sandy places of the Street of the Town of St. *Jago de la Vega*, after Rain, and in the sandy places of the Town *Savanna*.

One would wonder how Dr. *Plukenet* should come to think, in his *Mantiſſa, p.* 62. that this Plant might be the *Cyperus rotundus littoreus inodorus,* J. B. when their Figures and Descriptions are so vastly differing.

XLI. *Gramen cyperoides majus aquaticum, paniculis plurimis junceis sparsis, spicis ex oblongo rotundis spadiceis. Cat p.* 36. *Tab.* 76. *Fig.* 2.

This has a jointed Stem five Foot high, about the bigness of ones little Finger, being triangular, solid, gray, and each Joint, distant one from the other, four or five Inches. The Joint has always a Tuft of green Leaves, a great many together spring out alternatively, some two or three Foot long, others shorter, triangular, all being very green, serrated, and extremely cutting. The Panicles come out alternatively towards the top, at one Inches distance, standing on six Inches long Foot-Stalks, which are solid, and triangular, and at their ends have two clay coloured Membranes or Leaves, whose under part is as an *utriculus,* from which goes out the Panicle, or many longer, and shorter *Petioli;* at the ends of which stands one Spike, and round them, on other shorter *Petioli,* others, each of which is roundish, oblong, blackish, or dark rusty coloured, made up of many Scales, lying round, over one another, after the manner of others of this Kind.

It grew in the Fresh-Water River, above the Ferry going up to the *Laguna* near *Caymanes.*

Dr. *Plukenet* in his *Mantiſſa, p.* 98. thinks that this Grass may be the same with his *Gramen Cyperoides Madraspatanum, caule compreſſo, sparsa panicula junci Alm. p.* 179. *Tab.* 192. *Fig.* 5. This differs from it in the Stalks, being not compressed, but triangular, and in several other obvious differences.

XLII. *Gramen cyperoides sylvaticum maximum geniculatum asperius, semine milii folus. Cat. p.* 36. *Tab.* 77. *Fig.* 1.

This strange Cyperus Grass, has a slender jointed Stalk, rising to about fifteen Foot high, growing amongst the Bushes, and being supported by their help, though not turning round them. The Stalk is triangular, having three sharp, very rough edges, and a round hollow between them, like a three cornered Sword-Blade, and being of a very dark green colour, having here and there, at the Joints, Branches, which have Leaves at their Joints, being about a Foot long, narrow, having an eminent back, and being of a dark green colour, and withal much rougher than any of the other Cyperus Grasses, by the means of several small *Asperities* or Teeth on its edges. *Ex alis foliorum* rise small triangular three Inches long *Petioli,* on the top of which are Spikes about three quarters of an Inch long, on which alternatively grow three or four small blackish *Locuſtæ* or chaffie Heads. In each of these, between two black *gluma,* comes a roundish, large, whitish coloured Seed, like that of *Gromel,* a small Pearl, or that Barley, very much decorticated, call'd Pearl Barley.

It grew in *Moneque Savanna,* among the Trees.

'Tis evident this is not the *Gramen junceum elatius, pericarpiis ovatis Americanum, Pluken. Alm. p.* 179. *Phyt. Tab.* 92. *Fig.* 9. which Dr. *Plukenet* in his *Mantiſſa, p.* 98. thinks may be the same with it.

XLIII. *Gramen cyperoides majus, spicis ex oblongo rotundis, compactis ferrugineis. Cat. p.* 36. *Tab.* 77. *Fig.* 2.

This has a fibrous, reddish, brown Root, exactly like those of a Rush, the Leaves, and under part of the Stalk being covered with dry, reddish, foliaceous Sheaths, like them. The Leaves inclose the Stalk, by their under part, which is three Inches long, being usually two in number, and about three

Inches

Inches long the longeſt; very narrow, ſoft, and of a very green colour. The Stalk is cornered, ſtriated, compreſſed, two Foot long, on the top of which ſtand above two or three very ſmall, and ſhort Leaves, three or four oblong, round, compacted, ferrugineous Spikes, ſtanding ſome of them on half an Inch long *Petioli*, others on none, from whence the others proceed, and may be termed *prolifera*. Each of them is made up of Scales, lying one over another, between which are lodged yellowiſh, ſhining, cornered Seeds.

It grew in the Sea Marſhes near the Landing-place at Mr. *Delacrees* in *Liguanee*.

Dr. *Plukenet*, in his *Mantiſſa*, p. 97. queſtions if this Graſs be not his *Gramen cyperoides rarius & tenuius ſpicatum è Maderaſpatan, Phyt. Tab.* 192. *Fig.* 1. *alm. p.* 179. The Figures and Deſcriptions ſhew them plainly different.

XLIV. *Gramen cyperoides ſpica compacta alba, foliis ad ſpicam partim albis, partim viridibus. Cat. p.* 36. *Tab.* 78. *Fig.* 1.

From a fibrous and ſtringy Root, ſpring up ſeveral triangular blunt edg'd Stalks, of about a Foot in heighth. The Leaves are harſh to the touch. The Spike is compact, made up of many white Spikes, ſet cloſe in a Head, and has ſome long, harſh, graſſie Leaves cloſe under it, which for the firſt part, or half, are white, and towards the ends are green. The Seed is ſmall and yellowiſh.

It grows in thoſe places where water has in rainy times ſtood on the Ground, as in the Paſture beyond the *Angels* Ford going to ſixteen Mile Walk, in the *Caymanes* in ſeveral places, &c.

XLV. *Gramen cyperoides minus, ſpica compacta ſubrotunda viridi, radice odorata. Cat. p.* 36. *Tab.* 78. *Fig.* 2. *An Pee-Mottenga, H. M. Tab.* 53?

This has ſeveral fibers of a reddiſh colour, which united make a crooked, oblong Root, ſending up ſeveral graſſie, very green Leaves, of about four Inches long, the under part being reddiſh. The Stalk comes up in the middle of them: it is five or ſix Inches high, of the ſame colour with the Leaves, triangular, having at the top one pretty large, and on the ſides of it one or two ſmaller, oblong, compact Buttons, or little Burs, and under this Head ſome graſſie Leaves two or three Inches long.

The Root is very odoriferous.

It grew in *Coll. Nedham*'s Plantation, in ſixteen Mile Walk.

There is a variety of this, which is much larger in all its parts.

The whole Plant (if it be that of the *H. M.*) boiled in Rice, Water, and Oil *Marotti*, and green *Curcuma*, or in Coco-Nut-Oil and Cumin or long Pepper, makes a Liniment againſt the Liver Diſeaſe, *H. M.*

XLVI. *Gramen cyperoides minus, ſpicis compactis ſubrotundis flavicantibus Cat. p.* 36. *Tab.* 79. *Fig.* 1.

This has ſeveral long reddiſh, brown, ſtrings, which united make a ſmall knobby Root, ſending forth ſeveral harſh, green Leaves, ſix Inches long, two of them uſually incloſe a blunt triangular ſix Inches long Stem, at the top of which ſtands one oval, oblong Button, yellowiſh green, made up of green Chaff or *Locuſta*, between each of which lies a triangular ſhining Grain, having two others leſſer, ſtanding above it on one Inch long Foot-Stalk. The Spikes are inviron'd below on the end of the *Culmus*, by one five Inches long, and two ſhorter Leaves, like thoſe at bottom.

It grows on the ſides of ſandy Gullies round the Town.

Dr. *Pluke-*

Dr. *Plukenet* in his *Mantissa, p. 62.* doubts if this be not the *Gramen pulchrum parvum panicula lata compressa.* J. B. *tom.* 2. *p.* 471. There is scarce any resemblance between them, unless in the colour of their Panicles.

XLVII. *Gramen cyperoides minimum, spica simplici compacta, radice tuberosa odorata. Cat. p.* 36. *Tab.* 79. *Fig.* 2.

This small Grass has some brown Fibers, coming from a black tuberous body, covered with a brown Membrane, white within, and very odoriferous. From this Root rise very many one Inch and an half long, narrow, green, grassie Leaves, and from their middle small cornered five Inches high Stalks, of a pale green colour, having each his single, whitish, small, scaly Seed, between the Scales, are lodged black shining Seeds, like those of *Amaranthus.*

It grew in the Sea Marshes, near the landing place at *Delacrees* in *Liguanet.*

Dr. *Plukenet* in his *Mantissa, p.* 97. doubts if this be the same with his *Gramen cyperoides minus Virginianum spica simplici longiori. Alm. p.* 178. *Phyt. Tab.* 300. *Fig.* 6. It is very apparent they are two different Plants by their Figures and Descriptions.

XLVIII. *Gramen cyperoides minimum, spicis pluribus compactis ex oblongo rotundis. Cat. p.* 36. *Tab.* 79. *Fig.* 3. *An Gramen junceum perpusillum Capillaceis foliis Æthiopicum Pluken. Almag. p.* 179? *Tab.* 300. *Fig.* 5?

This had very many capillary, brown, Fibers for Roots, from whence rose many small, Inch long, narrow Leaves, reddish underneath; amongst which grew many triangular Stalks, about two Inches long, at whose top comes two or three grassie Leaves, very short, and small. Above these Leaves stand usually three small rusty scaly Spikes or Heads, two whereof have short *Pestoli,* and the others none, sufficiently discernible, from the others of this Kind, by their smallness.

It grew in the Island of *Jamaica.*

Dr. *Plukenet* questions, in his *Mantiss. p.* 97. whether this be not the same with his *Gramen cyperoides pumilum elegans Maderaspatan. Alm. p.* 179. *Phyt. Tab.* 191. *Fig.* 8. 'Tis very plain 'tis not that Plant, for their Figures and Descriptions are very different.

XLIX. *Gramen cyperoides polystachion, spicis ad nodos ex utriculis seu foliorum alis echinatis prodeuntibus. Cat. p.* 36. *Tab.* 80. *An Gramen multiplici spica Maderaspatanum, calamo secundum longitudinem aculeis horrido. Plukenet. Phyt. Tab.* 191. *Fig.* 2? *Almag. p.* 174?

This has several two or three Inch long Strings for its Roots, taking firm hold of the Earth, from whence rises a jointed Stalk, three or four Foot high. The *Culmus,* or Stalk is solid, triangular, or flat on one side, and round on the other. That part of the Leaf sheathing part of the *Intermedium* is rough or prickly, the other part is five or six Inches long, harsh, grassie, with a sharp back like the Cyperus Grasses, and about half an Inch broad, next the *Culmus,* where broadest. Towards the tops the Leaves, (which are always at the Joints,) are shorter, and more swell'd appearing like an *Utriculus,* having a row of prickles on its back, out of the *Ala* of which rise Branches below, and small Foot-Stalks above, sustaining one, two, three, or four Inches and a half long, green Spikes made up of many small Seeds, standing each above a very small, scarce discernible Leaf.

It grew in Fern *Savanna,* near *Guanaboa,* and Mrs. *Guys* Plantation in *Guanaboa* very plentifully.

Dr. *Plukenet*'s Title and Cut do not agree to this Plant.

L. *Jun-*

L. *Juncus cyperoides creberrime geniculatus, medulla farctus, aquaticus, radice vibra, tuberosa, odorata.* Cat. p. 36. Tab. 81. Fig. 1.

This Rush has a tuberous, red, knobbed Root, having a very grateful smell like that of *Calamus aromaticus*, covered with brown withered Leaves, as well as the under part of the Stalk, like other Rushes, and having several red Strings, going from the Root of one to that of another. The Stalk is round, green, three Foot high, smooth, having within it very strong, and frequent transverse Partitions or Membranes, making it jointed with a Pith between. At the top stand several brown chaffie Panicles, like those of Cyperus Grasses, the small, long, Spikes, being made up of several reddish Scales, lying over one another on the same Foot-Stalks, all coming from the Rushes top, as from a common Center.

It grows in the Rills of the *Savanna*, beyond two Mile Wood, about Mr. *Batchelors*.

This having a very gratefully scented Root, I question not but that it may be very successfully used in place of *Calamus aromaticus, &c.*

This agrees with *Hernandez*'s Description of *Phatziprandu*, and pretty well to his Figure, only it has no Leaves, which by the way are not like those of *Schænanth* in the Figure, and *Ximenes* says the Leaves, are like Leeks. In both these Authors may be largely seen the Vertues ascribed to this Plant.

LI. *Juncus cyperoides, culmo compresso striato, radice odorata tuberosa, capitulo rotundo compacto.* Cat. p. 36. Tab. 81. Fig. 2.

This Rush has a tuberous, red, knobbed Root, having a very grateful smell, like that of *Cyperus*. Each knob or joint is by long transverse Roots, of about half an Inch in length, joined to the nearest to it, so making a long Root, made up as it were of several Beads stringed. The Roots are covered over with brown, withered Leaves, as also are the under parts of the Stalks, which had some longer and shorter Sheaths, like other Rushes. From each knob rises a striated cornered Rush, something compressed, full of a spongy Pith, about two or three Foot high, on the top of which under a round Head, made up of many ferrugineous Scales or *Gluma*, standing close together, very compact, were some few very short Leaves, as in the Cyperus Grasses.

This has great resemblance, in manner of growth, with the *Vacembu*; *H. M.* p. 11. p. 99. or *Acorus verus Asiaticus radice tenuiore vel calamus aromaticus Garzia ab Horto. Comm. ibid.*

It was brought by a very curious Person from the Bay of *Honduras*, where he told me it grew among the Sand not far from *Truxillo*. It was used by the *Indians* in the Belly-ach, and I desiring the aforesaid Gentleman to bring me some Simples, used by the Natives of those parts whither he was to go, in the Cure of Diseases, he assured me he found none more celebrated than this.

This is perfectly differing from the *Gramen Cyperoides latiori folio aspero, panicula in summitate caulis conglomerata Plukenet. Alm.* p. 178. *Phyt. Tab.* 190. *Fig.* 8. as may easily appear to any who compares them, however contrary the Opinion of Dr. *Plukenet* may be in his *Mantissa*, p. 97.

LII. *Juncus aquaticus geniculatus, capitulis equiseti, major.* Cat. p. 37. Tab. 81. Fig. 3. *An scirpus Americanus, caule geniculato, cavo. Lign. Tournef. Inst.* p. 528?

From the same Root, made up of many Strings, as is that of other Rushes, rise five or six, two or three Foot high, hollow Rushes, within which are a great

many

many Membranes dividing it, and making it geniculated, and no Pith almost at all. It is of a fine green, shining colour, and at its top comes out a white, round, oblong, scaly Head, like that of the tops of Asparagus, *equisetum*, or the *Juncus capitulis equiseti*. C. B.

It grows after rainy Seasons in those holes of the low Lands or *Savannas*, where water has stood.

LIII. *Juncus aquaticus geniculatus, capitulis equiseti, minor. Cat. p. 37. Tab. 75. Fig. 2.*

It is in every thing like the former, only not over one third of its bigness or heighth.

It grows with the former.

LIV. *Juncus aquaticus capitulis equiseti minimus.*

This has several whitish fibrous Roots, and a very small Rush about two Inches long. On the top comes a scaly Head, between the light brown Scales of which, lie many small black Seeds.

It grew on the Banks of the *Rio Cobre*, under the Town of Saint *Jago de la Vega*, on the same side of the River.

LV. *Juncus lævis* ἁπαλος *secundus, vel juncus maximus sive scirpus major*, C. B. *Cat. p. 37. Scirpus palustris altissimus. Tournefort. Inst. p. 528. Scirpus altissimus, Ej. El. bot. p. 20.*

This grows plentifully in the great *Laguna*, near *Caymanes*, and in the boggy Fens in several places of this Island.

It is used to make bottoms of Chairs.

If boiled with Wine it stops Fluxes and the *Catamenia*.

Tied under the Arms People by them learn to Swim. *Lugd.*

The Pith dilates Sores and Fistulas.

The Men go naked. The Women (in 38°. N. Lat. in that part of *California* called *Nova Albion*) take Bull-Rushes, and kemb them after the manner of Hemp, and thereof make their loose Garments, &c. Which being knit about their Middles, hangs down about their Hips. *ibid.* Sir *Francis Drake, ap. Hakl. p.* 441. and *p.* 737. *p.* 3.

They are said by the Author of the *Mexican* Chronicle, to have grown in the Lake of that City, in the *Carrizales* of it, and to have made Beds for the new born Children to be wash'd and receive their name on. *Purchas, p.* 1102.

Chests of Bull-Rushes preserve the Wheat in *Azores, Mandelslo, p.* 221.

LVI. *Gramen Junceum aquaticum geniculatum, culmo nudo & folio non articulato. Cat. p. 37. Tab. 75, Fig. 3.*

This has several brown Strings, united in one common tuberous Root, from whence rises a green articulated Stalk, it being inwardly divided into many Sections by several Diaphragm's, growing smaller towards the top, within an Inch of which it has several long, slender, ferrugineous brown Heads, or Spikes making a sparse Panicle. At the Root are several graffie two or three Inches long Leaves.

It grows in the moister parts of the *Savannas* after Rain.

LVII. *Typha palustris major.* C. B. *Cat. p.* 37.

This grows on the brinks of *Black River* near Mr. *Byndloss*'s Plantation, and on the fresh water *Laguna*, near the *Caymanes*.

The

The Down is used in some places for Beds, and applied to kibed Heels Cures them, stamped with Swines Grease is good for burnings, *Diosc.* and given to a Dram, with other ingredients, Cures Ruptures, *Math.* This seems to be no good Medicine, but that other Ingredients mixed with it, are the causes of the Cure. *Lob.*

It is the φλόμος *Theophrasti* the flour or top used to be mixed with Clay or Lime, instead of Hair or Straw for Buildings, to keep the Mortar from breaking. The Leaves of the Male, before it flours or *Fæmina*, are good for binding any thing withal, and are now used to make Mats. *Herodotus* tells us of the *Indians* inhabiting the Marshes of the River, carrying *Vestem* φλοίνω, *Dod.*

The Leaves are used to cover Flasks, and for Chairs. *Math.*
The Leaves of this are cut for the *Papyrus* by *Dodonæus.* C. B.

Chap. V.

Of Herbs with less perfect or Stamineous Flowers.

There are in this Chapter some Plants which I confess I believe may by those more skilful be reduced to other Families, some of them having parts so extraordinary small as not to be easily visible, and perhaps others have perfect Flowers, which escap'd my observation.

Some of these are also very anomalous that I could not find any other place than by their Face or Leaves to reduce them to those of *Europe* placed by the more skilful here.

It may be objected to me that I have brought hither Nettles, or called Plants so, which have many of them tricoccous Seeds, which had better been with the *Ricini*; to this I answer that the *Hortus Malabaricus* has described several Plants under the name of *Schorigenam's*, with the same kind of Fruit, and no body found any fault with the Authors of that Book, or Mr. *Ray* for reducing them thither.

One thing in this Section falls very oddly, which is that in three or more sorts of such Plants as by their Face must of necessity be esteemed *Ricini* by all People, there are perfect *Pentapetalous* Flowers, and after these the Fruit follows contrary to those observed hitherto by others.

Before the Species of long Peppers are described, there is an observation relating to them to be taken notice of in this place.

1. *Urtica racemosa scandens, angustifolia, fructu tricocco. Cat. p.* 38. *Tab.* 82. *Fig.* 1. *An urticæfolia Jamaicensis tricoccos, Muf. Cotten. Plukenet. Alm. p.* 393 ? *Ricinus parvus urens urticæ filis Banister, Cat. Stirp. Virg.* ?

This has a woody, reddish, striated Stalk, something cornered, as those of Nettles, which turns about any Plant or Tree it comes near, and rises six or seven Foot high, having a strong Bark, something like that of Hemp. Towards the tops of this Stalk are many two or three inches long Branches, very thick set with Leaves without any order, each of which has a half Inch long, rough Foot-Stalk, is Inch long, and half as broad near the round Base, where broadest, and whence it by degrees lessens till it ends in a point, having the Margins deeply cut, and being all over very thick set with burning Hairs, as those of ordinary Nettles. Out of the *Ala* of the

leaves,

leaves, come an Inch long ſtrings of herbaceous Flowers, like thoſe of other Nettles, after which follows a tricoccous, rough Seed-Veſſel, each corner of which incloſes one round, reddiſh, brown Seed.

It grows among the Shrubs in the Town *Savanna*, going towards two Mile Wood, and elſewhere very plentifully.

If this be the *Valli Sthorigenam*, the decoction of the Roots drank, extinguiſhes the heat of the Liver, and is good for ſwell'd Bellies, with ſuppreſſion of Urine. The ſame beaten, and mix'd with Sugar and *Milk*, is good for the Itch. The Juice of it, beaten and drank, is good for thoſe piſſing viſcid and purulent Urine. *H.M.*

II. *Urtica urens arborea, foliis oblongis, anguſtis.* Cat. p. 38. Tab. 83. Fig. 1.

This Shrub I deſcribed, which ſeemed to be young, roſe to about eight or nine Foot high, by a round, ſtraight, woody Trunc, of the bigneſs of ones little Finger, covered with a ſmooth, browniſh Bark. It had towards its top ſeveral Leaves coming out alternatively ſtanding on an eighth part of an Inch's long Foot-Stalk, each whereof were about three or four Inches long, and about three quarters of an Inch broad in the middle where broadeſt, being even about the edges, unleſs one ſmall *Lacinia*, with a ſharp point, made the Leaf as it were eared towards the top (which ſome Leaves wanted, and ſo perhaps that was accidental) from the Foot-Stalks end it grew broader to the middle, and thence decreaſed, ending in a point. It was of a dark green colour, and had ſeveral Ribs, appearing on its under ſide, and on its ſurface and edges many long, ſmall prickles, which, as I was told, were very burning, and look'd ſo fierce that I was very loath to make the Experiment my ſelf, but very cautiouſly took the top of the Shrub, and dried it. I never ſaw either its Flower or Fruit.

I found it in the Woods on the Hills on the other ſide of the *Rio-Cobre*, near the *Angels*.

III. *Urtica racemoſa humilior iners.* Cat. p. 38. Tab. 82. Fig. 2.

This had a Root half a Foot long, very ſtrong, and deeply faſtned in the Earth, from which went ſeveral Filaments very far ſpread on every ſide, drawing nouriſhment to the Plant. The Stalk was two Foot high, ſquare, and covered with a reddiſh brown Bark, having ſeveral Branches, going out, oppoſite to one another. The Leaves ſtand towards the top, ſet oppoſite one to the other at half an Inch's diſtance, exactly like thoſe of Nettles, but they did not ſting. *Ex alis foliorum* comes an Inch long *Racemus* of green Flowers and Fruit, like thoſe of the *urtica major vulgaris*, J. B. only the *Apices*, or Seeds making up the *Racemus* were more ſolid, flat, and not cornered.

It grew on the Banks of the *Rio Cobre*, near Mr. *Fonſeca*'s Plantation.

IV. *Urtica iners racemoſa ſylvatica, folio nervoſo.* Cat. p. 38. Tab. 83. Fig. 2.

From a ſmall, ſtringy, brown Root, Springs a cornered, green Stalk, one Foot and an half high, having two Leaves at the Joints, ſtanding oppoſite one to the other, on two Inches long Foot-Stalks. They are four Inches long, and three broad in the middle, where broadeſt, rough, freſh green coloured, a little indented about the edges, having three Ribs running from the end of the Foot-Stalk, as from a common center, through the Leaf, with ſeveral tranſverſe ones, ſomething like the Leaves of Nettles. The Flowers ſtand at top in a bunch altogether, being browniſh, very ſmall, muſcoſe, reticulated, and making a pleaſing Figure.

It grows in the ſhady Woods near *Hope* River in *Liguanee*, and in the woody and ſhady Mountain near Colonel *Cope*'s Plantation in *Guanaboa*.

V. *Urtica*

V. *Urtica minor iners spicata, folio subrotundo serrato, fructu tricocco.* Cat. p. 38. Tab. 82. Fig. 3.

This has a Root five Inches long, big in proportion to the Plant, brown on the outside, and deeply fix'd in the Earth, which sends out on every hand several small Stems, lying along the Surface of the ground, two Inches in length, having small Leaves plac'd, without any order, on the Branches. Each leaf hath an eighth part of an Inch long Foot-Stalk, is one third part of an Inch long, and very near as broad at Base, where broadest: almost round, smooth, and snipt about the edges. The top of the Stem is a Spike of Flowers an Inch long, made up of a great many very small Purple Flowers, some little white intermingled. After these follow, in the same Spike, several tricoccous Heads, at first green, and then a little reddish, and rough on the outside. In each of these lie three roundish Seeds, every one is covered with a Membrane, and the three Membranes being joined, make up a tricoccous Seed, like that of the *heliotropium tricoccum*, only infinitely smaller.

It grows in the Town *Savanna*, every where in the sandy places thereof, especially towards two Mile Wood.

This cannot be the Plant mentioned by Dr. *Plukenet. Mant.* p. 190.

VI. *Urtica major racemosa humilior, pungens sed non urens.* Cat. p. 38.

This had a woody, reddish Root, about five Inches long, having several lateral Fibrils, by which it drew its Nourishment from the Earth. The Stalk was round, woody, green, and about a foot high, having near its top several Leaves, without any order, standing on Inch long Foot-Stalks. They were five Inches long, one broad, in the middle, where broadest, very much serrated, and rough like Nettle Leaves, of a dark green colour, the back side of the Leaf being lighter; these Leaves were very rough and pricking. In the Plant I examined, I could not observe any vesicles at bottom of the prickles as in Nettles; but the pricking of the Finger came from some very hard and white prickles was visible on the Leaf by the Microscope. By the Foot-Stalk, at the beginning of the Leaf, is a very short sustaining four small membranaceous Leaves or *Petala*, within which are five or six yellow *Stamina*, to which follows a roundish *Capsula*, made up of three pieces which open of themselves, throwing out several small oblong brown Seeds.

It grew on the Banks of the *Rio Cobre*, above the Town of St. *Jago de la Vega*, on the other side of that River.

'Tis plain by this description that this cannot be the *Urtica racemosa Canadensis Morif.* as Dr. *Plukenet* thinks in his *Mantiss*. p. 190. neither can it be the third described in this Chapter, as he there imagines.

VII. *Ambrosia elatior foliis artemisiæ, atrovirentibus, asperis, odoratis, non lanuginosis.* Cat. p. 38. *Katu-Tsietti pu. Hort. Mal.* part. 10. p. 89. Tab. 45. *sive Ambrosia Malabarica artemisiæ folio odoratissimo floribus flavis.* Comm. ib. *Bob.* part. 3. *Hist. Ox.* p. 4.

This rose to about three Foot high, by a strong, striated, woody, solid Stem, as big as ones little Finger, having at pretty large distances, Leaves, which stand on an Inch and a half long Foot-Stalks. They are cut and divided just after the manner of the Leaves of Mugwort, to which they are very like, only somewhat larger and stronger, being rough, but not hairy like the *ambrosia marina*, C. B. *pin.* and of a very dark green colour above, underneath more pale. *Ex alis fol.* and towards the top come Twigs, having smaller Leaves, and on their tops a great many small muscose Flowers,

Flowers, of a yellow colour, set close together, as in the others of this Kind. The Fruit is in an echinated Husk, just like the Fruit of *tribulus*, and the Seed like a Grape Seed.

The whole Plant has a very strong smell like the others of its Kind.

It grew in the stony places of dry River, in which sometimes water used to run, near Sir *Francis Watson's* in the seven Plantations, as well as in several other inland parts of this Island.

If this Herb be put under the sick's Pillow it foretels death if he sleeps not. Boil'd in *Zergelim* (i. e. *sesamum*) or *Tzit-elu* Oil, and burnt Wine, and applied to the part affected, it Cures *Empyema's*, and hid abscesses of the Stomach before they ripen, especially if the juice be drank with Honey. With Malabarick or *European* Horehound, made into a Plaister, and put on the pained place, it Cures the Cramp or *Spasm*. With Honey eaten fasting it Cures the Dropsie. The Root boil'd in Sergelim Oil, rub'd on a Cloath, and repeated every twelve hours, takes away freckles or spots. With Coco-Nut Milk boiled, it Cures eating Ulcers, and the Bark of the Root put on the affected part does the same. *H. M.*

VIII. *Ricinus Americanus fructu racemoso hispido. Cat.* p. 38. *Ficus infernalis. Catzeol. Muf.* p. 641. *Mexico* Seeds of *Med. Cur. Ricinus Americanus major rubicundus, Munt. Pa-tcu Cleyer.* p. 46. The Oil-Nut-Tree.

I have nothing to add to *Marcgrave's* description, but that 'tis generally nine or ten Foot high, and of two sorts, one with large Seed, which is made use of to make Oil, the other with lesser, which I do not see differing from the common *Ricinus*, and I am very apt to believe the red and green, as well as those with little and large Seed, to be only varieties.

They grow in *Bermudas*, &c.

It seems not to be different from the common *European Ricinus*, which is Perennial in *Spain* and *Crete*, where 'tis large, according to *Bellonius, Clusius* and Mr. *Ray.* The Seeds are commended by some for killing Worms in Human Bodies.

The Leaves dipt in Water or Vinegar Cure Tetters, and are generally made use of for dressing Blisters.

The Kernels after drying, and beating to Powder, are boiled in water, and an Oil swims at top, which is us'd for Lamps. It is good likewise against all outward and inward cold Distempers. It is hot in the third degree. Good against Wind. Opens Obstructions of the Spleen, &c. or gripings of the Belly and Mother. Applied outwardly to the Navel, as well as taken inwardly in Drops, it resolves Apostems, restores convuls'd or contracted Nerves, by distending them softly. Three or four Drops given by the Mouth or Glister, opens the Belly, and Cures Dropsies, and Diseases of the Joints. It Cures the Scab, and other external Diseases of the Skin. The Seed is a violent Purger, more than the Oil, both upwards and downwards, and therefore they want preparation, which is done by infusing them in Spirit of Wine, an ounce of which will bring off humours easily; seven of the Seeds are enough. The Oil is likewise drawn from these Seeds as from Almonds. *Piso.*

This Oil is called by *Monardes oleum Cicinum,* because *Dioscorides* and *Pliny* calls the Tree or *Ricinus Kiki,* and the Oil from it so. A drop or two given in Milk expels Worms in Children. If their Belly be anointed with it. It is good for Scald Heads, pains in the Ears, and Deafness. *Monard.*

The Root taken in Decoction, Cures pain from Wind in the Bowels, takes away the swelling of the Legs, and of the Kidnies and Belly, as also 'tis good for the Dropsie, Asthma, Gout and Erysipelas. The Leaves Cure the Head-ach. Beaten and applied, after they are heated at the Fire they are good in the Gout, as is also the vapor of the decocted Leaves received on the part. A Bath with it, is a good Diuretick. The Fruit is purging if in Powder it be fried and given with Sugar. The Seeds beaten ease the Kidnies if applied to them outwardly. The Oil is purging. Given with Milk it is good in the Collick, and inwardly applied it is good for pains. It stops Vomiting if the Bark be hanged about the Neck. *H. M.*

Clusius was mistaken when he thought this Oil was extracted out of the Physick-Nut, for 'tis commonly out of this which has a rough or murciated Fruit, whereas that has not, neither has the Physick-Nut a spotted Seed, although an Oil is drawn out of that too, but *Monardes* his description of the Tree to be like our *Ricinus*, must make it this Tree, and not the other which has a Face quite different.

'Tis good for clearing *Negroes* Skins, and for Lice on the Head. *Terre.*

Oleum Ricininum Dioscor. de Herva Arabb. is made by beating the Seeds, and pouring water, which is after heated on them, it swims at top, and is us'd for Lamps in *Egypt. Cord.* in *Diosc.*

These Seeds taken inwardly Purge upwards and downwards with great Anxiety. *Id. Hist. pl.*

It is sowed in Fields about *Milan* for Oil for Lamps. *Gesn.*

Thirty Grains beaten and drank Purges by Vomit and Stool. *Diosc. Fuchs. Lon. Trag.* The Leaves help Inflamations of the Eyes and Breast beaten with Vinegar, and applied. *Lon.*

The Inhabitants of *Pantalarie*, who are Poor, use the Oil made of this (if it be what he means) for Lamps and eating, and anoint their Hair therewith to make it grow. *Nicol.*

It is good for the same purposes that *Oleum de Cherva* is. *Mes. Fragos.*

The Leaves help St. *Antonies* Fire.

Five, seven, or at most fifteen Seeds are a Dose. *Mes.* Thirty endangers life.

The Oil is good for Plaisters and Ointments. This was what shaded *Jonas,* as *Arias Montanus* says. *Dod.*

A spoonful of the Oil Purges not, nor three spoonfuls in a Glister. The Leaves are good in the Head-ach. *Stubbs, Phil. transf.* N°. 36.

Green Oil is made so of Ebulus Fruit. *Math.*

The Oil is called *Azeyte de Cherva. Lac.*

Three of the Seeds of these of *Syria*, which are larger, are a Purge. *Com.*

An *Anonymus Portugall ap. Purchas, lib. 7. cap. 1. p.* 1368. tells us that in *Brasile* they dry them four days in the Sun, Stamp and Seeth them, an Oil comes at top, which is used for giving light, and anointing the Stomach to stop Vomiting, it also takes away Oppilations and Collick. *Ranolfo, p. 1. cap. 4.* relates that it is found about *Tripoli* in great abundance. *Belonius cap.* 18. that it grows to a Tree.

IX. *Ricinus Americanus major caule virescente, hort. Reg. Par. Cat. p.* 39. *Kiki Mirand. Sim. & Eut. p.* 894. *Lathyris major sive cataputia major,* i. *Ricinus. Ambros. h.* 36.

This differs only from the former in having its Stalk and Leaves green, whereas the other has them redish; wherefore I doubt it is only a variety of the precedent.

X. *Ricinus, ficus folio, flore penta-petalo viridi, fructu levi pendulo, Cat. p.* 40. *Ricinus Americus. Tab.* 2, *p.* 46. *Ricinus Americanus. Cont. m. p.* 1. *Den sive Dendt Arabum Culccil. muf. sp.* 588. *Dendt Muf. Muf. p.* 269. *Pinhos de Brasil*

Brasil Worm Muf. p. 211. *Noix Medecinales de Bonton,* p. 64. *Barbado* Seeds of *Med. Cur. Noix de medecine de Rochef. Tabl.* p. 25. *Ricinoides Americana gossypii folio. Tournef. Inst.* p. 566.

This Tree has deep brown Roots. It rises to about twenty Foot high, with a Trunc of soft Wood as thick as ones Leg, having a whitish coloured bark below, but towards the ends of its Branches the Bark is green, where are many Leaves set irregularly on all sides, having each a six Inches long Foot-Stalk. The Leaf its self is roundish, something like to a Fig-Leaf, only not so much laciniated, being soft, or a little woolly, having a very strong unsavoury smell. They, and the Bark yield a wheyish Liquor in pretty great plenty which stains Linnen. The Flowers come in Bunches on three Inches long Stalks near the ends of the Twigs. They are green *pentapetalous*, the *petala* being bowed back downwards, with yellow *Stamina* within, of a sweetish smell. After these follow the Fruit, several together hanging downwards from pretty long Stalks, bigger than Hasel-Nuts, first green, then yellow and brown, when ripe, having three obtuse sides, the outward Husk, or strong brown smooth Membrane, breaking in three places of its self, discovers three several Nuts or Kernels, each lodged in his own Cell, separated from the others by a partition, every one being almost round, oblong, like, but bigger than a Pine-Nut, covered with a rough or scabrous dark brown Shell, breaking or cracking in some places where 'tis whitish. The Kernel consists of two sweet, very white Lobes divided by a white Membrane, it is covered with a white thin *pellicule*, like that of the inside of an Egg-shell.

I take this to be the *Quauhayohuatlis.* 3. of *Hernandez* and *Ximenes,* as I have said already in my Catalogue p. 40. notwithstanding what is said by Dr. *Plukenet* in his *Mantiff.* p. 162.

Aldinus, or rather *Petrus Castellus,* has given a good figure of this Plant. These Seeds were put out of use by *Mechoacan. Frag.*

It is planted for Hedges at all times, and every where in *Jamaica,* being easily propagated by the slip. A quick grower, and good Fence; but of its self it grows on Banks near Rivers and Gullies every where.

I was very Sick, and so were all that were on Shoar (at *Sierra Leona*) with me, with eating of a Fruit of the Country, which we found on Trees, like Nuts, whereof some did eat four, some five, some six and more, but we Vomited, and Scoured upon it without Reason. *Ward ap. Hakl.* p. 3. p. 758.

Hughes, tells us p. 81. that three growing together are a Dose to weak Bodies, five, seven, or nine to others stronger. One eat thirty or forty, and came off, they work upwards and downwards, cleansing the Body of tough humours.

Ligon. p. 66. relates that in *Barbados* they likewise Plant it for Palisadoes, keeping it even with Rails and Brackets, and that no Cattle comes near it. The same tells us p. 67. that from five to three are a good Vomic, and that if you take out the Film it Vomits not.

Clusius says that half a Grain will Purge violently upwards and downwards, in which his Relator deceived him, or he meant of the *pinei nuclei Malucani Acosta* or *Granatilli.*

The Nuts are given from three to seven for a Vomit, and are counted hot in the third, and dry in the second degree.

It very often happens that new comers are deceived by these Nuts, being invited to eat them, by their pleasant taft. I have known several so deceived, but never any suffer more than a severe Vomit. Sometimes they are candied with Sugar into a Sweet-Meat, on purpose to deceive people, who eating of them unawares, are purged upwards and downwards with violence.

The

The great part of the Seed is divisible into two parts, but there lies in the middle in the hollow, a film which consists not only of two perfect Leaves, answerable to Lobes in other Kernels, but of these parts that become Trunc and Root. *Grew.*

In old Obstructions of the Bowels, four or five of these ripe Kernels, freed from their inward and outward pellicles or Membranes, then tosted and infus'd in Wine, are accounted a good remedy. They Purge upwards and downwards with that violence, that four will cause such symptoms sometimes as will force the use of Counter-Poysons or Opiats. They are likewise useful in affording an Oil for Lamps which the *Portugueze* use in *Brasil.* The Oil of this as well as of the *Ricinus* purges. *Piso.*

The inward Skin being taken out they purge more gently.

They sometimes Purge, but Vomit chiefly thick and tough Humours, therefore they are given in antient Distempers either five or seven, always odd, but they are first torrefied, and then when mash'd, they are steep'd in Wine or Water to be made milder. They are hot and oily. *Hernand.* His Figure is very bad.

The Dose is from three to six. *Tertre,* five *Rochef.* Rain dropping from them stains Linnen. *id.*

XI. *Ricinus minor staphysagriæ folio, flore pentapetalo purpureo.* Cat. p. 41. Tab. 84. *Ricinus Americanus lactescens, trilobato folio perennis flore nigricante* hort. Beaum p. 36. *Ricinus Americanus perennis floribus purpureis Staphidis agriæ foliis.* Commel. hort. Amst. p. 17. *Ricinoides Americana, staphis agria folio.* Tournefort. Inst. p. 566. Wild *Cassada.*

The Root of this is tapering, white, straight, two Inches long, having some few fibers drawing its Nourishment. The Stem seldom rises above two Foot high; it divides into Branches spreading themselves on every hand, they and the Stem being crooked, and having a light brown coloured Bark. The Leaves stand on long and rough Foot-Stalks, and are divided always into five points when old, at the tops, or when young, into three *Laciniæ*, which are of a reddish colour, like *Staphisacre,* or somewhat like *Mandihoca*, whence the common name of Wild *Cassada* in all our Plantations. They are of a dark, or very fresh green colour, and an unsavoury smell. The Flowers stand many together, on short Foot-Stalks, on the tops of the Branches, are *Pentapetalous* of a purple colour, with yellow *Stamina*, and very small. After these follows, a triquetrous Fruit covered with a dark brown muricated Skin, under which lie three hard clay coloured Shells, having a *sinus* or depression on their outmost corner or edge, and inclosing each within it an oblong Seed like the other *Ricini,* of an ash shining colour, with two Green Ears and a white Pulp.

It grows in the *Savannas* every where in *Jamaica,* as also in all the *Caribes,* and on the main Continent of *America.*

The Figures of the three Plants mentioned in my Catalogue, and cut by Dr. *Plukenet* in his *Phytographia,* agreeing with this Plant, which has sometimes in its Leaves five, and sometimes three *Laciniæ,* taken notice of by me in its Figure and Description, makes me believe they may be all the same, notwithstanding what he says in his *Mantissa,* p 161.

This is the most general remedy of the poorer sort in the dry Belly-ach, they take of the Leaves from seven to twenty one, and boil them, drinking the Decoction, and when nothing will move to Stool, they will. 'Tis not only used in our Plantations for this purpose, but on the main Continent of *America,* as I have been informed by those who practised Physick among the *Spaniards* inhabiting those parts.

L l The

The Seeds are very much coveted and eaten by ground Doves, whence they are used as baits to allure them into the *Clavanies*, or Traps made of Reeds, by *Negroes*, to catch them.

This seems to be the Plant *Hucipochotl* described and figured by *Hernandez*, nothing obstructing but the qualities, which he says, eaten as a boil'd Sallad restores the colour, when this purges very much, although those qualities may be consistent. The distilled water is very cordial, restoring even almost the dead. Five or seven of the Kernels, freed of their covering Membrane, Purge Flegm and Choler, both upwards and downwards, and that so gently, that the least thing taken stops them, if it be expedient. It is hot, and moist, or temperate. *Hernandez.*

XII. *Ricinus minor viticis obtuso folio, caule verrucoso, flore pentapetalo, albido, ex cujus radice tuberosâ, succo venenato turgidâ, Americani panem conficiunt. Cat. p. 41. Tab. 85. Jucca Mus. Swammerd. p. 12. Manihot* of *Ogilby. Afr. p. 556. An Hiuca sive Mizmaitl. Hernand.* p. 378 & *Mandihoca de Esquemeling. p. 55. Manyec de Rochef. Tabl. p. 52. d'Olivier Oxmelin, p. 74. Worm. Mus. p. 160. Cassavi vel Juca. Contant. p. 2. Magnec. de Bist. p.* 336. *Ginera Mus. Mosc. p.* 260. *Ricinus saxifera heptaphyllos, ex cujus radice venenata, placentas & panem conficiunt. Americani. Pluk. Mant. p.* 161. *Ricinus Americanus pentaphyllos, radiis foliorum integris, subtus glaucis, Cassava Barbadensibus dictus. Bob. hist. Ox. part.* 3. *p.* 348. *Manihot Thevett, Jucca & Cassavi.* J. B. *Tournefort. Inst.* 658.

This has an oblong tuberous Root, as big as ones Fist, having some fibers drawing its Nourishment, and being full of a wheyish venemous Juice. The Stalks are white, crooked, brittle, having a very large Pith, and several knobs sticking out on every side like Warts, being the remainder of the Leaves Foot-Stalks, which are dropt off. It usually rises six or seven Foot high, and has a smooth white Bark; the Branches, which come out on every hand, towards the top are crooked, and have on every side, near their tops, Leaves irregularly plac'd, which are finger'd like those of *Agnus Castus*, Hemp or *Lupins*, each Leaf having a four or five Inch long Foot-Stalk, and being cut, or divided almost to the centre or end of the Foot-Stalk into five or seven parts, each Section whereof is five or six Inches long, has one middle Rib, coming from the end of the Foot-Stalk, sending transverse Nerves to the sides, it is an Inch broad in its middle where broadest; smooth and of a dark green colour. The Flowers are pendulous, hanging down from the tops of the Twigs by three or four Inch branched Stalks, on the ends of which four or five hang together by small Foot-Stalks; being of a pale colour pentapetalous, with yellow *Stamina* in the middle, they are urceolated, or swell'd in the middle, like those of *Arbutus*. After these follows a triquetrous Fruit, about the bigness of a Hasel-Nut, covered with a thin green Skin having six Crests or Ledges on its Surface, and containing in several shells, three oblong gray Seeds, like those of the other *Ricini*, spotted with black spots.

It is planted every where, but more especially in the Low Lands, where it seldom Rains, this enduring the want of that better than any other of the Vegetables, whence Bread is made in these parts.

It is of the most general use of any Provision all over the *West-Indies*, especially the hotter parts, and is used to Victual Ships.

It is ripe twelve months after Planting, and ready to pull up for use. The longer 'tis planted the heavier is the Bread made of it, and it being sold by weight the more profitable. The new is Pleasanter for use in ones Family.

Martyr,

Martyr, by Relations from the *West-Indies*, gives it two years to ripen, to make of it the Bread *Cazabi*, which keeps two years, and says that they planted it in raised heaps nine Foot square, putting therein twelve pieces, and that they had a Tradition that one *Boitins*, a wise old Man, finding it on a Rivers brink brought it to his Garden, and that after several Trials, by which some were killed, they came to the knowledge of the Juices being the Poyson, and so came to prepare it right. The Juice when boiled is pleasanter, and as innocent as Whey.

Yuca, as *Oviedo* tells us in his *Coron. lib. 7. cap. 2*. Is of two sorts, the first with long and narrow Leaves, the second short and broad. The *Indians* make small round hillocks, almost touching one another, in which are put eight or ten pieces of the Trunc of the *Yuca*, which takes Root, and is weeded in the Continent. It kills not as that in *Hispaniola*, called *Boniata*, which is like the other, therefore they eat its Root, not made into Bread. *Cazabi* Bread is made by taking of the Rind of the Root with Shells, grating it, putting it into Bags made of Tree-Bark, pressing it, and putting it over the Fire. A Draught of the Juice Kills, but if it has three or four heats the Poyson vanishes, and the *Indians* eat it, but they eat not this when cold; because, though it be no Poyson, yet it is of bad Digestion. Evaporated it turns Sweet, and then Vinegar, but there's need of neither, because of the plenty of Limons and Sugar. It keeps a year, and the Root must be ten months old at least. The Root kill'd the *Indians* in two or three days, and the Juice presently. Fifty died together with the Juice.

The venemous quality lies not in the Juice only, separated from Earth, else the Juice alone would be Poyson, whereas the whole Root is so. The Venom lies chiefly in volatile parts, going away by Coction. *Aldinus*.

The Plant figur'd by *Hernandez* and *Terrentius*, under the name of *Hiuca five Mizmaitl*, seems not this *Cassada*, but rather a *Serpentaria* by its Figure, but I am notwithstanding apt to believe, considering *Hernandez's* Errors, this may be it.

The Juice of this Root (called *Manipuera* in *Brasile*) is poysonous, not only to Men, but all sorts of Cattle, unless they be used to it. I have seen Swine used to it, drink of it running from the Press, very sweetly, although to others, of the same kind, not used to it, it had been present death, by the relation of those knowing it.

The Symptoms following its being taken according to *Piso*, are swelling of the Body, pain and squeamishness at the Stomach, with *tenesmus* and belchings, dimness of the Eyes, Vertigos, constant Head-aches, cold in the extreme parts, Swoonings, and Death follows; wherefore those preparing this Root fortifie their Stomach and Heart, with some *Arnotto*, Roots and Flowers of *Nhambi*, against it, mixing them with their Meats and Drinks. The Juice kept some time corrupts, breeding Worms, called *Topuru*, of which powdered is made a greater Poyson, which, if given alone, in small quantities, kills presently, but if mix'd with *Nhambi* Flowers does it leisurely, whereby the Authors are undiscovered. The best Remedies are Vomits given presently, and Clysters mix'd with *Alexipharmacs*: then Diuretics, and such as cause Sweat, as *Malva d' Ises, pe de Galinha, Jurupeba, Urucu, &c*. Then the Root, and especially the Juice of the Pine-Apple frequently taken, and in large Draughts. If the Juice be boiled the Poyson evaporates, and the remaining part becomes nourishing. *Piso*.

It was in *Angola*, and there called *Hincea (Monardes*,) it Loves high dry Land. Wet spoils the Bread, *Farinha relada* is the name given the Meal when dry. The Juice evaporated over the Fire gives the *Tipioca* Meal, the Leaves beaten and boiled are eaten after the manner of *Spinach*, and called *Manicoba*. *Tipioca*, if well kept, is good against Dysenteries, &c.

No

The Natural History of Jamaica.

No increase is from the Root, it dying presently when out of the Earth. The Hills for it are made three Foot broad, and half a Foot high, that the water may fall off, the Juice boiled to the Consistence of Pottage is eaten. The Juice is sweet, but kills in two hours time, lets fall the Sediment, which dryed is called *Tipioca*, from whence Starch and several Medicines *Apimacaxera* is the only eatable *Caffada* Root, it is eaten rosted. *Marcgr*.

Gomera says that in 22 and 23°. beyond the Equinoctial, in going to *Magellan* Straits, they eat of Bread of *Madera Rallada*. cap. 91.

Christofile de Acugna, and from him Count *Pagan*, in his Relation of the River *Amazones*, says the Inhabitants of the Isles of that River make their Bread and Drink of this Root, and that they keep it under ground, well covered at top with Clay, from hurt by the Inundations there.

One sort of it on the Main eatable. *Gom*.

It came from *Brasil* to St. *Thomas*. If Boil'd it is innocent, swells much, whence a Law in the *Indies*, that it should never be put to Table without Wine or Water, left people should be suffocated with its swelling. It is not so wholesome as Wheat, and cannot be made light. *Reefs*. In their coming to St. *Thomas* gave over *Iams*, this being found firmer Nourishment. *id*.

Monardes and *Linschoten*, says that on the Continent the Root or Juice is not poysonous, but on *Sancte Domingo* it is, which is false, but there is one different kind of this Root in both those places, which is not poysonous, and which in *Jamaica* they Rost and Eat.

The *Brasilians*, either dried on the *Boucan* or Green, grate it on small sharp Stones set in a Plank, reduce it to Flower or Meal, having the scent of new making Starch, sift it, and put it into an earthen Pan, and set it over the Fire, ever moving or stirring it with Gourds cut in two, which they use for Spoons. That baked soft, is good for common use, the other for carrying to War. *Ouy-enta*, the name of this, which is hard and very much back'd, *Ouy-pou* the name of the other. It will not make Loaves, although it will rise with Leaven; the upper part drying, while the inward is meally. This by *Gomera in Gen. Hist. des Indes*, is called *Bois grate lib.* 2. *cap*. 92. Both Meals are good to make *Boulie* called *Mingaut*. The *Indians* eat this Meal instead of Bread, and throw it into their Mouths with their four Fingers never missing, which the *French* could not do, making their Faces meally. The Green Roots pressed yield a Milk or Juice, which set in the Sun till it thickens, and then put on the Pans makes good Meat, or Pancakes. *Aypi* is good if bak'd in the Ashes, but *Maniot* is Poyson, unless well bak'd in Bread. Pieces of these Stalks give great Roots in two or three months. *p.* 122. *Lery*. They to make their drink, slice the Roots, boil them in water till tender, and when cold chaw them, putting them out of their Mouths into Jars on the Fire, making them boil again, when boild enough they take it off and put it into other Jars, and there letting it work, cover it for use. A drink is made of Maiz boild and chawed the same way. *Id.* p. 132. called *Coenin*, No Men are to meddle with it in making. They drink it hot and muddy, three days and nights they are at it. They never mix drinking and eating. Eat when hungry, and never speak at meals. They slept on drinking our Sack two or three days. They try'd to make it without chawing, but it would not do. Water, their ordinary Drink, never did them harm. The chawing *Caouin* is not worse than Wine made with Feet. *Idem*.

It is in use for two thousand Leagues. *Thevet*. who makes that of the Continent differ from that of the Isles, but without reason.

The Bread is *Caçavi*, like a Target, it must be moistened in Water or Pottage, whereby it swells, to eat well, and neither in Milk, Cane Juice, or Wine will it swell. It has no tast but great Nourishment, and does not
Surfeit.

Surfeit. *Xauxa* is the finest fort of *Tuca* which is eatable, rosted or boiled, its Juice being no Poyson. *Tuca* is common in the Isles. *Acosta.*
The Stalks two Spans long, are buried in *tumuli*, called *Conúches.* It's ready in two years. The *Indians* clean it with Flints found by the Shore Press out the venemous Juice and bake it. It keeps three years, and has need of something of Broth to moisten it. *Benzo.*

XIII. *Ricinus minor vitleis angusto mucronato folio, caule verrucoso, flore pentapetalo albido, ex cujus radice tuberosâ succo venenato turgidâ, Americani panem conficiunt. Cat. p. 44*
I observ'd this Kind indifferently with the former, of which I suppose 'tis a variety.

XIV. *Ricino affinis odorifera fruticosa major, rorismarini folio, fructu tricocco albido. Cat. p. 44. Tab. 86. Fig.* 1. Wild Rosemary or *Spanish* Rosemary.
This Shrub has a Trunc as big as ones Arm, covered with a light brown, smooth Bark, rising five Foot high, having many white Branches, be set with Leaves in Tufts, at about an Inches distance from one another. Each of them is two Inches long, and a quarter of an Inch broad, green above, and very white below, standing on an eighth part of an Inch's long white Foot-Stalk, beginning narrow, and ending in a point. The tops of the Branches, for three Inches in length, are set thick with small white Flowers, Spike-fashion, made up of many *Stamina*, in a pentaphyllous white *Capsula*, or within five greenish white *Petala.* After this follows a tricoceous Fruit, sticking close to the Stalk, smooth and whitish, larger than that of the *Chameleæ tricoccos*, each of the three sides containing an oblong, brown, shining Seed.
The whole Plant smells very gratefully and strong.
It grows every where in great abundance on the Red Hills, and near *Passage* Fort, on the Road coming from thence to Town.
It is used very much in all sorts of Medicated Baths, and Fomentations for Hydropick Legs, &c.
In respect of the Fruit, this agrees with the *Pee-Tsherou-Ponnagem.* H. M.

XV. *Ricino affinis odorifera fruticosa minor, teucrii folio, fructu tricocceo dilute purpureo. Cat. p. 44. Tab. 86. Fig. 3. An Teucrii foliis frutex Curassavicus.* P. B. *p.* 380 ? *Pluk. Tab.* 228. *Fig.* 4 ? *Teucrio similis Indica fruticosa foliis crassiusculis, Hort. Beaum.* ?
This has several small woody Branches about four or five Foot long, sometimes rising upright, and at other times lying along the Surface of the Earth, having a gray Bark. The Twigs have Leaves at their ends, standing round them, about an Inch and one third in length, and of an Inches breadth, Oval, snipt about the edges, and of a very dark green colour, something like Germander, the Flowers consist of six greenish *Stamina*, coming from the same Center, standing in a pentaphyllous green *Capsula*, coming out *ex alis folis* by very short Foot-Stalks, to which follows a green, small, tricoccous Seed, which afterwards grows as big as that of *Heliotropium tricocceon* only 'tis smooth, and of a very pleasant pale purple colour.
The Leaves of this Plant bruised are very odoriferous.
It grew among the Trees of a Wood, between the Town *Savanna* and two Mile Wood.

Of Long-Pepper.

Although that which is the Long-Pepper of the Shops is not to be found in the *West-Indies*, yet there are many of that Tribe and Family. They have no perfect Flower, or at least it is so little, that 'tis not taken notice of, for there comes out a small *Julus*, which increases 'till the Fruit comes to its just dimensions, which is various, in differing sorts of it, but all of them are generally longer, and smaller than that of the Shops or *East*. They seem to come nearest the Nettle of any of this Tribe, which have a String with Seeds sticking to it round about, somewhat like those of this *julus*. All of them which are not herbaceous, have an hollow jointed Trunc, at every joint there is a Leaf, and opposite to it the *Julus* or Fruit, which is mostly sweet smelled, Aromatick, and biting. The Leaves are for the greatest part nervous, by which marks, they are easily distinguished from any other Plants. The Fruit, Tail, or *Julus*, is a String, on which are fastened round it many very small Seeds close to one another, as if they were united. The others, with these, are for Affinities sake added. The round black or white Pepper, seems in every thing to be the same, only to differ in having much larger *Atini*, and of this sort likewise is the Betle eat in the *East-Indies*.

Jaborandi gnacu Bras. Mentz. is of this Kinds.

XVI. *Piper longum arborum altius, folio nervoso minore, spica graciliori & breviori.* Cat. p. 44. Tab. 87. Fig. 1. *Ciribca Worm. mus.* p. 196. *quoad descr. Mecaxuchitl Ejusd.* p. 208. *quoad fig. An Planta Julifera aromatica in uliginosis crescens locis. D. Banister*: Long-Pepper of *Leigh ap. Purchas, lib. 6. cap.* 13. *p.* 1154. where 'tis proposed as a Commodity from *Guiana*?

This has several Stems rising twelve or fifteen Foot high, they are straight, green, smooth, jointed, and at every joint they are protuberant, distant from each other sometimes low near the Root, a Foot, hollow, whence the name of Elder. Upwards the joints are at less distance from one another. The Stalks have a Pith somewhat like our Elder. Towards the top stands the Leaves, one at a joint, they stand on one tenth part of an Inch long Foot-Stalks, are two Inches long, one broad near the round Base, where broadest, whence they end in a point. The Nerves, or Fibers, run the lengths of the Leaves from the Foot-Stalks end, as from a common Center, and have transverse ones, making a pleasant show in a very dark green, smooth Leaf. The scent of the Leaves, when rubbed, is very Aromatick. Opposite to the Leaf comes a *Julus* about one Inch long, slender, and of a yellowish pale colour, resembling Long-Pepper.

There is a variety in this, as to the length and shape of the Leaves, they being sometimes longer, and not so broad at Base.

Piso's Figure is good.

It grows in *Jamaica* near Colonel *Cope*'s Plantation, on a Hill above his House, and in *Barbados*.

The Leaves and Fruit of this is thought good against the Belly-ach, being boiled and eaten in their Pepper Pots or Pottages.

The Leaves and Root (the Fruit being not much used, because of so many other sorts of Pepper) are very hot, and esteemed extraordinary good, when dryed, to make all sorts of corroborating and strengthning Baths, against all manner of cold Distempers, in lieu of all other hot and strengthning Herbs. *Piso*.

The

The Natural History of Jamaica. 135

t. The Leaves of this Plant heal Ulcers of the Legs. The Root is very powerful against swellings. *Marcgr.*

Tlatlancuaye and *Acapatli* of *Hernand.* and *Xim.* are so very confusedly Figured and Described, that I can make nothing of them, though I believe there are two sorts common to *New-Spain* and this Island, whereof this is one.

This Wood is made use of to strike fire, if held between the Knees, and a hard stick with a tapering point turn'd round in it by the Hands. The Decoction of the Roots and Tops dissipates gross Humors, and heals Hydropick People. The Leaves applied, Cure bad Ulcers, and hot Baths are made with them for cold Fluxions. The Seed gives a good taft to Meat boiled with them. *Tertre.*

The Fruit, but especially the Decoction of the Root, is used in the Isles against the Stomach Evil, caused from intemperate cold, and moisture of the same, from too much Fruit, Drinks of the Country, or not being covered in a morning. They give two Glasses of it warm, and walk till they Sweat. *Plum.*

XVII. *Piper longum folio nervoso pallide viridi, humilius.* Cat. p. 44. Tab. 87. Fig. 2. *Arbuscula Brasiliensis piperis facie julifera* Raii hist. append. p. 1914. An *Tlatlancuaye seu piper longum pistorius quasi iconem Herna.* Jonst. dendr. 179 a *Plantula peregrina* Clus. Bmt. lib. 4. c. 11. Jonst. dendr. 181. Chabr. 127. *Piper longum angustissimum ex Floridis.* C. B. pin. 412. *Piper ex Florida.* J. B. t. 2. lib. 15. p. 187. Chabr. 126. *Arbor piperifera fructu longo Floridana.* Jonst. dendr. 180. An *Amolago* H. M. 6. 7. 31 ? *Spanish* Elder.

This Shrub has a round green Stem, four or five Foot high, having protuberant joints here and there, in that resembling Elder, whence, and from the pith or hollowness, the name. The Branches are likewise jointed, and there comes out Leaves, first on one side from one joint, and then from the other, another Leaf on the side opposite to that, and so alternatively always, one being at a joint. The Leaves have scarce any Foot-Stalks, are five Inches long, and two broad in the middle, where broadest, rough, having a great many large Nerves, running from the main one to the sides of the Leaves, of a yellowish green colour. At the joints on the side opposite to the Leaf, towards the top of the Branch goes out a Spike or *Julus*, standing on an half Inch long Foot-Stalk, of a pale greenish colour, and small aromatick biting taft, four Inches long, like that of Long-Pepper, being somewhat like a Rats Tail, and being generally crooked.

This agrees in every thing to *Piso*'s *Betre* or *Betys*, except in the white spots he marks on the Stalk, but being in every thing else it has so perfect an agreement with it, I take it to be the same.

It grows on the moist Banks of *Rio-Cobre*, near the *Crescent* Plantation, and in several other places of the Island, and in *Barbadoes*.

The *Juli*, which are like Long-Pepper, are of no use, the Root is very famous. It has something Aromatick, and in taft, colour, and smell, resembles Ginger, especially if it be fresh, and then it is not inferiour to it. It is very hot and dry. A Decoction of the Leaves, and Roots, Cures the Collick, and pains of the Limbs, eases the windy Belly, and takes away the cold tumors of the Feet. The same is done by Baths and Fomentations. *Piso.*

Piso's Figure is bad.

XVIII. *Piper longum arboreum foliis latissimis.* Cat. p. 45. Tab. 88. Fig. 1.

The

This agreed exactly with those of this Kind immediately preceding, in its Trunc, Joints, &c. only 'twas higher and larger; the Leaves were likewise very different, being much larger, they stood on one third part of an Inch long Foot-Stalks, were eight Inches long, and four broad, near the beginning where broadest, being pointed at the end, and roundish, at beginning smooth, of a dark green colour, and having few Veins in comparison with those preceding, and in proportion to the Leaf. The Spike or *Julus* was neither long nor big.

It grew in the Woods not far from *Rio-Nuevo* in the North side of this Island.

XIX. *Piper longum racemosum malvaceum. Cat. p. 45. Santa Maria Leaves.*

This has a strong Root composed of several very short blackish ones, which send up a straight Stalk, as thick as ones Thumb, jointed, of a gray colour, rough, round, striated, with some furrows in it, rising three or four Foot high, having towards the top Leaves alternatively, first on one, then on the other side of the Stalk, at the joints, where its four or five Inches long round Foot-Stalk encompasses the Stalk, leaving a mark when it falls off. The Leaves are cordated or like a Heart, or those of the Lime-Tree, only the Nerves run from the top of the Foot-Stalk, as from a common Center, through the whole Leaf, which is very soft, of a dark green colour, somewhat like those of Mallows, and of about seven and eight Inches Diameter. The Flowers and Fruit come out *ex alis fol.* being three or four *Juli* two Inches long, at first white, then green, standing upon a common half Inch long Foot-Stalk.

They grow in stony shady moist Woods, and by shady River sides, very plentifully.

The Leaves being very soft and large, are applied to the Head when it akes, or to any of the Joints in the Gout, and are thought to ease pain in every affected part, and therefore this is esteemed as a very rare Remedy, by all *Indians* and *Negroes*, and most part of Planters, but I could not find that this Leaf could do any more than Coleworts, only 'tis not so nervous, and so softer.

The Leaves are boiled and eaten in Pottage by the *Negros*.

If the *Juli* or Pepper be boiled in water, and exposed to the Sun, they grow stronger and more durable for all uses. The Root smells like Clover, and is hot to the third Degree, reckoned a Counter-Poyson, and of thin subtle, and therefore opening parts. If bruised and put like a Poultess to any diseased part, it ripens and cleanses. The juice of the Leaves, because cold, eases burnings; and the Leaves put into Clysters, have the same qualities with Mallows. *Piso.*

It is called in *Sancto Domingo Collet de Notredame*. *P'umier.*

This is not the *Serpentaria repens floribus staminiis spicatis bryoniæ folio ampliore pingui. Plukenet. pl.31. Tab. 117. Fig. 3. & 4.* as the Dr. thinks in his *Mant. p. 170.* Any person conparing that with this will find many differences.

XX. *Piper longum humilius fructu è summitate caulis prodeunte. Cat. p. 45. Macaxuchitl. Worm. Muf. p. 208. quoad descr. Piper longum Brasil. Eysd p. 214.*

This has a creeping jointed Root, striking into the loose Earth, several Tufts of hairy fibers at the joints. The Stalks are round, green, jointed, rising a Foot high, the Leaves are several, plac'd the length of the Stalk one by one, are four Inches long, and two broad, at the top, where broadest, having no Foot-Stalk, beginning narrow, and augmenting by degrees to the top, they are thick, succulent, smooth, of a dark green colour, having some

some few veins visible on their upper surface, like those of Water-Plantane, and sometimes a little defect or notch at the upper end of the Leaf. At the top of the Stalk is a jointed, red Foot-Stalk, sustaining a slender four Inches long Spike, *Julus*, or *Ligula*, like that of *Ophioglossum*, or some of the Long-Peppers, being sweet smelling, and sharp to the tast like them, and withal somewhat balsamick.

The Plant, if rub'd, smells very gratefully.

There is a variety of this, with smaller and more pointed Leaves, which I gathered in *Barbadoes*.

It grew in a Wood near Mr. *Batchelor*'s House, and in several thick and tall shady Woods of this Island, and in *Barbadoes* on rocky grounds.

It is hot in the fourth Degree, and dry in the third. It is drunk with *Cocoatlee*, and gives it a good tast, it strengthens the Heart, heats the Stomach, gives a sweet Breath, attenuates gross and thick Humours, resists Poyson, the *Iliac* Passion and Colick, is Diuretick, helps the *Catamenia*, and expels the dead Child, helps the Birth, opens Obstructions, and Cures pains from cold. It takes off the cold of Fevers, and such like. *Hermand*.

XXI. *Piper longum minimum, herbaceum, scandens, rotundifolium. Cat. p. 45. Serpentaria repens Americana, dictamni cretici villosis foliis apicibus nonnihil simulatis. Pluken. Alm. p. 343.*

This has a jointed, round, smooth, juicy Stalk, like that of Purslane. The Joints were an Inch and a half distant one from the other, and from each of these went several fibrils, into the Barks of Trees, or ground, thus drawing its Nourishment, and propagating its self for several Feet in length, by either climbing Trees, or creeping along the surface of the Earth. At each joint, was, on every side of the Stalk, one Leaf alternatively, standing on three quarters of an Inch long red Foot-Stalk, almost round, being more than an Inch long, and about three quarters of an Inch broad, smooth, of a yellowish green colour, juicy, and having Ribs like those of Water-Plantane. On the tops of the Branches *ex alis fol.* came several two or three Inch long, round, green *juli* or tops, small, having some brown spots on them, and being exactly like the tops of Mouse-Tail.

It grew on the larger Trees, and ground, in the Woods between *Guanaboa* and Colonel *Bourdens* Plantation.

I should not have thought Dr. *Plukenet*'s Title abovementioned, to have belonged to this Plant, had not he positively said it, *p. 177.* of his *Mantiss.*

XXII. *Yaruma de Oviedo Cat. p. 45. Tab. 88. Fig. 2. & Tab. 89. Ambayba* Fig-Trees of an *Anonymus Portugal* of *Brasile, Purchas, lib. 7. cap. 1. p. 1308.*

This is well described and figured by *Marcgrave* and *Piso*, and grows every where in the *Caribes* and *Jamaica* in the Woods, and is generally, being a quick grower, the first infesting clear'd grounds, being for the most part without Branches, though sometimes it has them.

It is used for the making of Bark-Logs or Floats, the Stalks being empty and light to lie under heavier Timber, and float it down the Rivers to the ⟨T⟩owns, where it may be useful, and *Peter Martyr*, as well as *Lopez de Gomara*, tell us that a *Lucaya Indian* Carpenter, fill'd one of these Trees (after having *Tuc*⟨h⟩owed it, and stopt the ends) with Maiz, and Gourds with water; and 'h another Man and Woman went to Sea, in order to go to their own Country (whence they had been carried, and made Slaves in *Hispaniola,*) and that they were unfortunately met and carried back when two hundred Miles onwards in their Journey.

138   *The Natural History of* Jamaica.

The hollow on the top of the Tree contains a white fat and juicy Pith, with which, as well as the young Leaves, the *Negroes* cure their Wounds and old Ulcers. This Wood is used in *Brasile* to rub fire with, making a hole in it, and turning round in it another harder till it takes fire.

The tender tops are adstringent, their juice is good against Fluxes, immoderate *Catamenia* and *Gonorrhæa*. *Piso*. It is good against the immoderate *Lochia*, if a Poultess of the Leaves be applyed to the Navel. *Idem*.

Peter Martyr, and *Lop. de Gom.* tell a Story of an old *Indians* curing a very great Wound, whereby the Arm was almost cut off from the Shoulder, with a Poultess of the Leaves of this Tree. But they say this Tree has an edible Fruit, which I never observed; at least that 'twas eat.

The tops cure new and old Wounds, and are Caustick, eating the proud Flesh, and generating new. *Oviedo*.

This being hollow, and very light, I believe may be what was made use of by the *Brasilians* to fish on, some pieces being joined together, and very fast tied, they called them their Piperis, in every thing supplying the use of Boats, as well as Bull-Rushes, or with the *Ægyptians* the *Scapi* of the *Papyrus*. *De Lery* tells us that they can never Drown on them, and that on these, they fish singly, and that they might be made use of here on Ferries.

The Fruit looks like Worms, and is wholesome and pleasant. *Lop. de Gom.*

Pigeons feed much on this Fruit, and the Wood is used by Turtlers for buoys to their Nets.

The Leaves are rough and polish Timber. The inner Rind, laid to fresh Wounds, with the outer tied on it, Cures them. *Purchas. Luet*.

That this is mentioned in two distinct places, as two several Trees, by *Johnston* in his *Dendrology*, is, I think, as plain as that they are the same, so that I much wonder at Dr. *Plukenet*'s remark in *p. 75*. and 76. of his *Mantissa*.

XXIII. *Fagopyrum scandens, seu volubilis nigra major, flore & fructu membranaceis, subrotundis, compressis. Cat p 46. Tab.90. Fig. 1.*

This Woodbind has round red succulent Stalks, by which it winds and turns its self round any Tree or Shrub it comes near, rising seven or eight Foot high, it has every Inch or half Inch towards the top, Leaves growing out of the Stalk alternatively. They have a quarter of an Inch long Foot-Stalks are grass green, juicy, smooth, thick, an Inch and a quarter long, and one Inch over at the base, being near upon of a triangular heart figure, or very much resembling those of the *Convolvulus niger semine triangulo*. C. B. *Ex alis foliorum*, towards the top, come out the Flowers, they are very many plac'd on three Inches long Spikes, by a very short *Petiolus*, they are round, flat, swell'd out in the middle, and green, and have a thin white membrane round them, looking like a Parsnip Seed. When these membranes are ripe Seed, they differ nothing from what they appear at first coming out when Flowers, but in being somewhat larger, and having their protuberant part in the middle, turn of a brown colour from a green.

It grows among the Trees near the Ruins of a Monastery by t[he] Town.

This, by its Description and Figure, appears to be quite different fr[om] the Plant, Dr. *Plukenet* thinks, (*Mant. p.* 74.) it may be, *viz. Fago tritic[o] majus volubile Virginianum, &c.*

XXIV. *Vo-*

XXIV. *Volubilis nigra, radice alba aut purpurea maxima, tuberosa, esculenta, farinacea, caule membranulis extantibus alato, folio cordato nervoso. Cat. p. 46. Ignafme de Bift.* p. 335. *Cambares de Flacourt.* p. 115. *An Omihares Ejusd. ib? racine noire de Maire.* p. 99. *An Rizophora Magna Virginiana bryonia nigra modo volubilis, singularis folio nervoso flexili, caule tetrageno ad angulos alato Plukenet. Alm.* p. 321? *Tames* of *Dampier?* p. 9? *An Rizophora caule alato rubente folio singulari Herm. par. Bat. Cat. p.* 111?

The Root is a Foot or more long, Brown on the outsides, with several long Fibers, running out on all sides to draw its Nourishment. It is very thick, sometimes roundish, being within viscid, before it be boiled, and of a white, or reddish purple colour, and very mealy when boiled, of the bigness of ones Leg or Thigh. The Stalk is of the bigness of a Goose-Quill, square, at each corner having a thin, reddish, extant Membrane, making it alated. It runs and winds its self round Poles by this Stalk, rising nine or ten Foot high, and putting forth Leaves at every three Inches distance, set opposite to one another, having two Inches long, green, square, alated Foot-Stalks. The Leaves are two Inches and a half long, an Inch and three quarters broad at the round Base, Almost of the shape of a Heart, and pointed, of a yellowish green colour, having many Ribs, taking their beginning from the end of the Foot-Stalk, as from a common Center, with transverse ones between. *Ex alis foliorum* come Inch or more long Strings, with small Flowers of a yellowish green colour, to which follow many dark brown, small Seeds, of an irregular shape, sticking sometimes to the String, and sometimes to its Branches.

Knox says that in *Zeylon* they grow wild in the Woods, and are there as good as those planted, only they are more scarce, and grow deeper, and so are more difficult to be plucked up, therefore are generally planted and very cheap.

Those of *Madagascar* plant them in Fields plain and untill'd, and after twelve months dig, and keep them in Magazines, They are planted by the people for their King in *November* in holes a Foot square, and at two Foot distance. *Id.*

They grow in many places of the *East* and *West-Indies. Dampier.*

They are eaten as Bread, being rosted under the Fire, or boiled, being very mealy. They are eaten likewise with Pepper and Oil, and for Bread in *Guiney.*

They are planted, having no Seed, by putting a small cut piece of one of the Roots into the ground. *Lery.*

The *Indians* say that a great *Caraibe,* or Prophet, brought these, and taught a young Woman how to plant them by pieces, from whence they are come common, and eat in lieu of Bread. *Thevet. Cosmogr.* These people lived on wild Herbs and Roots before. His figure is fabulous.

The Author of the *Hist. Lugd.* Confounds this and Potatoes, for these *Tams* are not planted by the Slip but Root.

They send them with Earth-Nuts, (*Arachidna,*) for *Lisbon,* from *St. Thomé,* to Victual their Slaves by the way. *Cluf.*

Roilf. doubts whether this Root, being like *Cassads,* it be not that innocent one of *Brasil,* may be eaten, but that is another, and particular kind of *Tuca.*

They grow in *Zeylan.* The Roots are eat by Swine and other Creatures. The Juice of the Leaves is good against Scorpions biting. The Root powdered with *Catu-panna-Kelengu,* is strewed on malignant Ulcers, with good success, and used in Fomentations. *H. M.*

XXV. *Ve-*

XXV. *Volubilis nigra, radice tuberosa compressa maxima digitata farinacea esculenta folio cordato nervoso. Cat. p. 46. Negro* Country *Yams.*

This has a great Root a Foot broad, and flatter than the former, almost palmated, or digitated after the manner of some *Orchis*'s, of a durty brown colour on the out side. The Stalks are not alated, but round, in all other things they are the same with the former.

They are both planted by cutting the Root in pieces, of about an Inch square, with a piece of the Rind on it, the larger the piece the bigger the Yam. Planting them in *March* or before, after *Christmas* they are at their full growth.

The Seed is not fruitful.

They being cut into pieces and boiled or rosted, are eaten by *Negros*, Slaves, or *Europeans*, instead of Bread, being a dry, mealy, pleasant, and very nourishing Root, and for this end are very much planted here.

Swine are fatted with these Roots in *Zeylan.* H.M.

XXVI. *Persicaria urens sive hydropiper* C.B. *pin. Cat.* p. 47. *An Eloquitic herba geniculata Hernand.* p. 210? *Persicaria Americana angustiori folio hirsuta fl. albis Plukenet. Alm.* p. 288?

This Arsmart sends out from every joint, touching the water or mud, a great many two Inch long Fibrils, shooting themselves into them, thence drawing their Nourishment. The Stalk is round, jointed at every Inch and an half, not straight, but inclining a little downwards, and two Foot long. At every joint there is a Protuberance, and at it upwards is an half Inch long Membrane, covering the Stalk. The Leaves come out at each joint alternatively standing on Inch long Foot-Stalks, they are eight Inches long, and two broad, where broadest, smooth, and in every thing like *Persicaria* Leaves. The Flowers stand on the tops of the Branches, Spike fashioned, like in colour, *&c.* to those of the ordinary *Persicarias*, and to them follows a black, flat, roundish, shining, smooth Seed, having two small Prickles or points at each end.

It grows by River sides, and in moist grounds all over the Island, and comes very near, if it be not altogether the same with our common *European Persicaria.*

A Fomentation of the Leaves of this, takes away old Aches and Colds of the Joints. Applied to the *Os pubis*, makes one Piss, if they be stop'd from a cold cause. It is a very good Caustick, and used by Chirurgeons in Putrid and Wormy Ulcers, for that cause. It takes away hardned tumours. *Piso.* It is Commended in this case by *Paracelsus.*

It is hot and dry, wasts Tumours, and dissolves congealed Blood or *ecchymosis*. Some of it put under the Saddle, and rub'd on a Horses Back refreshes a tired Jade. *Ger.*

Boiled in water, and applied, it carries away ill Humours from the Eyes. If beaten and applied with the Juice it helps purulent Eyes. It takes off Spots from the Body, if the Juice be rub'd on them Morning and Evening. It discusses hard Bruises and Swellings. *Dorst.*

Flies, Gnats, or Fleas, come not near this Herb or its Juice, and therefore 'tis very good for sordid Ulcers. *Trag.* and to keep these Vermine from Rooms strowed with it.

Its Juice kills Worms in the Ear.

The Leaves dried are used for Pepper. *Dioscorid.*

The Oil is good for a knotty Gout, which is made of the Juice of Arsmart, Lovage, and Shepherds-Purse of each *q. s.* Five Black Weathers Heads, and fifteen Frogs boiled in *balneo*, in two quarts of Oil, to the dissolution

solution of the Bones, then poured off and kept for use, it is good in Tophaceous and Oedematous Tumours. *Leb. Obf.*

The Juice of the whole Plant, 'baring the Root, boiled in Sergelim Oil, is a Cephalick Liniment, even to be applied in bleedings of the Nose. The tender Leaves *cum oryza lotura*, made into a Potion, diminishes the humour *Paddaur. H. M.*

This Herb Boiled with Oil makes a Liniment against the Gout. The Root being taken with hot water, loosens the Belly. The Leaves given in sower Milk aswage the swellings of the Belly. *H. M.*

Bruised and given with sower Milk, it takes away the griping of the Guts. *H. M.*

XXVII. *Potamogeiton aquis immersum folio pellucido, lato, oblongo, acuto.* Raii hist. p. 188. Cat. p. 48. *An potamogiton pellucidum nostras, foliis longis, perangustis, apicibus acutis,* Plukenet. Alm. p. 304?

This grows very plentifully in the salt and fresh water Rivers in the *Caymanes.*

XXVIII. *Amaranthoides humile Curassavicum foliis polygoni* Par. Bat. prodr. Par. R. p. 17. Cat. p. 48. Tab. 86. Fig. 2.

The Stalks of this Herb are round, reddish, tender, hairy, jointed, and spread on the Surface of the Earth for some Feet in length. Almost every Joint puts forth some small Roots or Fibrils into the Earth, as also some Leaves of a dark green colour, and smooth above, hairy underneath, bigger, but of the shape of those of *Polygonum*. At every joint likewise come out many white Tufts of Flowers, made up of many long white Membranes, dry, and not fading, laid *squammatim*, one over another, very close, and making in all a round Conglomerated Head. The Seeds are round, flat, and of a Chesnut colour.

It grows in the *Savannas*, near the Town of St. *Jago de la Vega,* very plentifully.

XXIX. *Amaranthoides humile Curassavicum foliis cepæ lucidis, capitulis albis,* par. Bat. prod. p. 15. Cat. p. 48.

From the Root of this are scattered, on every hand, a great many trailing Branches, lying on the surface of the Earth. They are a Foot long, round, red, jointed at every Inches distance, smooth, small, and having Branches set opposite the one to the other at every joint. The Leaves are almost round, green, one third of an Inch long, coming out opposite to one another on the smaller Branches, upon the ends of which, for the most part, and of the twigs at every joint, come the Flowers, being set round in a Head, pretty close together, like the Trefoils, each of them being long, tubulous, yellow within, and white above, having several yellow *Stamina,* the whole being a round Head, made up of many dry Leaves or Membranes laid close, *squammatim* one over the other, as in others of this Kind.

It grew near the Sea side among the Salt Marshes, at the *Canoes* near old Harbour.

It is somewhat in qualities like to Sampier, the short Branches and Leaves are a little boiled, and being covered with Vinegar, are kept as a Pickle to eat with Victuals, opening Obstructions, moving Urin, and exciting the Appetite. *Pif.*

XXX. *Amarantus fruticosus erectus, spica viridi, laxa & strigosa.* Cat. p. 48. *Tab.* 92. *Fig.* 1.

This Shrub has greenish, woody, and small Stalks, it rises to about two Foot and a half high. The Leaves are many, smooth, of a dark green colour, plac'd along the Branches without order, having half an Inch long Foot-Stalks. They are an Inch long, and three quarters broad, a little from the round Base where broadest, and whence they decrease by degrees to the point. The Flowers stand in Spikes at the tops of the Branches, about three Inches long, are not open, but made up of five Leaves, of a yellowish green colour, in the middle of which is a large blackish *Stylus*, which comes in some time to be a Seed-Vessel or Husk, containing several Seeds, each of which is scarce discernible to the Eye, shining, and of a brown colour, roundish, and hollow on one side like a Dish if viewed by the Microscope.

It grows by the Banks of *Rio Cobre,* near the Town.

XXXI. *Amarantus Siculus spicatus radice perenni Boccone.* Raii hist. p. 203. Plukenet. *Tab.* 260. *Fig.* 2. Cat. p. 49.

This Plant rises three or four Foot high, by a square, jointed, brownish green Stalk, having Branches set opposite to one another, *ex alis foliorum,* at about two Inches distance. The Leaves stand on one third part of an Inch long Foot-Stalks, the greater ones being three Inches and an half long, and two broad in the middle, where broadest, with one middle Rib, being of a dark green colour, woolly, smooth, and pointed. The Flowers stand in Spikes on the Branches ends, six or seven Inches long, being placed on every side of the Stalk, appearing at first nothing but short reddish Hairs or Filaments, after which follow rough, prickly, green, reflected *Capsula's* or Cones, each of which is divided into five points, containing in it one small oblong Seed, reddish, like Wheat, only smaller.

It grows in Ditches, and several places about the Town of St. *Jago de la Vega,* and in the Island *Madera.*

XXXII. *Amarantus, panicula flavicante, gracili, holoserices.* Cat. p. 49. *Tab.* 90. *Fig.* 2. *Amaranthus nodosus pallescentibus bliti foliis parvis, Americanus multiplici, speciosa, spica, laxa seu panicula sparsa candicante* Plukenet. Almag. *p.* 26. Phytogr. *Tab.* 261. *Fig.* 1.

This had weak, cornered, yellowish green, hollow, smooth Stalks, needing the support, though not turning round, its neighbouring Plants, rising to about three Foot high, being as big as a Goose Quill, and having few joints, and thereat Leaves, standing opposite to one another, on half an Inch long Foot-Stalks, they are about an Inch and an half long, and about half as broad near the round Base, where broadest, and from whence they end in a point, being smooth, and of a yellowish green colour. The tops of the Stalks, as well as Branches, growing *ex alis fol.* are branched Panicles, or branched Spikes of Flowers of a pale yellow colour, shining like Silk, as some of this Kind, only they are much smaller than any I know, otherwise like them in every thing else.

It grew in the Hedges of Lime-Trees, among the Sugar Plantations in *Guanaboa,* as well as in most of the *Caribes.*

XXXIII. *Blitum album majus scandens.* Cat. p. 49. *Tab.* 91. *Fig.* 2. *An Amaranthus Americanus, altissimus longifolius, spicis è viridi albicantibus.* Plum. Tournef. Inst. p 235 ?

This

This has a green Stalk as thick as ones Thumb, weak not able to support its self without the help of Shrubs or Trees, on which it leans, growing five, six, or more Feet high, putting out here and there Branches, having Leaves at about an Inch and an halfs distance, standing on Inch long Foot-Stalks. They are three Inches long, and half as broad, near the middle, where broadest, being narrow at the end of the Foot-Stalk, widening by degrees to near the middle, from whence they decrease to the end. They are smooth, of a dark green colour, and soft. *Ex alis fol.* come the several Spikes of Flowers, the Branches sustaining them, being two or three Inches long. At every half Inch is a Spike of Flowers about an Inch long. They are like those of the Blites or *Amaranti*, pale green or of an herbaceous colour, a great many together, each of which has five *Apices*, a yellowish *Stylus*, and *Stamina*: after these follow in green Seminal Vessels, or Husks, breaking horizontally, small black Seeds, shining, compress'd, and of the Figure of a Kidney.

It grew in a Wood by the Banks of the *Rio Cobre*, near the Town on the same side of the River, and on the Road between that place and *Passage* Fort, very plentifully.

XXXIV. *Blitum Americanum spinosum. Rai hist.* p. 199. *Cat.* p. 49. Red Weed of *Barbados*.

An oblong deep reddish Root, with some Fibers, sends up one, roundish, red, strong, striated Stalk, which has several Branches of the same colour, going out *ex alis fol.* The Leaves come along the Branches without any order, and are like those of the small Blites, and of a reddish colour, and usually under them are some sharp, short, small prickles. The Flowers come in long Spikes on the tops of the Branches, are of an herbaceous colour, and like those of the other Blites, and after them, follow, small, black, shining flat Seeds, like the others of this Kind.

It grows every where by the way sides in *Jamaica*, and the *Caribe* Islands.

The Leaves of this, as of others of its Kind, are eaten in the *Indies*. *Herm.*

XXXV. *Blitum minus album polyspermon folio subrotundo. Cat.* p. 49. *Tab.* 92. *Fig.* 1. *An Blitum Virginianum Polyspermon erectum viride, D. Sherard. Floken. Alm.* p 68. Caterpillers or Culilu.

The Root is large, strong, perpendicularly fix'd in the Earth, straight, reddish towards the top, and sending out round it several Branches on every hand, often trailing on the ground, and very rarely erect, two or three Foot long, striated, green, and succulent, along which come out several Leaves, in shape, &c. exactly like those of the small white Blite, only something longer, and now and then covered with a brownish *Farina*. The Flowers are Spike fashioned, very numerous, along the Branches, they are green, like those of this Kind, and to each Flower follows one Seed, round, compressed, black, shining, and little, very well inclosed in a pale green Membrane.

It grows every where in the low Lands, and Plantations, and is to be gathered very plentifully every where after Rain.

It is gathered, and when the Leaves are stript off, and boiled as a Sallet, is one of the pleasantest I ever tasted, having something of a more fragrant and grateful tast, than any of these Herbs I ever knew: whence likewise 'tis shred and boiled in Pottages of all sorts, and so eaten, is emollient, loosning, and provokes to a Stool.

It is used in Clysters in the Belly-ach, as the best and most common emollient Herb, this Country affords.

It is eat as Spinage, &c. for the same purposes. *Pis. Marg.*

Marcgraves Figure is good. The Figure here exhibited was taken in time of drough.

XXXVI. *Blitum polygonoides viride, seu ex viridi & albo variegatum, polyanthos.* Cat. p. 49. Tab. 92. Fig. 2.

Upon every side of a white, deep, and single Root, are spread several Foot long, green, round Stalks. The Leaves are set along the Branches, and just like those of the small, wild, green Blite, only sometimes they are variegated very pleasantly with a large white spot. The Flowers come out of the Stalk very thick on every side round it for near its whole length, but more especially *ex alis fol.* without any Foot-Stalks. Each of them is small pentapetalous, of a pale green colour, with a purple stroak on each of the *Petala*, and a green *Stamen* within, after each of which follows a round, compress'd, blackish brown shining Seed.

If any one will make this a *Polygonum* they shall have my leave, for it is somewhat *anomalous*, and because of its pretty large perfect Flower, ought to be neither Blite nor *Polygonum*.

It grows in hard Claiy grounds, and amongst Rubbish, every where about the Town of St. *Jago de la Vega*.

This is not the *Portulaca affinis polygonoides bliti folio & facie Maderospatensis* of Dr. *Pluken.* in his *Phyt. Tab.* 120. *Fig.* 3. as he thinks it may, *p.* 155. of his *Mantissa*.

XXXVII. *Blitum pes Anserinus dictum Raii.* Cat. p. 49. Goose-Foot or Sowbane.

I found this growing on the sides of the Streets, and by the High-way sides near the Town of St. *Jago de la Vega*. It seemed not to differ from that of *Europe*, only the Root was larger, the Stalks reddish, not so high, and the Leaves were whiter, being covered over with a whitish Meal, all which may come from the variety of Soils and Climates.

If it be eaten by Swine it kills them. *Lon. Trag. Fuchs.*

XXXVIII. *Parietaria foliis ex adversa nascentibus, urtica racemifera flore.* Cap. p. 50. Tab. 93. Fig. 1.

I could not observe this *Parietaria* here, in any thing different from that in *Europe*, only the Branches and Leaves stand opposite one to the other at joints, and the Stalks are square, green, smooth, and shining, sometimes reddish. The Flowers are racemose, having Strings like Nettles coming out *ex alis foliorum*. I am apt to believe on these scores it may be really different from the *Parietaria's* of *Europe*.

It grows on the sides of the shady Rocks going to sixteen Mile walk, and several such like places of this Island.

XXXIX. *Kali fruticosum, coniferum, flore albo.* Cat. p. 50. Another sort of Sampier.

This has several upright, woody, round, gray Stalks, about a Foot high, branched towards the tops, and having there many round, green Leaves, three quarters of an Inch long, succulent, and salt to the tast, set usually opposite to one another, and something like the Leaves of *Kali*, the Flowers are white, very small, and coming out from between the green Scales, of a small Cone, never, or very seldom bringing Seed.

It

It grows on the sandy, salt, marish Grounds at *Passage* Fort in the very Town, and in other such places in this Island.
It is pickled and candied in *Barbados*.

XL. *Herniaria lucida aquatica. Cat. p. 50. Tab. 93. Fig. 2.*
The Roots of this are many, small and hairy. The Stalk is green, round, erect, lucid, or almost transparent, about a Foot high, having on each side alternatively a small Branch, and opposite to it a tuft of Leaves, and out of the Branches after the same manner come Twigs, having very small, green, lucid Leaves, like those of *Polygonum*, only smaller in every part, very thick set, one against another. The Flowers come out *ex eorum alis*, on very small *Petioli*, either reddish, or green, which looked on by the Eye, arm'd with a Microscope, appear Tetrapetalous. The Seed follows as small as dust.
The whole Plant is astringent to the taste.
It grows on the Banks of most Rivers, and on the wet sides of Rocks.

XLI. *Corchoro affinis, chamædryos folio, flore stamineo, seminibus atris quadrangulis duplici serie dispositis. Cat. p. 150. Tab. 94. Fig. 1. An Corchorus Americanus minor carpini folio siliqua angustissima ex lateribus ramulorum proveniente. Br. pr. 2. p. 36?*
This has a very deep blackish coloured Root, which sends up a round, brownish, woody Stem, rising three or four Foot high, being divided into Branches on every Hand. The Leaves come out several together, some greater, some smaller, at half an Inches distance, on half an Inch long Foot-Stalks. They are half an Inch long, and a quarter broad, at Base where broadest, of a grass green colour, indented about the edges, and smooth. Opposite to the Leaves comes yellow Flowers, being stamineous, after which follows a two Inch long, dark brown Pod, or Seed-Vessel, something like those of the *Sesamum*, only having two round sides, instead of four, five, or six; in each of which two rows, are contained, a great many black, quadrangular, small Seeds, the rows or sides being separated from one another by a Membrane, dividing them. The Pod when ripe opens at the end, and scatters the Seed.
It grew on a rocky Hill, on which Colonel *Fuller*'s House was built, and in several rocky grounds near *Guanaboa*.

XLII. *Aparine paucioribus foliis semine lævi. Cat. p. 50. Tab. 94. Fig. 2.*
The Root is jointed, having at every joint several hairy fibers striking themselves into the Earth, sending up a jointed, greenish square Stalk, four or five Foot high, striated, a little rough and hollow, slender, and needing the support of neighbouring Plants. At the joints which are always protuberant and reddish, stand the Leaves opposite one to the other, on an half Inch long Foot-Stalks, being an inch and a quarter long, and half an Inch broad in the middle where broadest, of a pale green colour, a little rough. The Flowers come out *ex alis foliorum*, are pale green, Muscose, or made like *Juli*, each of which is a quarter of an Inch long. After these follow several brown Seed-Vessels disposed *Verticillatim*, two being always joined together, each being round on one side, or Semicircular, and flat on the other, and both make a Globe, each half containing one black Seed, of the same shape.
It grows plentifully amongst the Woods going from Town to *Guanaboa, &c.*

P p   CHAP.

CHAP. VI.

Of Herbs with monopetalous Flowers.

IN *Jamaica* the Tribe of Plants with monopetalous Flowers is pretty large, especially that elegant Sett of them called *Convolvuli*. They are most beautifull for the number and elegancy of the colours of their Flowers, far exceeding those of *Europe*. 'Tis somewhat odd, that whereas most of this kind of Plant are Purgative, a sort of Potato, first *Convolvulus* here mentioned, affords a Root, which by being an Aliment, goes a great way in affording Nourishment to the Inhabitants of the hotter parts of the World. It is also remarkable that there are more *Convolvuli* than one here, which have only one Seed following the Flower; whereas, generally speaking, there are in most *Convolvuli* three, and yet no person versed in Herbs, but who on reading the Description, or at first sight of these Plants, but must by the whole face of them, allow them to be true *Convolvuli*.

There are some of this Tribe that seem somewhat more anomalous than ordinary, all I can say to my reducing them hither, is, that 'tis the best place I could find, and that I shall be very glad any body of better Judgment will find a more proper.

There are none in this Tribe stranger than the Kinds of *Arum*, many of which are scandent, and several cultivated for Food, some for their Roots, and others for their Leaves, as shall be seen hereafter. *Galen* seems to take notice of *Colocaffia* Roots, and *Apicius* has many Receipts for dressing of them.

I. *Nicotiana major latifolia*, C. B. *Cat. p.*51. *Petun de Bouton. p.* 89. *Tabac. Ejusd. p.* 124. *Tabaco de Esquemeling*, p. 52. & 57. *Tabacum seu nicotiana major latifolia Eyst. Petun ou Tabacque ou Nicotien de Flacourt*. p. 146. p. 146. Tobacco of *Dampier. cap.* 12. *Tabac de Biet*. p. 336. *De Rochef. Tab.* p. 57. *De Olivier Oexmelin*. p. 76. *De Maire*, p. 80. *Tabacum Mirand. sin. & Eur.* p. 873. *Sana sancta Indorum sive Nicotiana Gallorum. Swert. part.* 2. *Tab.* 23. Tobacco.

This growing here agrees exactly to the Description given of it in Authors.

It is planted by most Planters in their Plantations, for the use of themselves, and in some for Sale. Before the *English* took this Island, the *Spaniards* had here as good as any was in the *Indies*, which they were careful of, and planted it by the dry River; but the *English* not taking care of their Seed, they lost the best sorts, and what they have now is planted along the Banks of this River.

Tobacco was most planted, used, and in esteem in the North parts of *America*, where probably (the want of Provisions being greater than in most parts of the World) it in some measure help'd the Inhabitants to pass their time without them. *Nicol* tells us five of his Company were lost in time of distress, who could not smoke it, though he confesses it did not Nourish. From the *West-Indies* it was propagated to the *East-Indies*, *Africa*, &c. and in 1586. 'twas by Sir *Francis Drake* brought into *England*. In all places where it has come, it has very much bewitched the Inhabitants from the more polite *Europeans*, to the Barbarous *Hottentots*.

This

This Plant having a Narcotick intoxicating quality, was used in the *West-Indies* by their Priests, to Dream and see Visions by, as appears by *Lopez de Gomara, Roulox Baro, Monardes* and *Morisot, viz.*

The Priests of *Espanola*, called *Bohitis*, who are Physitians likewise, do eat of this bray'd or made small, or the smoak of it is taken into their Nostrils, when they are to give answers, by which they see many Visions, being not themselves. The fury being over they recount for the will of God what they have seen. When they Cure they shut themselves up with the Sick, surround him, smoaking him with the same, suck out of his Shoulders what they say was his Disease, showing a Stone or Bone they kept in their Mouths, which Women keep as Relicks, thinking they facilitate Birth. *Lopez de Gomara.*

They cannot in *Brasil* Sacrifice without Tobacco, nor consult the Devil in their Affairs. The Priest offers a great Pipe of Tobacco, and Blesses them with its Smoak. *Roulox Baro, p. 225 and 238.*

In going to War and Weddings the Devil smoaks out of a Coco, and the Priest Incenses the People. *Morisot. 300.*

Priests, and *Indian* Inchanters, take the fume till drunk, and fall in an Extasie, giving an ambiguous answer, and then tell people of the success of their business. *Mon.*

The Oil or Juice dropt into the Ear is good against Deafness, and is used outwardly against Burns. *Ger.*

A Clyster being ready to be given with a Bladder, the small end of a Tobacco-Pipe was put into the Bladder and tied (but so that it might be drawn closer after the Pipe is pulled out) that had Tobacco in the Bole, which only blowing the Smoak thereof into the Bladder, and so given (as a Clyster) hath given perfect ease. *Park.* This Custom is yet continued of giving Tobacco smoak Clysters, with success in Colicks.

It was brought into *France* by one *Nicotius*, an Ambassador, who got it at *Lisbon* of a *Flandriquen*, who came from *Florida* about 1560. *C. B. pin.*

The Seed lies long before it rises from the ground, therefore it is sown in Autumn. *J. B.*

The Syrup of the Juice, or infusion in Wine, is a good Asthmatick Remedy. *Park.*

Four or five Ounces of the juice drank by one, a strong Man, in a Dropsie, purg'd him vehemently up and down, and then put him into a Sleep, after which he call'd for Meat, and was cur'd. *Dod.*

If Tobacco be Bruised and put on Wounds, it Cures them if small, if large they must be wash'd with Wine and stitch'd. This was taught our Men by the *Indians*, and did us much service in conquering Provinces. They Burn Shells, and mix the Powder with this Leaf, holding a Pill between their Teeth and under Lip, swallowing ever now and then the Spittle, it hinders the sence of Hunger or Thirst for three or four days. The Green Leaf put on the Spleen, helps its Obstructions, or a Rag dipt in its juice. If it be used likewise in Leaf, or in Substance chawed, it draws away the Flegm occasioning it. *Fragos.*

It is heating and drying, cleansing, resolving, binding, and is a Counter-Poyson, from thence it is called *Herba Santa* of the *Portuguese*. The fresh Leaves and their Juice, and Balsom, do not only check cancerous Ulcers, but heal poysonous Bites. The Water, or other convenient Liquor, in which it has been infused, kills Lice, and cleanses the Head of many of its Skins Diseases. The Ashes of it, when dry, kill Worms. Chaw'd it takes away Hunger and Weariness. By its spicy quality it strengthens the Stomach and Heart. It sometimes causes vomiting, at other times Sleep, and draws away Flegm from the Brain, if smoak'd or chaw'd, making sometimes those using it

Drunk,

Drunk. The Syrup of Tobacco is good against the *Asthma* and Dropsie, but must be given with Caution, because it sometimes Works very violently. *Piso.*

It is sharp to the taste, hot and dry in the third Degree. The Smoak makes one Spit much Flegm, it is good for the *Asthma*, and Wheesing, cures Hysterick Fits, strengthens the Head, brings Sleep, eases Pain, and takes away Weariness. A Leaf rubbed with an oiled Hand, and applied hot to the Stomach and Back helps Digestion, and Cures Surfeits; the same discusses a swell'd Spleen, eases pains from Cold, cleanses ancient and cancerous Ulcers, begetting Flesh, and cicatrizing them. The Juice dropt into the Wound, some dry Powder strowed on it, and the Marc put on after the same way Cures Wounds of the Head, if the Bones or Nerves be not touched. It Cures the Tooth-ach put into the hollow Tooth. The Powder taken at the Nose hinders one from pain, and the sense of Stroaks. As much as a Nut-Shell will hold of the Bark Inebriates, making them half dead and Mad, but those who use it much lose their Colours, have inflam'd Livers, squallid Tongues, and falling into *Cachexia* and Dropsie dye. It resists Poyson, some of the Powder, or Juice put into the wounded and poysoned part, this was the Remedy used by the *Cannibals*; and try'd on a Dog by the King of *Spain*. It Eases the Gout, and all pains from Wind. The Decoction sprinkled about the House kills Flies. Dry Leaves powdered ten parts, with Lime chawed one part, brings Sleep, eases pain, gives patience, Cures the Tooth-ach and pain of the Stomach. The Leaves are good for *Struma*, and are chawed for the Gout and fried with Butter, the Oil rub'd, Cures the Colick. A Syrup made by Infusion and Decoction is good against Worms, to two Spoonfuls. The Juice is given by some to four or five Ounces, but is too violent a Purger upwards and downwards. The distill'd water is good likewise. A Leaf used like a Suppository, Cures Quartanes. *Hernand.*

They mix the Seed with five times so much Ashes, Sow it, and cover the ground with Branches to keep of too much Sun, and replant it in a rainy time, at three Foot square distance; weeding it, and croping the top when going to Flower, as also the under Leaves, leaving ten or twelve Leaves on each Stalk, about two Foot high, clearing it of new Shoots every eight days. When the Leaf, by doubling breaks, they cut it, and string it, so as not to touch one another, and after fifteen days drying, take off the Leaves, out the Ribs, and twine it with Salt water into Ropes, to be made in Rolls. *Tertre.*

Ximenes tells a Story of one, who in fifteen days, with excessive taking Tobacco, at all Hours, Chocolate and Wine, had his Legs and Feet swelled, and was all broke out, and argues that it gives no Nourishment. Nor, if it be Physick, ought at all times to be used.

It was anciently used by the *Indians* for a Vulnerary, and only sent into *Spain* for its Handsomness for Gardens, but is now in use for its Faculties. Its name was *Piciels*, Tobacco was given to it by the *Spaniards*, from the Island of that name, where it grew very frequently. The Leaves Cure the Head-ach, being applied to it after being heated, and the Tooth-ach being put into the hollow Tooth. They ease all outward pains from the Stomach, Stone, and *Uterus*. The Smoak wakes an Hysterick Person. *Mon.*

Sir *Richard Greenfield*, on his discovery of *Virginia* in 1585. found the *Indians* used Tobacco in Clay-Pipes for their Health, whence he brought some Pipes, and they were made after the same Fashion in *England*, and thence used very much at Court. *Cluf.*

Tobacco was first showed in the *East-Indies* by the *Dutch* Seamen. *Vanderhagen.*

Seamen used to carry Pipes about them made of Palm Leaves, in which they smoaked to ease their weariness, bringing forth much Phlegm: It takes away Hunger and Thirst, *Lob.*

After the *Indians* have gathered it, and hang'd it up by small handfuls, and dry'd it in their Houses, they take four or five Leaves, and wrap them up in a great Leaf of a Tree, like a Paper made like a Funnel in which Spices are put, and put fire to the end, and draw it into their Mouths, which although the Smoak comes out again, yet by it, they subsist three or four days. When they go to War, or deliberate on it, they smoak and speak, if they take too much of it, it inebriates as Wine, and occasions great disturbance in those who take it. *Thevet.*

It is a Counter-Poyson. *Acosta.*

It is used in Leaves rolled up and smoak'd, they, who take it, lye senseless and stupefied most part of the day and night. Others take more moderately, having a Vertigo. I was forced out of the Houses by Smoak. Priests and Physicians are the same in all places where I was. They Smoak, and lie stupefied.

'Tis very likely the Powder mentioned by *Fernan Colon. p.* 125. to be suck'd from the Statue of *Cemi,* was this, for with it the *Indians* went out of their Senses like drunken Men, and likewise 'tis the *Cogioba* of *Roman* there. *p.* 132. and *Cohoba,* with which drawn by the Nose in Powder, the *Bohitis* are out of their Senses, and pretend Revelations. *p.* 144. They give likewise the Powder to purge the diseased with. *ib.* It makes them see People and Houses topsie turvy. *p.* 138.

It was in Juice used against poysoned Arrows, in place of Sublimat at *Porto Rico. Lugd.* and for Wounds and Ulcers in Cattle, and was eaten to prevent the Gout. *id.*

It is very vulnerary. *Oviedo.*

Upon the whole matter 'tis most certain, not only by the Eye Witnesses abovementioned, but many others, that Tobacco is a Plant of very extraordinary Vertues, not only for ill natur'd Ulcers, but even poyson'd Wounds. That chawed, snuff'd or smoak'd, 'tis good for Catarrhs, Headaches, Rheums, Defluxions, the Gout, Asthma, &c. 'Tis likewise very certain that the Priests or Physitians made use of this to intoxicate themselves withal, and afterwards to abuse the People, by telling them what they saw, or pretended to see, or foresee in such Extasies, would be the event of Wars, &c. From this Narcotick quality it is, that those who use it improperly, or in excess, turn yellowish, fall into Obstructions, and suffer almost the same Accidents as Persons that drink excessively of fermented Liquors, or take *Opium* in too great a quantity. From this Narcotick quality it is also not unlikely that it takes off the sense of Hunger, thereby calming the Mind, whereby 'tis hindered from fretting the Body.

II. *Gentianella flore cæruleo, integro vasculo seminali ex humidi contactu impatiente. Cat. p.* 52. *Tab.* 93. *Fig.* 1. *Gentianella utriusque Indiæ impatiens foliis agrati, Pluken. Phyt. Tab.* 186. *Fig.* 2. *Alm. p.* 167. Spirit-Leaf.

This has several brown, round, straight, an Inch and an half long Roots, almost finger'd like those of the *Oenanthe's.* From these rise two or three Stalks, four or five Inches high, at about one Inches distance jointed, and four square, and at the joints come out the Leaves. They are something like those of *Mercurialis,* of a dark blewish green colour. *Ex alis foliorum,* come the Flowers. They are large, monopetalous, like those of the *Convolvuli,* Bell fashion'd, and of a delicate blew colour, after which succeeds a four square, brown, Inch long Seed Vessel; containing a great many flat, brown Seeds,

When any wet touches the end of the Seed Vessel, with a smart noise, and sudden leap it opens its self, and with a spring scatters its Seed to a pretty distance round it, where it grows.

It grows under the Shrubs in the *Savanna's* about the Town, and is in perfection some time after a rainy season.

The admirable contrivance of Nature, in this Plant, to propagate its self, is most plain, for the Seed-Vessels being the best preserver of the Seed, 'tis there kept from the injuries of Air and Earth, till it be rainy, when 'tis a proper time for it to grow, and then it is thrown round the Earth as Grain by a skilful Sower.

This is a very good Wound-Herb, a very excellent Salve, being made with it and Suet boil'd together, and then strain'd.

It is us'd likewise applied on Issues to make them run.

I should not have taken this Plant to be mentioned by Dr. *Plukenet*, as above, had not he positively said so, *p* 167. of his *Mantissa*.

III. *Convolvulus radice tuberosa esculenta, spinachiæ folio, flore albo, fundo purpureo, semine post singulos flores singulo.* Cat. p. 53. Patatas de Elquemeling, p. 54. *Batatas Hispanorum Swert.* part. 2. Tab. 35. Patales de Bouton, p. 47. Patates de Biet, p. 334. *Rochef. Tabl.* p. 48. *Convolvulus angulosis foliis, Malabaricus radice tuberosa eduli. Plukenet. Almag.* p. 114. Potatoes of Dampier. p. 10. &c. *An Mawandres de Flacourt.* p. 116 ? Spanish Patatas.

The Root is tuberous, for shape and bigness very uncertain, but being for the most part oblong, as big as a Hen-Egg, from a swell'd middle tapering to both extremes, yellow, and sweet within when roasted, tasting like a boil'd Chesnut, and having many fibrils, by which it draws its Nourishment. The Stalks are green, a little cornered, and creeping for many Feet in length along the surface of the Earth, and putting forth Leaves and Flowers at every Inches distance. The Leaves stand on five Inch long green Foot-Stalks, they are almost Triangular, having two Ears, and a sharp point opposite to the Foot-Stalk. They are five Inches broad from Ear to Ear, and three from the Foot-Stalks end to the point, having under them purple Ribs, being soft, smooth, and of a yellowish green colour, something resembling the Leaves of Spinage. The Flowers come out *ex alis fol.* standing on a three or four Inches long, green Foot-Stalk, being monopetalous, Bell Fashion'd, not very open, purple within and whitish without, having in the middle some *Stamina*, and a *Stylus*. After each Flower usually follows one Seed, brown, and having several depressions in it. It is inclosed in a roundish, brown, membranaceous *Capsula*, under which stand five brown capsular withered Leaves, as in the other *Convolvuli*.

There is another Kind of this, the same in every thing, only the Roots are reddish, which is as common as the white, and grows indifferently with it.

They are every where planted after a rainy Season in the Plantations, for Provision, by the slip, a piece of the Stalk and Leaves, being put either into the plain Field after Howing, or into little Hillocks raised through the Field, in which they are thought to thrive better. In four months after planting they are ready to be gathered, the ground being fill'd with them, and if they continue therein any longer they are eaten by Worms.

Linschoten, in his Description of *America*, seems to make *Ages* and *Batates* two Roots, which are nevertheless the same.

They vary very much as to the figure and bigness of the Root, the colour of its Skin being sometimes red, and most commonly white. They are sometimes turbinated, at other times round, and most commonly biggest in the middle, and tapering to both extremes.

They

They are boild or roasted under the Ashes, and thought extraordinary good and nourishing Food, and because of their speedy attaining their due growth and perfection, they are believed to be the most profitable sort of Root for ordinary Provision.

They are used in great quantities to make the Drink called Mobby.

In dry times when Grass is scarce, or at any other, the tops are given to feed Rabbets.

They are windy. *Piso.*

Peter Martyr reckons up many sorts, only differing in the colour of the Skin, and *Benzo* speaks of *Haias* less, and more savoury, which I believe is only a variety of this.

They are very nourishing and provoke to venery.

Many conserves, like to Marmelade of Quinces, are made of them. *Linschot.* They are common at *Velez-Malaga,* whence ten or twelve Caravels are loaded with them every year to *Sevil.* They are temperate and loosning. *Mon.*

People feed on them in *Trinidado. Thevet.*

They are ripe in six months, and Breed Wind. *Benzo.*

These Roots were by *Colon* brought from the *West-Indies* into *Europe,* in his first Voyage, to shew the different Productions of the one and the other. *Lopez de Gom.*

They are planted by the Slip, the Root being necessary for Bread, which is a great Providence. *Ludg.*

They are best about *Malaca. Clus.* From whence they were brought to *Cadiz* and *Spain.*

IV. *Convolvulus radice tuberosa, esculenta, minore, purpurea. Cat.p.54. Batates Ind. Or. part. 6. p.85.* Red *Spanish Batatas.*

This has a Root four or five Inches long, as big as ones Finger, biggest in the middle, having a small lower end, and several fibrils drawing its Nourishment from the Earth. It is of a very deep red, or purple colour, and being broken, yields Milk very plentifully, which dyes of a purple colour. The Stalks were two or three Foot long, round, and green, putting forth at every Inch, or more, Leaves very like those of the precedent, only not so large, nor cornered, of a deep grass green colour, and thin, almost like those of a Violet, standing on an Inch, or a two Inches long Foot-Stalk. This here described was a very young Plant.

It grew at Colonel *Bourdens* Plantation, beyond *Guanaboa,* where it was planted.

It is used only to give Mobby a fine reddish colour.

V. *Convolvulus maximus, caule spinulis obtusis obsito, flore albo, folio hederaceo, angulofo. Cat.p. 55. Tab. 96. Fig. 1.*

This grows to a very great length, covering sometimes many Trees, or the Banks of Rivers for many Paces, having a round and reddish Stalk, arm'd with blunt, herbaceous, and short, variously shap'd Prickles, winding its self about any thing it comes near, or creeping along the surface of the ground. At unequal distances come out smooth Leaves, standing on six Inches long Foot-Stalks, they are three pointed, being four Inches long from the Foot-Stalks end to their points, and as broad from Ear to Ear at the base, there being a defect or *Sinus* from the Ears to the Point. The Flowers come out *ex alis foliorum,* standing upon an Inch long Foot-Stalks, having a four Inches long, green *tubulus,* from whence, by degrees, it opens its self into a white monopetalous Bell-Flower, of five Inches Diameter, a little situated, and having five green discernible streaks on its out side. The Seed-Vessel has

has five little, brown, short, capsular Leaves, standing underneath. It is an Inch long, pyramidal and brown, having four protuberances, and contains three irregularly figur'd large Seeds, very dark brown colour'd, hard and smooth.

It grows by the sides of Black River, and *Rio Cobre*, very plentifully.

The Leaves of this are very different from those of the *Convolvulus Americanus, subrotundis foliis, viticulis spinosis*. Plukenet. *Tab*. 276. *Fig*. 3. *Alm*. p. 115. So that they cannot be the same, though Dr. *Plukenet. p. 54. Mant.* thinks it may be the same.

VI. *Convolvulus major heptaphyllos, flore sulphureo, odorato, speciosissimo*, Cat. p. 55. *Tab.* 96. *Fig.* 2. *Spanish Arbor* Vine, or *Spanish* Woodbind.

This has a round tuberous Root, as large as ones Head, sending forth a brown, corner'd Stalk, which mounts, and turns round the highest Trees, and covers them with its numerous Branches. The Leaves come out at three or four Inches interval, they are finger'd like those of the *Hedera quinquefolia Canadensis Cornuti*, having two Inches long, green, round Foot-Stalks, and being divided into seven Fingers or Divisions, each Section having a middle Rib, beginning narrow, growing larger, and ending in a point. The Sections at base are shortest, and narrowest, each growing larger to the seventh, which is four Inches long, and one broad, thin, smooth, and dark coloured. The Flowers stand on three Inches long, round, green Foot-Stalks, coming out *ex alis fol*. being monopetalous Bell Fashion'd, of a very fine yellow colour, and smelling sweet. After these follow the Seeds, contained in a *Capsula* as big as a small Wallnut. The *Capsula* is thin, membranaceous, brown, and covered with foliose, or capsular Leaves, sticking to its Base. These Seeds are very large, being usually three, and triangular, having one round side, and being of a dark brown colour, looking like Sattin, by many brown Hairs, are on their surface.

It grows among the Trees by the Banks of *Rio-Cobre* near the Town, and is planted by Arbors to make shades, they covering them, and by their Leaves keeping out the Sun-Beams, better than any one of this Kind I know.

A Planter finding this tuberous Root in his ground, very plentifully, thought that he had found a new sort of Provision like *Batatas* for his *Negroes*; but he was mistaken for on boyling this Root as the others it would not at all relish with them.

This cannot be the *Convolvulus Americanus villosus pentaphyllus & heptaphyllos major*. *Icrm*. as Dr. *Plukenet* thinks, p. 55. of his *Mantissa*.

VII. *Convolvulus pentaphyllos, flore pallide flavescente, caule hirsuto, pungente*. Cat. p. 55. An *Convolvulus Americanus, pentaphyllos, folio glabro dentato, viticulis hirsutis*, Plum. *Tournef. Inst.* p. 84 ?

The Root of this Plant is oblong, and tuberous, of an Ash colour, from which rises a large purple Stem, branch'd out into others, very rough, and purple, taking hold, and climbing up by its Stalk any Plant or Herb it comes near. The hairy long prickles on the Stalks of this Plant are pricking and troublesome, like those on Cowhage. At about two Inches distance, come out the Flowers and Leaves, the latter stand on three quarters of an Inch long Foot-Stalks, being divided into five Sections from the Center of its Foot-Stalks, as *Lupin, agnus castus*, or *Hedera quinquefolia Canadensis Cornuti*. That Section opposite to the Foot-Stalk, is about an Inch long, the rest being shorter, in proportion to their being near the Base, being all smooth, and of a pale green colour, like the Leaves of Burnet. The Flowers come out *ex alis fol*. standing on two Inch long Foot-Stalks. They are small, monopetalous, Bell

The Natural History of Jamaica. 153

Bell fashion'd, of a very pale yellow colour, having a green pentaphyllous *calyx*. After the Flowers follows a round *Capsula*, divided into four Loculaments or Cells, in each of which lies one brown Seed, covered with a woolly Hair, triangular, round on one side, and flat on the two others, each Cell being by several membranes divided from the others.

It grows in the *Savannas*, and by the Rivers sides very plentifully.

I think this Plant agrees with Dr. *Plukenet*'s *Convolvulus quinquefolius, glaber, &c.* as I have said in my *Catalogue*, notwithstanding his being of another Opinion. *Mant.* p. 55.

VIII. *Convolvulus pentaphyllos minor, flore purpureo.* Cat. p. 55. *Tab.* 97. *Fig.* 1.

A small stringy Root, sends up a round purple Stalk, winding about any thing near it, and rising two Foot high, on which, here and there, stand purplish green, five pointed, smooth Leaves, the notches reaching almost to the purple Foot-Stalks end, somewhat like Papaw Leaves. The Flowers come out *ex alis foliorum*, they are monopetalous, of a purple colour, and very pleasant, after which follows a round, brown *Capsula*, membranaceous, and inclosing the Seed, as the other *Convolvuli*.

It grew among the prickly Pears in the Town *Savanna*, going towards two Mile Wood.

'Tis easie by comparing the Description and Figure of this, and the *Convolvulus quinquefolius, glaber, Americanus. Plukenet. Phyt. Tab.* 167. *Fig.* 6. That that Plant and this are different, though Dr. *Plukenet*, p. 55. of his *Mantissa*, thinks they may be the same.

IX. *Convolvulus major polyanthos, longissime latissimeque repens, floribus albis, minoribus, odoratis.* Cat. p. 55. *Tab.* 97. *Fig.* 2.

The Plant covers sometimes a great many Trees, and sometimes Pastures for a great breadth. It has a broad or compress'd, flat, long, cornered Root, of a brownish colour, from whence many strings go under the surface of the Earth, to draw Nourishment to the whole Herb. The Stalks are whitish, broad, smooth, having several round eminencies on their surface, and rise about any Tree they come near, to sometimes a very great heighth, at other times spread on the surface of the Earth to a great breadth, putting forth Branches adorn'd with Leaves at an Inch and an halfs distance. They are shap'd like a Heart, an Inch and a half long from the Inch long Foot-Stalk to the end of the Leaf, and an Inch broad at the round Base, where broadest, smooth, soft, and of a darkish green colour. The Flowers come out on the Branches in great numbers. They stand on an Inch long Foot-Stalks, are monopetalous, Bell fashion'd, white, with five greenish *Fascia*, little in respect of the Plant, and smelling very sweet. After each of these Flowers succeeds one large Seed, of an oval Figure, like brown Velvet, solid, inclosed in a brown membranaceous hairy Seed-Vessel, having five capsular brown Leaves, standing out on every side under it like the Rays of a Star.

It grows on the plain grounds near the River side, by the Town of St. *Jago de la Vega*, and in other places of the Island, very plentifully.

It Flowers in *May* and *December*, when the humming Birds are very busie about it, feeding on the *Farina* sticking to the Flowers.

It has a pleasant smell, when in Flower, much like that of the *Narcissus Medioluteus*, call'd *Pissaulitz* at *Montpelier*.

R r X. *Conv*

X. *Convolvulus Polyanthos, folio subrotundo, flore luteo. Cat. p. 55.*
This *Convolvulus* has a round, woody, green Stalk, by which it climbs, or winds its self round any Tree it comes near for many Feet high, putting forth Leaves at four Inches distance. They are cordated, more than two Inches long, and as broad at the round Base, where broadest, having some Nerves going from the end of the Foot-Stalk, through the Leaf, and some few transverse ones, smooth, of a dark green colour, and exactly in every thing like the precedent, standing on three Inches long Foot-Stalks. The Flowers are many, stand on four Inch long Stalks, they being at their end, as at a common Center, divided into several three quarters of an Inch long *Petioli*, sustaining several large monopetalous, Bell fashion'd, yellow Flowers, having some *Fasciae* within them, and a few slight Incisures on the edges with a *Stylus*, and some *Stamina*. After each of these follows a perfectly Spherical, brown *Capsula*, with five capsular dry Leaves, and within it three Velvet, or sattin'd, brown, triangular Seeds like the other *Convolvuli*.

It grew in a Wood, going towards the Ferry, near the *Crawle* Plantation.

I gathered a variety of this in *Barbados*, having hairy *Calices*.

I see no reason given by Dr. *Plukenet* in his *Mantissa. p. 54.* why I should think those *Synonima*, I have taken in my Catalogue to belong to this Plant, not to belong to it.

XI. *Convolvulus major, folio subrotundo, flore amplo, purpureo. Cat. p. 55. Tab. 98. Fig. 1.*

The round, green Sarments, or Stalks of this Plant mount about any Shrub, Tree, or Hedge, to a great heighth, cloathing them green with their many Branches and Leaves, which are two Inches and an half long from the Foot-Stalks end, to that of the Leaf opposite to it, and two Inches broad at the round Base, from one Ear to the other, the Leaves being fashion'd like a Heart, smooth, of a yellowish green colour, and standing on an Inch and a quarter long Foot-Stalks. The Flowers are of a pale purple colour, very large, monopetalous, and Bell fashion'd, after each of which succeeds a brown *Capsula*, having above five dry capsular Leaves, four round Protuberances, and in each of them a large triangular, smooth, solid, whitish brown Seed.

It grows every where on the Hedges and Ditches of the moister grounds.

The Decoction purges gross, and cold humours with Worms. It is to be taken in the morning, and made of the green Herb, otherwise it is not useful. *Hernand.*

XII. *Convolvulus folio lanato, in tres lacinias diviso, flore oblongo, purpureo. Cat. p. 55. Tab. 98. Fig. 2.*

This by its round, whitish, woolly Stem, turns its self round the Truncs of Trees, rising twenty Foot high, at every Inches distance, putting forth Leaves, standing on three quarters of an Inch long Foot-Stalks. They are something like the Elder Leaves of Ivy, being divided into three *Laciniae*, an Inch and an half long from the Center of the Foot-Stalk, to the point opposite to it, and as much or more from one Section at Base, to the other; they are of a very white green colour, soft, and covered over with short Wooll. The Flowers come out *ex alis foliorum*, standing on a quarter of an Inch long Foot-Stalks in a pentaphyllous, green *Capsula*, are monopetalous, an Inch and an half long in the *Tubulus* of the Flower, which opens it self Bell fashion, of a fine purple colour, with some yellow *Stamina* in the middle, and five paler Streaks or *Fasciae*. After these follows a brown membranaceous

Capsula,

Capsula, with four round Protuberances, under a thin membrane, containing three sattin'd Seeds, like the other *Convolvuli*.

It grew on the larger Trees, in the Road to *Guanaboa*, and at Colonel *Fuller*'s House in St. *Dorothies* near the Bridge, over Black River.

XIII. *Convolvulus folio hederaceo, anguloso, flore dilute purpureo. Cat. p. 56.*]
This sends forth several Stems from the same Root, which is oblong, deep and large, each of the Stalks being round, reddish, and about three or four Foot long, trailing on the surface of the ground, at every three or four Inches distance putting forth Leaves and Flowers *ex eorum ala*. The Leaves stand on two Inches long Foot-Stalks. They are an Inch and an half, or two Inches long from the Center of the Foot-Stalk to the opposite point, and as much from one end of the Section at Base, to the other, every Leaf being angular, having two Sections at Base or Ears, and a third sharper and longer, opposite to the Foot-Stalk, very like the elder angular Leaves of Ivy. The Flower has an half Inch Foot-Stalk, a pentaphyllous *Capsula* or *Calyx*, and a monopetalous, Bell fashion'd, pale, purple Flower, agreeing exactly with the other *Convolvuli* in all its parts.

It grows in very great quantities on the Red Hills near *Guanaboa*, and in cleared low Lands, as also on the Banks of the *Rie-Cobre*, below the Town of St. *Jago de la Vega*, on the same side of the River.

By the Figure of the Leaves, which agrees with those of what *Piso* says gives Jalap, I concluded this to be the same, but on tryal I found it had no such Root as I expected.

XIV. *Convolvulus folio hederaceo, anguloso, lanuginoso flore magno, cærulee, patulo, Cat. p. 56.* [*Habalnil. Avic.*

The Stalk of this is round, hairy, and pretty large, having Leaves standing at about two Inches intervals, on Inch long hoary Foot-Stalks. They are shap'd like those of Ivy, having three Angles or Points, whitish, hoary, woolly, soft, an Inch and an half long, and an Inch broad at Base, where broadest. *Ad alas foliorum*, come the Flowers, being several on the same Foot-Stalk. They are large, blew, monopetalous, and extremely pleasing to the Eye, in every thing agreeing with the other *Convolvuli*.

It grew in Mrs. *Guys* Plantation in *Guanaboa*.

Lobel tells us that it was in the Gardens of *Italy*, *France*, and the *Low-Countries*, and *Terrentius* in those of *Italy* in his time, and their Descriptions agree very well to this. *Dodonæus*'s Cut seems to make the Flower five pointed, which may come from the *Fascia* most *Convolvuli* have saying they are *per oras angulosi*, and so do other Authors, so there may be some doubt whether it be that.

It purges gross Humours. *Avicen.*

XV. *Convolvulus marinus catharticus folio rotundo. Cat. p. 57. An Convolvulus marinus Catharticus folio rotundo, flore purpureo. S. Patate de Mer. & Camaulroulee de Plumier. Plukenet. Alm. p. 113? Phyt. Tab. 324. Fig. 2?*

This had a very deep, white, oblong Root, and a great many long, round Stalks, as big as ones little Finger, green, spread on the surface of the ground for several Yards in length. The Leaves stood on them without any order. They had two Inches long Foot-Stalks, were almost round, only a very small notch, or defect, as if a little piece had been cut out with a pair of Scissors, at the end opposite to the Foot-Stalk, making the Leaf cordated, though sometimes it wants this defect. It is of two Inches Diameter, having several Ribs from the Foot-Stalks, and middle Nerve, smooth, of a yellowish green colour, and in its surface is like the *Caltha palustris*. The

Flowers

Flowers were monopetalous, Bell fashion'd, of a pale purple colour, and after them follow four rough, triangular Seeds, fattin'd like those of the other *Convolvuli*, each being set in a distinct Cell, and all of them in a round, brown, membranaceous pentaphyllous *Capsula*.

The whole Plant was milky.

It grew on the *Cayos* near *Port-Royal*, and on the sandy Sea shore at *Rio Nuevo*, in the North side very plentifully.

The Leaves are used in Bathes for the Dropsie, and to put on Issues to draw them.

Boil'd in water it makes a Fomentation to ease gouty Pains. The Leaves prepar'd into a Potion with Goats Milk are given for the *Hæmorrhoids*. *H. M.*

The Stalks and Leaves are temperately warm, and emollient, and therefore good in Baths, and to strengthen the Body, especially in cold Diseases. The Decoction of them are given inwardly, for the same purposes. *Pise.*

I have learned, of persons well experienced, that the inspissated juice is very purgative, and it is a kind of Scammony, and may be given as ordinary Scammony, from ten to twelve or fourteen Grains. It may be corrected with Sulphur, Creme of Tartar, or ordinary Quinces, or in want of them, with the Flesh of the Fruit *Guava*, or Almonds, or the cold Seeds. *Plumier.*

XVI. *Convolvulus maritimus major nostras rotundifolius. Morif. Cat.* p. 57.

I could not see any difference between the *European*, and this Herb.

It grew on *Gun Cayos*, a small Island off of *Port-Royal.*

It is very purging, especially of watery and hydropick Humours, and either given in Powder, or boil'd in Broths, but very strong, and not fit for weak Persons. *Ger.* who says likewise that it was used about *Hampshire* for Scurvy-grass.

The juice condens'd, either outwardly applied to the Belly, or inwardly given, is commonly, though not safely, known to help the Dropsie. *Lob.*

It is griping. *Dod.*

Three Drams of the Powder of the Leaves drank with Whey for some days, purges notably hydropical Humours. *Lac.*

XVII. *Convolvulus minor lanuginosus, folio subrotundo, flore cæruleo. Cat* p. 58. *Tab.* 99. *Fig.* 1.

This has a small, round, green Stalk, by which it winds and turns itself round any Plant it comes near, creeping a great length, and having very few Leaves, or set at great intervals, one from another. They stand on an Inch and an half long Foot-Stalks, are cordated, or shap'd like a Heart, an Inch and an half long, and an Inch broad at their round Base, of a very green colour, and covered with a little white Wooll. The Flowers come out *ex alis foliorum*, sometimes several together, and sometimes only one by its self. They are monopetalous, Bell fashion'd, and of an extremely pleasant, lively, blew colour, standing in a pentaphyllous, rough, hairy *Calix*, and on a quarter of an Inch long Foot-Stalk. After these follow, in a tapering or pointed, brown, membranous *Capsula*, three, almost triangular Seeds, brown, and like those of the other *Convolvuli*, being lodged in three several Cells, distinguish'd by so many membranes.

It

The Natural History of Jamaica. 157

It grows in great plenty flowring in *November*, in the open grounds, at *Guanaboa*, and among the prickly Pears, and other Bushes near the old Monastery of the Town of St. *Jago de la Vega*.

XVIII. *Convolvulus minor repens, nummulariæ folio, flore cæruleo.* Cat. p. 58. Tab. 99. Fig. 2.

From a small, stringy, and fibrous Root, spring, long, trailing Stalks, taking Root here and there, where they touch the ground, and putting forth, alternatively, at small, unequal distances, Leaves almost round, like those of the *Nummularia minor flore purpurascente*. They are three quarters of an Inch long, and an Inch Broad, having a snip, or small notch at the end, and a quarter of an Inch long brown Foot-Stalks. *Ex alis foliorum*, come the Flowers, standing on short Foot-Stalks. They are monopetalous, Bell fashion'd, of a light blue colour, after which follows a brown *Capsula*, containing two or three brown Seeds.

It grows very plentifully after Rain, in the Town *Savanna's*, and in *Barbados*.

XIX. *Convolvulus rectus minor, folio angusto candicante.* Cat. p. 58. Tab. 99. Fig. 3. *An Convolvulus Americanus, minimus, villosus, helianthemi folio. Plum. Tournef. Inst. p. 84?*

This has a long straight Root, which shoots forth several small, round Stalks, straight up, rising to scarce a Foot high, covered over with a hairy Down, the Leaves stand irregularly along the Stem. They are an Inch long, and not over a quarter of an Inch broad, smooth, and have a hairy Down on their underside. The Flowers come out *ex alis foliorum*, are monopetalous, Bell fashion'd, white, with some *Stamina* in the middle, and a *Calix* underneath, by which it adheres to the Stalk, to this follows a pentaphyllous *Capsula* containing several Seeds.

It grows plentifully in the *Savanna*, near the Town of St. *Jago de la Vega*, after Rain.

XX. *Rapunculus fruticosus, foliis oblongis, integris, villosis, ex adverso sitis, flore purpureo villoso.* Cat. p. 58. Tab. 100. Fig. 1.

This Shrub had many small Stalks, rising from the same Root, to about four or five Foot high, each whereof was square, about the bigness of ones little Finger, having under a smooth, clay coloured Bark, a hard, whitish Wood, and very large Pith, with many transverse dividing membranes. The Leaves were set opposite to one another at the Joints, which were two Inches, or an Inch asunder. Each of them had one third part of an Inch long, rough Foot-Stalk, was about two Inches long, and one Inch broad in the middle, where broadest: of a dark green colour, rough, or set all over with short whitish Hairs. *Ex alis foliorum* came the Flowers, standing on purplish, rough, one third part of an Inch long Foot-Stalks. They were large and monopetalous, difform, divided at the ends into several Segments, being tubulous, purple, and set very thick over, with strong short Hair. The bottom of the Flower swelling, there comes in it a *Capsula* or Seed-Vessel, in which lies much small, oblong, crooked, brownish Seed.

It grew in the mountainous Woods of this Island.

XXI. *Rapunculus fruticosus lenifolius, flore luteo specioso, foliis ex adverso sitis.* Cat. p. 58. Tab. 101. Fig. 1.

This rises to about four or five Foot high, being branched on every Hand. The Branches and Twigs are woody, set with Leaves opposite to one another at an Inches distance from each other. They stand on an eighth part of an Inch long Foot-Stalks, are an Inch and an half long, and about half as broad in the middle, where broadest, being smooth, and of a dark green colour. The tops of the Twigs are branched generally, and carry several Flowers of a yellow colour, very pleasant, being tubulous and monopetalous, the *Ora* divided into five Sections, and this standing on the rudiments of the Fruit, which augments into a pyramidal oblong Head, covered with some few small Leaves, and is made up of three several Cells, in each of which is great plenty of small, brown Seed.

It grew on the Red Hills going to Colonel *Cope*'s Plantation in *Guanaboa*.

XXII. *Ranunculus aquaticus, foliis Cichorii, flore albo, tubulo longissimo. Cat. p.58. Tab. 101. Fig. 2.*

This had a deep and thick Root, with which it was firmly fix'd in the Earth, and from whence rose a roundish Stalk about four Inches high, having many Leaves, going out alternatively at very short intervals, each being without any Foot-Stalk, about three Inches long, and three quarters of an Inch broad near the further end, where broadest, and whence they decrease, ending in a blunt point. They are rough, of a whitish green colour, and laciniated round their edges, after the manner of *Dens Leonis*, or Cichory, to the Leaves of which they are not unlike. Towards the tops come the Flowers, being of a very white or milky colour, and pleasant to look on. The *Tubulus* is the longest I ever saw, being about three or four Inches long. The *Ora* are divided into five points, and after these follows, in small Heads, much small Seed.

It grows on the moist Banks of the *Rio Cobre*, above and below the Town very plentifully.

XXIII. *Ranunculus folio oblongo, serrato, flore galeato, integro, pallide luteo. Cat. p.58. Tab.95. Fig. 2. An Rapuntium Americanum Altissimum, foliis cirsii, flore virescente. Pluk. Tournef. Inst. p. 193?*

This has a Stalk as big as ones Finger, rises thee Foot high, being green and smooth, and having very many Leaves set on it, without any order, each of which is ten Inches long, and two broad in the middle of a dark green colour, and indented about the edges. At the top are a great many Flowers, they are of a pale yellow colour, and galeated, having a long *Galea* turn'd up, and some *Stamina* coming out of the middle of the Flower. The Seeds were very small, and scarce discernible, brown like those of *hyoscyamus*, and contained in several Cells, in one *Capsula*, surrounded with four *foliola*.

It grew in the Woods by the Path going to sixteen Mile Walk, and other Woods about *Guanaboa*.

This is a *Rapuntium*.

XXIV. *Speculum Veneris majus, impatiens. Cat. p. 59. Tab. 100. Fig. 2.*

This rises to three or four Foot high, having a square jointed, rough, and a little hoary Stalk. The Leaves stand opposite one to the other, at every Inches distance, on half an Inch long Foot-Stalks, being an Inch and an half long, and three quarters of an Inch broad, near the middle, where broadest, hoary, and of a bluish green colour. *Ex alis foliorum* come out the Branches, two always set opposite, on whose tops stand Flowers without any Foot-Stalks. They have five capsular, long, green Leaves, and a long white *tubulus*, and on its top an open fine deep blew Flower, whose *Ora* are deeply divided into five Sections, to which follows a quadrangular, three quarters of an Inch long *Capsula*, containing

round,

The Natural History of Jamaica. 159

round, flat, brown Seeds, which are thrown out of the *Capsula* with violence, when 'tis either touch'd or wetted on the end.

It grew about the Town, in many places, amongst the Bushes.

XXV. *Flori Cardinalis five rapuntio affinis anomala, caule quadrato, flore coccineo, capsula pyramidali.* Cat. p. 59. *Euphrasia alsines majoris folio, flore galeato, pallide luteo Jamaicensis.* Plukenet. Almag. p. 142. Phytogr. Tab. 279. Fig. 6. *An Cara-Caniram.* H. M. part. 9. Tab. 56?

The Root was short, thick, and divided into three or four long, reddish, strong Branches. The Stalk was square, green, jointed, three or four Foot high: slender, and scarce able, to hold its self up, having Branches coming out at its joints, set opposite to one another, at every two Inches distance. At every Joint were large Leaves, standing on an Inch and an half long Foot-Stalks. They were two Inches long, and one broad, near the Foot-Stalk, where they were broadest. They were a little rough, had a point opposite to the Foot-Stalk, and were of a dirty green colour. The Leaves on the smaller Branches were lesser, but of the same shape. The Flowers stand in Spikes on the ends of the Branches, at half an Inches distance one from another, being of a very curious Scarlet colour, three quarters of an Inch long, tubulous, widening towards the top, where they were open, withsome *Stamina*. After these follow'd three or four round, flat, black Seeds, having a notch or defect in every one of them, and lying in a greenish *Capsula*, pyramidal, being round at top, and sharp at bottom.

It grew in a sandy place, near the *Rio Cobre*, just by the Town of St. *Jago de la Vega*.

XXVI. *Rapunculo affinis anomala vasculifera, folio oblongo, serrato, flore coccineo tubuloso, semine minuto, oblongo, luteo.* Cat. p. 59. Tab. 102. Fig. 1.

This Plant has several strong, short, blackish Roots, sending forth a round, woody Stem, having a clay colour'd Bark, with some *Sulci* in it, rising three or four Inches high, having at the top very many oblong Leaves, standing very thick, without any order, on a quarter of an Inch long Foot-Stalks, cover'd with a reddish Wooll like Moss. Each of these Leaves is seven Inches long, an Inch and an half broad, near the further end, where broadest, they beginning very narrow, widen themselves to near the end, where they straiten again, and end in a point, being much snipt about the edges. *Ex alis foliorum* comes out a small Stalk, divided into several Branches, having above a five pointed green *Calix*, an Inch long tubulous, scarlet colour'd Flower, something like those of *Periclymenum*, with some yellowish *Stamina*, after which follows, in a short, fungous, cornered Seed-Vessel, having no distinct Cells, but one cavity, a great many small, oblong, yellowish Seeds.

It grows in the Crannies of the steep Rocks, in the Road going to sixteen Mile Walk.

XXVII. *Stramonia altera major sive Tatura quibusdam.* J. B. Cat. p. 59. *Stramonium fructu spinoso, oblongo flore albo* Tournef. Inst. p. 119. *An Stramonium majus purpureum* Park. par. p. 36? *Stramonium majus album & vulgatius fructu oblongo spinoso.* Bobart. hist. Ox. part. 3. p. 607. Thorny Apples of *Peru.*

I could not observe any difference between this Plant found here, and that describ'd by Authors.

It grew just by the Prison going to the River. Whether it was wild, or came there accidentally, I know not; but I obferv'd it here, and in most of the *Caribe* Islands.

It is of great use in Surgery, as well in Burnings and Scaldings, as virulent and malignant Ulcers, Apoftems, &c. *Ger.*

The Seed came from *Conftantinople. Id.*

An Ointment is made of the Juice boiled with Hogs greafe, curing all Inflammations, Burns, &c. The Leaves boil'd in Oil till burnt, then ftrain'd and mix'd with Wax, Rofin, and Turpentine, doth, (made into a Salve,) cure Ulcers and Wounds, new and old, *Id.*

It is cold, the Decoction of the Leaves is ufed as a Fomentation, or Liniment in Fevers, efpecially Quartans. The Fruit and leaves againft pain in the Breaft, the Leaves infus'd in water into the Ears, cures Deafnefs. Put on the Pillows it brings Sleep to thofe who are awake. If too many be eaten they bring madnefs. *Herm.*

Four Grains make one Drunk. Two Drams kills, if not helpt with Vomiting with warm Water and Butter, or Oil, and a Bath to the Legs and Arms. *Math.*

The fixth part of a Dram Inebriates. An Ounce kills the fame day. *Lugd.*

Mixt and beat with *Sirgelim* Oil (or *Sefamum*) it is applied to humoral Tumors. The Juice of the Leaves mixt with Sugar (*Jagra de cana,*) and applied to an *Eryfipelas* cures it. Three Seeds are good in a cold Fever. The fame beaten with water wherein Rice has ftood, is fuccefsfully applied to tumified parts. *H. M.*

XXVIII. *Linaria minor, erecta cærulea. Cat. p.* 59. *Tab.* 103. *Fig.* 1.

This had a round, fingle Stalk, rifing to about two Foot high, on which were placed Leaves alternatively, being an Inch and an half long, narrow like the Leaves of *Linaria lutea vulgaris.* J. B. The tops of the Stalks were branched into feveral fix Inches long Spikes of blue Flowers, as the others of this Kind, after which followed fo many roundifh, turgid Seed-Veffels, each divided into two Cells, in which lie flat, brown Seeds.

It grew on the Inland *Savannas* of this Ifland.

This Plant is, by its Title, Defcription, and Figure, fufficiently diftinguifh'd from the other *Linaria cærulea,* as alfo the *linaria annua purpureo-violacea,* &c. mentioned by Dr *Plukenet* in his *Mantiffa, p.* 118. where he leaves out one of the Notes I give of this, *viz.* that it is leffer than the others that are blue, and then finds fault with my Title, as not defcribing it particularly enough.

XXIX. *Antirrinum minus anguftifolium, flore dilute purpureo. Cat. p.* 49. *Tab.* 103. *Fig.* 2. Balfam-Herb.

This had a great many hairy, red fibers for Roots, which fend up a fquare nine Inches high, brownifh Stalk, as big as thofe of leffer Centaury, on which ftood the Leaves oppofite to one another, having fmall Foot-Stalks. They were an Inch and an half long, and not over an eighth part of an Inch broad, of a very dark green colour. Towards the top the Stalk was divided into Inch long Branches, fet with feveral Flowers, like thofe of *Antirhinum,* of a pale purple colour. I faw not the Seed, but doubt not it ought to be referr'd to this Kind.

It grew on the rocky and woody Hills between *Guanaboa,* and Colonel *Bourden*'s Plantation, on each fide of the Road very plentifully.

It is accounted by the *Indians,* and thofe who came from *Surinam,* to be an extraordinary Vulnerary.

XXX. *Sefa-*

XXX. *Sesamum Veterum.* C. B. pin. *Cat. p.* 59. *Digitalis Orientalis Sesamum dicta.* Tournef. Inst. p. 163. *Hackalenah.*

It is frequently planted here by the *Negros* in their Gardens, and agrees to the Descriptions of Authors, having a Flower like *Digitalis.*

The Seed is very often beat up in Chocolate.

In *Æthiopia* and *Ægypt*, they use the Oil as we do Oil Olive: it is made by beating the Seds in a Mortar, and expressing the Oil. *Maregr.*

The Seed and Oil is hot, moist, emollient, and resolving, breeds gross Nourishment, and is hurtful to the Stomach, and is good in Diseases of the Ears. It is Emplastick. *Ger.*

The *Arabians* call this Oil *Good Oil*, by way of excellency, thinking it better than any other. *Vesling.*

A Decoction of the Plant is used for resolving *Ophtalmie*, if they be applied to the Eyes, for Coughs, Pleuresies, Inflammations of the Lungs, and hard schirrous Tumours. Women use it for hardness of the *Uterus*, it moving the *Menses.* They use the same for the Diseases of the Skin and Face from Spots, &c. The Herb and Seed boil'd in Honey, makes a Resolving Plaister for hard and sanguineous Tumors, and dried Nerves. The Decoction is good in Clysters. The Seed gives gross Nourishment, and fattens very much. The decorticated Seed fattens; the Oil more; and the Dregs (*which are eaten for Food in* Ethiopia) more than that. Women ordinarily drink the Oil to be fat, with the Dregs it is given to four Ounces in Plurisies and Pains, and in all defedations of the Skin, outwardly, as well as inwardly.

Zanoni, contrary to all others, seems to distinguish between the *Sesamum* of *Egypt* and *Africa.*

The Oil drawn from the Seed is used by the Poor in the *East-Indies. Lobes.*

All *Greece* uses it for Cakes, mixing it in making their Bread, and sprinkling it with water on the top of the Bread. The whole is hot in the first, and moist in the second Degree, Emplastic, and Emollient. *Dorst.*

It is planted for the Oil in *Bengale. Bern.*

It comes from *Greece* and *Peloponnesus*, its numerous Roots makes the ground poor. *Mash.*

The Oil takes off the roughness of the Throat, clears the Voice, mollifies hardned Apostems, and fattens very much. *Lac.*

It is burdensome to the Stomach by its Oiliness.

The Oil is made by bruising the Seed, and throwing of it into water. It is better for making odoriferous Oils than common Oil, because of its durability. *Ætius.*

The Oil is, if taken to four Ounces for many days, good against the Itch, hard breathing, Plurisie, *Peripneumonia, ad menses movendos*, and for pains in the Stomach, Womb and Guts. *Arab. Phys.*

The Oil of the Seeds is given for any pains. The *Indians* use it in colouring their Bodies with some Simples added. *H. M.*

The Seeds of this Plant are used in many places of the *East* and *West-Indies* for Food. The Oil in the *East-Indies* serves for all the purposes, whether Medical or Culinary, for which we use ordinary Oil: It is there called *Gergilim Oil.* It is used also as an excellent Remedy in shortness of Breath, as we now use Linseed-Oil. Lately Mr. *James Cunningham*, F. R. S. and my very good Friend, wrote to me from *China*, where he is Physician to the *English* Factory, that the Bean or *Mandarin* Broath, so frequently mention'd in the *Dutch* Embassie, and other Authors, is only an emulsion made of the Seeds of *Sesamum* and hot water.

T t   XXXI. *Aristo-*

XXXI. *Aristolochia scandens odoratissima, floris labello purpureo, semine cordato.* Cat. p. 60. Tab. 104. Fig. 1. *An Aristolochia Americana, folio cordiformi, flore longissimo atropurpureo.* Plum. Tournef. Inst. p. 163 ? *Contra-Terva.*

This has a long, round, geniculated Root, as thick as ones Finger, from whence rises a round green climbing Stem, taking hold of any Tree or Shrub it comes near, rising six or eight Foot high, and covering them with its numerous Branches at every two or three Inches putting forth Leaves *ex eorum alis.* The Leaves stand on the main Stalks or Branches, by an Inch and an half long Foot-Stalks. They are cordated or triangular, roundish at base, four Inches long, and three and an half broad at base, frome one round Ear to the other, of a dark green colour, smooth, and having Ribs running through its surface, taking their original from the end of the Foot-Stalk. The Flower stands on a three Inches long Foot-Stalk, is made like the Flowers of the *Aristolochia*'s, of a yellowish colour, the Label being covered with a purple *Farina.* After this follows a Fruit two Inches long and hexangular, containing, in six Cells so many rows of small, flat, brown Seeds, exactly of the shape of a Heart, the points lying in, and the bases making the angular Pod, which, when ripe, leaps open, the Seeds dropping out.

The whole Plant smells very strong, and very gratefully.

It grows every where in the Woods about the Town.

The Root is bitterish, hot in the third Degree, smelling sweet and rosiny. If put in form of a Poultess on Swellings it Cures them. It eases pain, and puts off the cold fit of an Ague. It strengthens the Heart, Stomach and Brain. Cleanses the Stomach and Breast, and stops Fluxes. *Hernandez.*

The Decoction of this Plant with Oil, is a good Liniment against Serpents bites. The Decoction of the Root, or the same mixt with water, is good in the same. It takes away cold Fevers, Headaches, Dropsies and *Dysuris.* If us'd in Lotion it Cures the Gout. The Juice, with Pepper and hot Water, Cures the Flux of Blood. *H. M.*

XXXII. *Digitalis, folio oblongo serrato, ad foliorum alas florida.* Cat. p. 60. Tab. 164. Fig. 2.

This grows by a rough, round, green, woody Stalk, to about two Foot high, having Leaves set on it without any order, ten Inches long, and two broad in the middle, indented about the edges, rough, and like to the Leaves of *Digitalis.* Towards the top *ex alis foliorum* come out four or five Inches long Stalks, being branch'd at their tops, and sustaining several yellowish green, difform, monopetalous Flowers, with dark spots within, after which follows, in a hard Seed-Vessel, many small brown Seeds.

The whole Plant has a strong smell.

It grew by the sides of the Path going to sixteen Mile Walk, and under the Town on the Banks of the *Rio Cobre.*

XXXIII. *Solanum lethale, fructu rubro, semine copiostori minutissimo.* Cat. p. 60.

This rises to five Foot high, by a quadrangular Stem, as thick as ones middle Finger, covered with a clay coloured smooth Bark, being jointed at every two or three Inches distance, and having two Leaves at every Joint, set opposite to one another, each of which has an Inch long, green, round, Foot-Stalk, is nine Inches long, and four broad in the middle, where broadest, being narrow at the beginning and end, a little indented about the edges, having one middle, and several transverse Ribs, being smooth, and of a dark green colour. *Ex alis foliorum* come the Flowers, several together, being on the same Inch long, green Foot-Stalk. They are yellowish tubulous, with

with several *Stamina* in the middle, after which follow so many red Berries, as big as a larger Pea, containing in a Pulp of the same colour, an infinite number of extremely small, brown Seed, almost like that of Tobacco.

It grows in sixteen Mile Walk-path, by the Rivers side, and under the Town of St. *Jago de la Vega*, on the Banks side, on the same side of the River.

XXXIV. *Zingiber* C. B. *pin. Cat. p.* 60. *Ind. or. part.* 6. *p.* 83. *Nieuhof. p.* 83. *Zinziber flore albo, à D. Steph. Swert, Sacaviro de Flac. p.* 126. *Boym. Flor. sin. lit. V. Theuen. p.* 26. *Gingembre de Feynes p.* 105. *De Rochef. Tab. p.* 60. *Zingiber angustiori folio, famina, utrinsque India alumna. Plukenet. Alm. p.* 307. Ginger.

The Root of this is flat, as broad as ones Hand, having several round *Apices* or ends, on every side of it, like the *Cymæ* or *Gemmæ* of some Plants, about the bigness of ones little Finger, of half an Inch thickness, it is of a whitish yellow colour, and has many fibrous Roots drawing its Nouriment. From this rises a Stalk about two Foot high, having several Leaves going alternatively from opposite sides of the Stalk, each whereof is in colour like the gladiolus Leaves, though in their Consistence and Veins they resemble those of the *Canna Indica*, being not so broad, and somewhat longer. The Flowers stand on a distinct Stalk, without any Leaves, taking their rise from the Root as those of our *Arum*. It is about a Foot high, jointed, there being at each Joint a Membrane, of a yellowish green colour, covering the Stalk to the next Joint. At and near the top come the Flowers, a great many together, set very close, and making a long spherical Head, which is made up of a great many single Flowers, each whereof has one large broad membranous Leaf, of a yellowish green colour, Concave above, and Convex below. In its hollow side are the Rudiments of the Fruit, *viz.* a round purplish Knob, standing on a small Foot-Stalk, and being coronated by a long Membrane of the same colour. In some time the round Knob, swells, and, I suppose, coms to a Berry as these others I take to be of the same Kind which are described hereafter.

The whole Plant smells gratefully sweet, and pleasantly.

It is planted in all parts of this Island, but was brought first from the *East-Indies*, it growing about *Malabar* in great plenty.

Great quantities of it are preserved here while it is yet green, and before it be grown full of Strings or Filaments. 'Tis boiled in several waters, and made tender before it be preserv'd, as likewise 'tis clear'd of its outward Skin. If it be too old it leaves Strings or Filaments in the Mouth, which comes from its Age, and no lye, whereby it was cured, as *Garcias* and others say.

A Limpid red transparent Oil, swimming on water, is by simple Distillation got out of these Roots, agreeing in smell and tast with Ginger, only more mild. At a year old 'tis taken out of the Earth, sun'd for fourteen days, then sprinkled with Ashes of Wood or Lime, and so, is kept, otherwise it is liable to Corruption and Worms. 'Tis sent dry'd, or preserv'd, every where. It does not like a sandy, or not cultivated Soil, and when 'tis above a year old 'tis fill'd with woody Filaments. It loses its Leaves in *January* and *February*. The chief vertue is in the Root, which, besides, taken in Victuals, remedies Asthmas, Coughs, tough Flegm, Squeamishness, being helpful to the evil disposed Stomach. Outwardly applied it Cures the Gout in the Feet and Hands. Beaten with water, and infused into the Eyes, it Cures the Vertigo that comes from Stoppages in Women

after

The Natural History of Jamaica.

after Childbirth, which, unless cured in time, brings Palsies, Epilepsies, Madnesses, and even Death its self in those Regions. *H. M.*

It is used against cold Diseases of the Stomach and Guts, being beaten and mixt with Coco-Oil, and applied to the part, and to venomous Wounds. A medicated Wine is made of it in the *Philippine* Islands. It is planted in *Mexico*, in fat, low, dunged, shady Lands, in the month of *March*, by pieces of the bigness of the Thumb, put five Inches asunder, and five Inches deep. It is covered with Earth, and watered after planting. It is to be watered afterwards twice a Week, or once if the Earth be moist. *Hernandez.*

In the *East-Indies* 'tis eat green as a Sallet, the Root being cut in small pieces and mixt with other Herbs. *Gar.* Or pickled with Salt and Vinegar, called Achar, or preserved. The best comes from *China*, little comes from the *East-Indies*, but most from St. *Thomas*, St. *Domingo*, *New-Spain*, and the hot *West-Indies*. *Linschotten* says they use to cure it by covering it with Potters-Earth, by which it was kept fresh, it is good for the Stomach. *Linschot.*

It gently loosens and heats, strengthening the Stomach. Candied it provokes to Venery. It is good against darkness of sight, and for every thing as Pepper *Ger.*

It is taken out of the ground, and a little dried in *December* and *January*, then covered with a little Clay, not to increase weight, but to keep it from Corruption. *Gare.*

It was first brought by *Francisco de Mendoza* to *New-Spain* from the *East-Indies*, and it thrives very much, being planted by Root or Seed. The dry is candied, being first buried in wet places, where Rushes grows, or frequently wash'd and laid in hot water till it be tender then with Sugar it is preserv'd as when green. *Mon.*

Lobel's Cut in the *Adversaria* is not good. That in *Pharmacopeiam Rond.* Truer.

It grew in *Balsora. Pedro Matteo apud Zanoni.*

There are two sorts, red and white, the red is rubb'd over with *Rubrica*, the white with Chalk to keep it from Worms, to which, when young and white, 'tis very obnoxious. *Bod. à Stapel.*

The first Ginger was brought to the *West* from the *East-Indies*, and multiplied then after such a manner, that in 1547. there was in the Flota 22053 Quintals. *Jos. Acosta.*

The Root grows in the *Maluccos. Gom.*

Wine with it, and *Cumin* boiled in it, is good against the Wind in the Guts and Stomach, occasioning pain. Taken to half an Ounce in a draught at night going to Bed, it expels ill humours by sweat. *Dorst.*

What grows wild is best. *Thevet.*

There are Roots white and red by the Druggists. The red is besmeared with *Rubrica*, the white with Chalk, to preserve them from Worms, call'd by *Amatus* red and white Ginger. *C. B. pin.*

There is the white and black, the first from the *East-Indies* is soft, the other from the *West-Indies* is really different. *Park.*

There is a difference of that from the *East* and *West*. It grows wild in the *East-Indies*, though it be not so good as that planted. They cure it when its Leaves dye. In *December* and *January* 'tis fit to Candy, 'tis first bark'd, and kept in Brine or Vinegar for an hour or two, then sun'd for so long, and then cover'd in a House, till all its humidity is gone, then they are candied with Sugar or Brine. Its Acrimony is lost by too much watering. It is good in Colicks, Lienteries, &c. *Pis.*

They

The Natural History of Jamaica. 165

They plant it in *Madagascar*, under the Eves for Rain. It is forbid, for hindering the sale of Pepper, by the King of *Spain*. The green Leaves are us'd in Sauce. The Roots are cur'd on the dying of its Leaves once a year. *Cauchi. p. 163.*

It was brought fifteen years since to Christendom from *Caire*, with other Spices, now it is carried thither. *Albert.*

It is forbidden to be brought from *Brasil* by the *Portuguezes*, because of the hindering the Vent of Pepper. *Pyrard. p. 2. p. 139.*

That Ginger which is gathered in *January* and *February* is best. They Give in *China* the Decoction with hot Water to Sweat. He that takes of it in a morning fasting, is preserved from Venom all that day. The Conserve is good for cold Diseases of the Stomach. *Boym.*

Ginger is very common in all the parts of the *West* and *East-Indies*. It came from the last of these places to the *West*, from whence most of what is brought into *Europe* comes.

Ginger when ripe is dug up. They cut off the Blades, scraping the Skin off, with Knives to kill it. *Negros* scald it to kill it, which makes it hard and black, the other method making it soft and white, with a cleaner and quicker tast. *Ligon. p. 79.*

XXXVI. *Zinziber sylvestre minus, fructu è caulium summitate exeunte. Cat. p. 61. Tab. 105. Fig. 1.* Wild-Ginger.

This, which was very like the *Zerumbet Garz.* grew to about four Foot high, the Stalk was round, and covered with Leaves from the Root upwards to the very top, each whereof was about four Inches long, and two broad, of an almost oval figure, and grassie, being thin, and having a great many Ribs run from the Center of the Foot-Stalk, through all parts of the Leaf, to the point, like to the *Indian* Cane. The Flowers stand at top of the Stalk in a Head together, which is composed of a great many set in a three Inch long Spike, pretty close together, of a pale purple colour, each whereof is made up of one membranous Leaf, Convex on its undermost side, and Concave above : in which stand, on an half Inch long Foot-Stalk, the Rudiments of the Fruit, being a round protuberance, or knob, on the top of which is a tuft of some long, withered, dry, membranaceous Leaves, or Flowers making a Crown. The round knob, after some time, swells into an oblong, oval, purple Berry. The Root is made up of a great many white, round, two Inch long, thick Strings, smelling like Ginger, spread on every side.

It grows on the North side of Mount *Diablo*, very pentifully, among the shady Woods, in the Path going to sixteen Mile Walk, and in most of the Inland woody parts of the Island.

The Root bruised, and applied as a Poultice in Cancers, *noli me tangere's*, &c. is reckoned a very extraordinary and admirable Medicine, and if one will give Credit to the Relations of *Indians* or *Negros*, is a never failing Remedy in those desperate Cases.

XXXVII. *Zinziber sylvestre majus, fructu in pediculo singulari. Cat. p. 61. Tab. 105. Fig. 2. Cardamomum Brasilianum latifolium, sylvestre, paucserosa Brasiliensibus. Breyn. pr. 2. p. 107. Canna Indica ad imum caulem racemifera, Fluken. Alm. p. 80.*

This differs in nothing from that immediately preceding, only the Stalk rises eight or Foot high, having much larger Leaves, and in nine lieu of having its Flowers and Fruit on the end of the Stalk it has a Stalk about three Foot high, immediatly springing from the Root like our *Arums* or *Orobanche's*, being jointed, and having each *internodium* covered

U u with

with a dry purplish membrane, coming from the under Joint, and on its top a four Inches long Spike of Flowers, exactly like the former, only larger.

It grew with the other on the North side of Mount *Diablo*.

The juicy Skin stains the Skin of a brown colour, as Black-Cherries, or Violets, and therefore is us'd for Ink. *Maregr.*

The Leaves, Stalk, and unripe Fruit, if rub'd, smell pleasantly, like Ginger, and therefore supply the want of *Spices*. Hot Baths are likewise made of them. *Piso*

Colon, in his discovery of *Hispaniola*, mentions *Gengevo* to grow *sponte*, by which, I suppose, he meant this or the precedent.

XXXVIII. *Cardamomum minus pseudo-asphodeli foliis. Cat. p. 61. Tab.* 103. *Fig.* 3.

The Leaves of this had more than an Inch long Foot-Stalks, were four Inches long, and more than one broad, in the middle where broadest, and whence they decreased to both extremes, ending in a point. They were even on the edges, thin, and like the others of this Kind. The Stalk was naked of these Leaves, having now and then some smaller, without any Foot-Stalks inclosing the Stalk, one of which was under every Flower, as in the *Orchides*. The Spike its self was large, about three Inches long, thick set with Flowers, the under part of which swelled out into a trigonous, oblong Head, in which, in several Cells, lay much very small downy Seeds, as in others of this Kind. The Stalk of this was not over seven Inches high, but I do not remember whether it was perfect.

I had it about *Guanaboa*.

XXXIX. *Arum maximum Ægyptiacum quod vulgo Colocasia.* C. B. *Cat. p.* 61. *Arum Ægyptiacum florigerum & fructum ferens, radice magna orbiculari. Pluken. Almag. p.* 51. *Tames* in St *Helena* of *Dampier. cap. ult.* Tayas.

These grow very commonly by the Rivulets, which come down from the Mountains, running the year long, and are planted likewise by some of the Inhabitants for Provision in their Plantations.

They have a Flower and Fruit, just like the *Arums*, which stands on a peculiar Foot-Stalk.

The Leaves are used to carry fresh Fruit, Cheese, &c. in *Madera*.

Cæsalpinus says that it grows in *Sicily*, and that the Root is eaten, and the Leaves, boiled in Vinegar.

The Roots are used in *Jamaica*, being boiled as *Yams*. They have a biting tast in the Throat, if not very well prepar'd, and therefore not very much covered, but only in scarce times.

Alpinus says he never could see it Flower, but I have seen it several times in Perfection, and therefore he very undeservedly blames *Dioscorides*. It is eaten by those indulging Venery very much, both raw and boiled, being thought helpful thereto. 'Tis common in the Markets, and cheap. *Alpinus Vestingius* describes and figures it flowering.

It is thought by *Alpinus* to be the *Faba, Egyptia, Colocassia & arum Ægyptium*, of the Ancients. *Theophrastus* says, both Leaves and Roots of *Arum* are eaten, *lib.* 7. *cap.* 11, 12. But the common *Arum* is sharp and opening, this Adstringent: *Cæsalpinus*'s spotted. *Alp.*

Bellonius found it wild by the Rivulets of *Crete* in great plenty.

This was sent from *Egypt* to *Rome*, there to be sold, whence *Martial*,

Niliacum ridebis olus, lanasque sequaces. Bod. à Stapel.

The

The Natural History of Jamaica. 167

The common *Arum*, Leaves and Roots, and *Dracontium* Roots were in ancient times boiled and eaten, as may appear from *Dioscorides* and *Galen*.

It is Adstringent, and good against Fluxes. It makes the People feeding on it pale. *Lon*.

Pliny says *Aron* is esculent. *Casalp*.

'Twas brought into *Portugal* from *Africa*, where the Slaves love it very much. In *Egypt* they feed on it as we on Turnips, putting a little of the Root into the River side. He saw it (*Rawolf*) grow about *Tripoli* and *Halepo*. *Clus*.

The *Hippopotamus* lives on its Roots in *Egypt*, where it grows wild, and in *Sicily*, and the Kingdom of *Naples*. It is eaten and praised by the Slaves and *Turks* at *Naples*. It is not the *Faba Ægyptia*. *Col*.

It is pernicious in the Root, and therefore must be three or four days macerated in water, being slic'd to wash off its Mucilage, which is deleterious, then the Roots are press'd, dry'd in the Sun, and made into Flower, afterwards made into Cakes. The *Jevous* of *Mataras* were killed by it, eating of it many Weeks, Besieging the *Dutch*, and wanting Rice. It kills by bringing the Dysentery. The first Decoction is to be thrown away. *Bont*.

XLI. *Arum minus, nymphæa foliis esculentum*. Cat. p. 62. Tab. 106. Fig. 1. *Choux Caraibes. De Bouton*. p. 47. *Sonzes de Flacourt*. p. 117. *Espece de Chou de Biet*. p. 335. *Rapunculus Brasiliensis tuberosus, seu Battata Tajaoba Brasiliensibus. Marcgr. Pis. Raii hist.* p. 1334. *Campanula tuberosa Indica convolvuli cordatis foliis denturibus radice esculenta. Pluken. Alm.* p. 78. *An Arum minus Bengalense colocasiæ foliis Ejusdem. Alm.* p. 51 ? *An Arum Zeylanicum minus colocasiæ facie prioribus punicantibus, & viridantibus. Herm. par. Bat. pr.* p. 314 ? *Parad.* p. 77. *Colocasia Brasiliana lactescens latifolia caule fusco. Tajaoba. Pis. & Marcgr. Ejusd. par. Bat.* p. 86. *An Arum humile Ceylanicum latifolium pistillo coccineo. Commel hort. part. 1.* p. 97. *Indian* Cail, or *Tajas* the lesser.

This has a small tuberous Root, bigger than a Walnut, and several Leaves rising from the same Root, about a Foot high from the Ground, the Foot-Stalks support a Leaf very like those of *Colocasia*, only smaller, being of a yellowish green colour, and very like in Figure, Colour, Ribs, &c. to the Leaves of the white water Lilly. The Flower and Fruit stand on peculiar Foot-Stalks, like those of the other *Arums*.

They are planted here very carefully, in most Plantations.

The Roots are eaten as *Batatas*, but their chief use is for the Leaves, which are boil'd, and with Butter eat, as Coleworts, and to the tast are extreamly pleasing.

They loosen the Belly. *Piso*. His Figure is good.

These of *Madagascar* eat this Root, in want of *Yams* or Rice, as we the Leaves in Pottage. *Cauché*.

XLI. *Arum minus esculentum, sagittaria foliis viridi-nigricantibus*. Cat.p.63. Tab. 106. Fig. 2. *An Arum montanum colocasiæ radice rotunda. Herm. par. Bat.* p. 78 ? *Colocasia Brasiliana lactescens latifolia, Caule viridi, Mangare penna. Pis. Ejusd. par. Bat.* p. 87.

This in every thing is like the preceding, only the Leaves are larger, narrower, and not so round, being liker to our common *Arum*: smooth, of a very dark green colour, soft, and somewhat corrugated on its surface, with a Welt round the edges.

It is planted as the former, and serves for the same purposes.

Piso in the first Edition says these Leaves are not eatable, but in the second that they are.

The Roots are boil'd, and tast musk'd or sweet. *Maregr.*

Piso's Figure is not good.

XLII. *Arum saxatile majus, foliis rotundioribus, fructu purpureo. Cat. p. 63. An Arum Americanum, folio amplissimo, flore & fructu rubro. Plum. Tournef. Inst. p. 259? vel An Arum Americanum, acinis amethystinis, parvo flore. Ejusd. ib. p. 160?*

Three or four Leaves spring from the same Root, standing on Foot long Foot-Stalks, being like the Leaves of Cuckow Pint, only larger, and rounder ear'd, being two Foot long, or from the Foot-Stalks end, to the roundest point opposite to it, and one Foot broad, from one round Base to the other, something of the shape of a Heart, and having great Ribs running from the Foot-Stalks end, as from a common Center. The Flower and Fruit, comes in every thing, after the same manner, as in the ordinary *Arum,* only the *Acini,* or Berries, are plac'd in a very comely order, on a Foot long Pestle, being larger below than at top, and are of a brownish or purple colour.

It grows on the Rocks in the shady Woods, on the red Hills near *Guanabos,* and near *Hope* River, in the Mountains of *Liguanee.*

XLIII. *Arum caule geniculato, cannæ Indicæ foliis, summis labris degustantes mutos reddens. Cat. p. 63. Canna Indica venenata, Ouvari forte pari. 8. Ind. Occident. p. 122. C. B. pin. p. 184. Plukenet. Almag. p. 79.* The Dumb Cane.

This rises to be about five Foot high, by a jointed, very green, succulent, solid Stalk, as big as ones Thumb, without any Leaves 'till towards the top, where come several, standing on large Foot-Stalks They are round at the Base, and from thence decrease to the point, being something like those of the *Canna Indica,* only much thicker, and of a darker green colour. Among these Leaves come out the Flowers, at the top of the Stalk, and after them the Fruit, being in every thing like those of the other *Arums.*

They grow in all moist low Lands and bottoms of this Island.

If one cut this Cane with a Knife, and put the tip of the Tongue to it, it makes a very painful Sensation, and occasions such a very great irritation on the salivary Ducts, that they presently swell, so that the person cannot speak, and do nothing for some time but void Spittle in a great degree, or Salivate, which in some time goes off, in this doing in a greater degree, what *European Arum* does in a lesser, and from this its quality, and being jointed, this *Arum* is called Dumb-Cane.

Pieces of this Stalk are cut, and put into Baths and Fomentations for Hydropick Legs, and are thought very effectual.

Strangers must be warned of these Canes, they looking like those of Sugar. *Laet.*

It is eat by *Indians* for want of better Meat. The Root is of more force than the Fruit or Leaves; besides, the first qualities, being of very small parts, and opening Obstructions, Fomentations are made of them against Inflammations and Obstructions of Hypochondres and Reins; and the Oil is good against those Evils, and supplies that of Capers, and *Lilies.* The Roots sliced and boiled in Wine, made into Baths, and used to the Feet, it is of great use against old and late Gouts. *Piso.*

XLIV. *Arum*

XLIV. *Arum maximum altissimus scandens arbores, foliis nympheæ laciniatis* Cat. p. 63. *An* A Vine with a Leaf pretty broad and roundish, and of a thick substance, of *Dampier, cap.* 16. *Dracunculus Americanus colocasiæ foliis laciniatis. Tournef. Inst.* p. 161.

This has a green jointed Stalk, which has Clavicles, exactly like the *Arum maximum trifoliatum altissimus scandens arbores, &c.* by which it sticks to the Trunks of Trees, and comes to be twenty or thirty Foot high, at its top having several Leaves like those of *Nymphæa* when young, but serrated, and sinuated about the edges when old, by which it may be sufficiently distinguish'd from others of this Kind. At the top of the Stalk come out Flowers, and a Fruit afterwards, like those of the other *Arum*'s. These come out from amongst the Leaves.

It is very ordinary in the larger Woods of the *Caribes*, as well as *Jamaica*.

If this be the same that Captain *Dampier* means, he tells us, that the Leaves pounded small, and boiled with Hogs Lard, make an excellent Salve for old Ulcers in Legs, and that one of his Men came to the knowledge of it by an *Indian* of the Isthmus of *Darien*.

XLV. *Arum maximum scandens geniculatum & trifoliatum foliis, ad basin auriculatis.* Cat. p. 63. *An Hiuca sive Mizmitl, Hernandez: Seu dracontium Mexicanum aromaticum. Herm. parad. Cat.* p. 92.? *Dracunculus Americanus, scandens triphyllus & auritus. Tournef. Inst.* p. 161.

This Plant has a round Stalk of half an Inch Diameter, green, very thick jointed, full of a spongy *Medulla* and milky Juice. From each Joint of this go out five or six Clavicles by them, it takes hold, and sticks very close, and fast to any Tree it comes near, and rises very often to twenty or thirty Foot high, being naked of Leaves, till near the top, where are a great many round it. The Foot-Stalks of the Leaves encompass the Stalk, leaving a mark when they drop off, making the Joint. Each of them is two Foot long, round, green, half an Inch thick, full of a spongy Matter, and having two *foliosæ fimbriæ*, one on each side, green and thin, running half of the length of the Foot-Stalk. The Leaf its self is very deeply divided into seven parts, or rather three Leaves with Auricles, the uppermost, or that Leaf or Section in the middle, being the longest and largest, *viz.* a Foot long, and half as broad, having one middle Rib, and some transverse ones, being of a very dark green colour, smooth, and sinuated as the Leaves of *Arum*. The two other Leaves or Sections *ad basin*, are lesser than the other, and besides have a large Appendix or Ear, at their outside Base, which makes up the five Leaves or Sections, and this Appendix or Ear has another like it self, only lesser, making up the seven Leaves or Sections. *Ex alis foliorum*, at the top come out Stalks, sustaining within a white monopetalous Sheath, Hose, or membranous covering, a white Pestle, in every thing like that of the *Arum*s.

Every part of this Plant is milky.

It grows on each side of the *Rio Cobre*, below and above the Town, on the sides of the larger Trees, and in the Woods in most parts of this Island.

XLVI. *Arum saxatile, repens, minus, geniculatum & trifoliatum.* Cat. p. 36.

This is in every thing like that above described, only lesser, and the Leaves without Ears, and each of the three Leaves or Sections equal to one another. The Flowers and Fruit are likewise the same.

It grows running along the Rocks, in shady Woods, in sixteen Mile Walk, and elsewhere.

XLVII. *Colocasia hederacea sterilis minor folio cordato.* Plumier. *Cat. p.* 63. *Arum Americanum scandens foliis cordiformibus.* Tournef. *Inst. p.* 159.

This has a green, juicy, jointed Stem, of about the bigness of a Goose-Quill, being round and smooth, climbing by means of its large Clavicles it has at joints to about thirty or forty Foot high, and hanging down again to the ground. At every three or four Inches, it puts out Leaves standing on two Inches long, green Foot-Stalks, the Leaf it self being cordated, or of the shape of a Heart, three Inches long, and two broad at the round Base, where broadest, and whence it decreases, ending in a point, being smooth, or equal on the edges, very juicy, green, and having several *Striæ*, or small superficial Veins appearing on it, being lucid and shining, and very pleasant to look on. What Fruit or Flower it bears I know not, never having seen them, though I have frequently observed it, both in *Jamaica*, and the *Caribes*, but believe with *F. Plumier* it is a *Colocasia*, and therefore have placed it here.

It grows in the woody shady places of *Jamaica,* and the *Caribes*.

CHAP. VII.

Of Verticillated Plants.

THere are very few verticillated Plants wild in the Island of *Jamaica,* at least I met with very few of that Tribe there, as will appear to any one who peruses the following Observations, and yet there are (I think) more of the *European Verticillata,* that there grow, and thrive well by Culture, than of any Tribe whatever; for there is Rosemary, Lavender, Marjoram, Pennyroyal, Thyme, Sage, Savoury, *&c.* in great plenty, whereas many other Kinds of as useful Plants are very hardly raised, or brought to perfection.

I. *Pulegium fruticosum erectum verticillis densissimis.* Cat. p. 64.

This has a four Inch long, reddish Root, with some lateral fibers, from which arises a square, woody, brown Stalk, three or four Foot high, branch'd towards the top, and thick set at the Joints, which are an Inch asunder, with long undivided Leaves, an Inch long, and not over the eighth part of an Inch broad in the middle, like those of Hysop, of a yellowish green colour, smooth, and having a smell like those of St. *John*'s Wort. The Flowers are many, set round the Joints in a large, round knob, are not galeated, but only have Lips divided into four parts, white, and set very close together, making a very large, round knob, and in the *Calyx* of the Flower follows brown, small, oblong Seeds, each of which has a small Furrow, or Canalure on one side, and is round on the other, in that resembling the ordinary Wheat.

It grows very plentifully in the Town *Savannas,* and Flowers the whole year round.

II. *Men-*

II. *Mentastrum maximum, flore cæruleo, nardi odore. Cat. p.* 64. *Tab.* 102. *Fig.* 2. *Erva Cidreira Lusitanis. Maregr. Citrago seu Melissa Citrata Brasiliensis. Raii Hist.* p. 1332. *Melissa Jamaicana odoratissima. Plukenet. Alm.* p. 247. *Phytogr. Tab.* 306. *Fig.* 3. *An mentha Americana melissæ foliis graveolentibus. Herm. par. Bat. pr. Plukenet. Alm.* p. 248 ? *Phytogr. Tab.* 306. *Fig.* 6 ? *Indian* Spikenard.

This near the Root has a red, round, rough Stem, which rises nine or ten Foot high, having towards the top a square Stalk, and opposite Branches. The Leaves stand on an inch long Foot-Stalks, are more than two Inches long, about one Inch broad in the middle, where broadest, are sinuated, or deeply jagged on the edges, hoary, and of a pretty dark green colour. The Flowers are standing round the Stalk *Verticillatim*, blue, small, tetrapetalous, in a quarter of an Inch long striated, furrowed, or cannulared *Calyces*, the top of which has five Hairs or Prickles. Each of these *Calyces* contains two black, almost round, flat Seeds, having a protuberance in their middle.

The whole Plant smells very strong, pleasantly, and like Spikenard, or somewhat like a Citron, whence the name given it by the *Portugueses*.

It grew on the sandy Sea Banks near old Harbour, in the Low Lands or *Savannas* near the Town, and in several places of this Island.

It is esteem'd a very great Alexipharmac, and is much used in outward Fomentations.

Piso us'd this in place of Balm, either in outward Fomentations, or as its distill'd water in the Collick, or other flatulent Distempers, as well as in Cordial Potions.

III. *Verbena folio subrotundo serrato flore cæruleo. Cat.* p. 64. *Tab.* 107. *Fig.* 1. Vervain.

This Plant has a long Root, very strongly fix'd into the Earth by several lateral fibers, drawing its Nourishment, and sending up several two or three Foot long, jointed four-square Stalks. The Leaves stand at the Joints opposite to one another, on short Foot-Stalks, are two Inches long, one broad, having many Ribs, the principal whereof are purple. They are deeply serrated, very smooth, and green, having smaller ones coming out *ex eorum ala*. The Flowers stand on the Branches ends Spike fashion. The Spike is long, the Flowers are thick set round it, without any Leaves between, are monopetalous, with the *Ora* divided into five parts, of a deep blue colour, one, two, three, four, or five Flowers, opening at a time, successively one after another from the bottom upwards. After each of these follows in a greenish brown *Calyx* or Husk one Seed or rather Husk, something like to that of Wheat in shape and colour, only in every thing smaller, being easily divisible into two, both being close covered with the same Membrane. The Seeds being in a Cavity on the side of the Spike, and are covered with three sharp, brownish, membranous Leaves.

This grows in clay grounds, in the *Savannas*, in the Road going to *Gua-naboa*, on the *Red* Hills, and in many other places in all the *Caribes*, but most plentifully near *Bridgetown* in *Barbados*.

It is very much us'd in Clysters for the Belly-ach; and by some in Poultesses, with Onions, for the Dropsie.

This and Lime-Roots boiled together the Decoction is said here to cure the Dropsie, or a Decoction of this after the manner of Tea, is of its self counted a powerful Sudorifick, lying warm after it. It is to be drank very warm.

It is very much in repute among the *Indian* and *Negro* Doctors for the Cure of most Diseases.

If

If this be *Monardes* his *Verbena* of *Peru*, as by the Vertues, it may, 'tis thought by him, and he tells several Stories, whereby he would make it appear to be one of the best Medicines against Worms, if the depurated Juice, with Sugar, be drank. It is bitter.

It is used in Ulcers, beaten and laid as a Poultess. A spoonful of the juice is given to those who have the Collick, Dysentery, *Cholera Morbus*, or any Bowel Disease. It is good against Charms. *Bont*.

IV. *Verbena minima Chamædryos folio.* Cat p. 64. Tab. 107. Fig. 2.

This Plant has a great many blackish fibrills coming from each side, of a long, reddish brown, deep Root. At its appearance out of the Earth it sends out on every hand several small, square, trailing, jointed Stalks, two Foot long, at the Joints striking some fibrous Roots into the Earth. At every two Inches distance are swelled, reddish Joints, where come the Leaves set opposite one to the other, on a quarter of an Inch Foot-Stalks. They are three quarters of an Inch long, and half an Inch broad, hairy, snipt about the edges, and like those of the *Chamædrys Spuria*. At the ends of the Branches come the Inch and an half long Spikes of small blue headed Flowers, each of which stands in a rough *Calyx*, and after them succeed several roundish Seeds, having Asperities and Depressions in them, and being of a light brown colour.

It grew near the Banks of the *Rio-Cobre*, below the Town of of *St. Jago de la Vega*, on the same side of the River.

V. *Verbena aut scorodoniæ affinis anomala, flore albido, calyce aspero, allii odore.* Cat p 64. *Guinea* Hen-Weed.

This Plant has a very strong Root, deeply fastned in the Earth, of a brownish white colour, from whence springs a very strong Stalk jointed, two Foot high, at whose Joints are Leaves an Inch and an half long, and half as broad in the middle, where broadest, smooth, and having many Nerves appearing in their surface. The tops of the Branches are, for a Foot in length, without any Leaves, set close on every side with white Tetrapetalous Flowers, in a very rough *Calix*, sticking close to the Stalk, without any Foot-Stalks, after which follow, inclosed in a very rough Seed-Vessel, one brown long Seed.

All the parts of this Plant have a very strong unsavoury smell, like to Wild-Garlick.

It grows in shady Woods, in the *Savannas*, every where.

The Roots of this Plant going very deep into the Earth, afford it Nourishment, when other Herbs and Grass are burnt up, and when Cattle can find no other Food, they feed on this. Hence Cows Milk in dry Seasons, in the *Savannas*, tast so strong of it as not to be savoury, and the Flesh of Oxen tast of it so much as scarce to be endured, and their Kidnies after a very intolerable manner. To avoid these inconveniencies, Grasiers, who feed Cattle for the Market, take them off such Pastures, and feed them with other sort of Food, and in about a Weeks time they are fitted for the Butcher, their Flesh having no tast of this remaining.

This tast in Milk or Flesh, is said commonly, and believed to be from the Calabash-Tree, on whose Fruit and Boughs Cattle likewise then feed, but 'tis perfectly the tast of this Plant, and not of the other.

A piece of this Root being put into a hollow Tooth, Cures the aching thereof.

VI. *Oci-*

VI. *Ocimum rubrum medium.* Cat. p. 65.

The Plant growing here, agrees exactly to the Description of *Ocimum Indicum*, *Cluf.* only it has neither spotted, nor serrated Leaves, and is not above half a Foot high, which perhaps may come from the variety of Soils, that described by *Clusius*, being the Plant raised from Seed sent from the *Indies*.

It grows every where in the moist places of the Low Lands, or *Savannas*.

It is reckoned a great Cordial, and therefore Distill'd, and us'd several other ways, especially by the *Spaniards*.

VII. *Hormino affinis, foliis angustis, glabris.* Cat. p. 65. Tab. 101. Fig. 3.

This had a square Stalk, hollow, with some Branches, standing opposite to one another, on which, at half an Inches distance, were placed the Leaves opposite to one another likewise, being an Inch long, and about a quarter of an Inch broad, slightly indented about the edges. The Flowers came close together at top in a Spike, being large, and standing on one third part of an Inches Foot-Stalk.

I found it in the North parts of this Island, whence I brought it to the South, and described it some days after, from whence comes the shortness of my Description, and imperfections of my Observations about it.

The indentures of the Leaves, Flowers set close at top, &c. distinguish this sufficiently from *Hormino accedens angustissimo folio Madraspatensis*, Pluken. *Phyt. Tab.* 194. *Fig.* 7.

VIII. *Nepeta maxima, flore albo, spica habitiori.* Cat. p. 65. Tab. 108. Fig. 1.

This rises to seven Foot high, though sometimes, in a different Soil, it may not come to half that heighth, having a square, brown Stalk. Towards the top are many Branches opposite to one another. The Leaves come out at uncertain distances, likewise opposite to one another, standing on an half Inch long, green, hoary Foot-Stalk, (in which they differ from the *Mentha Cataria angustifolia major*. C.B. *pin.*) They are two Inches long, and three quarters broad at near the Base, where broadest, whence they decrease, ending in a round, blunt point. They are hoary, snipt, or indented about the edges, of a whitish green colour. *Ex alis foliorum* come small Stalks, having sometimes Leaves, and sometimes none, but many white labiated Flowers, set close together round it, Spike fashion. All the parts of this Plant smell very strong, like Cat-Mint.

It grows in the Ditches about the Town of St. *Jago de la Vega*.

IX. *Prunella elatior flore albo.* Cat. p. 65. Tab. 109. Fig. 1.

This had several small, brown, two or three Inches long, fibrous Roots, The Stalk was quadrangular, jointed, green, two or three Foot high, being a little protuberant at each Joint, having towards the top Leaves and Branches, standing opposite to one another. The Leaves stood on very short Foot-Stalks, were an Inch and an half long, and three quarters of an Inch broad, of a pale green colour, with some Hair on them, being very like those of *Prunella. Ger.* The tops were short Spikes of white, and tubulous Fowers, like those of this Kind, and after them follow'd among the Leaves, *Capsula's* which were first green, and then whitish, containing several thin, round, membranaceous, black Seeds with a white Margin, lying on one another.

It grew on a rocky Hill, just over Colonel *Bourden's* House beyond *Guanabos*.

Y y X. *Pru-*

X. *Prunella flore dilute cæruleo pentapetaloides. Cat. p. 65.*
This is another fort of *Prunella*, which is in every thing the fame with the former, only not fo high. The Leaves are greener, the Flowers of a pale blue colour, monopetalous, the *Ora* being deeply divided into five Sections, or *Petala*.
It grows in most Woods of the Island.

XI. *Sideritis spicata scrophulariæ folio, flore albo, spicis brevibus habitioribus rotundis, pediculis insidentibus. Cat. p. 65. Tab. 109. Fig. 2.* Wild-Hops.
This has several small, white Roots, which, united, send up a four square, purplish Stalk, rising three Foot, having at every two Inches distance Leaves opposite to one another, exactly like those of *Scrophularia* in bigness, &c. Near the top, *ex alis foliorum*, goes out an Inch long Foot-Stalk, sustaining a Sphærical Head, about the bigness of a Hasel-Nut, made up of a great many galeated white Flowers standing close together, as in the *Trifolium pratense purpureum*. After this follow many small, black, shining Seeds, like to the smallest Gun-powder, contained in a hollow, long, brown *Calyx*, having five prickly *Apices* on its top. The whole Plant, when in state with its Heads, resembles very much a Branch of Hops, whence the name.
It grows on the Road to *Guanaboa* by a Gully near the *Red* Hills, near the Church in *Guanaboa*, and in several other places of this Island.
This Plant is somewhat anomalous, but this is the best place I could find for it.

XII. *Lysimachia cærulea galericulata, foliis angustis, longis, serratis, fœtidis. Cat. p. 66.*
A great many two Inch long white Roots, take very firm hold of the Earth, and raise an hexangular Stalk to one Foot and an half high, having here and there Joints and Branches. The Leaves stand three together, at a Joint, are long, narrow, jagg'd on the edges, covered with a white Wooll, or Down, and ending in a point from a broader beginning. The Flowers come out *ex alis foliorum*, are blue, small, like the other *Lysimachia galericulata*, having one Section turn'd up, and three hanging down, to which follow a great number of small, brown, red Seeds.
The whole Plant smells very strong.
It grows in clay low places of the *Savanna*, near the Town of *St. Jago de la Vega*, and other places where water has stood in rainy Seasons.
The Leaves of this being serrated, it cannot be the *Scutellaria Virginiana hyssopi angustis foliis, flore cæruleo. Plukenet. Alm. p. 338.*

XIII. *Scorodonia floribus spicatis purpurascentibus pentapetaloidis, semine unico, majori, echinato. Cat. p. 66. Tab. 110. Fig. 1.*
This has a great many four square, hollow Stalks, sometimes green, and sometimes purple, having here and there about an Inch distant from one another, at Joints, Leaves standing on an half Inch long Foot-Stalks, opposite to one another, being about an Inch long, and half as broad at Base, purple on the upper side, green underneath, rough and indented about the edges. The top is three or four Inches long, having on it a great many pale purple Flowers, standing in a green woolly *Calyx*, the *Ora* of each being divided into five Sections, after which follow so many rough, round *Capsula's*, brown, and inclosing an echinated, round, brown, large Seed.
It grows on the rocky, barren, clear'd Wood Lands, about *Guanaboa*.
The juice is counted a good vulnerary, healing green Wounds, after application giving some smarting pains.

XIV. *Scor-*

XIV. *Scordium maritimum, fruticosum, procumbens, flore cæruleo. Cat. p. 66. Tab.* 110. *Fig.* 2.

This Plant has small fibers going from the Stalks into the Sand, supplying the place of Roots. The Stalk is rough, four square, three or four Foot long, lying on the top of the Earth, at every half Inch's distance sending out Leaves, two opposite to one another, and sometimes Branches after the manner of other four-square stalk'd Plants. They are oblong, almost triangular, the Base sticking to the Stalk without any Foot Stalk, from whence it decreases to the point, being indented or snipt about the edges, hoary of a rank smell, and somewhat clammy. The Flowers come out *ex alis foliorum*, are blue, small, and tetrapetalous, with *Stamina* in the middle. After these succeed a great number of small, black, cornered Seeds, not discernible to the Eye, sticking to a middle Pillar, covered with long Skins or membranes, lying under each Leaf.

It grew among the loose Sand just by the Town of Old Harbour.

Chap. VIII.

Of Herbs that are leguminous, or have a papylionaceous Flower.

THIS Tribe of Plants is very numerous in *Jamaica*, and of great use to the Inhabitants, who feed much on the Kinds of Beans, Pease, or *Phaseoli* hereafter mentioned.

They are to be divided into two Kinds, such as live several years, which I call *Perennes*, and those which are sown every Season after Rain, rise and grow to their due maturity in very few months, if not weeks. 'Tis on these, many of which are *Redi*, or grow erect without climbing, that the Slaves are fed.

There are not wanting of this Kind such as are very large, beautiful, lactescent, &c. and other Kinds very differing from those of *Europe*.

There is use made not only of the Seeds, but also of the Leaves and Stalks of these Kind of *Phaseoli* for Provender for Cattle.

1. *Phaseolus maximus perennis, semine compresso, lato, nigris maculis notato. Cat. p.* 66. *Tab.* 111. *Fig.* 1. *Autre sorte des feves de Bies. p.* 335. *Phaseolus Barbadensis fruticosus septennii durationis.* Plukenet. Almag. *p.* 291. *An pois gros & plats blancs de Bouton. p.* 91 ? The great Bean.

This has a Stalk at coming out of the Earth as big as ones little Finger, angular, covered with a grayish Bark, turning and winding its self about any Perch, and rising to about seven or eight foot high, then falling, the Branches invicgle one another. At about two Inches distance, they send out Leaves, three always standing together on the same common, Inch long Foot-Stalk, each of them being almost round, of near an Inch Diameter, standing on very short Foot-Stalks, that in the middle, or opposite to the Foot-Stalk, having a *Petiolus* half an Inch longer than the others at Base. The Flowers stand several together, on an eighth part of an Inch long Foot-Stalk, being small papylionaceous, having a contorted *Stylus*, and some Leaves or *Petala*, of a white and purplish colour, with some green. After these follows a Pod at first green, then of a Clay or Ash colour without, and white within, crooked, more than two Inches long,

long, and an Inch broad in the middle, where broadest, containing one or more flat, or compress'd, broad Seeds or Beans, at first purple with black spots, then of the figure of a Kidney. They are black in the Circumference, and white towards the Center or *Hilus*, where they adhered to the Pod, and one Inch long, and three quarters broad in the middle where narrowest.

They are planted in most Gardens, and Provision Plantations, where they last for many years, bringing every Season a great many Beans.

They are eaten when green, and are fit for the Table in *December*. They are very good, as any of the *Legumina*, especially when the outward Skin is taken off.

They must have Poles or Perches to climb up, and sustain themselves by, they being otherwise too weak in their Stalks to support themselves, or bring Fruit.

It was first brought from *Africa*. C. B.

II. *Phaseolus perennis angustifolius flore luteo, semine lato, compresso, minore, rubro, maculis nigris notato.* Cat. p. 67. Tab. 111. Fig. 2. *An pois gras & plats rouges de Bouton.* p. 51 ? *Phaseolus Americanus longissimis & angustis foliis. Plukenet. Almag.* p. 291 ? The small red spotted Bean.

This *Phaseolus* has several angular green Stalks, by which it mounts about any Perches, or runs along the Hedges, at every three Inches distance, putting forth Leaves and Flowers, the first stand on two Inch long Foot-Stalks, always three together. They are an Inch and half long, and three quarters of an Inch broad at their round Base, where broadest, from thence decreasing, and ending in a Point, the odd one being three quarters of an Inch beyond those at Base. *Ex alis foliorum*, comes an Inch long Spike of Flowers, set round after one another. They are very small, papylionaceous and yellow. After them follow Pods an Inch and an half long, almost straight, with a sharp end, brown on the outside, white within, and containing two, or more red, flat Pease or Beans, something of the shape of a Kidney, scarlet coloured, with black Specks here and there on their surface.

They are planted as other of the *Perennial Phaseoli*, and give Fruit every year about the beginning of *February*, yielding a very great increase.

They are very good to be eaten any way, as any of the other *Phaseoli*.

III. *Phaseolus peregrinus octavus seu angustifolius alter, fructu ex albo & nigro vario.* C. B. *Cat.* p. 67.

This is likewise planted amongst the other Pease in *Jamaica*.

IV. *Phaseolus major perennis, floribus spicatis, siliqua breviori rotundiore, semine albo sphaerico.* Cat. p. 67. Tab. 111. Fig. 1, 2, 3. *Phaseolus tumidus minimus niveus siliqua brevi Virginianus* Raii, *hist.* p. 885. *An phaseolus hortensis fructo albo, minore, oviformis, venereus dictus. Hofm.* Cat. *Aldr ? Jamaica* Pease.

These are much the same with the former, only the Leaves are larger, and more pointed, standing on longer Foot-Stalks. The Flowers are more in number, standing Spike fashion, with an Eye of yellow in them, and the Pod is an Inch and an half long, and scarce half an Inch broad, containing three or four white, roundish Seed fastned by their middle to the Pod. They are not much bigger than our small Field Pea.

They are planted, and continue as the former.

V. *Pha-*

V. *Phaseolus maximus perennis, floribus spicatis, albis, speciosis, siliquis brevibus, latis, semen album hilo albido fere circumdante.* Cat. p. 67. Tab. 113. Fig. 1, 2, 3. *Phaseolus Jamaicensis & Barbadensis Ægyptiaco similis, semine ex toto eburneo nitoris.* Pluken. Almag. p. 291. Bonavist of Ligon. p. 22.

This *Phaseolus* has a very strong, round Stem at its Root, which sends out a great many round, green, striated, long Branches, climbing, and covering any Trees, Hedges, or low Houses they come near, sending out here and there Leaves and Flowers, which have a little knob at their Foot-Stalks, parting from the main Stem. The Leaves are always three, set on a long Foot-Stalk, large, woolly, and in every thing like the others of this Kind. The Flowers come out on each side of a Foot long Spike. They are white, papylionaceous, and make a very comely show on Hedges or Houses. After each of these follows a Pod two Inches long, and half an Inch broad, a little crooked, clay coloured, membranaceous, and containing four or five white, roundish, compress'd Pease, having a long white *Hilus* or Eye, almost surrounding the whole Bean.

They are planted here for Food as well as Ornament. They make a pretty appearance in Planters Gardens.

One Root will last a long time, and yield many Dishes of Pease, which, when eaten green, are not unpleasant.

These Beans rosted, as Coffee, and the Powder mixed with Rum-Punch, as Nutmeg-Powder uses to be, will intoxicate the Drinkers of it for some time.

VI. *Phaseolus maximus perennis, floribus spicatis albis speciosis, siliquis brevibus latis, semen russum hilo albido fere circumdante.* Cat. p. 68. Tab. 113. Fig. 4. *Lablab semine subnubro.* Alp. p. 74. Vesling. p. 24. *Leplap alterum rusum.* Clus. rar. pl. hist. lib. 6. p. 227. *Phaseolus Ægyptiacus sive lablab secundum semine rufo.* C. B. pin. p. 341. *Phaseolus Lablab spadiceus hilo longo albo.* J. B. tom. 2. p. 271. Red Bona Vist.

This *Phaseolus* seems not to differ from one I had from a Bale of Coffee, came from *Arabia* or *Egypt*, and therefore I take it to be the same with the *Lablab*, and am very apt to believe this, with reddish Beans, and the black to be only varieties of the same Plant. *Prosper Alpinus* describing this to have a long Pod, gave me a doubt about it, but since seeing one among Coffee, and the Pod by *Clusius* to agree with this, I think it the same with the above described of the *West-Indies*.

This is much the same with the foregoing, only smaller in every part, and the Beans or Pease are not white but red.

They are sometimes planted, though not so generally lik'd as the white Kind, because of their greater flatulency.

The *Ægyptians* use them for Food, and they are not less pleasant than our Beans to the tast. Women use their Decoction with Saffron, *ad excitandos menses.* It is likewise helpful for the Cough, difficulty of breathing, and to provoke suppress'd Urin. Alp.

VII. *Phaseolus maximus, siliqua ensiformi nervis insignita, & semine albo, membranula incluso.* Cat. p. 68. Tab. 114. Fig. 1, 2, 3. *Phaseolus Indicus siliqua magna falcata, quaternis in dorso nervis, cum eminentiis plurimis verrucosis, secundum longitudinem insignita, fructu amplo niveo, hilo croceo.* Pluken. Alm. p. 292. Horse Beans.

This sends out on every side of its Root, for some Feets distance, several Stalks, which are strong, and climbing about any thing they come near, at every Inches distance sending forth Leaves and Flowers. The Leaves are always

always three together, standing on the same two Inches long Foot-Stalk, which at parting from the Stem, has a swelling, each Leaf is larger, of a darker green colour than the other *Phaseoli*, smooth, having the Ribs appearing as those of the *Trifolium paludosum*. The Flowers stand several together on the same two Inches long Foot-Stalk. They are close papylionaceous, and of a bluish purple colour, after which follows a Foot and an half long Pod : strong, of a clay colour, a little crooked, or shap'd like a Scymiter, being an Inch broad, not altogether compress'd, but a little roundish, having on each Valve, within an eighth part of an Inch of their back Seam, where they are united, two small ridges or eminences running the whole length of the Pod. The Seeds are perfectly white, the largest of this Kind, being of a Kidnies shape, or something Oval at its narrowest or middle part, having a black *hilus* or speck, and each Seed is lodged in a distinct, very thin, white Membrane or Bladder, and five or six of these are contained in the same Pod.

I found this Bean first growing *sponte* at the upper end of the Town of St. *Jago de la Vega*, and afterwards in several Planters Gardens.

They are eaten as other *Phaseoli* by some, and counted good Food, though their greatest use is to fatten Hogs.

Nothing but the colour is different in *Clusius* his Description of this Lobe or Seed, which may be from the age or durtiness of the Plant.

VIII. *Phaseolus maximus perennis, folio decomposito, lobo maximo contorto.* Cat. p. 68. *Phaseolus utriusque Indiæ arboreus, alatis foliis, fructu magno cordiformi, lobis longissimis, nodosis, plerumque intortis* Pluken. *Alm.* p. 295.

This is very well described and figured in the *Hortus Malabaricus*.

They grow in the inland Woods of this Island, creeping up the Trees, and covering their tops for many Acres, as in the Thickets beyond Mount *Diablo*, going to St. *Anns*, on the *Moneque Savanna*, and in the Hills between *Guanaboa*, and Mountain River Plantations.

The Beans are pick'd up, and the mealy part being taken out at the *hilus*, they are tipt with Silver, and made into Snuff-Boxes.

The Beans Purge and Vomit, and are therefore Merchandise. *H. M.*

This sort of Bean is one of those found thrown up on the Shores in the North-West parts of *Scotland*, concerning which see the Philosophical Transactions. Numb. 222. p. 398.

IX. *Phaseolus Brasilianus frutescens, lobis villosis, pungentibus, maximus.* Herm. Par. Bat. pr. Cat. p. 68. *Phaseolus Brasilianus foliis mollis lanugine obsitis, fructu magno.* C. B. Muf. *Swammard.* p. 51. *An Lobus Cartilagineus ex insula.* S. *Mauritii.* Cluf. Brasil. Bester. *fasc?* Horse-Eye-Bean.

This has round, green Stalks, about the bigness of a Goose-Quill, by which it winds and turns its self round any Hedge or Tree it comes near. At about four Inches distance, it sends out Leaves, standing on two Inches long, green Foot-Stalks, three always together. The Foot-Stalk has a yellowish, and rough protuberance at coming out of the Stalk, as have the Leaves at parting from the Foot-Stalks. The Leaves are all equal to one another, that opposite to the Foot-Stalk, or in the middle, having a *Petiolus* three quarters of an Inch long. Each of the Leaves is three Inches long, and half as broad, on the upper surface smooth and green, on the under hoary and white, having some fibers from the Center of the Foot-Stalk, and others from the middle Rib, running through the Leaf. *Ex alis foliorum* hang down the Flowers by three Inch long Foot-Stalks, being eight or nine together, umbell-fashion, fastned to the end of the Foot-Stalk by a quarter of an Inch long *Petioli*. They consist first of a hoary, yellowish green *Capsula*, divided into four parts, within which is another yellowish and smooth capsular

capsular Leaf. Within these is an Inch long tubulous, papylionaceous, yellow Flower, with *Stamina*, and a *Stylus*. After these follow several Pods three Inches long, and half as broad, flat, having both Valves, at the opening, two waved eminent Lines, and all along their surface the same shorter waved eminences, very thick set with very sharp and small Prickles, both on the eminences and furrows. They are first of a green, and then when ripe of a blackish colour, and contain several round Beans, of about an Inch Diameter, flat, of a light brown colour, with a black ledge or *hilus* almost round them, looking something like a Horses Eye, whence the name.

They are eaten by the *Caribes*, and the juice of the Leaves is used by them for Dying Cotton Hamacs, of a black colour. *Plum*.

They grew on a Lime-Hedge, near Colonel *Copes* at *Guanaboa*, between his House, and the Mountains, and going down to the Ferry by the Rivers side over against *Atkins's* Plantation.

Snuff-Boxes are made of them. *Tertre*.

These Beans are very often to be gathered on the Sea Shore, cast up by the Waves, being dropt into some Rivers, or the Sea its self, whence they are again thrown up.

They are troublesome to Travellers, stinging them as they ride. *Piso* says they raise Pustles, and that he was not cured in eight days with Anodines and Coolers, they having hurt his Face and Hands in travelling.

They are thrown by the Currents of the Sea on the North-West parts of *Scotland*, concerning which the Philosophical Transactions before mentioned. *Clusius* saw them sent from *Barbary*.

Three of these Beans eaten cause death. The vertue of this Plant is in the Gout. The Leaves keep Women with Child from the Vertigo, for boil'd in Rice-water they dry superfluous humors. Its Bark, with dry Ginger, and the Fruit *Corimgola*, beaten, and boil'd in the Oil call'd *Fosale de enfermo & Bepu*, gives an Oil, which anointed on the Spinal Marrow, quiets Rheums and hurtful Catarhs. The Kernel, throwing away the inward Cuticle, boil'd with Milk and fresh Butter, is mixt to an Ointments Consistence against Pimples coming in Womens *pudenda*. With the Herb call'd *Felis oculus*, boil'd in Rice-washings and Butter-milk, it is a good Ointment in the Gout, and the same does the Root, with the Root of *Carimbola*, and *Capiram*, and the Leaves of *Munia* in form of Liniment, or if with the Bark of *Tamarind*, and dry Ginger, it be powdered and put into Whey, and then with the Oil *Bepu* it be boil'd and made into a Liniment. *H. M.*

They are used to be made into Buttons for Coats, sometimes tipt with Silver, sometimes not tipt.

X. *Phaseolus maritimus rotundifolius, flore purpureo, siliqua brevi cristata, semine fusco striato.* Cat. *p.* 69. *Phaseolus Brasilicus de Bry. florileg.* The Sea-Bean.

This has a deep, white, round Root, sending out on every hand several very long, small, white Filaments, running through the loose sandy Soil in which they grow to seek Nourishment to the Plant. The Stalks are many, lying on the surface of the ground for many Yards round, being about the bigness of a Swans-Quill, green, and a little cornered, putting forth at every three or four Inches Leaves alternatively, three always standing together on a common two Inches long Foot-Stalk, protuberant at its coming from the Stalk. The Leaves are almost round, that opposite to the Foot-Stalk, or in the middle, is the largest, standing an Inch beyond the other two at Base. It is two Inches long, and one and an half broad in the middle where broadest, having one middle Rib, and some transverse ones being of a Grass green co-colour, and smooth. The Flowers stand on an Inch long Foot-Stalk, are papylionaceous, and of a pale purple colour. The Pod is two Inches long, and three quarters of an Inch broad, straight, of a clay colour, swell'd out, or the Pease appearing in it before it be opened, having two crests, rais'd Ledges, or eminent Lines, one on each Valve, near the opening of it. The Pease are about six in number, each lying in a different Membrane, of the bigness of an ordinary Bean. They are Oval, brown, with clay coloured Spots on them, having a black Eye or *Hilus*, by which they are fastened to the Pod.

It grows on the *Cayes* near *Port-Royal*, and in all the *Caribes* by the sandy Sea Shores very plentifully.

They are dangerous to eat, therefore not gathered. *Tertre.*

Smith in his History of the Summer Isles, has a passage which I am apt to think relates to these Beans. The passage is this, *p.* 107. A kind of Woodbind, there is likewise by the Sea, very commonly to be found, which runs upon Trees, twining its self like a Vine. The Fruit somewhat resembles a Bean, but somewhat flatter, the which, any way eaten, worketh excellently in the nature of a Purge, and though very vehemently, yet without all peril.

This Plant is very well figured by *Ferrarius.*

XI. *Lobus oblongus aromaticus. Cluf. exot. Cat. p. 70. Vanillias piperis arbori Jamaicensis innascens. Plukenet. Alm. p. 381. Phyt. Tab. 320. Fig. 4. Volubilis siliquosa Mexicana foliis plantagineis. Raii hist. p. 1330. Siliqua Tlilxochitl. Worm. Muf. Swammerd. p. 16.* Vinelloes of *Dampier. cap. 3 & 8. Volubilis Americana capreolata; plantagineis foliis, siliquis longis moschum olentibus Bobart. hist. Ox. part. 3. p. 613. Lathyrus Mexicanus siliquis longissimis unctuosis Moschatis nigris. Ampan. char. p. 436.*

It is said by several that they grow in this Island about *Aqua Alta*, and that before the felling of Timber, and clearing ground, they were common in the shady bottoms of the inland parts of this Island.

They move Urin and the *Menses*. Comfort the Brain, Expel Wind, and Concoct crude humors. *Piss.*

It is mixed with Chocolate.

It grows about *Bocatoro Guatulco & Campeche*. It has a yellow Flower. It grows about the Trees, is yellow when Ripe, and laid in the Sun turns soft, and to a Chesnut colour, then they press it between their Fingers, and make it flat. The *Spaniards* sleek it with Oil. *Dampier.*

XII. *Phaseolus glycyrrhizites, folio alato piso coccineo atra macula notato. Cat. p. 70. Tab. 112. Fig. 4, 5, 6.*

This, by its Stalk, winds its self round any Shrub it comes near, rising by its help seven or ten Foot high. The Stalks are of the bigness of a Goose-Quill, round, woody, smooth, shining, having here and there alternatively Branches and Leaves. The Branches are about six Inches long, and at each Inches interval are winged Leaves, about four Inches long, beset by equal numbers of *Pinna*, opposite to one another, each of them being three quarters of an Inch long, and one quarter of an Inch broad, of a yellowish green colour, about eight or nine pair, growing on the same middle Rib. At the end of the Branch stand the Flowers very close in a Spike together, being in all two Inches long. Each of them is pale purple, and papylionaceous. After them follow as many short, broad, greenish brown Pods, broad at bottom, and sharp towards their ends, each containing three or four exactly round, scarlet Pease, having a black spot on that part where they stick to the Pod.

It

It grows in the Vallies or Low Lands, towards the Sea side in great plenty, flowering after the great Rains.

The whole Plant is sweetish to the taft, and therefore call'd wild Liquorish.

It is commended for a Remedy against the Belly-ach, if the Leaves be boil'd in Broth, and the Decoction given to the Sick.

Clusius was told, that these Pease string'd and wore as Bracelets, were used as Piony Roots against the Falling-Sickness; but he believed they were wore only as an Ornament, and I am of the same mind, they being us'd for that purpose to this day.

Clusius says, that they remain sometimes three years in the ground after planting, before they rise, which he ascribes to their hardness.

The *Egyptians* feed on them, being boil'd, but they are of all other Beans the hardest, and of worse Digestion, wherefore they are very troublesome to the Stomach, begetting much Wind. They are very hurtful to Hypochondriack People, very hardly to be digested, and beget bad Nourishment. *Alp.* They were brought from *Arabia Felix* into *Egypt*, and there Sown, and us'd more for Girls Ornament than Food. *Vesling.*

Best in his Description of the third Voyage of *Frobisher* to *meta incognita*, situated in above sixty degrees of North Latitude, says, that they found in the Tents of those people a *Guinea*-Bean of red colour, the which doth usually grow in hot Countries, thereby conjecturing they travelled into far distant Countries, or traded with people from thence.

They are commended by some (*I suppose in Necklaces, vid. Cluf.*) in teething Fevers, and Spasms of Children. *Cam.*

They are used for weights, *Ind. or. par. 6. p 35.*

The Leaves with *Calamus aromaticus*, boil'd with Oil, help Pains from sharp Humors. Their Powder by Insolation, mixt with Sugar, stays a sharp Cough. The Juice express'd, puts away Pains and Cold, and discusses thick and tough Humors, &c.

They are not eaten, but used as Weights in *Madagascar* and *China*. *Grew.*

These Beans are frequently gathered on the Shores of the North-West parts, and Islands of *Scotland*, with other Seeds which are before, and will hereafter be mentioned.

Upon being sometimes wetted, these Pease will turn of a black colour, whence I conjecture the two sorts mentioned by Dr. *Plukenet* in his *Alm.* p. 294. and *Mantissa*, p. 150. may be the same.

XIII. *Phaseolus sylvaticus flore patulo, dilute purpureo, siliqua tenui nigra, semine minore maculato.* Cat. p. 71. *Phaseolo affinis folio terebinthi duritie, Brasilianis flore purpureo maximo.* Pluk. *Alm.* p. 291.

This, by its tender Stalks, winds its self round any Plant it comes near. The Leaves are very thin, of a fresh green colour, and are always three, standing on an Inch long Foot-Stalk, each of which is about an Inch long, and three Inches broad, near the round Base where broadest, and whence they end in a point which is blunt, being thin and smooth, and of a yellowish green colour. The Flowers stand on a long Foot Stalk, are papilionaceous, very open, of a pale blue colour, with some purple Streaks, after which follows a small three Inches long, thin, black Pod, containing several oblong Pease, speckled with black spots. The Pods have two eminent raised Lines on their outsides.

There is a variety in the largeness of this Plant.

A a a   Dr. *Plukenet*

Dr. *Plukenet* in his *Mantissa*, p. 84. says that I, *more solito*, confound the synonimous names of Plants. I am sure he has done so in this. *Vid. Alm.* p. 157. and 291.

It grows above Mr. *Batchelor*'s House among the Honey-Comb-Rocks, and in *Barbados* very plentifully.

XIV. *Phaseolus sylvestris minor, flore minimo, siliquis longis, teretibus, alba lanugine hirsutis*. *Cat.* p. 71.

This has several hairy small Stems, by which it turns it self round any thing it comes near, and mounts to four or five Foot high, at every Inch and and an halfs distance, putting forth Leaves. They are always three together, on an Inch long Foot-Stalk, the odd one being two Inches long, and three quarters of an Inch in breadth, hairy, of a yellowish green colour, and plac'd one third of an Inch beyond the two Leaves at Base, which are every way smaller. Opposite to this Leaf comes the Flower, being of a pale green colour, and scarce discernable besides the Hair. After this comes a small Pod, green and hairy, two Inches long, of a dark brown colour, with a white Wooll on it, containing several oblong, round yellowish Pease, sticking to the Pod by a white Eye in the middle.

They grow very plentifully in the open Grounds on the Hedges in *Guanaboa*.

Concerning its Vertues. *vid. H. M.* Part. 8. p. 68. Where amongst others 'tis said to be good, being mix'd with Cows-Milk, either outwardly as a Liniment, or inwardly as a Potion, against the bitings of Scorpions or Rats, and against the swellings of the *Testes*.

XV. *Phaseolus minor lactescens flore purpureo*. *Cat* p. 71. *Tab.* 114. *Fig.* 4.

This by its round, small, woody, Stalks, turns its self round, and mounts about any Tree or Shrub it comes near, rising to fix Foot High. At every Inch or more, putting forth Leaves, three always together, standing on an half Inch long Foot-Stalk. That one of the three opposite to the Foot-Stalk, or in the middle, being the largest, is an Inch long, and three quarters of an Inch broad in the middle where broadest, being roundish or oval, of a dark green colour, smooth, hard and nervous. The Flowers come out *ex alis foliorum*, being many on the same very short Foot-Stalks, are three quarters of an Inch long, hollow, papilionaceous, and purple. After them follow two Inches long, and one fourth part of an Inch broad Pods, round, a little crooked, and sharp at the end. At first they are green, then brown, and contain several brown, small Pease.

All parts of this Plant are milky.

It grew on the Red Hills, on each side of the Road, among the Bushes going to *Guanaboa*, and other places very plentifully.

XVI. *Phaseolus minimus, foetidus, floribus spicatis è viridi luteis semine maculato*. *Cat.* p. 71. *Tab.* 115. *Fig.* 1. *An Phaseolus Americanus, fructu minimo, variegato. Plum. Tournef. Inst.* p. 415 ?

This has round, small and tender Stalks, rising by, and twisting its self round any thing it comes near, 'till it be six or seven Foot high, having here and there along the Stalks, Leaves and Flowers. The Leaves are always three, of a yellowish green colour, standing on the same common half Inch long Foot-Stalk, round, and about the bigness of those of the *Trifolium pratense album*. Its Flowers are many, standing on an half Inch long Foot-Stalk, Spike fashion, are very small, scarce opening themselves, papilionaceous, and of a greenish yellow colour. After these follow as many quarter

of

of an Inch long, black, rough Pods, containing one or more, small, oblong, blackish, green, speckled Pease.

The whole Plant has an unsavoury rank smell.

It grows in rocky places, where the Woods are clear'd, near Mr. *Batchelor's* House, and in several sandy places of the Town *Savanna*.

XVII. *Phaseolus erectus lathyroides, flore amplo, coccineo. Cat. p. 71. Tab.* 116. *Fig.* 1.

This has an oblong, large, white Root, going a Foot deep into the ground, from the top of which grow several trailing Branches, round and green, set pretty thick with Leaves three always together, on an half Inch long Foot-Stalks. Each of them is an Inch long, and not over the twelfth part of an Inch broad, smooth, of a blewish green colour. The Leaf opposite to the Foot-Stalk, is longer, and stands on the top of the Foot-Stalk by an eighth part of an Inch *Petiolus*. The Flowers stand severally on two Inches long, round, green, naked Foot-Stalks. They are papilionaceous, of a scarlet colour, the two biggest opposite *Petala*, being very large, to which follows a slender, brown Pod, containing several Pease.

It grows in the claiy parts of the *Savannas*.

XVIII. *Phaseolus erectus major, siliqua tereti, semine rubro. Cat. p.* 71. *Tab.* 115. *Fig.* 2, 3. *An Phaseolus Americanus hortensis affinis fructu minore rubro, Callavance Jamaicensibus dictus. Pluken. Almag. p.* 289 ? *Jamaica* Red-Pease.

This *Phaseolus* has a small Root, and strong green Stem, which, at about three Inches from the ground, shoots out several trailing Branches, at unequal distances sending forth Flowers and Leaves, of the latter there are always three, like the other *Phaseoli*, standing on a four Inches long Foot-Stalk. The Flowers are two, on an Inch and an half long, strong Foot-Stalk, one against another, papilionaceous, and of a pale purple colour. After these follow two Pods of a reddish purple, colour, having a swelling over every Pea.

It is planted as the former, and counted very good Provision.

XIX. *Phaseolus major erectus, caule purpurascente, siliqua tereti ventriosa longa. Cat. p.* 72.

This has a cornered, somewhat purple Stalk, slender, eight or nine Foot long, at every three or four Inches distance sending out Leaves, three being always on the same two Inches long Foot-Stalk, like in every thing to those of the other *Phaseoli*. *Ex alis foliorum* comes a two Inches long Foot-Stalk, and on it several pale papylionaceous Flowers, and after them follows a round, whitish Pod, ventriose, as big as ones little Finger, almost straight, four or six Inches long, containing about fifteen cornered, reddish brown coloured Pease, just like an *English* Rouncival, having in their Pods a thin membrane between every Pea.

They are planted as frequently, and set for the same purposes as other the *Phaseoli* of this Kind.

XX. *Phaseolus erectus minor, siliqua tereti ventriosa, pallide flavescente, semine albo. Cat. p,* 72. Clay coloured Pease, or six Weeks Pease.

From a small Root, with many fibers, rises a Stem, sometimes purple, having several Branches on every hand, about a Foot and an half high, and here and there, without any order, Leaves and Flowers, *ex eorum alis*, both standing on an Inch long Foot-Stalks. The Leaves are always three, woolly, almost round, two Inches long, and one and an half broad, made like the other *Phaseoli*. The Flowers are papylionaceous, of a pale blue,

blue, purple colour, or sometimes whitish. The succeeding Pods are about three Inches long, clay coloured, round, only swell'd more or less according to the number of Peas, every one having a swelling over it, and inclosing a great many white Peas, oblong, roundish, and shap'd like a Kidney, as other the *Phaseoli*.

There is a variety in this Pea, some being larger than others.

They are planted here in clear'd grounds after a small Rain or Season. They howe, or make superficial holes in the Earth three Foot distance one from another, and therein put three Pease, covering them with the Mould, one whereof they think will spring, the other two are supposed to be eaten with Rats, Mice, &c. In six weeks time they are ready to be gathered. If not gathered when Ripe, the first Shower of Rain shakes them, and they are lost, which shows the contrivance of Nature to perpetuate the Plant, Rain shaking the ripe Seed at a time when proper to grow.

They are reckon'd very good Provision for *Negroes*, white Servants, or to fatten Hogs withal, and very much planted for those uses.

XXI. *Phaseolus erectus minor, semine sphærico albido, hilo nigro. Cat. p. 72. Tab.* 117. *Fig.* 1, 2, 3. *Phaseolus Indicus fructu striato, albo minore nigra macula insignito.* Plukenet. *Almag. p.* 290. *An phaseolus Barbadensis erectior siliqua angustissima tinctorius.* Herm. *par. Bat. p.* 111 ? *Calavances*.

This sort of *Phaseolus* has a Stalk, putting forth several Branches, rising about a Foot high, along which come out Leaves three together, on a three or four Inches long Foot-Stalk. That Leaf which is odd, or opposite to the Foot-Stalk, is an Inch and an half broad at Base, and two Inches and an half long, plac'd three quarters of an Inch further than those at Base, which are smaller, they are very soft, of a yellowish green colour, and have their Ribs from the ends of their Foot-Stalks. The Flowers are white, papilionaceous, and stand on nine Inches long, strong Foot-Stalks, coming out *ex alis foliorum*. After them follow three or four Inches long Pods, almost round, clay coloured, and almost straight, containing very many almost round, white Pease, something resembling a Kidney, with a black Eye, not so big as the smallest Field Pea.

They are planted, and perish every Season, as some of the other *Phaseoli*.

They are accounted the sweetest, and best for Food of any of them.

XXII. *Arachidna Indiæ utriusque tetraphylla. Par. Bat. pr. Cat. p.* 72. *Mandobi fructus pisonis Muf. Swammerd. p.* 15. *An Terfex.* Ogilb. *Africa, p.* 22 ?

I found this planted, from *Guinea* Seed, by Mr. *Harrison,* in his Garden in *Liguanee*.

The Fruit, which are call'd by Seamen Earth-Nuts, are brought from *Guinea* in the *Negroes* Ships, to feed the *Negroes* withal in their Voyage from *Guinea* to *Jamaica*.

They are windy and Venereal. *Piso.*

If eaten much they cause the Head-ach. *Marcgr.*

An Oil is drawn out of them by Expression, as good as that of Almonds.

If they are beaten and made into a Poultess, they take away the pain of Serpents bites. *Du Tertre.*

This is the Nut *Clusius* speaks of, wherewith the *Portuguese* Victual their Slaves to be carried from St. *Thome* to *Lisbon.*

XXIII. *Hedysarum triphyllum fruticosum, flore purpureo, siliqua varie distorta. Cat. p.* 73. *Tab.* 116. *Fig.* 9.

This

This rises by a woody, brown coloured Stem, having several green, rough Branches, to four Foot high. The Leaves come out on every side, without any order, having two Appendices at parting from the Stalk something like a *lotus*, three always together on the same Inch and an half long Foot-Stalk, the upper one being longer, and having an Inch long Foot-Stalk proper to its self, or being plac'd an Inch further than the two under ones, which are rounder, and opposite to one another. They are all thin, smooth above, of a dark green colour, and rough underneath. The tops are long Spikes of Flowers thinly plac'd on an half Inch long Foot-Stalk, papilionaceous, of a pale purple colour. After these follow several Pods, slender, rough, jointed, and variously turn'd and distorted.

It grows in a Gully towards the Angels beyond the Town of St. *Jago de la Vega*, and in the Paths among the Sugar-Canes in several places of this Island.

XXIV. *Hedysarum triphyllum fruticosum minus.* Cat. p. 73. Tab. 118. Fig 1.

This had a very strong Root and Stem, from whence went several Branches about three Inches long, having here and there alternatively Leaves very like the former, three always together on the same half an Inch long common *Petiolus*, each whereof were about half an Inch long, smooth, of a yellowish green colour, and of an Oval shape. *Ex alis foliorum* rise Foot-Stalks, about three Inches long, sustaining some papylionaceous Flowers, and I guess Seeds in Pods, as the former, although I never saw them, and therefore cannot positively assert that it belongs to this place.

I found it in *Jamaica*, but do not remember where, neither can I give a more particular account of it.

XXV. *Hedysarum triphyllum fruticosum supinum, flore purpureo.* Cat. p. 73. Tab. 119. Fig. 2. *Onobrychis Americana floribus spicatis foliis ternis canescentibus siliculis asperis Plukenet.* Alm. p. 278. Phyt. Tab. 308. Fig. 5.

This has a long, small, woody Root, sending forth several Foot long Branches, lying along the ground, whose Stalks are reddish, rough, round, and woody, having at unequal distances, on half an Inch long Foot-Stalks, their Leaves, which are always three together, whereof that in the middle is the longest, and all are green above, and more pale beneath. The tops of the Branches are Spikes of purple, papylionaceous Flowers, to which follows geniculated, crooked Pods, forming a Semicircle, of a brown colour, each joint of which is fastned to that next it, by a very small Isthmus, whereby its adhesion to it is so easie, that by its roughness sticking to any Garment, they leave one another, whence the *Portuguese* Name *Erva d' Amor*. Every joint contains one pale yellow Seed.

It grows every where in the Woods, especially in those of the *Savannas*.

The Root is hot, and a Decoction of it in fair water, or other *Vehicle*, is one of the best Remedies against a cold Flux of the Belly. The fume or smoak of the Leaves received with a covered Head, cures the Head-ach which comes from Cold and Catarrhs. *Piso*.

XXVI. *Hedysarum minus diphyllum, flore luteo.* Cat. p. 73.

This Plant, from a small woody Root, puts out several three or four Inches long Branches, trailing, or lying on the surface of the ground, having several Leaves, two always standing on the same Foot-Stalk, of a yellowish green colour, and a little hirsute. The Flowers are yellow, and papylionaceous. The Pods are a little crooked, hirsute, or rough like a half
B b b   Moon.

186   *The Natural History of* Jamaica.

Moon. Each Seed or Pea being inclos'd in a Semicircular joint, every one of which is parted from the other by a small Neck or *Isthmus*.

It grows very copiously in the sandy and dry places of the *Savanna*, near the Town of St *Jago de la Vega*, after rainy Seasons.

An Apozem is made of this for cold Fevers. *H. M.*

XXVII. *Hedysarum caule hirsuto, mimosæ foliis alatis, pinnis acutis minimis gramineis. Cat. p. 74. Tab. 118. Fig. 3. An securidaca clypeata flore luteo lentis folio Zeylanica. Breyn. pr. 1. p. 46 ? Vel an Mimosa siliquis latis hirsutis articulatis. Herm. par. Bat. Cat. p. 10 ? An Onobrychis Brisnagarica mimosæ foliis, siliquis ad unum solummodo latus dentatis ex alis polyceratos. Pluk. Alm. p. 270. Phyt. Tab. 49. Fig. 5 ?*

The Branches of this Plant were about a Foot long, roundish, filled with a fungous Pith, set very thick on the outsides, with very large and fierce Hairs, or small prickles, of a white colour, as were also the Twigs, whose ends were set with alated Leaves, whose *Pinnæ* were very small, sharp, or pointed at the end, grassie or striated like grass Leaves, and numerous. The Flowers came out of a hairy, or echinated small Leaf, *tanquam ex utriculo*, being many standing on the ends of branched Foot-Stalks alternatively, and after them follows articulated Pods, like to those of the precedent.

I found it in the inland parts of the Island, but where particualrly I do not remember.

XXVIII. *Quadrifolium erectum flore luteo. Cat. p. 74. Tab. 116. Fig. 3.*

This rises to about a Foot high, being erect, branched, and having Twigs set thick with Leaves alternatively, on a three quarters of an Inch long Foot-Stalk, there being constantly, as far as I could observe, four on the same Foot-Stalk. Each of them are small, and have a small snip or defect on their further ends, where they are largest, being of a yellowish green colour, and smooth, having one middle Rib, eminent on the backside. *Ex alis foliorum* towards the top comes a yellow papylionaceous Flower, as in others of this Kind.

It grew in the *Savanna* near Two Mile Wood, and several other clay parts of the *Savannas*.

XXIX. *Loto pentaphyllo siliquoso villoso similis, Anonis non spinosa, foliis cisti instar glutinosis & odoratis. Cat. p. 74. Tab. 119. Fig. 1. An Anonis viscosa spinis carens lutea major nonnihil procumbens, medio tantum folio per extremum serrato, pediculis florum indivisis. Pluk. Mentiss. p. 15 ?*

This has a great many wooddy Branches from the same Root, rising to about two Foot high. The Stalks are round, shrubby, gray, and branch'd out into many Twigs, which are green and hoary, having without any order several Foot-Stalks half an Inch long, sustaining three Leaves, each smaller and longer than those of the *trifolium hæmorrhoidale*, and being purple on the edges, and having purple spots on their backs, and a down of the same colour. The Flowers are on the tops of the Twigs, several opening successively one after another, being yellow. After these follows a shining brown, single Seed, with a point on one side, and defect on the other, inclosed in a Husk or Pod; which is very short, thin shap'd like a Scymiter, having several appearing Lines or Nerves on it, in each of which lies only one single Seed.

The whole Plant is clammy, and smells like the ordinary *Cistus's*.

It grows in claiy or gravelly parts of the Town *Savanna*, and elsewhere in *Jamaica*.

It

In qualities it agrees with those of this Tribe that are in *Europe*. The vertues of the Root being heating, of subtle parts, and fit to cleanse the Bladder. *Pise.*

XXX. *Anonis non spinosa minor, glabra, procumbens, flore luteo.* Cat. p. 75. *Tab.* 119. *Fig.* 2.

This has a long, deep, round, brown, tough Root, from whence spring many round, Foot long Stalks, hairy, branch'd into others lying round on the surface of the Earth, being very thick set with Leaves, three always together on the eighth part of an Inch long, or very short, Foot-Stalk, each being smooth, and having many white Veins appearing on its under surface. Towards the top come the Flowers, they are papylionaceous, of an Orange colour, with a little purple in their middle. Afterwards follows a very small, rough, short Pod or Husk, inclosing one reddish Seed or Pea.

It grew near the River-side in a Field below the Town, and near the watering place by the River very plentifully.

Chap. IX.

Of Herbs whose Flowers are compos'd of two or three Petala or Leaves.

WHAT are chiefly remarkable in this Tribe, are some of the following *Viscum's*, which are a new kind of Parasitical Plants, differing from all those of *Europe*. Their way of growth, Flower, and Seed, are very particular, and extraordinary, and may be taken notice of in their Descriptions.

Arna overo ana Vareca di Padre Matteo, seems to be one of these *Viscums* described and figured by *Zan.* p. 29.

I. *Stellaria aquatica. Park. Cat.* p. 75.

It grows in most Rivers of this Island.

II. *Planta innominata prima Marcgr.* p. 8. *Cat.* p. 75. *Ephemerum Brasilianum ramosum procumbens bipetalon foliis mollioribus. Herm. par. Bat.* p. 145. *Phalangium Africanum helleborines folio non descriptum. Hort. Lugd. Bat. Rati. Hist. Dipetalos Brasiliana foliis Gentianae aut plantaginis. Rati Hist.* p. 1332. *Ephemerum Africanum annuum flore bipetalo Herm. Cat.* p. 231.

This sort of *Phalangium* is very common in all the moist places of this Island, as well as *Barbados.*

III. *Plantago aquatica. Fuchs. Cat.* p. 75. *Ranunculus palustris plantaginis folio ampliors. Tournef. El.* p. 241. *Inst.* p. 292.

It grows near Black River Bridge, going to Old Harbour, and in several other places of this Island.

This is thought to have the same qualities with Land Plantain, the Seed to be adstringent, and the Leaves good against Burns, and to be applied to Hydropick Legs. *Ger.*

The juice applied to Breasts is a great secret in clearing them of Milk. *Rossin.* J. B.

IV. *Sagitta*

IV. *Sagitta, Cast. Dur. Cat. p. 76. Ranunculus palustris, folio sagittato maximo, Tournef. El. p* 241. *Inst. p.* 292.

This Plant agrees exactly with *John Bauhin*'s Description, and seems to be the same Plant in every thing with that of *Europe*.

It grows in standing waters.

It is counted to have the vertues of Plantain, being thought Cold, Dry, and Adstringent.

I have seen this sent from the *East-Indies*, under the name of *Coolette Tella*, gathered near Fort St. *George*, where the Natives use the Root bruised to cure their sore Feet, which they often have in wet weather, going bare Foot.

V. *Plantaginis aquaticæ folio Anomala, flore tripetalo purpureo semine pulverulento. Cat. p.* 79.

This had several pretty large white Roots, two or three Inches long, which united send up several Leaves, four or five Inches long, green, succulent, rib'd like Plantain Leaves, an Inch and three quarters broad near the middle where broadest. In the center of these Leaves rises a purple jointed Stalk, a Foot and an half high, having a Spike of purple, or Carnation Flowers three Inches long, and at top three purple *Petala*, under which is a little swelling, which augmenting turns to a dust, and scatters with the wind out of a brown membranaceous Husk.

It grew in the Roads to Mountain River beyond Colonel *Cope*'s Plantation in *Guanaboa*.

VI. *Viscum Caryophylloides maximum flore tripetalo pallide luteo semine filamentoso. Cat. p.*76. *Philosophical Transactions.* Number 251. p. 114. Wild Pine.

A great many brown fibrils encompass the Arms, or take firm hold of the Bark of the Trunc of the Trees where they grow, not as Misleto, entering the Bark or Wood to suck Nourishment, but only weaving and matting themselves among one another, and thereby making to the Plant a firm and strong Foundation, from whence rise several Leaves on every side, after the manner of Leeks or *Ananas*, whence the Name of Wild-Pine, or Aloes, being foulded or inclosed one within another, each of which is three Foot and an half long, from a three Inch breadth at beginning or Base, ending in a point, having a very hollow or concave inward side, and a round or convex outward one, so that by all of their hollow sides is made within a very large Reservatory Cistern or Bason, fit to contain a pretty deal of water, which in the rainy Season falls upon the uppermost parts of the spreading Leaves, which have channels in them conveying it down to the Cistern, where it is kept as in a Bottle. The Leaves after they are swell'd out like a bulbous Root, to make the Bottle bending inwards, or coming again close to the Stalk, by that means hindering the evaporation of the water by the heat of the Sun. They are of a light green colour below, and like Leeks above. From the midst of these rises a round, smooth, straight, fresh, green coloured, three or four Foot long Stalk, having many Branches, when wounded yielding a clear white mucilaginous Gum. The Flowers come out here and there on the Branches. They are made up of three long, yellowish, white, herbaceous *Petala*, and some purple ended *Stamina*, standing in a long *Calyx* or *Tubulus*, made up of three green viscid Leaves, with purple edges. After these follows a long triangular *Capsula*, greenish brown, being somewhat like those of the *Cariophylli*, having under it three short capsular Leaves, and within several long pappous Seeds. The Seed it self being oblong-pyramidal, and very small, having very soft, downy hairs, or

Down,

Down, or *Tomentum*, much longer in proportion to the Seed, then any *Tomentum* I know, being as long as the Pod or *Capsula*.

It grows on the Arms of the Trees in the Woods every where, as also on the Barks of their Trunks, especially when they begin to decay, their Barks, receiving the Seed, and yielding then easily to the fibrils of this Plant's Roots, which in some time dissolves them, and ruins the whole Trunc.

The contrivance of Nature, in this Vegetable is very admirable and strange. The Seed has long, and many threads of *Tomentum*, not only that it may be carried every where by the wind, as papose and tomentose Seeds of *Hieracium*, *lysimachia*, &c. but also that it may by those threads, when driven through the Boughs, be held fast, and stick to the Arms and extant parts of the Barks of Trees. So soon as it sprouts, or germinates, although it be on the under part of a Bough, or the Trunc of the Tree, its Leaves and Stalk rises perpendicular, or straight up, because, if it had any other position, the Cistern (beforementioned, by which it is chiefly nourished, not having any communication with the Tree) made of the hollow Leaves, could not hold water, which is necessary for the Nourishment and Life of the Plant.

In the Mountainous, as well as dry low Woods, in scarcity of water. This Reservatory is necessary and sufficient, not only for the Plant its self, but likewise is very useful to Men, Birds, and all sorts of Insects, whither in scarcity of water they come in Troops, and seldom go away without Refreshment. *For the further account of this Plant and its Figure, as also concerning European Plants, somewhat analogous to it in some particulars, See the Philosophical Transactions.* Numb. 251.

Besides the Authors mentioned in my Catalogue of *Jamaica* Plants, p. 76. to take notice of this Plant, I find *Huldrich Schmidel*, cap. 46. p. 77. of his *Navig*. Printed 1599. 4to. to have the following passage, which I believe relates to this Herb.

Ex nostris autem hominibus multi siti moriebantur, licet ad hoc iter apud istos Caciqueos mediocri aqua copia nos instruxeramus. Inveniebamus autem in hoc itinere, radicem supra terram extantem, magna lataque folia habentem, in quibus aqua tanquam in vase aliquo manet, nec inde effanditur, nec etiam tam facile consumitur, capitque una harum radicum aqua circiter dimidian mensuram.

And Capt. *Dampier*, in his Voyage, *Vol.* 2d. of *Campeche*, p. 56. says thus.

The *Wild Pine* is a Plant, so called, because it somewhat resembles the Bush that bears the Pine: they are commonly supported, or grow from some Bunch, Knot, or Excrescence of the Tree, where they take root and grow upright. The Root is short and thick, from whence the Leaves rise up in Folds, one within another, spreading off at the top: They are of a good thick Substance, and about Ten or Twelve Inches long, the outside Leaves are so compact, as to contain the Rain Water as it falls, they will hold a pint and a half, or a quart: and this Water refreshes the Leaves, and nourishes the Root. When we find these Pines, we stick our Knives into the Leaves, just above the Roots, and that lets out the Water, which we Catch in our Hats, as I have done many times to my great Relief.

VII. *Viscum Cariophylloides maximum, capitulis in summitate conglomeratis.* Cat. p. 77. Tab. 120.

This has a great many long, dark, brown, small filaments, threads, or fibrils, which take fast hold of the Barks of Trees, to which it adheres, when all united making an oblong Root, and sending forth round about many Leaves, like those of white Lily, inclosing one another after the manner of Aloes, each of which is a Foot long, an Inch and an half broad, blunt or roundish. They are at the ends of a very dark green, and sometimes purple colour.

From

The Natural History of Jamaica.

From the middle of these Leaves rises a Stalk, about a Foot and an half high, on which, round about, are set several Leaves, smaller than those at bottom, on the top of which stand many pale, green, broad Leaves, having a glewy muciiage between them, and a great many reddish Leaves, making as it were a *Calyx*, and looking in all something like a Rose, in which are contained several Heads conglomerated, having Seed-Vessels, and Seed as the former.

It grows on old great Trees in the large Woods, in Sixteen Mile-Walk-Path, and sometimes on the ground when fallen.

VIII. *Viscum Caryophylloides majus, flore tripetalo cæruleo, semine filamentoso.* Cat. p. 77.

This by a great many small fibrils, wrapt together, catches hold, and surrounds the Branch of any Tree it grows on, for three Inches round. From thence rises a Stem, about which grow several Leaves, an Inch broad at their beginning, hollow or concave in the inside, and convex on the other, swell'd out, or prominent below, making a cavity able to hold a pretty deal of Rain water. The other, or upper part of the Leaf is narrow and grassie, something like those of Pinks, about nine or ten Inches long, bow'd backwards, and reflected, and so hanging down, of a whitish green colour, In every thing like the Leaves of the Wild-Pine in their contrivance. The Stalk is a Foot and an half high, jointed, at every Inches distance or joint, having a grassie Leaf, inclosing the Stalk at the joint. Near the top on each side, *ex alis foliorum* come the Flowers, which have three *Petals*, are blue with a few yellow *Stamina*, scarce appearing out of a green hollow Leaf. After which follow pappous Seeds, being small, oblong and brown, having many long, downy filaments, hanging from them, and all being inclosed in a first green, then brown triangular *Capsula*, supported by the aforesaid Leaf.

It grows every where in the Woods on the Branches of Trees, drawing its Nourishment from Rain-water, falling into a Cavity made by its own Leaves.

IX. *Viscum Caryophylloides minus, foliis pruinæ instar candicantibus, flore tripetalo purpureo semine filamentoso.* Cat. p. 77. Tab. 121. Fig. 1.

This Plant has several small fibers, warped, interwoven, or matted one within another, and wrapped about the Arms and Branches of Trees, from which, though sometimes it be on the under side of the Bough, rise straight up several Leaves, the under parts whereof inclose one another like Bulbs, making in their inward concave sides a cavity to hold Rain, thereby to nourish themselves and Stalks. The Leaves are long, narrow, grassie, exactly of the shape, make and contrivance with those of the Wild-Pine, something like the Leaves of a Gillyflower, only all cover'd over with a short, white shining Down, making the Leaves always look as if they were cover'd with a hoar Frost. In the middle of these rises a naked hoary, three Inches long Stalk, on the top of which stands a hoary half Inch long *Calyx*, on the end of which are three small, purple *Petals*, and afterwards follows a Seed-Vessel, exactly the same in every thing with that of the Wild-Pine, only in every part smaller.

It grows every where on the Branches and Arms of Trees in this Island.

X. *Viscum Caryophylloides minus, foliorum imis viridibus, apicibus subrubicundis, flore tripetalo purpureo semine filamentoso.* Cat. p. 77. Tab. 122. Fig. 1.

This

This is in every thing the same with the former, only larger and longer. The Leaves are very like those of Pinks in shape, their under parts are green, and tops reddish.

It is to be found on the Boughs and Trees on the Red Hills on *Guanaboa* Road, and near Mr. *Barnes*'s House there.

XI. *Viscum Caryophylloides tenuissimum, è ramulis arborum musci in modum dependens, foliis pruina instar candicantibus, flore tripetalo, semine filamentoso.* Cat. p. 77. Tab. 121. Fig. 2, 3.

The Stalk of this mossie Plant is about the bigness of a thread, consisting of a thin Skin, whitish, as if covered with a hoar Frost, having within that a long, tough, black Hair, like a Horse Hair. These Stalks (many of them being usually together) stick on any Branch superficially by the middle, and send down on each side some of the same Hairs or Stems, very often a yard long, hanging down on both sides from the Branches of the Trees they adhere to, being curled, or turning and winding one within another, and making the shew of an old Man's Beard (whence the name) or as if they were made to climb, which I never saw they did. These Stalks are branch'd, and the Branches which are two or three Inches long, are set with two or three Inches long, roundish, white Leaves, covered over as it were with a hoar Frost. The Flowers come at the ends of these Branches, have three *Petala*, and a Seed, with Seed-Vessel, &c. like the others of this Kind.

It grows on the Branches of the Ebonies, or other Trees in the *Savanna's* frequently, and looks very oddly.

It is us'd to pack up any thing in, which otherways may easily be broken, as Cotton is sometimes made use of with us.

The inward strong black Hairs of this Mosses Stalks, are made use of by the Birds called *Watchipickets*, for making their curiously contriv'd Nests, hanging on the Twigs of Trees.

This, by lying in the Air and Weather, or being by other means cleared of its outward Skins, has another appearance, whence Dr. *Plukenet* calls it *Cuscuta Americana super arbores se diffundens; Cuscuta trichodes lendiginosa*, &c. as I have observ'd, p. 221. of my Catalogue.

XII. *Ananas, Christ. Acost. Cat. p. 77. Fau-polo-mie seu Ananas fructus. Boym. lit. G. Thevenot. p. 21.* Pine-Apples of *Dampier*.

This Fruit is planted and us'd by way of desert, (having a very fine flavour and tast) all over the hot *West-Indies*, either raw, or when not yet ripe, candied, and is accounted the most delicious Fruit these places, or the World affords, having the flavour of Rasberries, Strawberries, &c. but they seem to me not to be so extremely pleasant, but too sowre, setting the Teeth on edge very speedily.

The Fruit ripened by the Sun is less esteem'd than that ripen'd in a Chamber. *Piso*.

It is clear'd of its outward Skin when ripe, and cut into slices, and so eaten, the middle fibrous or woody part being thrown away.

It is known when ripe by the colour of the tuft of Leaves at top, which then turn yellow, and will easily come off with the least pulling.

This Tuft, as well as young Sprouts or Succors from the old ones sides, are planted in any hot Soil, and seldom miss to prosper.

The slices are soak'd in Canary to take off the sharpness which commonly otherways inflames the Throat, and then they are eaten If the Wine in which they are soak'd be drank it inflames the Throat likewise, *Piso*. But I never found this which *Piso* speaks of.

It

It is a great Cordial to fainting spirits, and helps a squeamish Stomach. Its Juice and Wine is good for the suppression of Urin, and Fits of the Stone, as also against Poysons, especially *Cassada.* The same is done by the Root when the Fruit is wanting. The distill'd Liquor, by Fire from the Fruit, is yet more effectual if it be given in a small quantity, for if too much be given it is hurtful to the urinary passages. It is so injurious and corroding that it not only hurts the Tongue and Pallat, but leaves marks on the Knife that cuts it, especially if the Fruit be not ripe. And then 'tis as prejudicial as *Carawata* to Women with Child. It is hurtful to people in Fevers, to the wounded or ulcered, it is so hot as to be very injurious to them, therefore I wonder *Monardes* should reckon it cold and moist. *Christopher Acosta* saying better, that 'tis hot and moist, and begets inflammations if it be much used. *Piso.*

The juice is mix'd with water, and given to the sick as we give Mead. *Maregr.* It corrodes a Knife in a night, if it sticks in it. *Xim. Acost.*

Piso says that the old Inhabitants of *Brasile* told him that this was first carried thence to *Peru,* and the *East-Indies,* and *Linschotten,* that they came to the *East* from the *West-Indies.*

The juice takes spots out of Cloaths *Piso.*

It is cold and dry, it is given to those in Fevers to cool, and excite Appetite, though apt to turn to Choler. A slice held on the Tongue quenches thirst, and moistens the Tongue. *Hernand.*

Monardes was very much out when he describes this to have Seeds to be spit out when 'tis eaten.

The *Brasilians* use it in their sicknesses. *Thevet.*

It had its name from its likeness to a Pine-Apple, one was carried on its Plant to *Charles* V. but not lik'd. It is preserv'd in *New-Spain.* They are best on the Isles *Acosta.*

It is crowned to show its excellency, and that Crown planted gives a better Fruit than the Succors. It makes the Gums bleed. Its Wine is good, it spoils after three Weeks, but recovers again, both it and the Fruit cause Abortion. *Tertre.*

It is Cordial, and Stomachic, and is good in Gravel and Poisons. The distill'd water is good, but care must be had to Correct its Acrimony. *Roch.*

The Juice with Hony makes a drink in *Brasile.* *Morisot.*

The Leaves boil'd in Rice-water, mixt with *baleari* Powder, and drank Purges Hydropick Bellies. The unripe Fruit given with Vinegar expels the Child out of the *Uterus,* and eaten Corrects the swelling of the Belly from Wind. *Hort. Mal.*

This Plant went from the *West* to the *East-Indies,* where the Fruit is larger, and Leaves narrower. Large and ripe Fruit was in *Amsterdam* Garden for five years past. *Comm. ib.*

The Conserve of this Fruit does not preserve its natural taste. *Beym.*

CHAP.

The Natural History of Jamaica. 193

CHAP. X.

Of Herbs whose Flowers are compos'd of four Petala or Leaves.

I. Leucoium luteum, five Keiri minimum polygala facie. Cat. p. 79. Tab. 123. Fig. 1.

This has a great Root in respect of the Plant, being more than an Inch long, woody, of a white brown colour, and firmly fix'd in the ground. From hence spring two or three Stalks, two or three Inches long, having several Leaves exactly like those of *Polygala vulgaris*. The Flowers are at top of the Branches, yellow, and tetrapetalous, after which follows a small Pod.

It grows in the clay grounds in the Town *Savannas*.

This does not agree with the Title of *Polygala S. flos Ambervalis Viginiana floribus luteis in caput oblongum congestis Banisteri*. Pluk. Alm. p. 301. which Plant Dr. *Plukenet* thinks p. 153. *Mant*. may be the same with this.

II. *Sinapi foliis subrotundis, serratis, semine rufo*. Cat. p 79. Tab. 123. Fig. 2. Mustard.

This has a white large Root with many fibers. The Leaves at bottom round the top of the Root on the surface of the ground, are almost round at their end. They are six Inches long, narrow at their beginning, and growing broader towards their top, where they are round, and three Inches in diameter, very much serrated, and of a yellowish green colour. The Stalk is round, green, four or five Foot high, having some Leaves plac'd on it without any order, longer, smaller, and not so round as those at bottom. The Stalk has towards its top several Branches, beset with yellow tetrapetalous Flowers, having *Stamina* of the same colour, and in the Center a green *Stylus*. After which comes an Inch long Pod, swelled or protuberant on the outside over every Seed, round and containing two rows of small, round, smooth, reddish Seeds, with a thin membrane between.

It grows frequently in the cultivated places of this Island.

The Seeds, if prepared as our *European* Mustard-Seeds, make as good a Sawce.

III. *Nasturtium aquaticum vulgare*. Park. Cat. p 79.

This is very common in all not too rapid Rivers, chiefly near Springs, from whence they are brought down, and thrive in most Rivers of this Island.

This grows much larger than ordinary on the *Laguna* in the *Caymanes*, whence it is in great quantities carried to *Port-Royal* Market; but it seems to be no different Kind, but only a variety, and this variety in largeness, in deeper waters, is taken notice of by *Lobel* in his *Adversaria*.

It is very good against the Scurvy or Chronical Diseases, it is Diuretick, and very opening, being made use of for many days together, in Sallads or Broaths, especially the first.

The Seed breaks the Stone, and forces the *Catamenia*. *Dorst*.

It is not to be used by Women with Child, if the Child be not dead. *Lon*.

If Boiled in Milk or Wine, and so used, it cures the Scurvy. *Dod*.

It is very Diuretick even outwardly applied. *Cæsalp*.

D d d   *Lagunus*

Lacuna makes two Kinds of it differing in bigness, saying the last was hurtful, sometimes killing.

IV. *Sinapistrum Ægyptium heptaphyllum, flore carneo, majus spinosum. Herm. Cat. Plant. Jam. p. 80. Papaver corniculatum acre quinquefolium Americanum flore carneo majus spinosum. Pluken. Alm. p. 280.*

The Root of this Plant is deep, large, white, and firmly fix'd in the ground, by several smaller, going from the sides of the greater Root. The Stalk is very strong, round, hairy, green, rising to about four or five Foot high, spreading Branches on every side, having on their lower parts finger'd Leaves, standing on long Foot-Stalks, exactly resembling those of *Lupins*, or the *Pentaphyllum siliquosum, Prosp. Alp.* only each Finger is longer, narrower, hairy, and the whole Leaf, for the most part, seven pointed. At the Branches, and Leaves beginnings are usually two short, green, strong, straight Prickles. The Flowers come out on every side of the Branches tops. They are each made up of four long *Petala*, the first part of which is narrow, and towards the end broad, being shap'd like a Spoon, only not hollow, of a white colour, inclining to purple. In the middle of these *Petala* comes a great many long, purple *Stamina*. The Pods are small, round, and of a pale green colour, inclosing a great many very small, brown Seeds. The whole Plant is somewhat viscid, and has in every part of it a very grateful strong smell.

It grows every where in the Streets of the Town of St. *Jago de la Vega*.

V. *Sinapistrum Indicum pentaphyllum flore carneo minus, non spinosum. Herm. Cat. Plant. Jam p. 80. Papaver corniculatum acre quinquefolium Ægyptiacum minus flore carneo non spinosum. Plak. Alm. p. 280.*

This is commonly to be found near the Town of *Passage-Fort*, and other places as well of *Jamaica*, as the *Caribes*, and in *Egypt*.

Boil'd in Oil it remedies cutaneous Diseases, especially the Leprosie. The whole Plant beat with Juice of *Raca-palam*, and anointed cures Pustles. The Juice is snuff'd up to hinder Poyson from reaching the Head. The same does the Plant boil'd in water and drank. The Leaves provoke Appetite, are expectorating and comforting, dissipate Phlegm and Wind. The Root and all beat and applied in Balls under the Arms, cures or diminishes cold Ague Fits. The Juice with Oil helps Deafness dropt into the Ear. The Leaves beaten and applied to the Head cures its aching from cold. *H. M.*

VI. *Sinapistrum Indicum triphyllum flore carneo non spinosum. Cat. p. 80. Tab. 124. Fig. 1. Papaver corniculatum acre triphyllon Indicum floribus luteis viscosum Ramanissa Cochinensibus dictum. Plakenet. Alm. p. 280. An Aria-Veela. Hort. Malab. part. 9. p. 41. Tab. 23? Trifolium spicatum aliud Hernandez. p. 285? Vel Memoya de Tepozzlan Ejusd. p. 384?*

This has a four or five Inches long Root, small and white, with lateral Fibers, drawing its Nourishment. The Stalk is round, green, upright, two Foot long, without any Branches, but having Leaves thinly plac'd thereon, without any order, standing three always together on the same common Inch long Foot-Stalks. The uppermost of them is the largest, being an Inch and an half long, and half an Inch broad in the middle where broadest, and smooth. The top of the Stalk, is a Spike of tetrapetalous Flowers, each of the *Petala* being white, and standing round several long purple *Stamina*, inclosing a *Stylus* of the same length and colour, though of a different bigness, just like the others of this Kind, and after them follows a three Inches long Pod,

Pod, small, round, green, and ending in a point exactly like the Pods of the aforesaid Plants.

This grew on the Banks of the *Rio-Cobre* below the Town just by the water-side, on the other side of the River, and on a Gullies side near Colonel *Cope*'s Stables, in his Plantation by *Guanaboa.*

VII. *Iberis humilior annua Virginiana ramosior. Morif. Cat. p.* 80. *Tab.* 123. *Fig.* 3. *An Mexixquilitl seu nasturtium iberisve Indica. Nieremb. p.* 316? *Thlaspi Virginianum foliis iberidis amplioribus & serratis. Tournef. El. p.* 182. *Inst. p.* 213. Pepper-Grass.

This has a white Inch long single Root, with fibers from it, drawing its Nourishment. The first Leaves are spread round on the surface of the Earth, being about an Inch long, half of which is Foot-Stalk, and reddish, with three or four deep *Lacinie* or jags. At its beginning, towards the end of the Foot-Stalk, being Oval, half an Inch broad, indented about the edges, smooth, and of a dark green colour. The Stalk is round, whitish green, rising to a Foot and an half high, having longer, narrower, and deeplier jagged Leaves set thick about it without any order, the Branches also come out frequently on all hands, round which stand many Flowers, on the eighth part of an Inch long Foot-Stalks, being very small, white, and tetrapetalous, with green *Stamina* in the middle. After these follows a short Pod, round, with a notch at the further end, containing one oblong, reddish Seed in each Arch or Cell of the Pod.

It grows in all the *Caribe* Islands, and in this Island in most of the inland places.

The first Leaves being of a pleasing biting taft, supply the place of all the Cresses in Salleting.

VIII. *Eruca duodecima, sive maritima Italica siliqua hastæ Cuspidi simili.* C.B. *Cat. p.* 81. *Crambe maritima foliis eruca, capsula cuspidata. Tournef. El. p.* 180. *Crambe maritima foliis eruca angustioribus fructu hastiformi. Ejusd. Inst. p.* 212. *An eadem foliis latioribus. Ej. ib?*

I could not observe any difference between this Plant here, and that growing on the *Mediterranean,* if not in the colour of the Flower, which is white, but I look on that as only a variety.

It grew on House Cayes, a small Island off of *Port-Royal.*

Four Ounces of the distilled water, taken warm in the morning, helps Colick and Nephritick pains. *Lugd. Mycon.*

It Purges powerfully, but the Roots are useless. *Ang.*

IX. *Veronica fruticosa erecta dulcis, hexangulari caule, flore dilute cærulea. Cat. p.* 81. *Tab.* 108. *Fig.* 2. Another sort of Wild-Liquorice, or Sweet-Weed.

This has a strong Root, divided into smaller, two or three Inches long, crooked Roots, with several fibrils to draw its Nourishment. The Stalk is woody, covered with a gray Bark, having several Hexangular, green Branches, about a Foot and an half high, beset with Leaves towards their tops, three at a place, being without any Foot-Stalk, three quaters of an Inch long, and half an Inch broad near the end, where broadest, serrated about the edges, and of a grass green colour. *Ex alis foliorum* come the Flowers, standing on a quarter of an Inch long Foot-Stalks. They are tetrapetalous, whitish, with many *Stamina,* standing round on all hands within them, to which follows a roundish *Capsula,* or Head of two Leaves, light brown, membranaceous, no bigger than a great Pins head, containing very many, scarcely perceivable, small, brown Seeds, standing round a fungous substance

substance of the same colour. The Leaves of this Plant have a sweet tast like Liquorice, whence the name.

It grows near a Wood in the Town *Savanna*, towards two Mile Wood, by the River side going to the Ferry, and in several other places of this Island.

Three Spoonfuls of the expressed Juice of the Leaves of this Plant given Evening and Morning for three days, is counted an infallible Remedy for any Cough.

This, according to *Piso*, is very emollient.

X. *Veronica caule hexangulari, foliis saturcia ternis, serratis.* Cat. p. 81. Tab. 124. Fig. 2.

A great many white strings meet from every side, to make up one straight, oblong, woody Root, from whence arises an hexangular, woody, gray Stalk, spreading its self into green Branches about one Foot high. The Leaves come out three at a joint, at about half an Inches distance. They are long, serrated, small, and narrow, like those of Savoury, only of a pale green colour. This Plant has on an eighth part of an Inches long Foot-Stalk, a small whitish gray, tetrapetalous Flower, after which follows in a brown *Capsula*, some brown, angular Seeds. The Capsules stand round the Stalks *ex alis foliorum* on Foot-Stalks like to verticillated Flowers, each being made up of four membranes, they are not round as the former, but long, and pyramidal, and surrounded with four Leaves for its *Calyx*.

The Descriptions and Figures of this, and the foregoing, seem to differ, though Dr. *Plukenet*, p. 151. *Mant.* thinks them the same. Perhaps they may only vary.

It grows in the sandy *Savannas*.

XI. *Papaver spinosum.* C. B. Cat. p. 81. *Argemone Mexicana. Tournef. El.* p. 204. *Inst.* p. 239.

This Plant agrees exactly to the description given by Authors.

It grows every where about the Town of St. *Jago de la Vega*, on the road thither from Passage Fort, and in all the *Caribes* very abundantly.

The Leaves of this Plant boil'd promote sleeping.

A Thimble full of the Seeds are reckoned a very violent Purge.

The Seed powdered and taken to the quantity of two Drams, purges all Humours, especially Flegm from the Joints. The Milk, with a Womans Milk that bore a Female, dropt into the Eyes, Cures their Inflammations. It is good against intermitting Fevers. The Flower applied Cures the Scab. The Tast is bitter, and it is hot and dry. Its distill'd Water, with the tops of *Mizquitl* takes spots out of the Eyes, and eats Proud Flesh, takes away pains of the Head, and helps other such Diseases. *Hern.*

The Seed came from *England*, under the Title of *Figo del inferno. Banh.*

The prickly Head is long and round, somewhat like a Fig, and whosoever should have one stick in his Throat, quickly goes to Heaven or Hell, from thence 'tis called *Ficus Infernalis. Park.*

XII. *Chelidonium majus arboreum foliis, quercinis.* Cat. p. 82. Tab. 125.

This Shrub riseth to ten or twelve Foot high, having a straight Trunc, as big as ones Arm, covered with a white, smooth Bark, being branch'd towards the top, the Branches ends, having a great many Leaves set round them without any order. They are of the shape of Oak Leaves, have an Inch long Foot-Stalks, are seven Inches long, and three broad at the blunt top where broadest, being narrow at the beginning, and having on their sides some deep sinuations, one great middle, and several transverse Ribs, and

being

being of a yellowish green colour on the upper side, and whitish underneath. The tops of the Twigs, beyond the Leaves, are a Foot long, and branched out into very large bunches of many Flowers each standing on a short Foot-Stalk, and being made up of two green Leaves or Lobes, within which are many *Stamina* of a yellow colour, and a *Stylus* which grows roundish, big in the middle, and tapering to both extremes, and in its middle contains a small, brown oblong Seed.

All parts of this Plant yield on breaking a yellow juice, like that of Celandine.

It grows in a Gully near Mr. *Elletson*'s House in *Liguanee*, on the Road going to Collonel *Bourden*'s Plantation from *Guanaboa*, and several other places.

It is hot and dry in the fourth Degree, with some Adstriction. The Twigs bark'd take off spots and marks from the Eyes. The juice consumes Wind, cures Tetters as well as the Fruit, and eases pain from cold Causes. The Leaves cure old Sores, being applied to them. They take off Warts, especially those of the *Præputium* and *Pudenda*, which has been found by most certain Experiment. It is likewise called *Quanhchilli*, from being as sharp as *Indian* Pepper, and was planted by the *Indian* Kings in their Gardens, *Hæenandez*.

XLI. *Tithymalus dulcis parietariæ foliis hirsutis, floribus ad caulium nodos conglomeratis.* Cat. p. 82. *Tithymalus Americanus, humi fusus, serratus, floribus in capitulum alis adhærens, congestis.* Plum. Tournef. Inst. p. 88. *Tithymalus botryoides minor Americanus foliis hirsutis.* Pluken. Alm. p. 373.

This from an oblong rough Root, sends out several small, round, red Stalks hairy, and about a Foot long. The Leaves are set at the joints of the Stalks, they are sometimes red, and sometimes whitish green, almost like those of *Parietaria*. Out of the joints come likewise the Flowers, towards the under part of the Stalk having a Foot-Stalk, but towards the top none. They are very small, many being close set together in the same Head, or conglomerated, of a white or purplish colour, and after them follow tricoccous Seeds, as small as those of *Chamæsyce*.

It grows every where in *Jamaica*, and other Islands.

Its Vertues are thought many. Here 'tis very much commended in Claps as an *Antivenereal* Medicine, and by some it is given in the Belly-ach.

Piso says 'tis one of the best outward or inward Antidotes, that being fresh chaw'd, or beaten, applied to a Serpents biting, it not only takes away the pain, but draws out the Poyson, and cures Wounds, and likewise, that if it be dry and powdered, and given in a convenient Liquor, to the quantity of a Pugil, it corroborates the Heart, and restores the strength decayed by Poyson. He farther says, that scarce any who is prudent go's into the Woods in *Brasile* without either this Herb or its Juice, which drank in a good Draught cures the Poyson of Serpents.

This is the greatest Antidote against Serpents biting, being bruised, and applied to the Wound: if it have reach'd the Heart, a little of the Powder cures being inwardly taken. One drop kills a small Serpent. *Tertre*.

A Bath is made of this against Serpents Poyson. The Leaves with the Juice of the Bark *Lana*, levigated and anointed, Remedies Carbuncles and Phlegmons. *H. M.*

XLIV. *Tithymalus erectus, acris parietariæ foliis glabris, floribus ad caulium nodos conglomeratis.* Cat. p. 82. Tab. 126. *Tithymalus Americanus, erectus, serratus, floribus in capitulum longo pediculo insidens, congestis.* Plum. Tournef. Inst. p. 88.

Chamæsyce

An Chamæfyce Americana major floribus glomeratis cynocrambes folio glabro. Breyn. pr. 2. p. 31 ?

This has several reddish, stringy, crooked Roots, which united send out one strong red Stalk dividing its self into several green, round Branches, rising ten Inches high. The Leaves come out at the joints of the Branches, about half an Inch from one another, they are two and two, set one against the other on very short Foot-Stalks, are about an Inch long, and one third part of an Inch broad at round Base where broadest, and whence they decrease, ending in a point, being cut snipt or indented about the edges, and smooth. *Ex alis foliorum*, come for the most part Inch long Foot-Stalks, sustaining many white, or purple, tetrapetalous very small Flowers, set very close together one by the other, or conglomerated, after which follow as many small tricoccous Seeds, like the others of this Kind.

The whole Plant is milky.

It grows in laboured or cleared Grounds in the Plantations.

The Leaves, or any part of this Plant are poysonous, if eaten by Hogs.

The Milk of this Plant rubbed on Warts cures them.

XV. *Peplis fruticosa, maritima, geniculata Cat. p. 82. An sanamunda Africana. Park. p. 204?*

This small leav'd Sea Spurge, had a four Inches long, red, wooddy Root, from which rose several red Foot and an half high Stalks, straight, and jointed at every quarter of an Inch, having at the joints Branches, and on them several Leaves, standing side ways opposite to one another, at the joints likewise. They are very small, being not over one third part of an Inch long, and a quarter of an Inch broad, near the middle where broadest, smooth, of a very pale green, or glaucous colour, standing on very short Foot-Stalks. On the tops of the Twigs comes out a very small, tetrapetalous, pale, or whitish yellow Flower, after which follows a small tricoccous Seed, as in others of this Kind.

The whole Plant is very milky.

It grew on *Gun Cayes*, near *Port Royal*.

XVI. *Chamæsyce. C. B. Cat. p. 83. Tithymalus exiguus glaber nummulariæ folio, Tournefort. El. p. 74. Inst. p. 87. Tithymalus minimus ruber rotundifolius procumbens. Bob. Hist. Ox. part. 3. p. 342.*

I could not observe any difference between the *Chamæsyce* growing about *Montpelier*, and here in *Jamaica*, and therefore I reckon them the same Plant, and if there be any difference, 'tis, that in *Jamaica* it is larger than in *Europe*.

It grows in dry sandy places, about the Town of St. *Jago de la Vega*, very plentifully.

It takes off all Warts, they being rub'd with it. Boil'd with Victuals or Sallet, it loosens the Belly: the juice does the same. It is good for taking off spots from the Eyes, as well as Dimness and Clouds, being rub'd on them, mix'd with Honey. *Lon.*

It has all the vertues of the other *Tithymals. Dod.* Writing with its juice is not discovered but by Ashes. *Lac.*

The *Greeks* were not acurate sometimes in giving their names, as may appear by this, having no quality of the Fig-Tree only it has Milk, and so has Lettice. *Lac.*

This Plant grows in most parts of the World.

XVII. *Plantago Cæsalp. Cat. p. 83.*

'Tis

'Tis common on this side the Ferry going to *Liguanee* by Land, and in the North-side of this Island in several places.

'Tis reckoned one of the best Adstringents, therefore its Juice or Decoction, and all parts of it are thought to be very vulnerary, stopping all manner of Fluxes, either of Blood or Matter, and that taken several ways.

It is Cold and Dry, drying Wounds and Ulcers, abounding with too much moisture, taken as a boil'd Sallet, it cures the Dysentery. The juice heals Ulcers of the Ears, and stops the *Catamenia*, a Cloath dipt in this juice being made use of by way of Pessary. It stops bleeding of Wounds, being mix'd with white of Eggs. *Dorst.*

XVIII. *Aceris fructu herba anomala, flore tetrapetalo albo.* Cat. p. 83. Tab. 117. Fig. 1, 2.

This Plant has very small, brown, fibrous Roots, by which it creepeth along the Earths surface, sending up now and then a round, red, jointed, and juicy Stalk, about a Foot and an half high, having Leaves coming out at the joints alternatively, at an Inch and an half's distance, standing on an Inch long, round, and red Foot-Stalks. They are three Inches long, half as broad, rough, of a yellowish green colour, indented on the edges, being of an irregular Figure, having as it were a defect on one of the sides of the Base by the Foot-Stalks end, and the other side produced lower like an Ear, from whence it decreases, ending in a point. The Flower comes at top, is tetrapetalous and white, two of the *Petala* being large, and set opposite to one another, and two of them being much smaller, set in the same manner with yellow *Stamina*. After this follows one Seed, which is quadrangular, large, brown, oblong, inclosed in a two wing'd, triangular, extant membrane which surrounds it.

It grows in the woody and shady Paths going to sixteen Mile Walk by the *Rio Cobre*, in the shady Woods by *Hope* River, and other such places in several parts of the Island.

This in many things resembles the *Tsjerianarinampuli*. *H. M.* p. 9. Tab. 86. p. 167.

XIX. *Clematitis prima sive sylvestris latifolia.* C. B. *foliis ternis.* Cat. p. 84. Tab. 128. Fig. 1.

I did not observe any difference between this Plant here, and ours in *Europe*, but in every part found a perfect agreement, only the Leaves were not five as ours, but for the most part always three on the same Foot-Stalk, which may be the variety of Soil. C. B. says of his, that *Mire in foliorum divisuris ludit.*

It grows in the Woods going to *Guanaboa* by the Road side, and over the *Rio Cobre* near the *Angels.*

The Stalks are used for Withs.

The Root heated in Water, and mix'd with two Glasses of Wine, diluted with Sea-water, Purges Hydropical People.

The Juice and Flowers beaten, or boil'd, rub'd on the Skin, takes out its spots. It is very Hot. *Trag.*

Pliny tells us that these Sticks rub'd against one another they fire. Which is the way used by the *Indians* to make Fire at this day.

XX. *Solanum racemosum Americanum.* Raii. Cat. p. 84. *Phytolacca Americana.* Tournef. El. p. 146. *Phytolacca Americana fructu majori.* Ejusd. Inst. p. 299. *Blitum maximum caule rubente Virgin. seu Amaranthus Americanus baccifer.* Schuyl. p. 14. *Solanum Virginianum rubrum maximum racemosum bac-*

cis terulis canaliculatis. Bob. Hist. Ox. prt. 3. p. 522. An Cuechiliz tematl. Hernandez. p 374? The Great *Virginia* red Nightshade.

I could not observe any difference between *Parkinson's* Great red *Virginia Solanum*, and this growing here, and therefore will not give the description of it, it seeming to me to be the same.

It grows on the more mountainous parts of this Island, as in *Liguanee*, on the Mountains above Mr. *Elletson's* Plantation, on Mount *Diablo*, in going to the North side, and several the more cool places of this Island.

It is used by the *Indians* in *New-England* to dye their Skins and the Barks wherewith they make their Baskets. *English* People in *Virginia* call it Red Weed. *Virginia* Nightshade is a familiar Purger in *Virginia* and *New-England*. A spoonful or two of the juice of the Root works strongly. The dried Root has not, upon trial, been found to have that effect. *Park*.

Tsjeru-Caniram. H. M. Is of this Kind.

XXI. *Solanum racemosum Americanum minus. Cat. p. 85. An Heliotropium Caraſſavicum Scammonii foliis mollibus subhirsutis. Herm. par. Bat. prod ? Amaranthus baccifer Circea foliis. Hort. Amst. part. 1. p. 127.*

This has a small, oblong, fibrous Root, which sends up one green, round Stalk, two Foot high, having many Branches. The Leaves stand on the Stalks without any order, are almost like those of *Circea Lutetiana*, or the foregoing, only lesser. The Flowers come in a Spike on the tops of the Branches, which have some large hairs or soft prickles, they have very short Foot-Stalks, are many, white, and tetrapetalous, after which follow some very small, at first green, then red Berries.

It grows every where in the Woods of this Island, and in the *Caribes*.

De Tertre tells us of a small *Solanum*, and a Plant like *Circea*, which, with the juice of its Root, Cures the Tooth-ach. Perhaps that like *Circea* was this.

XXII. *Ghandiroba vel Nhandi-roba Brasilienſium. Marcgr. Cat. p. 85.*

This grew very plentifully on a Lime-Hedge near Mrs. *Guy's* House in her Plantation in *Guanaboa*, as also in Mountain River Woods, and in several other places. The Fruit, or inward Nuts, are carried down fresh Rivers into the Sea, and thence are thrown upon the Banks by the Waves again very frequently.

The Inhabitants of *Brasile* make an Oil of the Kernel of this Fruit, which they use in their Lamps, being very clear, good, and withal slow in consuming. It cannot be us'd for Victuals being bitter, as is the whole Fruit. *Marcgr.*

This Oil is good against Aches from Cold, it being Hot. I remember to have seen whole Families of those of *Brasile* Sick with the Night Air, restor'd with this Oil. *Piso.*

XXIII. *Clematis baccifera, glabra & villosa, rotundo & umbilicato folio. Plumier. Cat. p. 85. An Capela. Fif. Worm. muf. p. 158 ? An Volubilis seu baccifera Virginiana. osseulo compresso lunato, caudice lento, foliis hederaceis nasturtii Indici more umbilicatis. Banist. Pluken. Alm. p. 393 ?* Velvet-Leaf.

This has a round, whitish, wooddy Stalk, with which it turns about, and catches hold of any Tree it comes near, and rises to its top, and thence falls down again, putting forth all along some small Branches, having several Leaves, like a Heart, or almost round, of about an Inch in Diameter, the edges being undivided. They are of a white yellowish colour, very thick set with a whitish down, or soft hair, feeling to the touch as Velvet, whence its name. The Flowers come out among the Leaves are hanging on a two Inches long Strings, as those of Nettles, each Flower being very small,

small, tetrapetalous, of a pale green colour, with one black spot in the middle. I could never find any Seed, though several times I examined the Plant.

It grows every where in the Town *Savanna*, amongst the Shrubs.

The Leaf either applied whole, or bruised, to a wounded place, cures it very effectually.

It is a good Remedy against poisoned bites, the Leaves being bruised and applied. The Root is excellent against the Stone. *Marcgr. Pis.*

XXIV. *Lysimachia lutea non papposa erecta major, foliis hirsutis, fructu caryophylloide. Cat. p. 85. Tab. 127. Fig. 3.*

The Stalk is usually brown, strong, four or five Foot high, and has several hairy, red, angular Branches, on every hand very thick beset with long, narrow, hairy nervous Leaves, several of which come out together, some larger, some smaller. The larger are three Inches long, and scarce one broad, in the middle where broadest, of a light green colour, downy and soft like Velvet. The Flowers come out *ex alis foliorum*, are large, yellow, standing on an half Inch long Foot-Stalk, pentapetalous, very open, or spread with yellow *Stamina*, having under them four or five green, small Leaves, standing on an oblong, large, four or five square *Capsula*, or Seed-Vessel set on to the Stalk by an eight part of an Inch long Foot-Stalk, and containing in several Cells very much small, yellowish Seed, when full ripe.

The Descriptions and Figures of this and the *Lysimachia corniculata maritima sinuatis, & pubescentibus foliis Virginiana, Plukenet. Alm. p. 235. & Phyt. Tab. 203. Fig. 3.* shew these Plants to be very different one from the other. This I take notice of, because Dr. *Plukenet* in his *Mantissa, p. 123.* doubts if it be not the same.

It grows every where on the wet or low Banks of the *Rio Cobre*, and in other wet places of the Island.

XXV. *Lysimachia lutea non papposa erecta minor, flore luteo pentapetalo, fructu caryophylloide. Cat. p. 85. Tab. 128. Fig. 2, 3.*

This rises about ten Inches high, from small fibers it sends out from its joints, into the neighbouring mud, thereby drawing its Nourishment. Its Stalk is green, round, succulent, smooth and brittle, on which are plac'd by a quarter of an Inch long red Foot-Stalks, several Inch long Leaves, half as broad in the middle where broadest, smooth and shining. *Ex alis foliorum* come the Flowers. They are pentapetalous, very large, yellow, and making a fine show. The Seed I did not observe, though by its Stalks and Flower it should be of kin to, if not a Purslane.

It grows in Marshy places near Black River Bridge, &c. where water stands shallow most part of the year, among the Mud, into which it strikes its fibrous Roots.

XXVI. *Cuscuta inter majorem & minorem media, filamentis longis & fortibus latissime super arbores vel campos se extendens. Cat p. 85. Tab. 12. Fig. 4.*

This has very long, and strong filaments, by which it stretches its self over very large Trees, and whole Fields and Pastures, rising no higher than the Plants on which it feeds. The Flowers are white, and conglomerated on one side of the Stalk, as the others of this Kind are. The Filaments are larger than those of the *Cuscuta minor sive epithymum. C. B. Pin.* and lesser than those of the *Cuscuta major. C. B. Pin.*

F f f

The Natural History of Jamaica.

This is not the *Cuscuta Lendiginosa, &c. Plukenet. Tab.* 26. *Fig.* 6. as the Doctor conjectures, *Mantiff.* p. 61. as may be seen above p. 191. his *Cuscuta Lendiginosa,* being the inward part of the *Viscum cariophylloides, &c.* there described, the little knots in it being at the places of the Stalks where the Branches were placed.

It grows on the Palisadoes, Trees, Shrubs, and Fields near the *Rio Cobre,* about and below the Town of St. *Jago de la Vega* very plentifully.

CHAP. XI.

Of vasculiferous Herbs with pentapetalous Flowers.

I. *Cistus urticæ folio, flore luteo, vasculis trigonis. Cat p.*86. *Tab.* 127. *Fig.* 4. 5.

This Shrub has a Stem as big as ones Thumb, covered with a reddish brown, smooth Bark, rising three or four Foot high, and having many Branches towards the top, spread thinly on all hands. They are beset with Leaves and small Sprigs. The Leaves are some greater, some smaller, those largest stand on an half Inch long Foot-Stalks, being two Inches and an half long, and one Inch broad in the middle where broadest, notch'd about the edges, hoary or woolly, soft, having one middle Rib, and several transverse ones. They are of a yellowish dark green colour, and they have a scent like *Cistus*; and somewhat resemble the Leaves of *Cannabis spuria.* On the tops of the Twigs and Branches come the Flowers standing in a pentaphyllous *Calyx.* They are pentapetalous, yellow, or Orange coloured, very large, and like the Flowers of *Cistus,* after which follows a small Head, made up of three strong cartilagineous membranes, in which are lodged the Seeds, which are large.

It grew between *Guanaboa* and the Town on the Red Hills very plentifully, in the Road going to the *Angels* near the Gate, and in several other places of this Island.

II. *Chamæcistus urticæ folio, flore luteo. Cat* p. 87. *Tab.* 127. *Fig.* 6.

This has a small, straight, white, short Root, having some small, white fibers on each side, from whence springs one round, rough Stalk, two Inches high, having Branches opposite to one another, and Leaves which are three quarters of an Inch long, and one third part of an Inch broad, very deep cut in on the edges, and rough, like those of *Cannabis spuria.* The Flower stands at top, it is very woolly or hairy, surrounded by three or four Leaves. It is of a deep yellow colour in the inner part, pentapetalous, large, and making a fine show. After this follows a small Head, made up of three cartilagineous Membranes, inclosing some large Seeds, of a white colour, very pleasantly striated, and like those of Mallows.

It grows in the Town *Savanna* after Rain.

III. *Chamæcistus caule hirsuto, folio oblongo, angusto, sinuato, flore luteo, pediculo insidente. Cat.* p. 87. *Tab.* 127. *Fig.* 7.

This has a wooddy, oblong Root, with several lateral fibrils, which sends up sometimes a single, and sometimes two or three wooddy, round Stalks, about three or four Inches high, being covered with a reddish coloured hair, or small Prickles, which although they look very fierce, yet I never observed

to

to prick. The Leaves come out *alternatim*, being about an Inch long, and one fifth part of an Inch broad in the middle where broadest, and whence they decrease to both extreams, being somewhat hairy, of a dirty brown colour, and sinuated about the edges. *Ex alis foliorum*, and at the tops come the Flowers each standing on an half Inch long Foot-Stalk, being yellow, and after this follows a Seed-Vessel, or Head made after the same manner, and containing Seeds as the former.

It grew in the *Savanna* about the Town, with the former.

IV. *Anagallis cærulea, portulaca aquatica caule & foliis. Cat. p. 87. Tab.* 129. *Fig.* 1.

This small repent Plant, has round, smooth, green, juicy Stalks, which at joints strikes into the Earth, small, white, hairy fibers, whereby it draws its Nourishment, and likewise small, green, succulent Leaves, almost like those of the *Portulaca aquatica*, being roundish, thick, green, smooth, and very small, without Foot-Stalks, standing opposite to one another. Towards the ends of its small Twigs, *ex alis foliorum*, come out half an Inch long Foot-Stalks, and on them in a *Calyx*, consisting of two green Leaves, a pentapetalous Flower, of a pale blue colour, having some whitish *Stamina* within. After this follows a great number of very small, flat, brown Seeds, inclosed in a hard, brown *Capsula*, covered by some first green, afterwards brown Leaves, which were the *Perianthium* or *Calyx* of the Flower.

It grew on the Sea Marshes near the landing place at *Delacrees* in *Liguanee*.

V. *Alsine Americana nummulariæ foliis. Herm. parad. Bat. prod. p.* 306. *par. Bat. p.* 11. *Cat. p.* 87.

This sort of Chickweed had several very small, white Strings for its Root, from whence, on every hand, went several round, small Stalks, a Foot long, lying on the surface of the Earth. They were green, and upon stretching, like other *Alsine's*, the Bark breaks, the inward part of the Stalk holding. At more than an Inches intervals distance come the Leaves. They are two, one opposite to another, standing on short Foot-Stalks, are almost quite round, having Nerves, like Plantain, running from the Foot-Stalks and through the Leaf, each of which is of about one third part of an Inch's Diameter. The Flowers come out *ex alis foliorum*, several together standing on an Inch long Foot-Stalk. They are made up of many white, long *Petala*, as the other *Alsine's*, to which follows several small Seeds, like those of other Chickweeds.

It grew on the low, moist Banks of *Rio Cobre*, below the Town of *St. Jago de la Vega*, and several such moist shady places, in and towards sixteen Mile Walk, very plentifully.

VI. *Alsini affinis foliis bellidis minoris, caule nudo. Cat. p.* 87. *Tab.* 129. *Fig.* 2.

This had a crooked, white Root, with many Hairs to draw its Nourishment, about an Inch long. The Leaves lay on the surface of the Earth, spread round the Root, being about an Inch long from a narrow beginning, increasing by degrees to its round end, and very like the Leaves of the lesser Dasie, only not so thick nor succulent, and of a darker green colour. From the middle of these Leaves rises a four or five Inches long Stalk or two, without any Leaves, being branched towards the top, which Branches are divided into small Twigs, each whereof sustains a small Head, inclosed by four whitish capsular Leaves, there being within them a round, whitish, thin Seed-Vessel, full of roundish black Seeds, very small. I never found it in Flower. It

It grew on the fandy places of the Town *Savanna*, towards two Mile Wood.

VII. *Portulaca latifolia feu sativa.* C.B. pin. Cat. p. 87.

This grows every where in cultivated Grounds, without being Sown, and agrees in every thing with its Defcription in all Authors.

This name *Porculaca* is thought to come from its feeding Swine.

This with Fern and *Basilicum*, are common to *Europe* and *Brasile*. *Lery*.

It has fometimes a notch in the Flowers, or they are *bifid*.

The whole Plants are boiled, dried in the Sun, and ufed for Winter-Difhes. *Matth*.

The Stalks are pickled after drying with Fennel, Verjuice, and Salt. *Lugd*.

This is the only Herb common to *Maragnan* and *France*. *Abbev*.

It is Cold in the third, and Moift in the fecond Degree. It is eaten in Sallads, yields little cold, moift, and grofs Nourifhment, killing Worms, thickning and cooling the Blood, therefore ftopping all its Fluxes. It is good for the Bladder, extinguifhing Venery. It takes away the pain of the Teeth fet on edge, and thefe things are done likewife by the Seed. It helps the Inflammations of the *Uterus & anus*. *Ger*.

This beaten with *lotura oryzæ* remedies Carbuncles. *H. M.*

Applied outwardly it cures the *Eryfipelas* Inflammations, and procures fleep. *Lon*.

VIII. *Portulaca angustifolia five sylvestris.* C. B. pin. Cat. p. 87. *Pourpier de Bouton.* p. 48. *Taicombelahe de Flacourt*, p. 126.

It grows in more barren fandy, and fometimes wet Grounds, and therefore I doubt whether it may not be a variety of the former.

Mr. *James Lancafter ap. Hakl.* p. 119. p. 2. relates that being left on the Ifland *Mona*, near *Espanola* in the *Weft-Indies*, in great want, the beft relief they found was in the Stalks of Purfelain boiled in water, and now and then a Pompion found in the Garden of an old *Indian*, and that for twenty nine days they lived fo.

IX. *Portulaca aizoides maritima procumbens, flore purpureo. Cat.* p. 88. *An Portulaca Caraffævica angusto longo lucidoque folio procumbens floribus rubris. Pluken. Alm.* p. 303 ? *Comm. hort. Amst.* p. 9 ? Sampier.

This has many thick, juicy, round, red, frequently jointed Stalks, lying on the furface of the ground on every hand. The Branches and Leaves come at the joints, the latter being an Inch and an half long, like the *Chryfanthemum aizoides*, triangular, very juicy, faltifh in taft, like Sampier, not unpleafant, and very fucculent. *Ex alis foliorum* come the Flowers, which are pentapetalous. The *Petala* are thick, of a purple colour varying fometimes to white, ftanding Star-fafhion, with purple *Stamina* in the middle, and looking very pretty. After thefe follow many fmall, roundifh, comprefs'd Seeds, having a little defect in the middle, being of a fhining black colour, like the grains of Gunpowder, or other the Seeds of *Portulaca*, inclofed in a hexaphyllous Cup, with a clay colour'd cover to it, breaking horizontally, as other the *Portulaca*.

It varies very much in bignefs in feveral places.

It grows on the falt marfhy Grounds near Paffage Fort, old Harbour, on the *Cayos* off of *Port Royal*, and other fuch places very plentifully.

'Tis pickled, and eat as *Englifh* Sampier.

The Leaves are preferv'd with Vinegar and Brine, as Sampier, *Soldanella, &c.* and are eaten as a Pickle. *Pif.*

X. *Portulaca aizoides maritima erecta. Cat. p.* 88.

This has several Branches spread round, reddish, and as thick as ones little Finger, cornered, and set with Twigs, at a third part of an Inches distance round the Stalk, mounting upright a Foot high. The Leaves are almost triangular, two Inches long, and one twelfth part of an Inch broad, pyramidal, very-green, tasting like others of this Kind. I did not observe its Flower or Seeds, but believe them like the former.

It grew on the sandy Shore near Passage Fort.

XI. *Portulaca erecta sedi minoris facie, capitulo tomentoso. Cat. p.*88. *Tab.*125. *Fig.* 3.

This has small, white, stringy Roots, which send up several round, smooth, reddish Stalks, four Inches high, having many small, round Leaves, like those of the *Sedum minus foliis teretibus,* about half an Inch long, and having at their tops seven or eight Leaves, encompassing several Flowers in a downy Wooll, after which follow, in a *Capsula,* like that of Purslan, several small, black Seeds, like those of that Plant.

It grows in the dry *Savannas* after Rain.

This differs very much from the *Portulaca Curassavica lanuginosa, &c. Plukenet. Phyt. Tab.* 105. *Fig.* 4. *Alm. p.* 303. though the Doctor thinks them the same in his *Mantissa, p.* 154.

In the Figure the Leaves on the Stalk are not taken notice of.

XII. *Portulaca facie maritima fruticosa erecta americana. Cat. p.* 87. *An portulaca Americana erecta floribus albis. Parad. Bat. Commel. hort. Amst. part.* 1. *p.* 7 ? *An portulaca marina latifolia, flore suave rubente. Plum. Tournef. Inst. p.* 236 ?

This in every thing was like Purslan, only larger, and growing more upright. It had a long, deep, round Root, and several lateral fibrils went into the sandy, loose Earth, thence drawing its Nourishment. The Stalk was reddish, as thick as ones little Finger, succulent, rising to at least two Foot high, having Branches thick set with Leaves, without any order, about an Inch and an half long, and three quarters of an Inch broad near the end where it was broadest, and round, being at the beginning narrow, and from no Foot-Stalk increasing to that breadth. It is juicy, has one middle Rib, and a very unsavoury bitterish tast. What was its Flower or Seed I did not observe, and so cannot be positive that it belongs to this place.

It grew on Gun *Cayos* near *Port Royal.*

XIII. *Portulaca affinis folio subrotundo succulento, flore pentapetalo dilute purpureo. Cat. p.* 89. *Portulaca Curassavica procumbens Capparidis folio, flore muscoso, capsula bifurcata. Herm. par. Bat. p.* 213. Horse-Purslane.

This has trailing, juicy, round, smooth, green Branches, three Foot long, lying on the surface of the Ground, round the top of the Root, being jointed, swell'd at each joint, and branch'd every Inch and half, there being two little Branches set against one another, on which stand the Leaves opposite to one another, on an Inch long Foot-Stalks, being roundish, of about an Inch Diameter, and not altogether so succulent as those of Purslane. The Flowers are pentapetalous, of a light purple colour resembling those of Mallows. The Seeds are something flat, and round, being black, and contained in a green, oblong *Capsula.*

It grows by the Kings House, and in several places in the Streets of the Town of St. *Jago de la Vega,* after Rains.

This differs extreamly from *Patulaca similis planta India orientalis, &c. Plukenet,*

Plukenet. Phyt. Tab. 206. *Fig.* 3. though the Doctor in his *Mantiss. p.* 155. thinks they may be the same.

XIV. *Linum scandens flore dilute purpureo semine triangulari. Cat. p* 89. *Tab.* 130. *Fig.* 1. *An Linum sarmentosum, seu volubile Jamaicanum, flore cæruleo. Pluken. Alm. p.* 224 ?

This has a round, hoary, small Stalk, turning and winding its self about any Plant it comes near, and rising three or four Foot high, having Leaves at every Inches distance, standing on an eighth part of an Inch long Foot-Stalks, being three quarters of an Inch long, and half an Inch broad near the Base where broadest, and whence they decrease, ending in a point, a little hairy, and of a dark green colour. *Ex alis foliorum* come out several Flowers, standing on very short, crooked Foot-Stalks. They stand Star fashion, are pentapetalous, and of a pale purple colour, after which follows a small, brown, roundish Head, like that of Flax, divided into several Cells, and containing Seeds almost triangular, being round on one side, flat on the others, and of a light brown colour.

There is another sort of this, or rather variety, being much larger in Flowers, &c. They being of a white colour.

It grows, though rarely, in the *Savannas*, amongst the prickly Pears; but is frequent on the Red Hills in the way to *Guanaboa.*

XV. *Apocynum erectum fruticosum, flore luteo maximo & speciosissimo. Cat. p.* 89. *Tab.* 130. *Fig.* 2. *Savanna* Flower.

This rises three or four Foot high by a wooddy, branch'd Stalk, covered near the Root with a brown, smooth Bark, like that on Birch-Twigs. The Twigs at top have green Bark, and several Leaves, two always standing opposed one to another, on one third part of an Inch long Foot-Stalks. They are Oval, yellowish green in colour, smooth, two Inches long, and three quarters of an Inch broad. *Ex alis foliorum* towards the tops of the Branches come out several long, wooddy Foot-Stalks, supporting each a yellow monopetalous Flower, whose margin has five deep notches in it. After these follow two Pods set like Bulls Horns, very slender in respect of the others of this Kind, and long, containing many brown Seeds with much Down, resembling in every thing the other Species of this Kind. The whole Plant is very Milky.

It grows in the *Savannas* every where, and is in Flower most part of the year, making a very pleasant sight.

XVI. *Apocynum erectum folio oblongo, flore umbellato, petalis coccineis reflexis. Cat. p.* 89. *Tab.* 129. *Fig.* 4, 5. *Apocynum Curassavicum fibrosa radice floribus aurantiis Chamænerii foliis angustioribus. Herm. par. Bat. prod p.* 213. *par. Bat. p.* 36. *An Apocynum petræum ramosum, salicis folio venoso, siliqua medio tumente Virginianum Pluk. Almag. p.* 26 ? *Phyt. Tab.* 261. *Fig.* 3 ? *Apocynum Hyficanense erectum, salicis foliis angustioribus vel latioribus, Hort. Beaumont ?*

This has strong and deep Roots several Inches long, and by the many fibers, they have, draw copious Nourishment to the Plant, which has several Stalks streight, three or four Foot high, jointed at every Inch, four square, the Leaves coming out at the joints opposite to one another. They are long, and of a dark green colour. At the top is a Foot-Stalk, sustaining many Flowers umbel-fashion'd, every one being pentapetalous. The *Petala* are turn'd down reflected, or bow'd back. The *Stylus* standing up in the middle of them is yellow, the *Petala* are of a very fine scarlet colour. The Pods follow, which are three Inches long, roundish, as big as ones Finger in the middle, and tapering to both extreams. The membrane is first green, but turns

brown,

The Natural History of Jamaica. 207

brown, and contains, when opened, a great many flat, round, thin, brown Seeds, with much Silk Cotton sticking to their ends, in which they lie, fastened to a middle, white, soft membrane, dividing each Pod. All the parts of this Plant are very milky.

It grows very plentifully in the Road between Passage-Fort and the Town, as also in most Ditches and Pastures about the Town.

It is very troublesome to Planters, fowling their Pastures, and spreading its self therein much against their will, and to their damage by means of its winged Seed.

The Figure of this in Dr. *Herman*'s *Paradisus Batavus* is not very good.

XVII. *Apocynum fruticosum scandens, genistæ Hispanicæ facie, floribus lacteis odoratis.* Cat. p. 89. Tab. 131. Fig. 1.

This has near its Root a Stem as big as ones little Finger, covered with an ash-coloured Bark, having some superficial black *Sulci* in it. The Stalk is divided into very many round, green Branches, like to *Spanish* Broom, which turning round Trees, or leaning on any thing they come near, rise to sometimes thirty Foot high. At every two or three Inches distance towards the top, it has two small two Inches long Twigs set opposite one to the other at a little joint, each of which has Leaves set likewise opposite one to the other, standing on an eighth part of an Inch long Foot-Stalks, being Oval an Inch long, half as broad, a little hairy or like velvet, and of a Grass green colour. Six or eight Flowers stand at top umbell-fashion together, their Inch long Foot-Stalks coming from the ends of the Twigs, as from a common Center. They are each of them pentapetalous, the edges of the *Petala* being hairy, the *Stamina* are five, standing up in the middle, whose round ends are reflected inwards, all of a milk white colour, smelling sweet, and looking very pleasantly. The Seeds are many, brown and flat, lying in a silk down like other the *Apocynums*, all of them being lodged in an Inch long pyramidal Sheath, Pod, or *Capsula*, which opens on one side, letting the Seeds fly out: two of these Pods being usually set opposite like Horns, as the others of this Kind.

The whole Plant is very milky.

It grows on some Palisadoes about the Town of St. *Jago de la Vega*, and on the Trees of the sandy Banks of the *Rio Cobre* below the Town, on the same side of the River.

It is very plain, by this Description and Figure, that this cannot be the *Apocynum scandens angusto rorismarini folio, &c. Plukenet. Alm. p. 37. Phyt. Tab. 261. Fig. 2.* The Doctor, p. 17. of his *Mantissa* doubts if it may not be the same.

XVIII. *Apocynum scandens majus folio subrotundo.* Cat. p. 89. Tab. 131. Fig. 2. An *Apocynum scandens Malabaricum fruticosum floribus nerii Cariophyllos redolentibus. Herm. par. Bat. p. 62?*

This has a round, dark brown, and deep Root, sending out one or more round and green Stalks, which wind and turn themselves about any Plant or Shrub they come near, rising to six Foot high, and at about each three or four Inches distance are joints, out of which come Leaves, and sometimes Branches set one against another. The Leaves have a quarter of an Inch long Foot-Stalk, are almost round, two Inches Diameter, a little curled, of a very fresh green colour and smooth. Between them and the Stalk, *ex alis foliorum*, comes a half Inch long Foot-Stalk, sustaining a great many pentapetalous Flowers, each Flower standing in a contorted, greenish *Calyx*, which is long, its *ora* being divided into five parts. There is between each of the *Petala* or Sections, a small distance, they standing sparse from

one

one another, and each of them beginning very narrow, are at their ends round and broad; a little indented about the edges. After these follow large taper Pods standing like Horns, agreeing in every thing with those of the other *Apocynum*'s.

It grows in the edges of the Woods in the Town *Savannas*, and in several other places of this Island, as well as the *Caribes*.

XIX. *Apocynum minus scandens, flore albo, singulis petalis, viridi stria notatis.* Cat. p. 89.

The Stalk of this is near the Root, round, as big as a Hens-Quill, of an ash colour, turning and winding its self round any Tree or Shrub it comes near, and rising four or five Foot high; towards the top on its Branches are many Leaves set in Tufts on opposite sides of the Stalk. They have a quarter of an Inch long Foot-Stalks, are an Inch long, and one quarter of an Inch broad, towards an Oval shape though pointed, of a deep green colour, and smooth. The Flowers come out *ex alis foliorum*, being several, taking original from the same Center, *viz.* an eighth part of an Inch long Foot-Stalk. They are pentapetalous, long, small, and white, with a green streak in the middle of each of the *Petala*, and yellowish *Stamina* in the middle. After which follows in a pyramidal Husk or Seed-Vessel made up of one membrane, Seeds, lying in a silken Down both Seeds and Seed-Vessels, being exactly like those of the other *Apocynums* only smaller.

All parts of this Plant are very milky.

It grew in the Town *Savanna* among the Ebonies.

XX. *Apocyno affine, Gelseminum Indicum hederaceum herbaceum tetraphyllum, folio subrotundo acuminato.* Cat. p. 90. *Pseudo apocynum semine compresso & alato secundum, claviculatum tetraphyllum Barbadiense, flore luteo tuberoso inodoro brevioribus foliis.* Bob. hist. Ox. p. 3. p. 612. *Bignonia Americana, capreolis aduncis ornata siliqua longissima.* Tournefort. Inst. p. 164.

This had a wooddy Stalk, about the bigness of a Hens-Quill, covered with a smooth Bark of an ash colour, climbing any Trees or Hedges it came near, at about three or four Inches distance putting forth Leaves standing opposite to one another on an half Inch long Foot-Stalks. The Leaves were shining, roundish, about an Inch and an half long, and near as broad near the middle where broadest, smooth, and sometimes, though rarely, indented about the edges. At the ends of the Foot-Stalks came out small *capreoli*, by which it took hold of any thing it came near.

It grew in the Island of *Barbados* very plentifully every where.

XXI. *Nummularia minima flore albo.* Cat. p. 90. Tab. 131. Fig. 3.

This little Plant has several creeping Stalks, and many trailing Branches, with which it covers large spots of Ground. The Branches are round, brown, small, and take hold by fibrous Roots of the surface of the Earth. The Branches ends are set very thick with Leaves alternatively, they are round, and like those of *Nummularia minor flore purpurascente*, C. B. only smaller. The Flowers come out *ex alis foliorum*, stand on a short Foot-Stalk, are whitish, pentapetalous, with some *Stamina* in the middle, in a green oblong *Calyx*.

It grows very copiously in the *Savanna* near the Town of St. *Jago de la Vega*, in those places where Rain has stood after a wet Season.

XXII. *Nummularia saxatilis minima repens, foliis crenatis villosis, floribus albis.* Cat. p.90. Tab. 131. Fig. 4. *An Vitis Idaea palustris Americana foliis parvis circinatis, elegantissime crenatis ex insula Jaimaicana.* Plukenet. Alm.p. 392 ?

This

This small repent Plant has very small Hairs for Roots, which it strikes out at its joints. The Stalks are small, round, hairy, jointed at every half Inch, from whence come out on an eighth part of an Inch long Foot-Stalks, round, small Leaves of a pale green colour, rough, of about a quarter of an Inch Diameter, snipt about the edges. Towards, and at the tops come the Flowers in a tuft together, being very small and white, but whether they belong properly to this place or no, I know not.

It grows on the sides of Rocks, which it covers, among the Mountains near *Hope* River in *Lignanes*, by Mr. *Elletson*'s Plantation.

This is extreamly different from the *Nummularia major rigidioribus & rarius crenatis foliis, &c. Plukenet. Alm.* p. 254. though the Doctor in his *Mantiss.* p. 136. thinks they may be the same.

XXIII. *Tribulus terrestris major, flore maximo odorato. Cat.* p. 90. *Tab.* 132. *Fig.* 1.

From a pretty straight and deep Root spring a great many Foot and an half long trailing Branches, spread every way on the surface of the Ground, from the top of the Root, as from a Center. The Stalks are round, succulent, brittle and thick, from whence go several Branches set with winged Leaves. The *Pinna* are generally six, or three pair, of a dark green colour, the furthermost pair being largest. The Flowers come out towards the ends of the Branches, are of a pale Orange, or yellow colour, pentapetalous, very large, and smelling sweet. After which follows a small prickly Head, with a long process something like to the *Geranium* Seeds, only these are set with very strong prickles, though not very sharp on the largest side, or that part of the Seed nearest the Stalk.

It grows in the Streets of the Town of St *Jago de la Vega*, and in rocky or gravelly Grounds in most Plantations in the island.

A Salve is made of this Herb with Suet, good for the Ringworm, a frequent Distemper in this place.

The *European* Kind is adstringent, and good for all Inflammations. *J. B.*

This seems to differ much from the *Tribulus terrestris major Curassavicus. Herm.* in not being so large, nor having so many Wings in its Leaves.

XXIV. *Urtica folio anomala, flore pentapetalo purpureo, fructu pentacocco muricato. Cat.* p. 90. *Tab.* 132. *Fig.* 2.

From a redish, round, deep, oblong Root, come several round, green, rough Branches, about six Inches high, along which come out several small Leaves oval, snipt or deeply cut in on the edges, smooth, and standing on a small Foot-Stalk. Between them and the Stalk comes out a small, pentapetalous, purplish Flower, standing on a very small, reddish Foot-Stalk, and having one large *Stylus*, which in some time grows to be red, large, and afterwards rough and brown, it is pentacoccous, or divided into five *Cellulæ*, containing each a blackish Seed, and all are pendulous, or inclining towards the Ground.

It grows among the Grass in the Town *Savanna*.

XXV. *Gratiola affinis frutescens Americana, foliis agerati seu veronicæ erectæ majoris. Breyn. prod.* 2. p. 54. *Cat.* p. 90. *Capraria Curassavica & Cabritta vulgo. Herm. par. Bat.* p. 110. *An Tijeru parva. Hort. Mal. part.* 10. p. 105? *Tab.* 53? *West-India* Thea.

Several small two inches long brown Roots united, send up a Stem three or four Foot high, wooddy, covered with a smooth clay coloured Bark, and having several Branches, which are very thick set towards their tops with Leaves round their Stalks, without any order. Each Leaf is an Inch long, and not over an eighth part of an Inch over at top where broadest, having no Foot-

H h h   Stalk,

Stalk, but beginning very narrow, and augmenting to that breadth: Succulent, of a deep green colour, smooth, having notches or incisures into the Leaves like those of *Senecio*, or great Daisie. *Ex alis foliorum* comes the Flowers standing on a short Foot-Stalk, being small, white, and the margin very deeply divided into five parts. After which follows a great many very small brown Seeds, standing in an oblong, cylindrical, four cornered, brown *Capsula*, divided into two parts. Of this there is a variety, being smaller.

It grows in the *Savanna*, and about the Houses of the Town of St. *Jago de la Vega*, very plentifully.

Why some give it the name of Thea, I cannot imagine.

It grows in *Greece* wild, and is call'd *Cusferment* by the *Turks*. *Wheeler apud D. Plukenet*. It is call'd *Cabrita* by those of *Curasao*, because Goats feed on it. *id.*

CHAP. XII.

Of Herbs which are of the Kindred of Umbelliferous Plants.

I. *Valerianella folio subrotundo, flore purpureo, semine oblongo, striato, aspero. Cat. p.* 91. Hogweed.

The Root is single, very strong and deep, sending forth many Branches, lying along the surface of the Earth, on every side for a Foot or two in Diameter. The Stalks are red, round, succulent, like those of Purslane, jointed. At every joint are two Branches, and Leaves opposite one to another, set on one third part of an Inch long red Foot-Stalks. They are almost round, three quarters of an Inch Diameter, succulent, green with purple, and now and then curled edges, and smooth, on the tops of the Branches are a great many purple or scarlet Flowers, set close by one another in the same Head, and after them succeed so many brown, oblong, striated, and very rough Seeds.

It grows in gravelly Low Lands about the Town, and in most gravelly Soils in all the Plantations of the Island.

Hogs feed on this Herb with much delight: whence 'tis gather'd, and brought to them to their Styes to fatten them.

II. *Valerianella alsines folio scandens, floribus pallide luteis pyxidatis in Umbella modum dispositis, semine aspero Cat. p.* 91. *An Valerianella Curassavica femine aspero viscoso? Herm. par. Bat. prod. p.* 382. *par. Bat. Plukenet. Phyt. Tab.* 133. *Fig.* 7? *An Pseudovaleriana Curassavica semine aspero viscoso. Bob. Hist. Ox. part.* 3. *p.* 105?

The lower part of the main Stem of this Plant is as big as ones Arm, having a furrowed white Bark. The Stalk takes hold of any Palisadoe or Tree it comes near, and branches at the top, it rising seven or eight Foot high. The Branches are many, round, red or green, and brittle, hanging downwards. The Leaves come out at the joints, and are in every thing like those of our greater sort of Chickweed. The tops of the Twigs send out several Raies or Foot-Stalks, as from their common Center, like the *Umbellifera*, sustaining each one small, greenish yellow Flower, like a small Cup, being round, undivided, and almost like *Muscus pyxidatus* in shape. After which comes a small, long, brown Seed, almost like those of some *Umbellifera*, growing larger from the beginning to the top, and being a little rough.

It

It grows every where about Palisadoes, and among Shrubs in the *Savanna's*.

I think every body will agree with me, that this is rather a *Valerianella* than a *Solanum bacciferum*, notwithstanding Dr. *Plukenet*'s contrary Opinion in his *Mant. p.* 173.

III. *Dentellaria Lychnioides sylvatica scandens flore albo. Cat. p.* 91. *Tab.* 133. *Fig.* 1. *Plumbago Americana viticulis longioribus sempervirentibus ex Vera Cruce. Hort. Reg. Hampton. Bobart. part.* 3. *Hist. Ox. p.* 599. *An plumbago (forte) Americana ex conjectura D. Sherard Plukn. Phyt. Tab.* 312. *Fig.* 1 ? *An Chilmeatl seu Tcha. Hern. p.* 140.

The Stalks of this Plant are round, jointed, shining, green, wooddy, crooked, rising three or four Foot high, when its weak Branches are supported by Shrubs, among which it usually grows. The Leaves come out on the Branches without any Foot-Stalks, being thin, smooth, for the most part roundish, and about two Inches over, though sometimes oblong, like the Leaves of the common *Dentellaria*. The tops of the Branches are set with Flowers Spike-fashion, without any Foot-Stalks, but join'd to the Stalk by a rough, or almost prickly, viscous, striated, green *Calyx*, in which is plac'd a white pentapetalous Flower, like the Flowers of the *Lychnis sylv. flore albo*, after which follows one large four-square, brown Seed, in a rough, viscid *Capsula*.

It grows in the Woods, or among Shrubs every where, and very plentifully on each side of the Road, between *Passage* Fort, and *St. Jago de la Vega*.

IV. *Admirabilis Peruana rubro flore. Cluf. rar. pl. hist. Cat. p.* 91. *Jalapa flore purpureo. Tournefort. Inst. p.* 129. The four a Clock Flower.

It is frequent every where in the Woods and Plantations that are often watered with Rain, and usually opens its Flower about four a Clock, whence the name. It has in this Island, for the most part a scarlet or purple Flower.

At *Barbados* I was told by a person knowing in these matters, that its Root was *Mechoacan* or *Jalap*; but this I suppose came from its purging water, which it does successfully, and cures the Dropsie, as *Piso* relates.

The Flowers yield a tincture for painting Women withal. *Piso*.

Cortusus hath found out that two Drams of the Root, doth very notably purge waterish humours. *Cluf.*

The Roots are moist and cold, wherefore they are eaten, and outwardly applied to cool. Some say the Root of that Kind, with variously colour'd Flowers to two Drams, purges water. *Hern.*

The red Flower'd ones Seed sown brings always red Flowers. *Cluf.*

Plumier and *Lignon*, told Mr. *Tournefort* that the *Jalap* of the Shops was not different from his *Jalapa Officinarum Fructu rugoso*.

V. *Agrimonia lappacea inodora, folio subrotundo dentato. Cat. p.* 92.

This rises to four or five Foot high, being divided into several smaller Branches towards the top, which are beset with several Leaves without any order, standing on half an Inch long rough Foot-Stalks. They are almost round, though a little pointed, with two *Lacinia* or sinuations, being an Inch long and three quarters of an Inch broad at Base, where broadest, indented round the edges, woolly, of a deep green colour above, and paler below, having some eminent Ribs going from the Center of the Foot-Stalk through the Leaf. The Flowers stand on the tops of the Branches in a

Spike,

Spike, are yellow, the *Petals* being long and like those of our *European* Agrimony, only narrower. After these follow on a crooked Foot-Stalk, several brown round Burs, thick set with hooked prickles, sticking to any thing like the Seeds of Agrimony.

It grows about the Town of *St. Jago de la Vega*, and in several other places of this Island.

In my Catalogue I plac'd this amongst the Plants of the Kindred of the *Umbelliferæ*, as most of the best Botanists have done before me, so that Dr. *Plukenet* might have sav'd himself the trouble of his long Paragraph upon this occasion in *p.* 112. and 113. of his *Mantissa*.

VI. *Cotyledon aquatica. Cat. p.* 93. *Ranunculo affinis umbelliferis accedens in palustribus folio peltato repens Americana & nostras. Pluken. Alm.* p. 314. *Valerianella cognata folio cotyledonis. Herm. par. Bat. Cat.* p. 13. *Hydrocotyle Vulgaris. Tournef. Inst. p.* 328.

Out of every half Inch of a round, small Root, creeping under the surface of the Earth, at the joints, are a great many very small hairy, blackish fibers, by which the Plant is nourished, and from the same places are sent up the Leaves and Flowers Foot-Stalks, which are sometimes one, sometimes more, round, greenish, and four Inches long. The Leaves are round, thick, sinuated on the edges, smooth, of an Inch Diameter, very green, their Foot-Stalk entring in their very Center, from whence some Nerves are sent to the Circumference. The Flowers stand close together round their Foot-Stalks end. They are many, all near join'd to one another on almost no Foot-Stalks, and are of a greenish colour. In a short time appear the Seeds, being broad, of the shape of Parsnep-Seeds, striated, and standing on Foot-Stalks, taking their beginning from the great Foot-Stalks end, like the *Umbelliferæ*.

I can observe no difference between this and *Cotyledon aquatica Joannis Bauh.*

It grows in several places along the moist Banks of the *Rio Cobre*, in most Marshes or wet Grounds in the *Caymanes*, and other places of this Island.

This Plant is sharp to the tast, and has been taken by some of the Planters in place of Scurvygrass, by using its distill'd water as Spirit of Scurvygrass.

It has subtle and hot parts, pleasant and Aromatick to the tast. Its chief Vertues are in the Roots, which are opening. They open Obstructions of Liver and Reins, and help a hot Constitution, so that no Remedy is more proper. The juice of the green Leaves is esteem'd by the Inhabitants a famous Antidote, and they procure Vomit with it, as we do with the juice of *Asarabacca* Leaves. *Piso.*

It is us'd to take away the marks called *Os Figados* by the *Partuguese*, which I suppose are Liver marks. *Marogr.*

Gerard says that it was called Sheeps Bane by the Husbandmen, for that it kill'd the Sheep feeding on it, and therefore is angry with Apothecaries for using it for Wall-Penniwort alledging it would be more pernicious to Men than Sheep.

Lobel and *Dod.* tell us that they used it for *Cotyledon* in the *Ung. Populn.* both for ignorance and want of the true *Cotyledon.*

It is sharp and exulcerating, therefore not to be used, or called *Cotyledon,* but *Ranunculus. Col.*

It is called *White-Rot,* because it kills Sheep feeding on it. *Park.*

CHAP.

CHAP. XIII.

Of Plants that are rough leav'd, called Asperifoliæ.

I. **H**Eliotropium arboreum, maritimum, tomentosum, Gnaphalii Americani foliis. Cat. p. 93.

This rose to a Man's heighth, had a ſtraight Stem or Trunc, wooddy, firm, and ſolid, as big as ones thumb, with a pretty large Pith, a Bark all covered over with Down or *Tomentum,* ſmooth and white. Towards the top it had many Branches going out on every Hand, which are very thick ſet with a great number of Leaves round them, being each of them three Inches long, and not over an eighth part of an Inch broad near the round top, where broadeſt, being at the beginning narrow, and increaſing to that place. They are ſucculent, thick, and covered over with very much white Down, looking ſomething like the Leaves of the *Gnaphalium Americanum.* The tops are branched out into ſeveral Spikes of white Flowers, contorted like a Scorpions Tail, or the *Heliotropes,* to which in every thing they are like.

It grew on a ſandy Bay to the Eaſtward of *Bridgetown* in *Barbados,* and on the Sea ſide between *Paſſage Fort* and *Old Harbour* in *Jamaica.*

II. *Heliotropium maritimum minus, folio glauco, flore albo.* Cat. p. 94. Tab. 132. Fig. 3. *Heliotropium Americanum minus glabrum folio angulo glauco* Breyn. prod. 2. p. 55. *Heliotropium Americanum procumbens facie ilini umbilicati.* Herm. par. Bat. p. 183. *Heliotropium monoſpermum Indicum procumbens glaucophyllon floribus albis.* Plukenet. Phyt. Tab. 36. Fig. 3. Almag. p. 182. *An Tateey Xochiub. Hernandez. p. 432?* Wild Sampier.

This has ſeveral three or four Inches long Roots, white, and when united ſending up three or four Branches of the ſame length. The Stalks of it are round, green, juicy, and white, on the ſurface. The Leaves are ſet very thick in Tufts, ſome ſmaller, and others bigger, being an Inch long, and a quarter of an Inch broad in the middle where broadeſt, juicy, pale green, covered over with a white Meal, glaucous, being like the Leaves of *Echium maritimum,* P. B. or *Cerinthe.* Towards the tops of the Branches ſtand the Flowers. They are many on the upperſide of the Stalk, white, and turn'd like a Scorpions Tail, or thoſe of the other *Heliotropes.*

It grows in Salt marſhy Grounds near the Sea-ſide, by the *Canoes,* old Harbour and *Paſſage Fort.*

When I printed my Catalogue of *Jamaica* Plants, I did not think Doctor *Plukenet* had any where figured this Plant, neither ſhould I now, (his figure being not very exact) did he not ſay ſo in his *Mantiſſa, p.* 100. where he blames me for taking the ſynonimous names I have in my Catalogue from his *Phytographia,* whereas he has none there but his own name, and one out of *Breynius's* ſecond *Prodromus,* which was not then taken notice of by me, and therefore is not mentioned.

III. *Heliotropium Americanum cæruleum, foliis hormini. Dodart.* Cat. p. 94. Wild Clary.

The Stalk of this Plant is large, green, round, wooddy, crooked, and rises to a Foot high. The Branches are small, and hairy, sustaining Leaves just like those of Clary, whence the name. The Flowers are many, pentapetalous, of a pale blue colour, set in a double row on the upperside of the Branches, and turn'd like a Scorpions Tail, like the other *Heliotropes*. After which follow several cornered, brown Seeds.

It grows about the Town of St. *Jago de la Vega* very plentifully in *Jamaica*, and in the *Caribe* Islands on dry Grounds.

It is cleansing, and having a consolidating quality, is good against Wounds and Ulcers; it is likewise good against most hot Cutaneous Diseases. *Piso*.

If this be the *Bena-patsa. H. M.* the Authors of that Book say that boiled in Coconut-Oil, it cures the Disease called *Pitao* by dying, and is given in the poysonous bites of the great Fox *Jakhulsen*.

IV. *Heliotropium Caraſſavicum hormini foliis anguſtioribus. Hort. Beaumont.* Cat. p. 94.

This was larger than the precedent, having Leaves on its Stalk, usually taking their original opposite to one another, being narrow at the beginning, and pointed at their ends, not so rough as the precedent, nor blunt as it, but more glabrous or smooth. The Spikes sometimes come opposite to the Leaves, and are slenderer than those of the precedent.

It grew in one of the *Caribes*, where I found it.

V. *Heliotropium minus, Lithoſpermi foliis.* Cat. p. 95. Tab. 132. Fig. 4.

This Plant resembles very much the *Lithoſpermum arvenſe radice rubra.* C. B. only 'tis smaller. It has a small straight root, dark brown colour'd, from whence springs up one Stalk, three or four Inches high, divided into so many Branches, which are bowed or reflex'd like a Scorpions Tail, or the other *Heliotropes*. The Leaves come out alternatively, and are like those of *Gromwil*, only smaller. The Flowers are white, pentapetalous, and plac'd like those of the other *Heliotropes*. After which follow the Seeds, four being always together, each of which is almost triangular, of a dark brown colour, and irregular form.

It grows very plentifully every where in the Town *Savanna's* after Rain.

CHAP.

Chap. XIV.

Of Herbs commonly accounted to have many naked Seeds.

FOR the Tribe of Mallows 'tis very large, and to be divided here into three distinct Divisions. The first that of Mallows, where the Seeds have sticking to them their outward membrane very close. The other, or second, where this membrane is something laxer, or a Follicle, which are properly call'd *Abutils*. The third contains the *Alcea*, where these Follicles are not disjoined as in the others, but contiguous to one another. All these come near to the Multi-siliquose Tribe.

Mallows, according to some, are properly roundish leav'd, *Alcea* those whose Leaves are cut, and *Althea* those whose Leaves are hoary.

I. *Malva arborea, folio rotundo, cortice in funes ductili, flore miniato maximo liliaceo.* Cat. p. 95. Tab. 134. Fig. 1, 2, 3. The Mahot or Mangrove-Tree.

The Roots of this Tree are many, round, white, and long, entering the surface of the Earth, and not running very deep. Several Trunks as big as ones Thigh, rise up to about fifteen or twenty Foot high. Their outward Bark is very white, and almost smooth, the inward is yellowish when fresh, and red when dry, and then very tough, and serves to make Ropes. The Leaves stand, at the ends of the Branches, on four Inches long Foot-Stalks, being cordated, or almost round, of about five Inches Diameter, a little indented on the edges, soft, smooth, of very dark green colour, and having the Veins apparent, running from the Foot-Stalks end, as from a common Center, through the several parts of the Leaf. The Flowers come out of the ends of the Twigs, standing in a pentaphyllous, green, hoary *Capsula*, they are pentapetalous, each of the *Petals* being two inches long, and of a red colour, inclosing a red Pestle or *Stylus* of the same length, on which are many *Stamina*, the whole Flower looking like a red Lilly. After these follow brown Seeds, of the shape, and plac'd like those of Mallows, only much larger. Tab. 134. Fig. 1. shews the Leaf, Fig. 2. the Flower, and Fig. 3. a piece of the inward Bark, whereof are made the Ropes.

It grew in Colonel *Bourdin*'s Plantation, as well as in many of the inland moist parts of this Island.

'Tis chiefly useful by its Bark, which is peel'd off, and made into Ropes of all sorts, for the use of the Island.

The Bark is taken off the Branches with Knives, then beat with a Mallet 'till the first gross one be separated from the second: of the more gross is made Cords of the other Britches, for the *Negroes* and Slaves. *Casche*.

II. *Malva arborea maritima, folio subrotundo minore acuminato subtus candido, cortice in funes ductili, flore luteo.* Cat. p. 93. Tab. 134. Fig. 4. *Mahault de Bouton.* p. 68. *An Ampoufantchi de Flacourt.* p. 144? *Malva arborea Indica, abutili foliis argentea, sub externo cortice tenediophoros, & summis ramis radicosa.* Plnk. Mant. p. 75. *An Arbor Americana Taeniophoros, vittas & taeniolas plurimas transversicas seinvicem incumbentes, longitudinaliter sub externo cortice ferens.* Ejusd. Alm. p. 41. *?Mahot* of Ogilby. Amer. p. 348. & 377.

Maho-Tree, or Shrub of *Dampier, cap.* 3. The Sea Mahot or Mangrove-Tree.

'Tis the same in every respect with the former, only the Leaves are smaller, of a whiter green colour, and a little pointed. The underside being very white. The Flowers are of the same shape but yellow, and the Seeds are the same.

It grows by the Sea-side near Captain *Draxe*'s Plantation, in the North side of the Island in St. *Anns*, very plentifully.

The Bark is of the same use with the former.

The Ropes made of the Bark of this Tree, are used to tye the Human Bodies they in *Brasile* design to kill for Sacrifices. *Lery Linschot*.

The Leaves and Flower feed great Lizards. Its Bark is used for Tobacco in rolling it; for Matches for the *Spaniards*, and for sowing the Reeds together for the Houses to cover them. *Tertre*.

In our return to the North Sea, we cut and made Piperies or Floats of four or five of these Trunes, being light and floating; after barking they being tied together with wild Withs instead of Cords, two or three, or more of them, according to the bigness of the Pipery. *Raveneau de Lussan*.

Why Doctor *Plukenet* should call this *summis ramis radicosa* I know not.

III. *Malva arborea, folio oblongo acuminato, veluto, dentato & leviter sinuato, flore ex rubro flavescente. Cat. p. 95. Tab. 135. Fig.* 1.

This rises to about twelve foot high, having a woody Trunc, and several Branches, whose Twigs are hairy, and have some few Leaves set alternatively, at about an Inches distance towards their ends, each being placed on an Inch and an half long Foot-Stalks, they are about four Inches long, and near as broad near the round Base where broadest, and whence they decrease, ending in a point, being very soft, woolly, and having some slight sinuations, as well as indentures about the edges, and several Ribs running through the Leaf, taking their Original from the Foot-Stalks end, as from a common Center, being much more pointed than either of the foregoing. *Ex alis foliorum* towards the top comes out the Flowers, standing on large hirsute five Inches long Foot-Stalks, in a *Calyx* made up of many *Foliola* in two *Series*'s, the outward *Series* being the narrowest. Within this is a large Flower, of a yellowish Carnation colour, like in every thing to the others of this Kind. After which follows large brown Seeds, placed like those of Mallows, to which they are very like in every thing, only larger.

It grew on the Red Hills over against Mr. *Batchelor*'s House very plentifully.

IV. *Malva arborea, folio oblongo, acuminato, glabro, dentato. Cat. p. 56. Tab. 136. Fig.* 1.

This had several wooddy Branches, with a smooth ash-coloured Bark, white Wood, and large Pith. The Leaves came out on the ends of the Twigs, being set on two or three Inches long Foot-Stalks, longer than the precedent, and not so broad; serrated, but not sinuated about the edges, being smooth, and not hairy or hirsute as that, but in every thing else very like it. The Flowers come at top *ex alis foliorum*. They are lesser, as is also the Seed, both standing in a smaller *Calyx*, and on a slenderer, as well as shorter Foot-Stalk.

It grew in *Jamaica*.

The

The Natural History of Jamaica. 217

V. *Malva aspera major aquatica, ex hortensium seu rosearum genere, flore minore luteo, semine aculeato.* Cat. p. 96. Tab. 137. Fig. 1.

This Plant rises to about four Foot high, having one upright, strong, round, green Stalk, with a very rough, or almost prickly hair on it, with Leaves and Branches coming out every Inch and half without any order. They are like those of the Holyoks or *Lappa minor*, of a fresh green colour, smooth, almost round, being four Inches long, three and an half broad, angular, having two or three points, sinuated, and indented about the edges, and standing on large Inch long Foot-Stalks. *Ex alis foliorum*, on the same rough one quarter of an Inch long hairy Foot-Stalk, stand several Flowers, having each one green, ear'd, rough Leaf under it. The Flower is made up of five large yellow *Petala*, with yellow *Stamina* in the middle. After the Flowers follow five or seven triangular, whitish brown colour'd Seeds, being pointed on the under part, and round in the upper, plac'd like those of the Mallows, inclos'd in a small *Calyx*, and having that ear'd Leaf, formerly described with the Flowers, now turn'd brown, for their cover.

It grows in watry places of the *Savanna's*, and by the River sides in *Jamaica*, and the *Caribes*.

If this be that described by *Piso* he us'd the Leaves of it in Clysters, as the European *Althea*, and found no difference.

Any body who peruses the Description of this Plant, and looks on its Figure, and compares them with the Figure of *Althea abutili foliis, fructu hispido Americana Pluken. Alm.* p. 25. *Phyt. Tab.* 132. *Fig.* 5. will not (with Dr. *Plukenet.* p. 10. of his *Mantissa*) think they may be the same.

VI. *Malva minor supina betonica folio, flore coccineo, seminibus asperis,* Cat. p. 96. *Tab.* 137. *Fig.* 2.

This has a long whitish Root, very deeply fixed in the Ground, drawing Nourishment to the Plant by many fibres. The Stalks are round, tough, wooddy, whitish, and spread on the surface of the Earth, round the top of the Root, as from the same Center, for a Foot in Diameter. The Leaves come out along the Stalks here and there, though in greater abundance at the Stalks ends, several being there together. They stand on one third part of an Inch long Foot-Stalks, are half an Inch long, and one quarter broad, serrated about the edges, and of a dark green colour. The Flowers come out of the ends of the Stalks three or four together, they are pentapetalous, and of a purplish red colour, and after these follow several Seeds, set round a common Center, like those of this Kind, included in some hairy capsular Leaves, each Seed being almost triangular, and having its outside made rough with several small whitish prickles.

It grows in the dry places of the Town *Savannas*, and Flowers after Rains.

The Description and Figure of this Plant shew it to be very different from the *Alcea pusilla supina geranii exigui maritimi folio, &c. Pluken. Alm.* p. 14. *Phyt. Tab.* 132. *Fig.* 4.

VII. *Malva minor erecta betonica folio, flore luteo, semine duplici rostro seu aculeo predito.* Cat. p. 96. *Malva Indica foliis subrotundis. Tournef. Inst.* p. 96.

The Root of this is stronger, and every way larger than the former, the Branches more woody and erect, of a darker brown colour, rising a Foot high, the Leaves are broader at bottom, deeplier serrated, and for the most part purple about the edges, something like Betony Leaves, the Flowers are of an Orange colour, but in every thing else it agrees with the former,

K k k   having

having two prickles on the ends of the outermoſt Coat of the Seeds, which lies like the others of this Kind in a pentaphyllous *Calyx.*

It grows with the precedent, and in the *Caribes.*

The whole Plant boiled in water is Diuretick. The Root powdered, and taken with Pepper takes away the cold Fever, and from it is made an Oil. *H. M.*

VIII. *Malva erecta minor, carpini folio, flore luteo, ſeminibus ſingulis ſimplici aculeo longiori donatis. Cat. p. 96. An Alcea Carpini folio Americana frutescens, floſculis luteis, ſemine duplici roſtro donato. Commelin. hort. Amſt. p. 3? Malva ulmifolia ſemine roſtrato Tournef. El. p. 81. Inſt. p. 96.*

The Roots of this are the ſame with the former, as are the Stalks. The Leaves have a quarter of an Inch long Foot-Stalks, are deeplier ſerrated, larger, of a freſher green colour, and ſhap'd like the Leaves of the Horn-Beam-Tree, or Gooſeberries. The Flowers are yellow, and like the others, only larger. The Seeds are more in number, and each has a ſingle long prickle going out of one of its ends.

It grows with the two foregoing Kinds, and in the *Caribes.*

IX. *Althæa flore luteo. Cat. p. 96. Tab.* 136. *Fig.* 2.

This grew to about three or four Foot high, the Stalks being round, rough, hoary, having ſeveral Branches beſet with Leaves cordated, or of the ſhape of a Heart, ſerrated, ſomething like the Leaves of our Marſh-mallows, but rounder, without ſinuations. They ſtood on an Inch long Foot-Stalk, were very ſoft like Sattin, of a yellowiſh green colour. Towards the top *ex alis foliorum* came the Flowers ſtanding in a green ſat-tin'd pentaphyllous *Capſula,* being pentapetalous, and Orange colour'd, as the other *Althea,* to which this is very like in Face, Stature, and other particulars.

It grew on the ſandy Sea Banks, near *Old Harbour* very plentifully.

This differs extreamly from the following Plant notwithſtanding Doctor *Plukenet* makes them the ſame in his *Mantiſſa, p. 9.*

X. *Malva Americana, abutili folio, flore luteo, ſpicato, foliis hirſutioribus & craſſioribus. Herm. par. Bat. prod. Cat. Jam. p. 97.*

I found this by the Road ſides going to *Guanaboa* in *Jamaica. Breynius* had it from *Curaçao.*

XI. *Althæa ſpicata, betonica folio villoſiſſimo. Cat. p. 97. Tab.* 138. *Fig.* 1.

This riſes to about three Foot high with a woody Stalk, cornered, very rough, covered with a dark brown Bark, on which, towards the top, was much Wooll or Down. On this, at about an Inches diſtance from one another, come the Leaves alternatively, on oppoſite ſides of the Stalks ſtanding on one third part of an Inch long very hairy Foot-Stalks, each of them being about an Inch and an half long, and about an Inch broad at round Baſe, where broadeſt, and whence they grew narrower to their round ends, being ſerrated about the edges, and covered all over with a long yellowiſh hair, making the Leaf look of a dirty green colour. Towards the top, *ex alis foliorum,* come out the Flowers which are plac'd like theſe of the *Planta Verticillata* round the Stalk, or rather moſt on one ſide, each of them being incloſed at bottom by ſmall Leaves, rough, and very hairy, making a *Calyx.*

I found it on Mount *Diablo* in the middle of the Iſland.

XII. *Althæa spicata betonicæ folio, flore luteo, habitiori spicâ. Cat. p. 97. Tab.* 138. *Fig.* 4.

This has a three or four Inches long deep Root, sending up a pale green Stalk two or three Foot high, branch'd out into several Branches, and Twigs which are beset at three quarters of an Inches distance, with Leaves almost round, not so long as the precedent, standing on three quarters of an Inch long Foot-Stalk. They are an Inch and a quarter long, and three quarters of an Inch broad at roundish Base, where broadest, indented about the edges, of a pale green colour, and smooth. The tops of the Twigs and Branches for an Inches length, are thick set spike fashion, with Orange colour'd Flowers standing in five leaved *Calyces*, which are very hirsute. After them follow a great many brown Seeds, in situation, shape, &c. like those of the other Mallows.

'Tis very common in clear'd, barren, rocky Lands.

XIII. *Abutilon arboreum spicatum, betonicæ folio incano, flore minore purpureo. Cat.* p. 97. *Tab.* 138. *Fig.* 2, 3. *An Althæa Jamaicensis oblongo mucronato glauco folio profunde venoso margine undulato crispo. Pluken. Alm.* p. 25? *Tab.* 259. *Fig.* 6? *Vel Ricinus (forte) althea folio Jamaicensis glauco profunde venoso margine undulato crispo. Ejusd. ib. Alm.* p. 321? *vel betonica arborescens Maderaspatana villosis foliis profunde venosis. Ejusd. ib Tab.* 150? *Fig.* 5, & 6? *Vel Althea Jamaicensis, arborea, foliis oblongis mucronatis glabris floribus amœne rubellis, Ejusd. Tab.* 259. *Fig.* 3? *Malva vera crucis frutescens incana hirsuta flore parvo carneo purpureo Herm. par. Bat. pr.* p. 350.

This Tree riseth to about ten Foot high, by a Trunc as big as ones Leg, covered with an almost smooth, reddish Bark. It has several Branches towards the top, equally spread on every hand, whose Twigs have at their ends several Leaves standing on an eighth part of an Inch long Foot-Stalks. They are three quarters of an Inch long, one third part of an Inch broad, near the Base, where round and broadest, snipt about the edges and white, or of a very pale green colour, and pointed. *Ex alis foliorum*, on the tops of the Branches, stand several pentapetalous purple coloured Flowers set *racematim* in Branches or Spikes, about yellow *Stamina*, in a few green Capsular Leaves. After which follow Seeds in Heads just like those of the *Alcea arborea*, each Seed being lodged in a distinct Cell, five of these making up the Seed-Vessel, each of the five being pointed, and having two *Alæ extantes*.

It grows in several wooddy places about the Town.

There is a variety of this, which is larger in its Leaves, and not so white above as below.

XIV. *Abutilon fruticosum, foliis subrotundis serratis, floribus albis pentapetalis ad alas foliorum conglomeratis. Cat.* p. 97. *Tab.* 135. *Fig.* 2. *An Malva Americana, ulmifolia, floribus conglobatis ad foliorum alas. Plum. Tournef. Inst.* p. 96?

This Shrub riseth to about five Foot high, by a round Stem, covered with a brown smooth Bark, it has several long Branches, having Leaves going out at uncertain intervals from the opposite sides alternatively. They stand on one third part of an Inch long Foot-Stalks, are an Inch and an half long, and half as broad, are round at the beginning, and broadest very near the Foot-Stalks end, and from thence grow less to the end. They are thin, of a very green colour, and have Ribs conspicuous through the whole Leaf, smooth, and cut about the edges like Nettle Leaves. *Ex alis foliorum* come the Flowers, being several, standing in a knot almost round the Stem, like those

those of verticillated Plants, without Foot-Stalks. They stand in a reddish *Calyx*, are pentapetalous, and of a whitish colour. After these follow several roundish Heads, each being made up of five Follicles or Cells, standing close by one another, in each of which is lodged a small irregularly figur'd triangular black Seed.

It grows round the Town amongst the *Savanna* Bushes, and in the *Caribes*.

This seems to be quite different from the *Malva orientalis elatior, &c.* Plukenet. *Phyt. Tab.* 44. *Fig.* 5. though the Doctor, *p.* 10. of his *Mantissa* thinks it may be the same.

XV. *Abutilon herbaceum procumbens, betonicæ folio, flore purpureo.* Cat. *p.* 97. *Tab.* 139. *Fig.* 1.

This has wooddy, tough, round, red Stalks, two Foot long, spread, and lying on all sides, on the surface of the Ground, having many Branches, set with very few Leaves, on one quarter of an Inch long Foot-Stalks. They are an Inch long, one quarter of an Inch broad, at round Base, where broadest, and whence they diminish till they end in a point, being smooth, a little indented about the edges, having several Veins on the backside, and somewhat resembling Betony Leaves. The Flowers stand on the ends of the Branches, they are purple with yellow *Stamina* in the middle, like those of other Mallows. After each of these follows a five pointed pyramidal *Capsula*, in which are five Cells, in each of which lies a cornered, small, brown Seed.

It grew in gravelly low Ground near Mr. *Batchelor*'s House.

The Description and Figure of this Plant shew it to be very different from the *Alcea pusilla supina geranii exigui maritimi folio, &c.* Plukenet *Alm. p.* 14. *Phyt. Tab.* 132. *Fig.* 4. though in *p.* 6. of the Doctor's *Mantissa* he thinks it may be this.

XVI. *Abutilo affinis arbor altheæ folio, cujus fructus est styli apex auctus, quatuor vel quinque siliquis hirsutis, funis ad instar in spiram convolutis, constans.* Cat. *p.* 97.

This Tree is about fifteen Foot high, has a Trunc as large as ones Leg, a smooth white Bark, and is leaning or inclining towards the Ground. Its Leaves are two Inches and an half long, and one Inch and three quarters over from a round broad Base, ending in a point snipt, or serrated about the edges, having several pretty high Ribs on its under side, being soft, of a yellowish green colour, downy, and like *Altheæ* Leaves. At the tops of the Branches come the Flowers, standing in a rough green *Calyx*, they are white, standing like those of the *Digitales*, only smaller, out of the Center of which comes a long *Stylus* or String, having a roundish hirsute Button at the end, which augments and becomes its Fruit, and consists of four or five round, small, brown *Siliqua*, Ropes, or rather long Follicles, hairy, dark brown colour'd, very hard wreath'd, or roll'd spirally one by another, and containing within them great plenty of round, brown Seed, which falls out of the end of each of these Pods, Follicles, or Ropes, which open themselves for that purpose.

It grows near Mr. *Batchelor*'s Plantation, on the Red Hills every where on the Road to *Guanaboa*, and other places of this Island.

The Leaves are us'd in Decoctions for Clysters with Oil and Salt, as those of Mallows.

It seems rather to belong to this Tribe, than to the *Siliquose* one, because the Follicles of *Abutilon*, differ only in shape from these.

It has in the juice of the Root great vertues in the *Empyema*, and Stomach Diseases. The Root applied outwardly in Measles, Whitlows, and other such like Diseases, is very good. *H. M.*

XVII. *Alcea Arborea althææ folio, florum petalis luteis, deorsum reflexis.* Cat. p. 97. Tab. 140. Fig. 1, 2, 3.

This Shrub or Tree rises to seven or eight Foot high, the Stalks being covered with a whitish Bark, having Leaves alternatively, standing on two Inches long white Foot-Stalks. They are an Inch and an half long, and an Inch broad near the Foot-Stalks end, or at Base, where broadest and round. The Nerves run through the Leaf from the Center of the Foot-Stalk. They are soft, and of a whitish brown colour, much whiter underneath than above, and like the Leaves of Marsh-mallows. The tops of the Branches are divided into several Stalks, sustaining each a yellow pentapetalous Flower, the *Petala* being bow'd back, seemingly not divided, having under them five green whitish, soft, capsular Leaves. After these follow Seeds lying in several flat, broad, compress'd Follicles, ending sharp, joined to one another, making up the same round Head, having many Cells, in every thing exactly like the others of this Kind, there lying in each of the Cells some large, woolly, gray Seeds.

It grows about Colonel *Fuller*'s House, on a small rocky Hill near Black-River-Bride in St. *Dorothy*'s.

The Flowers of this are very yellow, and not at all purple, so that it cannot be (as Dr. *Plukenet* doubts in his *Mantissa* p. 10.) *Althæa betonicæ folio villoso floribus ex luteo purpurascentibus deorsum reflexis ad oras Coromandel.*

XVIII. *Alcea fruticosa aquatica, folio cordato, scabro, flore pallide luteo.* Cat. p. 97. Tab. 139. Fig. 2.

This Shrub has Roots made up of several long, and very white fibrils like Thread, hanging down into the Water or Earth. The Stalks or Stems are many, about the bigness of ones Finger, straight, five Foot high, covered with a white smooth Bark. The Twigs are several, and set about, towards their ends, with cordated Leaves, standing on an Inch and an half long Foot-Stalks. They are two Inches long, and one and an half broad, near the round Base, where broadest, ending in a point, are indented here and there about the edges, of a yellowish green colour, harsh, and having Veins going from the Center of the Foot-Stalk through the Leaf. The Flowers stand spike-fashion, set round on the tops of the Branches, for four Inches of their length, each being half an Inch distant from another, and standing on three quarters of an Inch long Foot-Stalk, having five pale yellow, half an Inch long *Petala*, with purple Veins, a large Pestle like *Stylus*, with *Stamina* on it, all within five green capsular Leaves, under which are twice as many more, very narrow Leaves, of the same make. The Seeds follow, being five, contained each in his Cell. The *Capsula* is large, five cornered, roundish, having at every corner an obtuse *Apex*, as in some others of this Kind.

It grows on each side of the Salt River, in Cabbage-Tree-Bottom very plentifully, and in some places on the Fresh River going up to the *Laguna*.

Any person who compares the Description and Figure of this, with that of *Althæa Indica latiori folio cordiformi ad summum sinuato Pluken. Alm.* p. 26. *Phyt. Tab.* 9. Fig. 2. will find them very different, though the Doctor, p. 10. of his *Mantissa*, thinks they may be the same.

XIX. *Alcea populi folio incano integro. Cat. p. 98. Tab. 139. Fig. 3. Alcea Jamaicensis, abutili facie, floribus exiguis flavis, folio vix crenato, prona parte molli & tenuissima lanugine canescente. Pluken. Alm. p. 17. Phyt. Tab. 254. Fig. 5.*

This has a brown, wooddy, branched Root, a round, wooddy, smooth, brown Stalk, three Foot high, from whence come several Branches, having towards their ends Leaves and Flowers *ex eorum ala.* The Leaves stand on one quarter of an Inch long Foot-Stalks, are extremely white, and soft with Down, an Inch and an half long, and half as broad at the round Base, where broadest, and whence they diminish till they end in a point. The Flowers stand on half an Inch long Foot-Stalks, are pentapetalous, and of an Orange colour. After each of these follows a Seed-Vessel, made up of five parts, like to these of *Fraxinella,* in each of which are contained Seeds, very like the Seeds or *Acini* of Grapes, which leap out of the open'd *Capsula* by the Suns heat.

It grows every where in the *Savanna's.*

There is a variety of this having the Leaf longer, broader at Base, not so woolly on the upperside, and fewer Flowers, growing indifferently with the precedent.

XX. *Alcea populi folio villoso, leviter serrato. Cat. p. 98. Tab. 139. Fig. 4.*

This has many green round Stems, rising two or three Foot high, having several Twigs, on the ends of which are two or three Leaves, standing on near an Inch long green Foot-Stalks. They are near an Inch long, and half as broad at round Base, where broadest, of a green colour, soft, hoary, sinuated, and snipt about the edges, ending in a point, having several Veins running through it from the end of the Foot-Stalk, and some from the middle Rib. *Ex alis foliorum* comes the Flower, standing on three quarters of an Inch long Foot-Stalk. It stands in a pentaphyllous *Calyx,* is pentapetalous, Orange colour'd, and after it follows a six cornered Seed-Vessel, first green, and then brown, or rather so many Follicles lying by one another, and inclosing the Seed like the *Abutila.*

It grows in several places of this Island.

XXI. *Alcea maxima, malvæ roseæ folio, fructu pentagono, recurvo, esculento, graciliore & longiore. Cat. p. 58. Tab. 133. Fig. 2. Alcea Indica Quigombo & Ochros dicta, siliquis prælongis planis quinquefariam divisis, ex insula Barbadensi. Pluken. Alm. p. 16. An Ketmia Ægyptiaca, vitis folio, parvo flore. Tournef. El. p. 83? Inst. p. 100? vel An Ketmia Indica vitis folio, fructu corniculato. Ej. ib. Bammia Calecolar. Mus. p. 520?* Ocra.

This has a round green Stem, which rises straight up to ten or twelve Foot high, being here and there divided into Branches, which are beset without any order, at about an Inches distance, with Leaves standing on seven Inches long Foot-Stalks, each whereof is divided by deep *Lacinia* into five Sections or Divisions, being six Inches long from the Foot-Stalks end, to the end of that division opposite to it, or in the middle, which is the largest, and four broad at Base, from Ear to Ear. They are somewhat rough, of a dark green colour, beset with short inoffensive Prickles, having five middle Ribs, taking beginning from the Foot-Stalks end, and running through every of the five divisions, being proportionably large to their bigness, and the whole Leaf resembling those of our Holyokes. The Leaves when they fall off leave a white knob on the Stalk. *Ex alis foliorum* come the Flowers, standing on an eighth part of an Inch long Foot-Stalk, having nine Inches and an half long, narrow, hoary Capsular Leaves, and one large green one, standing about a round, smooth, green knob. The *Petala* are five, large as those

of

of a Rose, yellow, with a purple bottom, in the middle of which is an half Inch long *Stylus* of the same colour, rough, and having a purple spot on its top. After these follows a three Inches long, crooked, green Pyramidal Pod, nor so big at bottom as ones thumb, having five Ridges or Eminencies towards the top, a little rough, when ripe growing brown, and containing in ten several Cells, so many rows of Seeds, each of which is of the shape, &c. of Mallows, only much larger.

I see no reason why I might not have doubted in my Catalogue of *Jamaica* Plants, whether this was not taken notice of by Writers before me, notwithstanding what is said by Dr. *Plukenet* in his *Mantissa, p. 7*.

XXII. *Alcea maxima, malva rosea folio, fructu decagono, recto, crassiore, breviore, esculento*. *Cat*. p. 98. *Tab.* 133. *Fig.* 3. *Alcea Brasiliana fructu maximo pyramidali sulcato*. *Plukenet. Alm* p. 16. *An siliqua magna decagona seu multicapsularis altheæ sinensis*. *Morif. Fig?* *Alcea Americana annua, flore albo maximo fructu maximo pyramidali*. *Commel. hort. Amst.* p. 37.

This is the same in every thing with the former, only the Stalks of it are not so green, being reddish. The Pod is not over two Inches long, and at bottom, being pyramidal, is of an Inch Diameter, and has ten ridges on its surface, containing, in a great many Cells, Seeds like the former.

If this be not the same, 'tis very near to the *Atlatzopillin sive aquosa herba appensa* of *Hernandez*.

They are both *(viz. the former and this)* very carefully planted by *Europeans*, as well as Slaves in their Gardens, and the unripe Pods, which are in use, are common in Markets.

The Pods of both are gathered when green, and before ripe, and being cut into thin transverse slices are boil'd with Pottages, *Oglio's*, and Pepper Pots, and are thought to be extreamly Nourishing, and very Venereal, being very viscous or mucilaginous. They are so used in *Egypt*, as says *Veslingius*.

The Leaves are us'd after the same manner, for the same purposes.

The tender Fruit is boil'd, and eat with Oil, Salt, and Pepper. *Maregr.*

The *Egyptians* use the Seeds as Beans, Pease, and other *Legumina*, in Victuals. The Leaves and Fruit are cold and moist in the first Degree, resolving and emollient, they are good for the Breast and Lungs, and loosning to the Belly, applied as a Fomentation or Plaister. They resolve Inflammations, ease Pains, soften Tumours, ripening them, Women use them the same ways, and in Baths for hardness of the *Uterus*. A Decoction with white Sugar, taken by the Mouth is very good in Inflammation of the Kidnies and Stone, used for some days. It is good in Decoction against *Ophthalmia*, used as a *Collyrium*, and with it in Pleurisies they foment the affected Side. *Alp.*

It is used to promote Venery *Cæsalp*.

It came from *Ethiopia*, or the Cape of *Good Hope*. *Cluf.*

XXIII. *Alcea hirsuta flavo flore & semine moschato. Maregr. Cat.* p. 98.

This is well describ'd and figur'd by *Piso*, and is to be found in every Plantation, every Seed dropt thriving very well in any Soil, especially stony. It is much more sinuated than that of C. B. &c. in the Leaves.

The Seeds are gathered by Children, and made into Beads, being covered because of their smell.

Drugsters usually adulterate Musk with these, which Sophistication is known by its small continuance. The Leaves are used as those of Mallows in Fomentations and Clysters, being of the same quality, *Piso*.

Young, green Locusts, covet these young Leaves very much. *Maregr.*

The

The Seed is counted Cordial, and powdered with Coffee, is good for the Stomach, Head and Heart. *Vesling.* and is therefore given in Palpitations of the Heart, powdered to a Dram. *Alp.*

It is in great price, and used by the *Egyptian* Women in Baths, but to what purpose they will not reveal to Christians. *Bell. ap.* J. B.

The Confectioners in *France* use the Seed to give a good smell to their Comfits. *Tertre.*

XXIV. *Aloca acetosa, trifido folio India orientalis.* Breyn. prod. Cat. p. 99. *Aloea Acetosa Indica gossypii folio pericarpio coccinei coloris, capsis siliquam annulante.* Plukenet. Alm. p. 15. *An sair Indorum Oxalis Cannabina fruticosa fructu coccineo coronato.* Triumf. MSS. Cupan. Hort. Cathol. p. 194. *Ketmia Indica, gossypii folio, acetosæ sapore.* Tournef. Inst. p. 100. French Sorrel.

This rises higher than a Man, it has a thick round Stalk, covered with a red Bark, with a Pith within it, and a great number of Branches coming out on all sides, without any order, three or four Foot long, having Leaves standing on red, Inch long Foot-Stalks, being about three Inches long, and two broad at the Base, and divided into three great *Laciniæ*, each of which has a Rib going from the Center through the Leaf, which are all red, they are also indented, about the edges green, the whole Plant having a sowr tast like Sorrel, whence the name. *Ex alis foliorum* come the Flowers standing on short Foot-Stalks, of a greenish yellow colour, with purple in the bottom, large, and pentapetalous, with *Stylus* and *Stamina*, as in this Kind, standing within five Capsular, red, small, long Leaves, inclosed by twice as many narrower of the same sort, being very red and thick, When the Flowers fall off there follows a membranaceous, oblong, round, five cornered, pointed, sharp *Capsula*, in which are five Cells, containing so many rows of roundish, compress'd, light, brown Seeds, like those of Mallows, only much larger.

It is planted in most Gardens of this Island.

The Capsular Leaves are made use of for making Tarts, Gellies, and Wine, to be used in Fevers, and hot Distempers, to allay Heat, and quench Thirst.

The Root given to two Drams, purges easily the Stomach and Guts. *Hernandez.* Whose Description agrees, the Stature excepted, but his Figure is very faulty.

The first Leaves are sometimes whole, commonly divided into three *Laciniæ*, and sometimes like *Ellebrastrum*, into seven Sections.

The Leaves either alone, or boiled with other Herbs, are eaten by *Indians*. The Stalks are, as Hemp with us, spun into Ropes and Yarn, therefore set in their Fields and Gardens. *Herm.*

XXV. *Caryophyllata foliis alatis.* Cat. p. 99. *Caryophyllata Campestris elatior Brasiliana foliis acuminatis.* Pluk. Almag. p. 87.

This is well describ'd and figur'd by *Piso*, so that there needs no more be said of it, but that 'tis very common in the Woods of this Island.

It is very Hot and Dry, it attenuates, Cuts, Cleanses and is Adstringent, and therefore is not only good to corroborate the Bowels, but likewise to cut tough Humours. *Piso.*

CHAP.

Chap. XV.

Of Herbs that are Bacciferous or Pomiferous.

I. **C**ucurbita longa folio molli, flore albo. J. B. Cat. p. 100. Cucurbita longa. Muf. Swamm. p. 13. Gourds, whereof they make Conferves of an *Anonymus Portugal* of *Brafile, Purchas, lib. 7. cap. 1. p. 119.* The fweet Gourd.

I could not obferve any difference between the Plant I faw here in Gardens, and thofe in *Europe*, only they feem'd to be larger, and had a whitifh Skin, which is only, I fuppofe, an accidental variety.

It was planted by Mrs *Aylmer* in her Garden at *Guanaboa*.

The Pulp of this is edible and cooling, whereas that of moft of the other Gourds growing here is very purgative.

This is what the *Jews, Spaniards,* and *Portuguefe* make into Sweet Meats, in the Ifland of *Madera* and elfewhere.

If crude it is not grateful to the Stomach.

Applied to the Head by way of Poultefs, it cures inflammations of the Eyes, and pains thereof. The Seeds are Diuretick and made into Emulfions temper the Urin, taking off its Acrimony.

Bodæus à Stapel tells us that he had one of thefe Gourds which came from *Indian* Seed, as large as a Human Body.

This is cut into long flices, and candied with Sugar by the *Spaniards* and *Portuguefe*, and fent into the North to fell, it is call'd *Carbafade*, and is very delightful to Dry or Feverifh People. *Lob.*

The Seed of this is one of the four greater cold Seeds.

It is eat with Onions boil'd after the manner of Coleworts, being good for Lean People. The diftilled Water is good in Fevers. *Lon.*

Children in Fevers are cured, if one of thefe, as long as themfelves, be put to Bed to them. *Dod.*

Being of no taft they imbibe any. *Lac.*

The long, round, and comprefs'd Gourds come of the fame Seed from different parts of the Fruit. *Durant.*

II. *Cucurbita fpharica maxima. Cat p.* 100. *The largeft round Gourd.*

In Leaves, &c. it agrees with the white, long, fweet Gourd, only the Shell is yellow, or Cinamon coloured, hard, fmooth, fhining, having a bitter Pulp within it, and fmaller and darker coloured Seeds. One of them is able to contain many Gallons of Melaffus, &c. for which they are us'd inftead of Bottles.

The Shell gives a purging quality to any thing infufed in it for fome times.

They are planted on ftony Hills.

They are fo large, that cut in two they make Panniers to carry any thing in. *Acofta.* And others are leffer for eating and drinking in.

III. *Cucurbita tertia feu lagenaria, flore albo, folio molli.* C. B. *Cat. p.* 100. *An Kabach, cucurbita lagenaria Herbar. Olear. p.* 231?

This grows with the former likewife.

This is falfly taken for *Colocynthis* though it Purges, and gives that quality to any Liquor ftanding in its Shell.

The *Genouese* cut them into long slices, dry them in the Sun, keeping them sweet with a little Sugar. *Lugd.*

IV. *Cucurbita lagenaria minima, collo longo recurvo.* Cat. p. 101. Tab. 141. Fig. 3.

This is like the others, only the round part is no bigger than a Tennis Ball, the Neck being four Inches long and crooked.

V. *Cucurbita lagenaria, longa, maxima, recurva.* Cat. p. 101.

This is two or three Foot long, round, about four Inches Diameter, smallest at both ends, and crooked.

This and the former are planted as the others.

VI. *Cucurbita lagenaria, longa, recta, minor.* Cat. p. 101. Tab. 141. Fig. 2.

This is just like the others in every thing. The Fruit is Pyramidal or Oval, six Inches long, and two in Diameter, very round and polish'd.

They are made into Bottles and carried about for Dram Bottles in ones Pocket.

Gourds are lighter, and not so brittle as Glass, wherefore covered to carry Wine and other Liquors in, in *France*, *Spain*, and *Italy*. *Lob*.

VII. *Pepo maximus Indicus compressus.* Lob. Obs. Cat. p. 101. *Melo pepo fructu maximo, albo.* Tournef. Inst. p. 106. Pompions.

They are commonly planted here in most Gardens, both by Poor and Rich, and the Fruit being boil'd as Turneps, are very savoury and cooling.

The juice, with a little Muschat, takes away weakness. *H. M.*

Marco Polo tells us, those in *Sapurgan* are the best Pompions in the World, being sweet like Honey. *Purchas, lib.* 1. p. 73.

VIII. *Anguria prima, Citrullus dicta.* C. B. Cat. p. 101. *Anguria Indica seu melo aquaticus.* Tournef. El. p. 89. *Melo sacharinus anguria folio, Virginianus, fructu magno cortice viridi, carne liquescenti albo vel incarnato.* Plukenet. Alm. p. 247. Banist. Cat. Stirp. Virg. *An cucumis peregrinus major sativus, anguria foliis horto Comptoniano è semine natus.* Pluk. Alm. p. 123. *Uva* of *Ogilby America.* p. 313. Water-Melons of *Dampier.* cap. 10.

This is commonly planted here, and is of two sorts, that with whitish green, and that with red Pulp, the Seeds of the latter being red, those of the first black.

They are used here by way of desert, are very much commended and every where planted, especially in dry, sandy, or rocky Grounds. They are Diuretick, counted very good in Fevers, extremely good against hot Livers, and Kidnies, very cooling, and therefore often eat with Wine.

The Seeds are us'd for Emulsions, and provoke to Sleep.

It grows better if Dung be mixed with the Sand, where it is planted *Maregr.*

Gerard seems very much deceived, when in his Figure and Description, this is sulcated or furrow'd, which it is not.

Hieronimo de Lobos tells us, in his Relation of the River *Nile, &c.* that about the Red Sea, were the best he ever tasted.

These Melons are much wholesomer than ours at *Paris*, raised by Dung and Water. *Thevet.*

They beget bad and venemous Humours, bringing Autumnal Fevers, they are of bad Digestion, wherefore not us'd as Meat. *Dorstein.*

Nicolas says they grow about *Argiers*, and are there eaten, being very cooling.

Casol-

The Natural History of Jamaica. 227

Cæsalpinus mentions two sorts of this Melon, saying *caro interna aliis candida, aliis rubens*; and that the latter is the best, he takes likewise notice of the Seed, that 'tis *colore Nigro, aut rubente aut fulvo*. The Pulp next the Bark is the less sapid. *Cæsalp.*

In *Alexandria* they make holes in the outward sides of this Fruit, and draw out the Liquor, and sell Glasses of it to those who are hot. *Id.*

The unripe Fruit kept in a heap of Wheat ripens. *Lugd.* Who mentions one reddish *semine ex Asia misso*.

The Seed of the *Indian* is larger and smoother. *J. B.*

The Seed from *New-York*, called *Maracock*, proved on sowing to be this. *Lact.*

IX. *Cucumis anguria folio latiore, aspero, fructu minore candido spinulis obtusis muricato.* Cat. p. 103. *Anguria Americana, fructu echinato eduli.* Tournef. Inst. p. 107.

This has a deep, white, oblong Root, sending forth several long, trailing Branches. The Stalks are four-square, and rough, five or six Foot long, at about every four inches distance of which come out the Leaves, Clavicles, and Flowers. The Leaves are divided into five Sections, they are curled, sinuated, and rough, the undermost, or Sections at Base, being smallest every way. The fifth Section is three Inches long, and has two notches in it. Their Foot-Stalks are four Inches long, and rough. *Ex alis foliorum* come the Clavicles, which are an Inch long; from thence also come the Flowers, several on two Inch long Foot-Stalks, which are yellow, monopetalous, their *ora* being divided into five Sections. The Fruit is of a pale green colour, Oval, as big as a Walnut, having many short, blunt, thick Tubercles, sharper than those of other Cucumbers, and within a Pulp, a great many small Seeds, like those of other Cucumbers.

It grows every where about the Town, and in most Plantations, as well as in the *Caribes*.

This Fruit is eaten very greedily by Sheep, and all manner of Cattle, and they are thought to thrive extremely by feeding on them.

The Fruit is likewise eat in lieu of our *European* Cucumbers, are very cooling, and equal, if not exceed them in every thing.

Piso in his first Edition says nothing of this, and *Marcgrave* mentions it, whereby *Piso* takes occasion in his second Edition to speak of it, and reasons upon it out of his own imagination, for the Fruit of this is no way like that of our *Cucumis asininus*, neither is it bitter, or yields any thing like *elaterium*, but is very much coveted, and eaten by Cattle of all sorts, and Men themselves, whence one may make a conjecture concerning several things of that Author, that they were his own conjectures, or the product of his own Brain, rather than his own Experience, or that of others.

The Leaves of this in *Jamaica* are rough, whatever Dr. *Plukenet* may have found in *English* Gardens, or say to the contrary, p. 59. of his *Mantissa*.

X. *Cucumis minima fructu ovali nigro lævi.* Cat. p. 103. Tab. 142. Fig. 1. *Bryonia Canadensis, folio angulato, fructu nigro.* Tournef. Inst. p. 102. *Cucumis fructu minimo, viridi, ad maturitatem producto nigricante.* Plukenet. Alm. p. 123. *An bryonia alba lævis Americana cortice albo nitente; forte Tzaetzavaalic Mexicensibus seu herba glutinosa & candens.* Hernandez apud Reccum, lib. 8. f. 283. Pluk. Alm. p. 71. Phytogr. Tab. 272. Fig. 3?

This Cucumber has a very slender Stalk, which mounts by its Clavicles, or runs along the Ground for five or six Foot in length, catching hold of any thing it comes near. The Leaves, Flowers, and Clavicles, come out here and there together. The Leaves stand on an Inch long Foot-Stalks, are of a

roundish

roundish triangular shape, a little auriculated, being an Inch and an half from Ear to Ear, and as much from the Foot-Stalk end, to the point or end of the Leaf opposite to it. They are rough and harsh to the touch, something sinuated and indented a little about the edges, and of a dark green colour. The Clavicles are very tender. The Flowers stand on a small Inch long Foot-Stalk, are yellow, monopetalous, though at the Ore divided into five Sections. The Fruit is of the shape of a Nutmeg, or Oval, though not so big as an ordinary red Gooseberry, smooth, blackish when ripe, and full of small white Seed, like that of other Cucumbers, lodged in a thin insipid cooling Pulp.

It grows in cleard low Grounds, by Hedges and Ditches.

The Fruit is eaten either pickled as unripe Cucumbers use to be, and are good, or when fully ripe, and are thought extremely cooling.

XI. *Colocynthis bryoniæ albæ folio, in quinque lacinias dentatas profunde sectò, aspero, cathartico* Cat. p. 103. Tab. 142. Fig. 2. Belly-ach-weed.

The Root of this was whitish, oblong, and deep, having several Stalks creeping along on the Ground, which at about two or three Inches distance send out Leaves and Clavicles. The Foot-Stalks of the Leaves were two Inches long, the Leaf its self being smaller than others of this Kind, rounder, and more deeply cut in on the edges. It is always divided into five Sections, the Section opposite to the Foot-Stalk being three Inches long, which is much longer than the two Sections next it, and they than those at Base. Each of the Sections had one middle Rib, and was jagged without any order. It was rough on its surface, and of a yellowish green colour. Its Clavicles were not very long, but caught hold of the Stones it came near, and would creep a great many Feet from its Root.

It grew at Mr. *Abraham's* Plantation in the Northside of the Island, upon some stony Hills, near the place called *Ocho Rios*, corruptly *Chirires*.

By what I saw, I question not but this is a *Colocynthis*, although I saw neither Flower nor Fruit, neither could any inform me any thing about them.

This is counted an extraordinary Medicine against the Belly-ach They take a handful of the Leaves, boil them in water, and give the Decoction, which usually Vomits and Purges, but more certainly the first.

It is accounted a very good Remedy against the Dropsie, and is taken the same way.

It is also used for the same purposes in Clysters.

XII. *Cucumis puniceus,* Cord. hist. Cat. p. 103. *Momordica vulgaris.* Tournef. El. p. 86. Inst. p. 103.

This is very much us'd for the Belly-ach.

The Leaves are accounted very vulnerary as well as the Fruit, and both taken inwardly, or their Oil outwardly applied are thought to ease Pains, and cure the Colick.

It consolidates Wounds, and eases pain of bad natured Ulcers, the Fruit being beaten and boiled in Oil. The Fruit is call'd *Charantia,* for being like a Lemon. *Lob.*

If an Ear be cut off of a black Horse, and fastened to that of a white Horse, likewise cut off, they will be healed with this Oil; but I am not certain that 'tis so. *Trag.*

The Oil cures Burns, and takes away Scars. *Dod.*

The Oil of the Seed is commended by some for the best Remedy. *Cam.*

XIII. *Flos passionis major pentaphyllus.* Cat. p. 104. *An Passiflora foliis latioribus citius florens.* Pluk. Alm. p.281 ? *Clematis passiflora pentaphylla, angustifolia.* Muns p. 53 ? Tab. 165 ? *Flos passionis pentaphyllus major, angustifolius semper virens.* D. *Kiggalaer.* Ib ?

This has a green, almost round Stalk, about the bigness of a Goose-Quill; mounting about any Tree it comes near, to twenty (sometimes more or less) Feet high, at every two Inches distance, putting forth Leaves, Clavicles and Flowers. The Leaves stand on an half Inch long, crooked, green Foot-Stalks, being divided even almost to the Foot-Stalks end, into five very long, and narrow Sections, that in the middle, which is opposite to the Foot-Stalk, is the longest and narrowest, being three Inches long, and about three quarters of an Inch broad, every Section having one middle, and some transverse Ribs of a dark green colour, and smooth. These Sections at Base are a little ear'd, and the whole Sections are entire and not serrated. Opposite to the Leaf is the Clavicle two or three Inches long, and taking hold of any thing it comes near. The Flower comes out in the middle between both, and stands on an Inch long Foot-Stalk, is very large, and of the same make with others of this Kind.

It grows in the Woods going from the Town to *Guanaboa*.

Piso commends the Fruit of this in hot Diseases, to cool and be used instead of Currans or *Berberis*.

The first of these was eaten in *New-Spain*, being thought cooling, and they have in their Flowers, if one be aided by Piety, the Figures of the Instruments of the Passion. The Fruit is not Savory to those who eat it at first. *Tertre*.

The Fruit is good in Fevers for the Spirits and Appetite, and is pleasant, and without nauseousness in those using it. *Rochef.* The Rind and Flowers are preserv'd likewise, and us'd as the Fruit. *Rochef.*

XIV. *Flos passionis folio hederacco anguloso, fœtido.* Cat. p. 104. *Flos passionis albus reticulatus Herm. par.* Bat. p. 273.

This has a round woolly Stalk at every Inches interval, sending forth pretty strong Clavicles, and reaching many paces in length. The Leaf has two Ears, or is three-pointed like Ivy Leaves, downy and soft, like those of *Althæa*, and of a very rank and offensive smell. The Flower is like others of this Kind, of the same colour with the ordinary. After these succeeds a round, hollow, reddish Fruit, in the Cavity of which are Seeds, sticking to the inside in rows, each Seed being inclosed in a white Skin, and the Fruit having a Fringe, Filament, or Leaves under it, like the Seed-Vessel of *Nigella Romana*.

It grows in Gullies near Mr. *Elletson*'s in *Liguanee*, Colonel *Crew*'s in St. *Dorothies*, the Banks of the *Rio Cobre* below the Town, and in several other places of this Island very plentifully.

It flowers in *May*.

XV. *Flos passionis, folii mediâ laciniâ quasi abscissâ, flore minore, carneo.* Cat. p. 104. *Granadilla flore suaverubente, folio bicorni.* Tournef. El. p. 206. Inst. p. 241. *Passiflora Americana flore suaverubente folio bicorni.* Pluken. Mantiss. p. 146.

The Stalk of this Plant is striated, redish green coloured, and mounts very high by its Clavicles about any Tree near it, or creeps along the Ground, covering it for some considerable breadth, putting forth alternatively Leaves, Clavicles, and Flowers. The Leaves are large, of a whitish green colour, having two points, and as it were a want or defect in room

Nnn

of the third, as if one had by Art cut of the third point, resembling in this the Leaves of the Tulip-Tree, or the *Acetofa rotundifolia repens Eboracenfis folia in medio deliquium patiente Morif.* The Clavicles are ftrong, and the Flowers very fmall, of a pale red, or flesh colour, in every thing else like the others of this Kind. The Fruit is oval, having fix red Lines on it, and contains several, long, black Seeds, each of which is inclofed in a white mucilaginous membrane, and plac'd as the others of this Kind.

All the parts of this Plant are without fmell.

It grows on the Banks of the *Rio Cobre*, about, and below the Town of St. *Jago de la Vega*, and elfewhere very plentifully.

XVI. *Flos Paffionis perfoliatus, five periclymeni perfoliati folio. Cat.* p. 104. *Tab.* 142. *Fig.* 3, 4.

This Paffion Flower has a round, purple Stalk, which has at every two Inches diftance, as it were two Leaves join'd together on one fide by a Seam, on the other between two Ears, it comprehends or catches hold of the Stalk, and feems to be perfoliated like the *French* Honyfuckle. Each Section or Leaf is two Inches long, very fmooth, one Inch broad at Bafe where broadeft, having one Rib from Bafe to end. Both Sections or Leaves are join'd to the Stalk by a fhort, thick, crooked Foot-Stalk. Out of the *Ala* of this comes a pretty long Clavicle, which catches hold of any thing near it, fo rifing or creeping as the others of this Kind. The Leaves come out alternatively on each fide of the Stem, where are the Flowers towards the top of the Stalk, having an Inch long Foot-Stalks. They are purple in colour, made up of five large, and as many narrow *Petala*, of about an Inches length, ftanding horizontally. It has fome ftrong, green Filaments, with orange colour'd tops, and a green *Stylus*, every way like the other Paffion Flowers.

It grew in a wooddy, rocky Mountains fide in *Liguante*, near Mr. *Ellet fon*'s Plantation.

This is not the *Paffiflora fcaphoides, &c. Pluken. Alm.* p. 282: being neither that defcrib'd by *Plumier*, nor *Hernandez*, mention'd by him in that place.

XVII. *Flos paffionis minor, folio in tres lacinias non ferratas profundius divifo, flore luteo. Cat.* p. 104. *Flos paffionis trifido folio, flore minimo pentapetalo viridi, fructu minimo, nigro, molli. Breyn. prod.* 2. p. 47. *Flos paffionis flore & fructu omnium minimis, Par. Bat.* p. 177. *Granadilla folio amplo tricufpidi, fructu olivæ formâ. Tournef. El.* p. 206. *Inft.* p. 240.

This has a three or four Inches long Black Root, about the bignefs of ones little Finger, throwing into the Ground feveral long Threads or Strings very deep. From the top of the Root, are fcattered on the furface of the Ground, or climbing any Plant for four or five Foot high, feveral round, green, rough Stalks, befet alternatively at an Inches diftance with Leaves fet on one third part of an Inches long Foot-Stalk, being an Inch and a quarter long, and about an Inch from the end of one divifion at Bafe, to that of the other, being divided into three parts or Sections, cut even almoft to the Foot-Stalk, each of the divifions having Ribs from the Foot-Stalk. They are of a dark green fhining colour, and fmooth. *Ex alis foliorum* come the Flowers, they ftand on three quarters of an Inch long Foot-Stalk, are very tender, and of a greenifh yellow colour, in every thing refembling thofe of the other Paffion Flowers, and after them follows the Fruit, in every thing like the others of this Kind, only fmaller.

It grows in all the rocky Banks and Sides of Hills in this Ifland.

This

The Natural History of Jamaica. 231

This is not the *Flos passionis folio augusto, flore amplo decapetalo, &c.* of *Breynius prod.* 2. *p.* 47. as I once question'd, but is mentioned by that Author by the name above recited. This is the *Clematitis Indica folio hederaceo, &c. Plum.* notwithstanding Dr. *Plukenet's* contrary Opinion, *p.* 146. of his *Mantissa.*

XVIII. *Flos passionis minor, folio in tres lacinias non serratas, minus profundas, diviso. Cat. p.* 104. *Passiflora hepatica nobilis folio parvo non crenato, flore ex luteo viridante. Pluk. Alm. p.* 282. *Granadilla pumila, flore parvo, luteo, Alexandri Badam. Geo. à Turre. Cat. p.* 55. *An Clematis seu flos passionis flore luteo, Munucuja Mexicana. Belluc. hort. Pis. p.* 182. *Alia Munucuja Species Marcgr. p.* 72 ? *Alia Munucujæ species foliis hederæ scandentis. Raii hist. p.* 656 ? *Clematis seu flos passionis flore viridi. Hort. Reg. Paris. p.* 53 ? *Clematis passionis hederæ folio floribus parvis herbaceis, fructu minimo quando maturus nigro. D. Banist. apud Raium, hist. pl. app. p.* 1874 ?

This is in every thing the same with the precedent, only the Leaves are not so deeply divided, the end of the Leaf having only two great notches, or defects for divisions.

It grows on the rocky Banks or Hills, every where through the Island.

XIX. *Smilax, aspera, fructu nigro, radice nodosa, magna, levi, farinacea, China dicta. Cat. p.* 105. *Tab.* 143. *Fig.* 1. *An Smilax viticulis asperis Virginiana, foliis angustis, lævibus nullis auriculis præditis. Plukn. Phyt. Tab.* 110. *Fig.* 4. *Alm. p.* 349 ? *China Root of Dampier, cap.* 15.

This has a Root as big as the thickest part of ones Leg, though sometimes it be no bigger than ones Wrist, having several swellings, being crooked and jointed, having some short membranes here and there, and a thin reddish brown Skin, within which is a friable light red coloured substance, more mealy than fibrous, yielding a reddish tincture to water. The Stalk is round, as thick as ones little Finger, very thick set with short and sharp prickles green, running its self round any Tree or Plant it comes near, and rising with their help to fifteen Foot high, putting forth Leaves, the lowermost of which are very like those of the *Smilax aspera*, being cordated, smooth, of a very dark green colour, with Nerves running through the Leaves like those of Plantain. The Twigs go out towards the ends of the Branches, having where and there Leaves, smaller, narrower, and thicker set than the others which are on the Stalk, not so broad at Base, and in greater numbers. The Flowers come out on the ends of the Twigs being several together, standing on an half Inch long *Petioli*, taking their Original from the same common Center, umbel-fashion, each of which has six *Petals*, or very small and green *apices*, standing round a green short *Stylus*. After these follow so many blackish Berries, being round, of the bigness of those of Ivy, containing, within an unsavoury purple Pulp, one round purple Stone, as big as that of Haws.

It grows on the red Hills very copiously, the moister the ground is wherein it grows the greater the Roots, and the worse to be cured.

The Leaves of this are both ear'd and long, so that I doubt concerning it, whither it may not be the same with both Dr. *Plukenet's* mention'd here, and in my Catalogue.

'Twas first known in the *East-Indies* from *China*, in 1535. *Lugd.*

This is used for *China* Roots, and yields a much deeper Tincture in *hanabar* of the *East-Indies*, whence I think it much better for the purposes to which it is employed, than that which is Worm eaten coming from *China*, although *Piso* seems to be of another mind.

It yields a Gum called *Testeli*, which the *Indians* chaw to strengthen the Teeth. The Decoction is good against Chronical Distempers, against the *French Pox*. Oriental *China* has a lighter, not so sirm, tenderer, and less
Adstringent

Adstringent Root, and yet this Kind does the same with it, *Sarsa* or *Guajacum*, if the same methods are followed. It is Cold, Dry, and Adstringent, but bitterish, and of subtle parts, strengthning the Stomach, expelling Wind, voiding by the Pores Melancholick Humours and Flegm, which eludes other Medicines, thereby giving ease. *Hernand.*

In the *East-Indies* they us'd *China* very much for the cure of Diseases, boiling one Ounce with two Drams of Smallage in sixteen Pints of water to ten: they drank a draught warm in Bed, lying two hours after, and another two hours before Supper, and cold other times of the day, making it fresh every day; and many use to take two Drams of the powdered Root morning and evening in Wine, or its own Decoction when going about their Business, or failing, with happy success. It is best preserv'd with broken Pepper. Its distilled water is good in Ruptures, Headaches, *Callus's* and Ulcers of the Genitals. It is also proper in Venereal Diseases, but the Decoction better, *Acosta.*

Saris, apud Purchas, lib. 4. *cap.* 2. *p.* 394. tells us that *China* Roots are a Commodity for *Japan.*

China was in use to cure a sort of Pox in *East India*, called *Afmaphoa* or Stink, it came to the knowledge of the *Latins* in the year 1535. by two *Chinese* Merchants, *Nacmoch* and *Makal*, Trafficking in *Africa*, Don *Martin Alfonse* was cured of an otherways incurable Distemper with the Pox by it, which was known to the *Arabians* before us, and at first sold for its weight in Gold. The Country Mountaineers eat of it raw, or rosted as Turneps, it is boil'd in Pottage by some, *f.* 417. *Thevet. Cosmograph.*

The White is the best, whence that from *New-Spain* is not good, which is so red that a little of it makes red water, neither is it so efficacious as the other. It is when fresh so tender as to be eat either raw, or boiled as Turneps here with Flesh. *Fragos.*

Canes are made in *Virginia* of them. *Pl.*

It appears by *Lane*, *Harlot* and *Laudonniere*, in the Places mention'd in my Catalogue, that this Root is used for Bread in *Virginia*, from these *Tsinaw* Roots, (says *Harlot*) new or fresh, chopt into small pieces and stampt, is strained with water, a Juice that maketh bread, and also being boiled a very good Spoon-Meat, in manner of a Gelly, which is much better in tast if it be tempered with Oil. and *Laudonniere, p.* 55. *ap. Hakl. p.* 344. says that beat in a Mortar it makes Meal, which boiled in water is eat in scarcity in *Florida.*

XX. *Bryonia racemosa foliis ficulneis. Plum. p.* 83. *Fig.* 97. *Cat. p.* 106.

This has a striated yellowish green Stalk, being jointed at every two Inches distance, not so big as a Hens-Quill, having at every joint one Leaf, standing on an Inch and an half long Foot-Stalk, each Leaf being deeply cut into three parts or *Lacinia*, like to ordinary white Bryony. The two Sections at Base have a round Auricle, which is serrated. The Section in the middle is from the Foot-Stalks end to the sharp point, two Inches and an half long, and 'tis three quarters of an Inch over in the middle. 'Tis likewise two Inches from one Auricle to the other, every Section has a middle Rib, furnishing transverse ones. The Leaf is somewhat rough, and of a yellowish green colour. Opposite to the Leaves stand Clavicles with which it mounts the neighbouring Trees or Hedges, as others of this Kind.

It groweth every where in *Barbados.*

XXI. *Bryonia alba triphylla, geniculata, foliis crassis, acidis.* Cat. p. 106. Tab. 142. Fig. 5, & 6.

The Root is two or three Foot long, reddish, and sends forth other Roots, smaller, pretty long and jointed, creeping under the Surface of the Earth. The Stalk is jointed, reddish brown, as big as ones Finger, every joint making a very Obtuse Angle with that next it, it sticks to Palisadoes or Trees by Clavicles, and on its upper parts towards the top, is beset with a great many Leaves, three always together on a pretty long Foot-Stalk. The Leaves are thick, juicy, sower, and cut in pretty deep on the edges like a Saw. The Stalks towards the tops are always red. The Flowers stand towards the tops of the Branches in bunches or *Corymbi*, like Ivy, and are of an herbaceous colour. After these follow several, little, round Berries, no bigger than a small Pea, of a black colour, standing on an half Inch long green Foot-Stalks. The Pulp is of a blackish green colour, and contains one small round Stone, having a white Kernel within it.

It climbs the Trees in the Woods in several places, and grows near the River sides, and about the Palisadoes, near the Town of St. *Jago de la Vega*.

It Flowers in *May*.

The Juice of this is used in Sauces as that of Sorrel.

Whosoever pleases to compare the Figures of *Bryonoides trifoliatum Indicum, &c. Plukenet. Phyt. Tab.* 155. Fig. 2. and his *Chamædrifolia scandens, &c. Phyt. Tab.* 81. Fig. 5. and has seen this Plant, will be ready to think the last only a Figure taken from a dry'd Plant, from which the Leaves are drop'd off, and the first from the same Plant, growing in a Garden, notwithstanding what that Author says, p. 33. of his *Mantissa*.

XXII. *Bryonia alba geniculata, viola foliis, baccis è viridi purpurascentibus.* Cat. p. 106. Tab. 144. Fig. 1. *An Bryonia alba lævis Americana, cortice albo nitente* D. *Plukenet. Phyt. Tab.* 272. Fig. 3. Almag. p. 71 ? *Vel an Bryonia Americana fructu aureo, cerasi parvi magnitudine tetrapyrene venenato, Barbadensibus nostratibus* Poison Wythe, *vocata. Ejusd. Alm. p.* 71. Tab. 151. Fig. 4 ?

This has jointed or geniculated Stalks as big as ones little Finger, and rises, catching hold of any thing it comes near by its Clavicles to seven, eight, or even thirty Foot high, being round, greenish, every joint an Inch long, making a very Obtuse Angle with that next it, and putting forth Leaves, Clavicles and Flowers. The Leaves stand on an half Inch long Foot-Stalks, are four Inches long, three broad at the round Base where broadest, being of a dark green colour, and somewhat resembling the Leaves of Violets. The Flowers are many, small, of a pale yellow colour, standing many together as the others of this Kind. After these follow Berries of a purplish green colour, like those of the other Bryonies or Ivy.

It grows near the Town about old Palisadoes and Trees, as likewise near the Bridge over Black River, on the Trees growing there, very copiously.

XXIII. *Bryonia alba triphylla maxima.* Cat. p. 106. Tab. 144. Fig. 2.

This has a many-cornered Stem about the bigness of a Goose-Quill. It has, at every Inches or more interval, Leaves, three always together at a crooked joint, and a Clavicle opposite to that, three Inches long, and catching hold of any Tree they come near. The Leaves stand on an Inch and an half long Foot-Stalks; that opposite to the Foot-Stalk, or in the middle, being an Inch and an half long, and an Inch broad near the further end

O o o where

where broadest. They are smooth, and of a yellowish green colour. What the Flower and Fruit were I did not observe, though I am apt to think, by its Leaves and Clavicles, it may be a Briony, of the Kindred of those describ'd before, and therefore I put it among them.

This is not the *Bryonia sideritidis folio multiplici, dispermos, flore cærulæ. Prom. Bon. spei. Plukenet. Phyt. Tab.* 152. *Fig.*1. *Alm. p.* 71. as Dr. *Plukenet* suspects in his *Mantissa. p.* 33.

It grew on the larger Trees in the Road to *Guanaboa*, and at Mrs. *Guys* Plantation.

XXIV. *Bryonia nigra fruticosa, racemi ramulis variè implicitis, atque cauda scorpionis instar, in se contortis, baccis albis una vel altera nigra macula notatis. Cat. p.* 106. *Tab.* 143. *Fig.* 2.

The Trunc of this Plant is near the Ground, as thick as ones Arm, woody, running about any Trees it comes near, smooth, hard, and of a dark brown colour, branch'd into a great many Twigs, crooked, and turning round any thing they come near, rising seven or eight Foot high, towards the ends of which are plac'd alternatively the Leaves. They stand at half an Inches distance from one another, on an eighth part of an Inch long Foot-Stalk, are an Inch long, and half an Inch broad in the middle where broadest, smooth, and of a dark brown colour, and a little bowed back. At the tops of the Twigs come out several two or three Inches long crooked Branches, very variously turn'd, twisted, and bow'd one within another, each of which is turn'd like a Scorpions Tail, or the Heliotropes, and sustains on none, or very small Foot-Stalks, a great many very small, five-pointed, herbaceous Flowers. After these follow so many Berries as big as Pepper Corns, round, first green, but when ripe, white, markt with one or two black specks, and containing, lodged in a Pulp, one or two black, round *Acini* or Seeds.

It grows about the Palisadoes in the Town of St. *Jago de la Vega*, and in the Woods of the *Savannas* about any Trees or Shrubs it comes near.

I have perused what Dr. *Plukenet* says in his *Mantissa, p.* 187. concerning this Plant, and remain of the same Opinion as when I wrote my Catalogue, that it is what he calls *Virga aurea Americana frutescens glabra, foliis subtus cæsiis, comis ad summitatem in ramulos brachiatos implicatis. Phyt. Tab.* 235. *Fig.* 6. *Alm. p.* 389.

XXV. *Bryonia nigra fruticosa, foliis integris ex adverso positis, flore luteo, racemoso, fœtido. Cat. p.*106. *Tab.* 145. *Fig.* 1.

This Plant has a Stem as large as ones little Finger, covered with an almost smooth, white Bark, creeping amongst the Trees it comes near, and putting forth here and there Foot-long Branches adorn'd with Leaves set at an Inches distance by pairs opposite to one another, standing on an eighth part of an Inch long Foot-Stalks. They are three Inches long, and an Inch and a quarter broad, near the middle where broadest, being of a very green colour, smooth, thin, having one middle, and several transverse Ribs. The tops of the Branches are a great many bunches of Flowers, each one standing on a weak Foot-Stalk, in a green *Calyx*, being long, tetrapetalous, yellow, with yellow *Stamina*, and of a very unsavory scent.

It grows on the Road side, among the Woods, going to *Guanaboa* very plentifully.

XXVI. *Bryonia nigra fruticosa, foliis laurinis, floribus, racemosis, speciosis. Cat. p.* 106. *Tab.* 145. *Fig.* 2.

The Natural History of Jamaica. 235

This had roundish, reddish brown Stalks as big as a Goose-Quil, by which it climb'd and turn'd its self round the Truncs and Branches of Trees, putting forth here and there Leaves, standing on one third part of an Inch long Foot-Stalks. They were four Inches long, and near twice as broad in the middle where broadest, being narrow at the beginning, and obtuse at the end, whole, very smooth, and having an eminent middle Rib, running through the middle of the Leaf. At the top of the Branch are Flowers standing in a bunch together, being many very beautiful and small.

I gathered it, if I rightly remember, in St. *Maries*, near *Cabeça del oro* in the North-side of this Island amongst the Woods.

XXVII. *Solanum bacciferum primum, seu officinarum.* C. B. *Cat. p.*106. *Solanum fructu nigro, Rudbeck. Brom.* p. 108. *An solanum Indicum vulgari simile, floribus albis parvis. Pluk. Alm.* p.349 ?

This has a green Stem, as big as ones little Finger, having some very narrow membranes which make it look angular, rising two or three Foot high, either straight up, or amongst other Shrubs. The Branches are inveigled among one another, spreading themselves on every hand, and having Leaves standing on a quarter of an Inch long Foot-Stalks, being an Inch and an half long, and half as broad near the middle where broadest; very much sinuated on the edges, soft, of a dark green colour. Towards the tops of the Branches come the Flowers, several together from the top of an half Inch-long green Foot-Stalk, each whereof is made up of five white, or pale yellow reflected *Petala*, with Orange colour'd *Apices*, standing up in the middle making an *Umbo*. After these follow round Berries, as big as Pepper-Corns, smooth, black when ripe, containing in a thin Pulp a great many very small, roundish, flat, white Seeds.

It grew near the Church-yard back-side, and near Mrs. *Guys's* House in a Gully in *Guanaboa*.

The Leaves of this are boil'd and eaten by *Negros* in their Pepper-Pots or Potages.

The Leaves being applied cure Wounds of the Shins, Fissures of the Nipples, and are good against all Inflammations. They are Anodine, and good against the Inflammation and heat of the *Anus*, the Juice being put up; but because of its very cooling and adstringent quality, it must be cautiously used. The Bark bruis'd and put into water, intoxicates Fishes so that they may be easily taken. *Piso.*

I only doubted whether the *Jamaica* and *European* Nightshades were the same, on account of the Leaves, being eaten there, and the *European* commonly accounted not wholesome, but when I found by *Cordus, Dorst.* &c. that the common *Solanum* was anciently sowed for Meat, I am confident 'tis in every thing the same, especially since the colour of the Berries differs in the *European* Plant.

The Leaves cool, being applied to St. *Anthony's* Fire. It is good for a hot Stomach. Beaten with Salt and applied, it discusses the *Parotides, Fuchs.*

It is good against hot Aposthems. *Dorst.*

The distill'd water outwardly or inwardly used is cooling in Inflammations, outwardly the Leaves being applied to the Head helps Phrensies, and all Inflammations. *Lon.*

Tragus seems to be confused as to this Plant, and says that 'tis not edible.

The Leaves are cooling and adstringent. It is hazardous to apply it outwardly, it being *Repercussive*, and not to be us'd in *Erysipelas's*, &c. *Dod.*

The Juice, with Oil, is good in the Phrenzy, &c. *Math.*

The Leaves eaten are adstringent and cooling. *Gal.*

236 *The Natural History of* Jamaica.

The Boys eat the raw Berries. It contemperates sharp and biting humours. *Schwenckf.*
The Juice is useful in Cancers. *Cam.*
The distill'd water is good in Fevers. *Lob.*

XXVIII. *Solanum bacciferum, caule & foliis tomento incanis, spinosis, flore luteo, fructu croceo, minore.* Cat. p. 107. Tab. 144. Fig. 3. *An solanum spinosum Jamaicense glabrum foliis parvis minus profunde laciniatis.* Pluk. Alm. p. 351. *Phyt.* Tab. 316. Fig. 5 ? *An solanum Americanum, tomentosum folio verbasci parvo, fructu flavescente.* Plum. Tournef. Inst. p. 150 ?

The Stalks of this Plant are very thick set with short crooked prickles, the points downwards, woolly, round, and about three Foot high, having Leaves set alternatively every Inch and halfs distance, about six Inches long, and five broad, beginning narrow, and ending in a point, very much sinuated on the edges, very rough, downy, of a whitish green colour, and having one middle Rib, and some transverse smaller ones, in every thing like those of the *Solanum spinosum fructu rotundo.* C.B. Pin. only lesser, both beset with prickles. In the middle space between the Leaves come out the Flowers, two or three together on the same half Inch long, hoary Foot-Stalk, monopetalous, though the *Oræ* be divided into five *Petala*, reflected back, of a yellowish colour, and having in the middle many *Apices* standing up together, making an *Umbo*, as the other *Solanum's*. After each of these follows Spherical Orange coloured Berries, as big as Field Pease, having five green Capsular Leaves under them, and being full of small, white, compressed, irregularly figur'd Seeds, lying in an Orange colour'd Pulp.

It grows on the sides of the Streets of the Town of *St. Jago de la Vega*, near the old Monastery, and on the sandy and gravelly Banks of the *Rio Cobre*, below the Town on the same side of the River.

The Leaves and Juice are good to temper and cleanse Wounds and Ulcers, and although it be bitterish, yet 'tis not hot. Their Roots are very bitter, and of thin parts, and excellent Vertue, especially the Male. Its Decoction is Diuretick. They open Obstructions of the Liver and Prostates, being us'd instead of the opening Roots. It is very much esteem'd both by the learned and unlearned. *Piso.*

The Leaves of this *Jurepeba*, which are not prickly, are commonly us'd to heal Ulcers of the Legs. *Marcgr.*

Margrave making mention of this, but of no distinction of Sex, I am apt to believe *Piso* was impos'd upon; especially since he does not explain himself sufficiently about it.

The Root given to half an Ounce purges all Humours downward. *Hern.*
The Decoction of the Root is good in burning Fevers, and with Honey in Catarrhs, and in the Strangury with some *Cardamoms.* It is proper for windy Guts. The Juice of the Roots and Leaves is good for Concoction, and the Juice with Sugar is good for the soreness of the Breast. The Decoction of the Leaves is good for the Itch with Sugar and Lime. *H. M.*

XXIX. *Solanum fruticosum bacciferum spinosum, flore cæruleo.* Cat. p. 108. Tab. 145. Fig. 3. *An solanum Americanum fruticosum, persica foliis aculeatum.* Plum. Tournef. Inst. p. 149 ?

This Shrub by crooked woody Stems, as thick as ones little Finger, covered with a whitish coloured Bark, and having here and there some sharp, small, and short prickles, rises three or four Foot high, having several Branches and Twigs of a green colour, and, without any order, Leaves standing on an half Inch long Foot-Stalks, they are two Inches and an half long, and almost one broad, somewhat rough, and like those of the

Antonum

Amomum Plinii. Ger. Towards the ends of the Branches stand several purple pentapetalous Flowers, having five Orange-coloured *Stamina*, *Apices*, or points standing straight up in the middle, making an *Umbo*, the *Petala* being bow'd back or reflected. After which follows a round, red Berry, like those of the *Amomum Plinii*, having many small flat Seeds, white, and of the shape of a Kidney.

It grows in the Sand by the Sea-side, near *Old Harbour*, on the Road from the Town to Colonel *Cope*'s Plantation in *Guanaboa*, and in the *Caribe* Islands.

XXX. *Solanum pomiferum quartum, sive fructu oblongo.* C. B. *Cat.* p. 101. *Melongena fructu oblongo*, *Tournef. El.* p. 126. *Melongena fructu oblongo violaceo, Ej. Inst. sp.* 151. *Mala insana. Swert. An Nila-barudena. H.M. part.* 10. p. 147. *Tab.* 74 ? Mad Apples.

This has several white Roots, sending up a branched purple Stalk, three or four Foot high. The Branches ends have Leaves many in number, standing on four Inches long purple Foot-Stalks. They are six Inches long, and three broad, sinuated, of the shape of those of the others of this Kind. The Ribs are all purple, and the Leaves somewhat hairy and rough. The Flowers come out several on the same Foot-Stalks, they being monopetalous, the *Ora* divided into five Sections, blue, and in every thing like those of the other Nightshades. The Fruit is oblong, pear-fashion'd, from a narrow round Stalk, ending larger, first purple, then yellow, containing in a fungous Pulp, a great number of flat, roundish, brown Seeds.

The Seed was brought from the Main Continent of *America*, where it was planted by the *Jews*. Mr. *Harrison* planted it in *Liguanee*.

The Fruit boil'd, and drest as Turneps, is very much commended, and eaten by the *Jews*.

The *Arabians* used it in their Meats, but it breeds vitious Juices. Its tast is sweetish, flat, and bitterish. It is only for show sown in Gardens or Pots. *Cord.*

It is eaten as Mushrooms. *Ruellius.*

It is boil'd and eat as Mushrooms. *Cæsalp.*

Rosted under the Embers, boiled or fried, every day 'tis eat in *Egypt. Bellon.*

It is eaten to promote Venery. *Math.*

Ranolf apud Lugd. ap. says that the black is different from the others.

They are susceptible of any tast, much used about *Toledo*, and boil'd first, then fried with Fat. They cause Obstructions, darken the Complexion, breed Melancholy, &c. *Loc.*

They are boil'd with Wine and Pepper, and tast like Artichoaks. They are Diuretick. *Bont.*

This agrees in every thing with *Nila Barudens H. M.* p. 10. p. 147. *Fig.* 74. only that has prickles, this has none. *Commelinus* makes it *Focky focky & mala insana.*

These are the *Melongena* of the *Arabians, ap. Purchas, lib.* 9. *cap.* 9. p. 1499. where we find *Mahomet* affirm'd he had seen this Plant in Paradise, and measur'd the quantity of Mens Wits by their eating store hereof.

This seems to be one sort *Melanzana, Melongena* or *Beudengian*, called *Bathleschain*, oblong of *Rawolfe, lib.* 1. *cap.* 6. which grows about *Aleppo*, and is there eaten boiled as *Averrhoes* mentions.

XXXI. *Solanum secundum racemosum cerasorum forma: vel cerasa amoris racemosa rubra.* C. B. *prod. Cat.* p. 109. *Tab.* 146. *Fig.* 1. *Lycopersicon fructu cerasi rubro. Tournef. Inst.* p. 150. *Lycopersicon fructu cerasi ejusd. El.* p. 125. Tomato Berries.   This

This grows in several places about the Town of St. *Jago de la Vega*, and in *Guanaboa*, near Mrs. *Guy's* House, in her Plantation, but I cannot be positive that 'tis wild. It grows likewise in the *Caribes*.

They are eaten by some here, are thought very naughty, and yielding little Nourishment, though they are eaten either boil'd or in Sauce by the *Spaniards*.

They are good to eat, being cooling, give a relish to Sauces, and take off the ill effects of *Indian* Pepper, which is too heating. *Jof. Acofta.*

The Fruit is innocent, and not bad smell'd. *Gefn.*

This came first from the *Peru* Isles. The Fruit boiled or sun'd in Wine is good against the Scab. *Cam.*

The Juice is good for Eyes with Defluxions, against *Eryfipela's*. The Chymical Oil is good for Burns, and to procure sleep rub'd on the Temples. One slept in a frenzy with these under him, and in his Hand. *J. B.*

The Fruit boil'd, or infus'd in Oil, is good against the Itch. *Park.*

XXXII. *Solanum veficarium trectum folani vulgaris foliis. Cat. p.* 110. *Alkikengi Virginianum fructu lutto. Tournef. El. p.* 126. *Inft. p.* 151. *An folanum veficarium Virginianum procumbens annuum folio lanuginofo. Bob. hift. Ox. part.*3. *p.* 527?

This has a four-square hollow green Stalk, rising three Foot high, branch'd out on every side. The Branches have several Leaves set on an Inch long Foot-Stalks, they are two Inches long, one broad, ending in a point, of a dark green colour, and like the Leaves of the ordinary *Solanum*. The Flowers are on the tops of the Branches, on half an inch long Foot Stalks, monopetalous, with five notches in them of a yellowish white colour. After these follows red Bladders, in all things like those of *Alkekengi.*

It grows by the *Rio Cobre* in wet places above the Town.

The Fruit is eaten, and tasts like *European* Winter-Cherries. A Decoction of the Herb serves to wash the Legs. *Marogr.*

The Fruit is of several parts, and takes off the Obstructions of the Liver, and Kidnies, differing only from that of *Europe* in that its Diuretick quality is more remiss. *Pifo.*

The Root is hot, cures Surfeits, resolves Wind, provokes Urin, eases pain in the Belly mix'd with *Indian* Pepper. The Decoction drank, or the Root to a Scruple in quantity, stops all Fluxes from hot Causes, and applied to the Breasts drys up the Milk. *Hern.*

The *Indians* and *Chinese* eat it with *Capficum*, and love Apples. The green Juice is used in hot Diseases of the Bladder and Kidnies, and in Venereal *Gonorrheas*, it is a great Remedy easing pain. *Bont.*

The Juice diminishes the Tumours of the Testicles. The Roots help the burning Fever, and fried with Oil helps the Dropsie called *pitao. H. M.*

XXXIII. *Pifum decimum five veficarium fructu nigro alba macula notato.* C.B. *Cat. p.* 110. *Corindum ampliore folio fructu majori, Tournef. El. p.* 342. *Inft. p.* 431. *Cor Indum fructu majore Pluken. Alm. p.* 120.

This has a woody, cornered, rough Stalk, taking hold of any Tree or Shrub it comes near by its Clavicles, and mounting to eight or nine Foot high, the tops then falling down, and covering the Tree or Shrub it Climbs. At about every three Inches distance it puts forth Leaves, Clavicles and Flowers, at the same place. The Leaves stand on two Inches and an half long Foot Stalks. They are very much divided or laciniated, cut always into nine Sections, standing three together on the same common *Petiolus*, coming from the end of the Foot-Stalk, that division of the three opposite to the end of the *Petiolus*, or in the middle is the biggest, being two Inches

long,

long, and one broad where broadest, deeply notch'd, or cut in on the edges, of a dark green colour, being very smooth, soft, and thin, the other two at Base, being of the same shape, &c. only smaller. The Clavicles stand opposite to the Leaf, being five Inches long. *Ex alis foliorum* come the Flowers several together, standing on three Inches long Foot-Stalks, being white, pentapetalous, and very open. After the Flowers follow three cornered, oblong Bladders, having in each of them three distinct Cells, and in every one of these lies fastened to a membrane, a round, dark, brown, or black Seed, about the bigness of a small Field Pea, having three Triangular Lines, meeting at the Center of a clay coloured, or whitish triangular or cordated spot, which is at the place where 'tis join'd to the Bladder, or its *hilus*.

It grew on some Bushes, on the Banks of the *Rio Cobre*, a little below the Town, and very copiously among the Shrubs on each side of the way, riding through the *Thickets*, near the *Moneque Savanna*, in the North side of *Jamaica*, and in the *Caribes*.

There is a variety of this, which is in every thing less, but I believe it no distinct Kind.

From the likeness of the division of the Leaves of this, to those of Parsley, it got the name of wild Parsley in the *Caribes*.

This is no *Solanum*, because it has no Berries. *Gesn.*

It's not *Isopyrum*. *Col.*

Its thought good for Diseases of the Heart. *Schwenef.*

The Seeds occasion greater Sleep than *Opium*. *Cord.*

The Plant beaten with water, and applied cures the Gout, and coldness of the Joints with stiffness. The Juice of the Leaves with *Zit Avanacu* Oil Purges. The juice warm is good for the gouty, with black Cumin Seed 'tis good for the Heartburn, the beaten Leaves are good for the Cough mix'd with Sugar, and boil'd in Oil it is good for the Eyes. *H. M.*

XXXIV. *Pisum cordatum non vesicarium. Cat. p.* 111. *Leguminosa Brasiliensis fructu ovato costa folii appendicibus aucta. Raii hist. p.* 1347. *An cordis ludi folia & facie frutescens Curassavica latifolia. Horn. par. Bat. prod. p.*328? *Plukenet. Phyt. Tab.* 164. *Fig.* 6?

This by round, smooth, brown Stems, rises to a great heighth, mounting by its Stalk sometimes twenty Foot high, more or less, according to the Tree or Shrub it Climbs, having Leaves at every half Inch's distance, standing on one third part of an Inch long Foot-Stalks, being usually nine Leaves plac'd by threes, on the same common, small Foot-Stalks, those three in the middle, or standing opposite to the Foot-Stalk, being the largest, and that Leaf of the three in the middle the largest Leaf, being more than an Inch long, and about half as broad near the further end where broadest, smooth, of a yellowish green colour, augmenting from the Foot-Stalk to near the end, and thence decreasing to the point. On the tops of the Branches come the Fruit, standing in Bunches, or many together on branched Twigs, having small Clavicles. Each of them is a Triangular Head, having three plain sides, and three very small extant membranes, or sharp Corners, red or black, when opened containing three, large, black, shining, almost round Seeds, or Pease, with a white *Hilus*, Eye or Spot, at that place where they were join'd to their *Capsula*.

It grows between *Passage Fort* and the Town, on each side of the Road, on the Trees in the Woods, and on the *Red Hills* very plentifully.

The Fruit bruised and put into water intoxicates Fishes. The green Leaves bruised, or their Juice, is good for Wounds, being vulnerary, and cleansing them. *Pif.*

XXXV. *Cap-*

XXXV. *Capsicum minus fructu rotundo, erecto, acerrimo. Cat. p.* 111. *Capsicum siliquis surrectis eteraß forma. Tournef. Inst. p.* 135.

From a woody, brown, strong Root, spring several woody Branches, two or three Foot high. The Stalks, Leaves, Flowers, and other parts of this Plant are in every thing like those of the other *Capsicums*, only every way smaller, and the Fruit is plac'd on the tops of the Twigs. It is almost perfectly round, no bigger than a Field-Pea, and full of such small Seeds as are common to others of this Kind.

It grows in shady low Woods, whether Parrots and other Birds resort to it by natural Instinct, to forward the Digestion of those Fruits they meet withal and feed on in the Woods.

It grows not only in the Island of *Jamaica*, but in the *Madera*, and all the *Caribes*.

This sort is counted the sharpest and most biting of any of its Kind, and is much used by *Indians*, *Negros* and *Europeans*, who have liv'd here any time.

The Birds, *Toucan* and *Saviath*, feed of this, and their Dung produces it, being Sown, as if the whole Fruit were Sown. *Thevet. Cosm.*

Powdered with Salt, 'tis a portable Sauce in little Room, and agreeing almost to every Dish and Pallat, being mix'd with Gravy or Vinegar.

It is counted very good against the Belly-ach.

The Savages cure Fevers by drawing over the Eyes of the Sick a Thread dipt in this, by making them forget it. *Tertre.*

There are two other sorts very like this, common in most Planters Gardens, which may perhaps be only varieties of it.

This was brought by the *Spaniards* in *Colons* first Voyage from the *West-Indies*, to show the rarities of them, and was then admired. *Lopez. de Gomara.*

Merchants brought it from *Brasile*, though it be not so good as *Malagette. Thevet.*

The Leaves of this burnt with Mother of Pearl Oisters shells, and wet with water till 'tis white, makes an Ointment which rub'd on Teeth makes them black, and free from aching. *Ben.*

This was call'd *Caribe* by the *Indians*, which signifies *sharp* and *strong* in their Language, and because they of *Espanola* found the *Canibals*, or those inhabiting the *Antilles*, to be sharp and strong like this, they gave them the name of *Caribes. Martyr.* So that we need not trouble our selves so much about the Derivation of that Word as *Rochefort* has done.

This sort, if tasted, the sharpness cannot be got out of the Throat in some days. It grows in *Brasile* and *Portugal. Cluf.*

One must have a care after touching these *Capsica* not to touch the Eye with the Hand it occasioning great pain, which is remedied by cold water.

They are reckoned much wholesomer than the *East-India* Peppers, they give a good tast, being cut and mixed with Fish or other Meat. The small whole Pepper, being swallowed helps Digestion, Corroborates the Stomach, and expels Wind. The same is done with Vinegar in which it has been infused whole, or Salt, and it powdered and mix'd together. It hurts the Breast and Body, and occasions a Cough, if it be put on Coals, and the Fume received by Mouth or Nostrils. The Leaves and Roots are the First Ingredients of hot Baths. *Pifo.*

XXXVI. *Capsicum minus fructu parvo, pyramidali, erecto. Cat. p.* 112. *Tab.* 146. *Fig.*2. *An solanum mordens Americanum perenne Berberidis fructu surrecto* Berberry Pepper. *Barbadensibus vulg. Pluk. Alm. p.* 354 ? *vel Solanum mordens*

The Natural History of Jamaica. 241

mordens foliis majoribus surrectum fructu parvo, oblongo, Ejusd. ib? vel solani mordentis siliquis surrectis rotundis, alterum genus fructu parvum acuminato. Ejusd. ib? Capsicum siliquis surractis & oblongis exiguis. Tournef. Inst. p. 152. Capsicum sive piper Barbadiense fructu Berberidis acerrimo. Bob. hist. Ox. part. 3. p. 530.

This differs only from the foregoing, in that 'tis not perfectly round, but pyramidal, and a little pointed.

It grows commonly with it.

XXXVII. *Capsicum oblongum minus recurvis siliquis. Park. Cat. p. 113. Capsicum siliquis recurvis, minus. Tournef. Inst. p. 152. Piment qui semble du corail de Biet. p. 334. An Solanum mordens oblongum minus pendentibus recurvis siliquis puniceis. Pluken. Alm. p. 353? Vel Solanum mordens propendentibus siliquis oblongis recurvis. Ejusd. ib?*

This is in every part like the other larger *Capsicums,* only its Fruit is two Inches long, a little crooked, round, about the bigness of a Goose-Quill, but towards the end decreasing, and ending in a blunt point, of a very fine scarlet colour.

It is planted by the Inhabitants very carefully in their Gardens, for its use in Pottages, &c. and is us'd indifferently with the Fruit of those of the other Kinds.

This powdered with Salt is made into Loaves, after every Morsel, some of which is taken up between the Fingers, and eat by the *Brasilians.* It is the only Simple of *Brasile* described by *Mathiolus, Lery.*

This is figured in *Cordus* with two other sorts on the same Shrub, from *Tragus,* and in *Loniceras* with one other. In many of the old Herbals many of these *Capsica* are figured together on the same Stalk.

XXXVIII. *Capsicum siliqua lata & rugosa. Park. p. 114. Tab. 146. Fig. 3. Solanum mordens fructu magno petasoide,* Bonnet Pepper, *nostratibus vulgo. Pluk. Alm. p. 353.* Bell-Pepper.

This rises four or five Foot high, is in every thing like the other *Capsicums,* only the Fruit is large, turbinated, conoidal, or somewhat shap'd like a Bell, whence the Name, hanging down towards the Ground, the sides of it being deeply sinuated or furrowed here and there, especially towards the point.

It is sweet smell'd. *Greg. de Reg.*

A little put on Coals, the Fume entring the Nose and Lungs, excites a troublesome Cough, not to be remedied but by a Handkerchief wet in Vinegar. *Tertre.*

It is the most commonly planted of any of the *Capsicums,* and used extremely by *Indians* and Blacks.

It is very often pickled by cutting off the largest part next the Stalk, and clearing it of its Seed, and putting it into Pickle of Vinegar and Salt.

It is us'd in every thing as the others. The *Indians* and *Negroes* make it the proper Correctice for all sorts of *Legumina,* and Sallets, and will scarce abstain from it in hot Diseases.

These Peppers ought not to be inwardly us'd, having something venemous and malignant in them. *Dod.*

It is used all over Spain for Pepper. *J. B.*

XXXIX. *Capsicum siliqua lata non rugosa. Cat. p. 114.*

It differs only from the other in making a lesser not furrowed Fruit, which is very shining and polisht. This seems to be only a variety of the former.

Q q q XL. *Capsi-*

XL. *Capsicum cordatum propendens. Park. Cat. p,* 114. *An Solanum mordens fructu dependente subrotundo crasso. Pluk. Alm. p.* 353? *Vel Solanum mordens siliqua cordiformi pendula, ejusd. ib? Piper Indicum cordatum majus siliqua plana & propendente Hort. Reg. Paris. p.* 142.?

This *Capsicum* is like the others in every thing, only the Leaves come out in opposite Tufts along the Branches at an Inches distance. The Fruit is Conoidal, and instead of being sharp at point, is blunt, very shining, polish'd, and smooth, exactly like that of a Womans Nipple.

It is planted as the other Kinds.

A Pessary is made of this kind of Pepper, *ol. laurin.* Gentian and Cotton, which *Vulvæ ori impositum, purgationes Menstruas deperditas revocat.* Taken to a Scruple it cures Pains of the *Uterus,* if it be boil'd in Wine, and with it the *Hernia aquosa* be bath'd it cures it. If the Fruit be infused in Wine it takes away stinking Breath, and mixt with *Hydromel,* us'd as an *Errhinum,* it takes away the smelling of the Nose. If infus'd in *Aqua Vitæ* it helps the parts grieved with the Palsie if rub'd with it. Mixt with Vinegar it resolves Apostems and hard Spleens. Drank with the Decoction of Bay-Berries it cures the Colick. Chaw'd with Raisins and *Stavisacre,* it draws Flegm from the Head. Boil'd in water the Decoction cures the Tooth ach. Drank with Wine it cuts rough Flegm in the Breast, Lungs, or Guts. It takes away the cold Fits of Agues if mix'd with *ung. de Alabastro,* if it be rub'd on the Back-Bone. With Hens Grease it resolves Apostems and Buboes, it cures the Gout and Nodes, and voids Hydropick waters. Mixt with *pil. aloephang.* with Decoction of Mallows it is Diuretick. It takes away the Flegm in the Kidnies, being drank with *Saxifrage* water. Given to a Scruple with Broath every morning, it warms the cold Stomach, discusses Flegm, and dissipates Tough Humours of the Stomach. Taken three days together with *Decoct. Pulegii,* it expels the dead Child. *Gregorius de Regio.*

This is one of the three Fruits figured by *Tragus* on the same Stalk, and from him in *Cordus & Tab. ernamontanus.*

It raises Blisters. *Park.*

Those using it are troubled with their Stomach and Yellowishness. *Roch.*

The Root or Fruit bruised and applied to the Bite of a Mad Dog, cures it. *H. M.*

This Pepper is us'd all over the *West-Indies,* it is hot in the fourth, and dry in the third Degree. It excites Flatulency, and Venery. It purges with griping in those not used to it, from its Acrimony, but easily in others; it helps the *Catamenia,* and is Diuretick; it is good against the Sciatica made into a Plaister with Honey. The *Indians* help Hecticks with pricking their Bellies or Loins with Needles or Pins dipt in this; but immoderately used it causes Inflammations of all sorts: it is eat by the *Indians* for ordinary Food, by others for Sauce. *Xim.*

Infused in Spirit of Wine it is Diuretick, and cures Palsies if the part be often washed with it. Mix'd with Hens Suet it resolves Apostems that are cold. Inward Apostems and Abscesses are caused by *its immoderate use. Piso.*

Five Grains of this Pepper makes pleasanter Potage than twenty of the other from the *East-Indies. Martyr.*

It kills Dogs if eat by them. *Chabr.*

Hernandez and *Ximenes* are so confused in their Descriptions, and Names, that although the Figures are good, yet the Descriptions are so very bad that I cannot make any thing of them.

It was the only general Spice of the *West-Indies*, and a Merchandise in esteem amongst the *Indians*, some strong, others eatable, it is only strong in the Seeds and Veins. It is corrected by Salt or *Tomato*-Berries: it is too hot for young Men, and Venereal, *Acosta*.

The Decoction of the Fruit brings away the Dropsie water. *Cam.*

It is planted very much in *Spain*, for use in their Kitchens, either fresh or dry'd 'tis in use. *Cluſ.*

The Roots, Stalks, Leaves and Flowers, are not in use, only the fleshy part of the Pod and Seed, is planted for Ornament or Use. It raises Pustles wherever applied. In opening the Pod, and taking out the Seed, a subtle Vapor penetrates the Nose, going to the Brain, and draws out much tough Matter, causing Sneesing sometimes, and entering by the Mouth it causes Cough and Vomits. The Hands are so inflamed in holding it, that they must be speedily wash'd for their burning. *Cluſius* touching his Eyes, when watering, with this Vapor from the Seed, a great Inflammation came, which had almost cost him his Eyes but they grew well with frequent Lotion with cold water: burnt they raise a very stinking Smoak. Three of the *Siliqua* are dry'd over the Fire, cut in small pieces, mix'd with a pound of Flower bak'd like Bisket, then sear'd and kept as the best Preparation, and gives an Apperite; it is good against old Coughs. *Greg. de Reg.*

It dissolveth *Struma*, cures the Sciatica and Quinsie, and Freckles, applied to the Face with Honey. *Ger.*

Most of the Fruits of these *Capsica* are first green, then turn purple, and afterwards scarlet.

The Fruit was used by the *Indians* as a punishment for Vagabonds, who were forc'd for their Faults to receive the fume of it.

It is used by the *Spaniards* in their Chocolate as well as Potages.

All these sorts of *Capsica* differ little from one another in Vertues.

XLI. *Periclymenum rectum herbaceum, gentianæ folio, folii pediculo caulem ambiente. Cat. p. 115. Tab. 147. Fig. 1.*

This has a green, round, smooth, jointed Stalk, rising about a Foot high, at every joint having a Leaf, whose Foot-Stalk encompasses the Stalk at the joint, making a hollow *Tubulus* wider then the Stalk about half an Inch above the joint, which may be able to hold some water. The Leaves are five Inches long, and two broad, smooth, thin, and something like the Leaves of *Phalangium Diperaſon*, or Gentian. Towards the top over against every Leaf comes a five Inches long, jointed Foot-Stalk, on the top of which stand above two green, small Leaves, several, small, white Flowers. After these follows several, round, pretty, large, black *Acini*, clustered very close together, making one Berry. In each of the *Acini* lies one black Seed within a very thin Pulp, which usually dries away.

It grew in the Woods by the Path going to Sixteen Mile Walk, very copiously, and in the Woods going to the North-side, and elsewhere.

XLII. *Viola folio baccifera repens, flore albo pentapetaloide, fructu rubro dicocco. Cat. p. 115. Pyrola affinis Malabarica Karima-Kali. H. M. Pluken. Almag. p. 309. Periclymeno accedens planta utriusque Indiæ, foliis periclymeni rotundioribus fructu bipyreno. Bob. hist. Ox. part. 3. p. 535.*

This Herb has a small, round, creeping Stem, putting forth at its joints many, small, fibrous Roots, and having small Branches, at about one Inches distance from one another, each of which is about an Inch and an half long, having roundish Leaves, standing opposite the one to the other, on an Inch long, reddish Foot-Stalks, in every thing resembling those of Violets, only smaller and rounder. The Flowers come out at the tops of the Branches, they

244   The Natural History of Jamaica.

they are white, and divided in their Margins into five Sections, and to them follows several red, smooth Berries, round, as big as a small Pea, containing in an Orange colour'd Pulp, two oblong, brown Seeds, each of which is flat on one side, and rais'd on the other, with a sharp or more eminent ledge on it.

It grows in shady, dark, and moist Woods in the Path going to Sixteen Mile Walk, and elsewhere.

Boil'd in Whey it cures the Flux. Boil'd in Oil it cures bloody Eyes. *H. M.*

Chap. XVI.

Of Herbs with bulbous Roots, those of their Kindred, and of Herbs with Flowers that have six or more Petala or coloured Leaves.

I. *Narcissus totus albus latifolius polyanthos major odoratus, staminibus sex è tubi ampli margine exiantibus. Cat. p. 115. Autre sorte de Lys. Rochef. tabl. p 112. Narcissus Americanus, flore multiplici albo odore balsami Peruviani. Tournef. Inst. p. 358.* White Lilly.

This has a tunicated bulb as large as ones Fist, made up of one white thick Coat over another, as Onions, and having at base many white fibers, by which it draws its Nourishment. The Leaves are two Foot long, about three Inches broad, channel'd, or being a little concave in the inside, very green, juicy and smooth. In the middle of these rises a flat Stalk, four Foot high, not hollow, but fill'd with a fungous Matter, when cut dropping water, about one third part of an Inch thick, and being sharp at the edges. At the top of this are six or seven, or more white Flowers, standing each on a six Inches long Foot-Stalk, each of which has a white large *Tubus*, having six Ribs, very long *Stamina*, or *Lingula* with long *Apices*, dividing it into so many parts, and standing up above the Flower on its edge or margin two Inches long, being there of a green colour. A *Stylus* of the same colours, is in the middle. There are six white, five Inches long, very narrow, divided *Petala*, which stand between the aforesaid *Lingula*.

The Flowers of this Plant have a very fragrant smell.

It grows in the *Savanna*'s beyond the *Black River* in the Low-Land Woods every where in *Jamaica*, in the Woods in St. *Christophers*, and by the Roads in *Barbados*.

It is not only coveted as as ornamental and pleasant in Gardens, but likewise the Roots are us'd all over these Islands, in lieu of White-Lilly-Roots for Maturating Cataplasms.

II. *Lilio-narcissus polyanthos, flore incarnato, fundo ex luteo albescente. Cat. p. 115. Lilium Americanum, puniceo flore, bella donna dictum. Herm. par. Bat. p. 194. Lys des Antilles pareilles à nos Lys jaunes ou Orangers. Rochef. Tabl. p. 112. Lilio-narcissus Americanus puniceo flore Bella donna dictus. Plukenet. Alm. p. 220.*

The Root of this is no larger than that of a great Onion, or the half of ones Fist, a little oblong, made up of many white Tunicles or Coats, inclosing one another, after the manner of Onions, having under its Base many whitish fibers drawing its Nourishment. The Leaves are one Foot long,

long, an Inch and an half broad, juicy, of a very fresh green colour, blunt, round, or obtuse at their ends, channel'd or furrowed towards the Stem, or inwards. The Stalk rises from the Leaves, being one Foot and an half high, hollow, of about one quarter of an Inch Diameter, sustaining on its top several Flowers going out of, or inclosed in a membranaceous Sheath or Follicle bow'd back, or hanging down by two Inches long Foot-Stalks. Each of the Flowers is wide open, of a yellowish and white colour in the middle, and of a Carnation, or pale red the rest, having in its Center several reddish and yellow *Stamina*.

It is planted along the Walks sides for Ornament in Gardens, and comes from *Barbados*, where it is wild. It is said likewise to grow wild in the Gullies here, and to come from *Surinam*.

III. *Aloe Dioscorid. & aliorum.* Col. min. cognit. stirp. Cat. p. 115. *Aloe Musf. Moscand.* p. 289. *Aloe vera vulgaris Munt.* Phyt. cur. p.20. Tab. 96. Sempervive.

This grows every where, where it is or has been planted, but I never saw any that I thought was spontaneous, in *Jamaica*, though enough about old ruin'd Plantations. It is planted both here, and chiefly in *Barbados* to make Aloes to send into *Europe*.

In speaking of this Plant, *Piso* says he never found it Purge, but only that it was cleansing if used by Chirurgeons, which is manifestly contrary to Experience, whence may be almost plainly gathered that he took out of *Maregraves* Notes only, which in this Chapter he calls his own, whereas what *Maregrave* there speaks of seems not to relate to this Aloes, but to all of them in general, and to the *Caraguata-guasu* in particular. What they say of the Metle, must belong to that not this, which is the true Aloes.

The Figures of these Plants in *Piso*, are so transpos'd that I cannot make any thing of them.

It purges and fortifies the Stomach against crude Humours. *Dal.*

It purgeth Choler, Flegm, Worms, opens Womens Obstructions, and the Hæmorrhoids. It is good against Surfeits of Meat or Drink. Wash'd it fortifies more, and purges less. It is hot in the first, and dry in the third Degree. It preserves Carcasses, Heals bad Sores, stops Bleeding, is good for the Eyes, being drying, &c. *Ger.*

The *Indians* have a Medicine made of Aloes and Myrrhe, call'd *Meethar*, which they use in curing Horses, and wormy Wounds. A Decoction of three Ounces of the Leaves, with two Drams of Salt, being boiled over a gentle Fire, then strain'd over night, and given the next morning to eight Ounces, gives four or five Stools. If it be bruised with Milk, and given to those troubled with an Ulcer of the Kidnies or Bladder, it cures them. The same cures Birds broken Legs, and they are us'd in *India* to ripen Swellings. There is great Controversie between the *Greeks* and *Arabians* concerning Aloes, the last saying that it strengthens the Stomach, and opens the Hæmorrhoids, the others denying it, who were certainly mistaken. The best Aloes is that which is solid, without empty spaces, and not mix'd carelesly with Sand. The best formerly was accounted that coming from *Alexandria*, which is the same now coming from *Socotora*. It is made likewise in *Cambaya* and *Bengale*, but less esteemed. *Garcias ab Ort. Acosta.*

Dioscorides says it is very bitter and strong smell'd.

The Leaves are to be cut transversly, not long ways to cut the Veins, which drop a yellowish Juice, of which Aloes may be made, it dropping on a glazed Tile. *Col.*

The Natural History of Jamaica.

There are three sorts of Aloes, *viz. Lucida, Citrina,* and *Hepatica,* which is the worst, the first and second are almost the same or succotrine Aloes, the last is black and harder to break. 'Tis odoriferous as Myrrhe, not ill smelled, unless not good or fresh. *Bod. à Stapel.*

'Tis insipid with us, but grows bitter in Stoves, being hang'd up, the *Hepatic* is best. *Cord.*

They take it up in *Cyprus* before Winter, and hang it up, planting it again in the Spring. *Gesn.*

It does not purge less mix'd with Honey, and is not hurtful to the Stomach, *Fuchs.*

It is used, the Leaves being beaten, to consolidate fresh Wounds, and for Corns. *Casalp.*

Aloe is of two sorts, the *Caballina,* or *Arenosa,* and the *Secatrina,* from that Island or *Hepatica,* from its consistence it is friable in Winter, and softish in Summer. *C. B.*

It grows wild about *Lisbon, &c.* on Walls. *Clus.*

This is the true Aloe, from whence comes the Aloe of the Shops. It is made by cutting the Leaves obliquely, that a yellow juice may come from the Veins running its length, which is evaporated a little, and makes a mass agreeing to the Aloe of *Dioscorides, &c.* being all dissoluble in water, and friable to the Fingers. It is purgative, and outwardly applied stops Blood. The Leaf, the outward Skin taken off, with the juice applied, cur'd a Palm of the Hand where Nerves and Tendons were hurt. *Col.*

Pieces of the green Leaf are given to Horses for the Worms, as also to Children for the same Disease, with great success.

It is hang'd up to be ready in Houses to apply to fresh Wounds. *Park.*

IV. *Aloe secunda seu folio in oblongum aculeum abeunte. Morison. Cat. p.* 117. *An Aloe Americana spinifera angustis foliis radice bulbosa cujus folia ad pannum conficiendum sunt apta. Pluken. Alm. p.* 19? *Aloe mucronato folio Americana Major. Mont. Phyt. cur. p.* 19. *Tab.* 91.

I can add nothing to the Description extant in Authors.

It grows frequently on the rocky small Hills, in several places of this Island, in *Brasile, New-Spain, &c.* and Flowers generally about *May,* afterwards dying down, Root and all.

The Stalk and Flowers being very straight, and twenty Foot high, are planted before Houses for May-Poles.

The Leaves are us'd to scour Rooms, Plates, or any thing withal, instead of Soap, having a viscid Juice. It occasions a great pricking and tingling in the Hands of those which are besmeared with it.

Cloath is made of this, little inferior to Linnen Cloath, and Nets to Fish withal, both being made of the fibers of the Leaves.

The Root or Leaves being fresh, and bruised, and thrown into water, kill the Fish, that they can easily be taken with the Hand. *Marcgr.*

The Wood is as good as Touch-Wood to kindle Fire, *Acosta,* or with another harder rub'd on it to beget Fire, and to hang Hamacs by. *Marcgr.*

They put forth new Leaves, like *Sedums,* being hang'd up in a Room.

Piso's Figure is very faulty.

It is somewhat doubtful whether this be the same Plant grows in *Mexico,* call'd *Maguei* or *Metl. Columna* thinks this differs from it. However most of the properties of that of *Mexico* agreeing to this, I have set down the uses they in *Mexico* make of it, *viz.*

About

The Natural History of Jamaica. 247

About *Mexico*, and other places in *Nova Hispania*, there groweth a certain Plant called *Maguies*, which yieldeth Wine, Vinegar, Honey, and black Sugar, and of the Leaves of it dried they make Hemp, Ropes, Shoes which they use, and Tiles for their Houses, and at the end of every Leaf there groweth a sharp point like an Awl, wherewith they use to bore or pierce through any thing. *Chilton. ap. Hakl. p. 3. p. 462.*

There is much Honey, both of Bees, and also of a kind of Tree, which they call *Magueiz*. This Honey of *Magueiz* is not so sweet as the other Honey is, but it is better to be eaten only with Bread than the other is, and the Tree serveth for many things, as the Leaves make Thread to sow any kind of Bags, and are good to cover and Thatch Houses, and for divers other things. *Hawks ap Hakl. p. 2. p.464.* where he tells us it grows about *Mexico, p. 405.* The *Indians* are given much to drink both Wine of *Spain*, and also a certain kind of Wine which they make with Honey of *Magueiz*, and Roots, and other things which they use to put into the same. They call the same Wine *Pulco*, they are soon Drunk, and when so, are given to Sodomy, &c. wherefore all Wines are forbiden by a Penalty on Buyer and Seller *ib.*

There (in the way from *Panuco* to *Mexico*) also groweth a strange Tree, which they call *Magueiz*, it serveth them to many uses: below by the Root they make a hole whereat they do take out of it twice every day a certain kind of Liquor, which they Seeth in a great Kettle, till the third part be consumed, and that it wax thick, it is as sweet as any Honey, and they do eat it. Within twenty days after that they have taken all the Liquor from it, it withereth, and they cut it down, and use it as we use our Hemp here in *England*, which done they convert it to many uses, of some part they make Manties, Ropes, and Thread: of the ends they make Needles to sow their Saddles, Pannels, and other Furniture for their Horses, of the rest they make Tiles to cover their Houses, and they put it to many other purposes. *Hortop. ap. Hakl. p. 3. p. 492.*

The *Magueit*-Tree or *Cabuya*, yields Wine, Vinegar, Honey, Beds, Threads, Needles, (out of the prickles of the Leaves) Tables, and Hafts of Knives, besides many medicinable uses. *Pedro. Ordonnes ap. Purchas, lib. 7. cap. 4. p. 1421.* speaking of *New-Spain*.

Oviedo in his *Coronica de las Indias, lib. 7. cap x.* tells us that they make of this and *Heniquen*, or Silk-Grass, good Ropes. The Leaves are laid in Rivers, and covered with Stones, as Flax in *Spain*, for some days, then they dry them in the Sun, after clear them of filth, with which they make many things, especially *Hamacas*, some of this is white, others reddish. The *Indians* with these Threads have broke Prisons, and Chains of Iron several times; nay, on the Continent cut Anchors in pieces, rubbing it in the same place with this Thread, and putting now and then some small Sand, taking a new firm place of the Thread as it breaks.

Hernandez, whose figure is not good, says this Plant alone is sufficient for Fields and Gardens. The Leaves are good Thatch, the Stalks Beams. The fibrous or nervous part supplies the uses of Flax, Hemp, or Cotton, to make Thread or Cloath; the prickles are good for Pins, Needles, Nails, Bodkins, and Piercers to make holes in the Ears. The *Indians* likewise us'd them to do Pennance on their Bodies, neither were they unfit for instruments of War. If this Plant be Lopt, or the Trunc cut off, there issues out forty or fifty *Arrobas* (each of which is thirty two Pounds) of Liquor from each Plant, out of which is made Wine, Vinegar, Honey and Sugar. The Liquor is sweet of its self, and drinkable, growing by boiling thicker, turning first to Syrup, then to Sugar. They mix Water with the Juice, and some Orange and Melon Seeds, adding likewise some intoxicating Ingredients, with which they love to be Drunk. Vinegar is made by mixing the Sugar with Water,

and

248   The Natural History of Jamaica.

and expofing it nine days to the Sun. The Juice brings down the *Catamenia*, is Diuretick, opens the Belly, cleanfes the Kidnies, Bladder, and Ureters, breaks the Stone, it is likewife vulnerary. The rofted Leaf, the Juice fqueez'd out with a little Nitre, takes away Scars if they be yet new, and anointed therewith. The Leaves and Trunc are, when bak'd, eatable. The Leaves, rofted and applied, remedy Convulfions, and take away Pains, efpecially if the hot Juice be drank at the fame time, even if they come from the *French* Difeafe.

This Aloes after ftanding one hundred years in *Avinion*, flowr'd with them, growing to its full Stature in forty five days, as may be feen by *Fontanus* in his Epiftle, in *Clufius*'s *Cura Pofteriores*.

You muft not expect to make Aloes of the Shops of this. *Col.*

The *Mexican* Hiftory, *apud Purchas*, tells us that the Children of eight or nine years of Age, were chaftifed with the Thorns of this thruft into their Bodies, for Difobedience or Negligence, and the Priefts corrected Novices with them, and brought blood for Sacrifices. The Wine made of this had its inebriating quality from the Root, *Tepatcli* mix'd with it. *f.* 998.

This is what the Women Cloath themfelves with in *Sibola*. *Lop. de Gomara*.

It is ufed for a Fence, andcalled *Cardon*; Shirts and Harnaesare made of them. The Fume from it boiled, cures the Pox, caufing very much fweat. *Cluf.*

Laet tells us of *Nequen*, a Cloath for the meaner fort in *New-Spain*, made of *Henequen*.

This bruis'd and fleep'd makes Flax, of which they make very white Cloaths. *C. B.* and of this I believe *Clufius*'s Thread was made, mentioned *Exot. p.* 6.

The Leaves boil'd yields Thread. The Root or Leaves bruifed put in a River, gives a Juice intoxicating the Fifh that they may be taken by Hand. The great Trunc, dried, burns like a match, efpecially being rub'd againft with another. *Du Tertre.*

The tops and tender Leaves make Conferve, the Leaves are fit for Parchment, or Cards to write on, and the Vapour from large peices cures the Pox. *Duret.*

V. *Caraguata-acanga Pif Cat. p.* 118. *An Ananás fylveftris Brafiliana Kirbita vulgo. Herm. par. Bat. Cat. p.* 3 ? Yellow *Penguins* of *Dampier. cap.* 9. *Ananas Americana fylveftris altera minor Barbados & Infula Jamaica Penguin dicta. Pluken. Phyt. Tab.* 258. *Fig.* 4. *Penguins.*

I cannot add any thing material to the Defcription extant in feveral Authors.

It grows very plentifully in the *Caribes*, and *Jamaica*, between *Paffage Fort* and the Town, as likewife towards the Sea-fide by the Salt Ponds.

The Fruit is very acceptable by reafon of its grateful acidity, but it not only fets the Teeth fpeedily on edge, but likewife brings the Skin off of the Roof of the Mouth and Tongue. It quenches Thirft extremely, and on the landing of the *Englifh* Forces on *Hifpaniola*, in their want of water, was thought to fave many Lives by that its quality.

A Spoonful of the Juice with a little Sugar given to Children, cures them of Worms, and the Thrufh or Ulcers of the Mouth. It is good in Fevers. It is very Diuretick, and brings down the *Catamenia* very powerfully, even to too great a quantity if the Dofe be not moderated. It caufes Abortion in Women with Child, of which Whores being not ignorant make frequent ufe of it to make away their Children. An excellent Wine is made

of

The Natural History of Jamaica. 249

of the Fruit, the Juice being squeez'd out and kept for use; but because 'tis strong, it intoxicates and heats the Blood. *Piso.*

Piso's Figure is bad, as is *Hernandez*'s, if this be it he means, to which the Description in every thing agrees, but in the shortness of the Stalk. *Jo. de Laet* thought it the same.

The Fruit helps Ulcers in the Mouth from heat. *Hern.*

It is antiscorbutick, and good in Fevers. *Laet.*

VI. *Aloe Yucca foliis. Cat.* p.118. *An Aloe Americana non spinifera Yucca foliis Domini Bobart.* Pluken. *Alm.* p.19? *Yucca Virginiana foliis per ambitum apprime filatis. Ejusd. ib.* p.396? *Yuccafolia filamentosa & bulbosa. Ejusd. ib.?* *Aloe Pita dicta. Herm. par. Bat. cat.* p. 3? Silk-Grass.

This has long arundinaceous Leaves, and grows in the *Caymanes.*

This I suppose to be what *Lery* tells us the Savages in *Brasil* made use of for Fishing-Lines, and Bow-Strings, and *Du Tertre* says, is made into Stockings and Hamacks. They make a running Knot fast to a Tree, and so draw the Leaf through, first one way, and then another, and keep the Flax in their Hand. *Tertre.*

In the *Spanish* Galeons that were taken at *Vigo*, was a good quantity of a kind of this, or Hemp, or Flax, the fibers whereof were three Yards long, and very strong. It was of a grayish colour; and I am told is brought to *Spain* to be wrought, either there or in *Italy*, into Point called *Punta da pita.*

In the *East-Indies*, in the Kingdom of *Orixa*, near the River *Ganges*, grows a Plant, wich yields fibers as Flax or Hemp, or this Aloes. The fibers are whiter than those of the Aloes, and finer, and of these mention is made in *Haklyt*, in the two following places.

Cloath of Herbs, which is a kind of Silk, which groweth amongst the Woods, without any labour of Man, and when the Bole thereof is grown round as big as an Orange, then they take care only to gather them. *Frederick ap. Hakl.* p.2. p.230 ?

In *Orixa* is Cloath made of Cotton, and great store of Cloath which is made of Grass, which they call *Teroa*, it is like a Silk, they make good Cloath of it, which they send for *India*, and divers other places. *Fitch. ap. Hakl.* p. 237.

VII. *Aloe visci in modum arboribus innascens. Cat.* p. 119.

The Leaves of this are very large, and like those of Aloes, always ready, and fit to retain the Rain water, several Stalks rise up from among them about three Foot high, their tops being Cones or Spikes a Foot long, beset very thick on every hand with an Inch long Foot-Stalks, sustaining each several Heads, sometimes lower, and sometimes higher. They are oblong, roundish, in the middle biggest, and have each a red top, and three prickles going out thereat.

It grows on the large Arms and Trunes of great Trees especially those decaying through Age.

In scarcity of Wells or water in dry Countries, Travellers come to this for relief, it being capable to hold much pure water, able to extinguish their Thirst. The best Polypody grows on this.

The Thread lies on the Surface, whereas in the other Aloes it lies within. *Tertre.*

This seems to be mention'd by *Knivet*, *viz.* Thus seeing my self at the last cast, I espied a great Tree, in which grew a thing of thick long Leaves, called by the *Indians Caravala*; as big as the Nest of an Eagle, I got me into that, &c. *Knivet, ap. Purchas, lib.* 6. *cap.* 7. §. 2. p.1210. where he hid himself in it from the *Indians* shooting at him.

S f f   VIII. *Orchis*

VIII. *Orchis elatior latifolia asphodeli radice, spica strigosa. Cat. p.* 119. *Tab.* 147. *Fig.* 2.

The Root of this was double, fungous, two or three Inches long, being somewhat of the shape of those of the *Asphodel's* or *Ornanthe's*, and not so round as those of the *Orchides Testiculata*. The Stalk was about a Foot and an half high, being slender, jointed, and beset with Leaves alternatively, which had Foot-Stalks of about an Inch long, by their under part next the Stalk incompassing it, and making a Sheath for it. The Leaves are about three Inches long, and one and an half broad in the middle where broadest, whence they decrease to both extremes, ending in a point, being nervous, and something like the Leaves of *Saponaria*. The top of the Stalk, about two Inches in length, is a slender Spike of Flowers, under each of which is a small membranous Leaf: the *Petiolus* of the Flower is crooked, the Spur blunt, the *Labellum* small, and the *Galea* large, and divided as others of this Kind.

It grew in the Woods of Mount *Diablo*.

IX. *Viscum radice bulbosâ majus & elatius, delphinii flore ferrugineo guttato. Cat. p.* 119. *Tab.* 148. *Fig.* 1. *An Tzauxochitl. Hernandez. p.* 433? *vel Amazautli. Ejusd. p.* 349? *Uracatu Marcgr. p.* 35?

This grows on the Truncs and Arms of Trees, as Misletoe, or others of this Kind, and is the largest of all those I have met with of its sort. The Roots are large, and the Leaves many, long, narrow, smooth, of a dark green colour, and somewhat like those of our common White Lillies. The Stalk is round, tough, brown, crooked, rising six Foot high, and join at every eight or nine Inches distance, where are Branches standing straight out with several Flowers, whose Foot-Stalks are an Inch long. The Flowers, themselves are of six or seven *Petala*, each of which is narrow at the beginning, and round towards its end, being of the shape of a Spoon, only not hollow, of a ferrugineous colour, and spotted, except one difform, hooded *Petalum*, which is in the middle of a white colour, and within which are several Orange colour'd *Stamina*.

It grows on the Truncs and Arms of Trees, between the Town and the Salt-Ponds.

X. *Viscum radice bulbosâ minus, delphinii flore rubro speciosa. Cat. p.* 119. *Tab.* 121. *Fig.* 2. *An viscum arboreum seu epidendron scilla foliis Barbadensium. Pluk. Alm. p.* 390?

This has a great many white, thick fibers or Roots, like the fibers of Leeks, or *Capreoli* of Ivy, taking firm hold of the Trees Bark whereon it grows, and being matted, or interwoven one within another. When united they send up one thick greenish, almost round, a little compress'd bulbous or tuberous Leaf or Root, of an Inch Diameter, cover'd with some brown wither'd Filaments. From the top of this comes two smooth, striated, hollow, hard, light colour'd green Leaves, three Inches long, and one broad, between which springs out a naked, brown, jointed, round, smooth Stalk, about a Foot high; near, and at the top of which stand several long, reddish purple Flowers, very beautiful, made up of six *Petala*, five whereof are broader, and shorter than the others of this Kind, standing round, and inclosing in their Center a sixth large difform one, or inward Flower, like the Flower of Lark-Spurs, which is in the inner part thereof yellowish, with purple streaks.

It

It grows on the Ebonies, and other Trees in the *Savanna* Woods, very plentifully, as also on the Palisadoes inclosing the Gardens of the Town of St. *Jago de la Vega*. *Terrentius ap. Hernand.* his Description is as exact as can be from the Figure of this Plant.

XI. *Viscum delphinii flore minus, petalis è viridi albicantibus angustioribus radice fibrosâ. Cat. p.* 120. *Tab.*121. *Fig* 3. *Orchidi affinis planta parasitica folio crasso sulcato. Par. Bat.* p. 187. *Epidendron Curassavicum folio crasso sulcato vulgò. Ejusd. ibidem.*

Several *Caprioli*, a little longer, but of the colour and bigness of those of climbing Ivy, warp and knit themselves one within another, sticking very close on every side to the Bark of the Tree, or Palisado, or even into the Body of the Tree, (for the most part rotten) on which they grow. From those Roots come out several purplish, round, jointed Stalks, from the uppermost joint, about two or three Inches from the Root, (each of the under *Internediums* being very short) stands one pointed Leaf, which is very thick or almost round, three or four Inches long, of the bigness of a Goose-Quill, the two insides flatted, purplish in colour. Out of the inside of this, upon a three, four, or five Inches long, green, round Foot-Stalk, or top of the Stalk, come several Flowers, having a long green *Calyx*, with five greenish white, narrow *Petala*, standing Star-fashion, and in the middle is one white hooded, large, difform *Petalum*, of a very odd shape, and to these follow an angular Tricapsular knob, very like those of several of the bulbous Tribe, in which is contained a white, very small *Farina*.

It grows on old Palisadoes and Trees, about the Town of St. *Jago de la Vega*.

XII. *Viscum delphinii flore albo guttato minus, radice fibrosâ. Cat.* p. 120. *Tab.* 148. *Fig.* 2.

This from a matted Root like the others of this Kind, sends out several Leaves three Inches long, and not one quarter broad, almost triangular, and of a yellowish green colour, from the midst of which comes a Stalk in every thing like the former, only the Flowers are more, and different, *viz.* each is made up of four little white *Petala*, spotted with brown, and one large one with fewer spots, on which is a small yellow Hood, as in the Flowers of Larks-Spurs, and opposite to it, one like it of a blue colour, all standing on Inch long Foot-Stalks, round the top of the Stalk.

It grows on the Ebonies every where, especially on the way towards the *Angels*.

XII. *Viscum delphinii flore minimum. Cat.* p. 120. *Tab.* 148. *Fig.* 3.

This was for Roots, manner of growth, *&c.* exactly the same with the preceding, only much lesser. The Leaves were striated, green, carinated, and long, the Stalk not over three Inches high, having two or three lesser Leaves on it. On the top of the Stalk were four or five Flowers, standing on crooked, large Foot-Stalks, which are the Rudiments of the Fruit. The Flowers were so small that their *Petala* and parts were not easily to be distinctly discerned, but I suppose they were the same with the others.

It grew on the Trunks of Trees by the way going to *Guanaboa* on the *Red Hills* and other places.

This is very like *Tsjeron-Mau-Maravara. H. M.* p. 12. p. 11. *Tab.* 5. but lesser.

XIV. *Nym-*

XIV. *Nymphaa alba major.* C. B. *Cat. p.* 120.
I could not obferve any difference between that here and in *Europe*.
It grows in a Pond near the *Angels*, on the frefh water *Laguna* in the *Caymanes*, and in the way to it very plentifully. We are told it grows in *Java*, by *Bont. p.* 129.

It is for qualities the fame with the *European* being dry. The Leaves applied cure hot Ulcers, Inflammations, and the *Erifypelas* of the Legs. The Oil of its Root is likewife moft excellent againft both inward and outward hot Diftempers, the Root boiled is accounted an Alimentary Medicine, by way of Sallat, and cures Fluxes of Blood. *Pifo.*

It extinguifhes the Appetite to Venery both inwardly and outwardly ufed. *Ger.*

The Infufion of the Flowers in water for a night, drank in the morning, is ufed by the *Turks* to keep them from the Head-ach. *Dorft.*

In *Ferdinando de Soto's* Expedition into *Florida*, written by a *Portugal* of *Elvas, p.* 54. *ap. Purchas, p.* 1533. The *Indians* being furrounded in a Lake by the Chriftians, they endeavour'd their efcape in the night, with thefe Water-Lily-Leaves on their Heads.

XV. *Nymphaa Indica flore candido folio in ambitu ferrato, Commel. Cat. p.* 120. *Nymphaa Indica crenata flore pleno, candido, Pluk. Alm. p.* 267. *Nymphaa Ægyptiaca alba folio crenato radice tuberofa. Bob. hift. part.* 3. *p.* 513. *An Nymphaa Malabarica alba, crenatis foliis, radice fibrofa floribus ex albo refaceis. Ej. ib?*

This differs from the former by its indented Leaves, which are deeply cut in on the edges, agreeing with the Defcription and Figure of this Plant extant in Authors.

It grew on the Frefh River going up to the *Laguna.*

The *Egyptians* eat the Stalks in the Heats. They ufe the Leaves and Flowers, as likewife the Juice for all hot Pains, Inflammations, Burnings, Ulcers, &c. as likewife the Oils, which are ufed in want of Sleep, The Seed and Roots are ufeful in Dyfenteries, *Diarrhæas, Gonorrhæas*, and the *Fluor albus*; but it makes People frigid, therefore 'tis us'd by Hermits. *Alp.*

The *Egyptians* make their *Sarbet Nufar* of Sugar diffolv'd, from which the water is evaporated till it Candy's, then they put to it fuch a fmall quantity of the depurated Juice of this Plant, fo as not to hinder its Concretion. *Vefling.*

This was carried to the *Indies* by way of Merchandife. *Bod.*

Salmafius mended *Pliny* putting the word *Refidentibus* for *Refedentibus*; for it grows on the top of the water. *Pliny* tells us it was ufed for Bread by the *Egyptians*, and that when hot it was good, never occafioning Loofenefs or *Tenefmus.*

Diodorus Siculus, mentions it among the Edibles of *Egypt.*

It extinguifhes Venery very much. *J. B.*

Its Root is ufed as Meat. The Root is alfo given in Decoction for the *Dyfuria*. The Seeds candied with Sugar take off the heat of the Bones. The Leaves, beaten together with thofe of *Ottel Ambel*, and boil'd with Butter, makes an *errhinum* which is good for pain'd Eyes. *H. M.*

The *Ambel* of the *H.M.* and the *Lotus Ægyptia* of *Alpinus*, feem to me to differ in very little from each other.

XVI. *Nymphæa minoris affinis Indica flore albo pilofo. Commelin. Cat. p.* 121. *Nymphaa Indica fubrotundo folio minor, flore albo fimbriato. Plukenet. Alm. p.* 267. *Nymphoides Indica flore albo fimbriato. Tournef. Inft. p.* 154.

This

This had a Leaf somewhat like Coltsfoot, which floated on the top of the water like the Leaves of Water-Lilies, each Leaf was roundish, and about two Inches Diameter, having a defect towards the Foot-Stalk, being thick, of a yellowish green colour, and smooth without any appearing Nerves in it. The Foot-Stalks of each Leaf were about a Foot long, or reached to the bottom of the water, round, and brownish, and out of them, just under the Leaf its self, came the Flowers, which were several, some on Inch long, others on shorter Foot-Stalks, being enclosed in a *Calyx*, made up of several small Leaves, like the former, and containing in rotten Heads, some pretty large Seeds.

It grew on the surface of standing waters in the *Savannas*, where they were not deep.

The whole Plant, bruised and boiled with Butter, taken inwardly, is an Antidote against the biting of the Snake called *Cobra Capella*. *H. M.*

XVII. *Nymphææ affinis palustris, plantaginis aquaticæ folio, flore hexapetalo stellari cæruleo. Cat. p* 121. *Tab.* 149. *Fig.* 1.

This has a great many white fibers, like those of the Roots of a Leek, and several Leaves, the Stalks of which inclose one another, and are full of Cells or Membranes, as other watry Plants. These Foot-Stalks are about seven Inches high, and about their middle, like to the Figures of *Gramen Parnassi*, send out an Inch long Foot-Stalk, sustaining a blue, hexapetalous, starry Flower, after which follows a great many small, flat, blue Seeds. The Foot-Stalks have at their ends green, roundish, nervous Leaves, like those of the lesser Water Plantains.

It grows in the *Savannas*, in places where water has stood, most part of the year.

This is of the same Kind with *Carim-Gola. H. M. P.* 11. *p.* 91. *Tab.* 44. only lesser and fewer Flowers come out together.

XVIII. *Canna Indica. Riv. Cat. p.* 121. *De Bry Florileg. Canna Indica flore rubro. Swert. part.* 2. *Tab.* 32. *Cannacorus latifolius vulgaris. Tournef. El. p.*295. *Inst. p.* 367. *Canna Indica sylvestris fructu saxea duritiei, & gypsi adinstar, mansa sub dentibus scrupose*, Wild Plantine *Barbadensibus dicta. Pluken. Alm. p.*80.

It grows in the Lower Grounds very commonly, having scarlet coloured Flowers.

The Leaves are cold in the second Degree, and cleansing. They are useful against many cutaneous, inward and outward Distempers. Applied to the right *Hypochondre*, with White Water Lily, and *Aninga* Oil, they cure an over-heated Liver, or Spleen. The Gum coming out of this Plant does the same. *Piso.*

The Seeds are made into praying Beads. *Maregr.*

This Plant repels Tumours, for the Root which is used is glewy, of a sweet taft, and cold and moist. *Hern.*

It seems to wrap up, Gum *Elemmi. Bauh.*

It grows very well under the water Spouts in *Portugal. Clus.*

Out of the Fruit, a little rosted, a Juice is drawn, which put into the Ears eases their pain. Of the same, and Sugar is made a Mass, which applied to the Navel cures the Diabetes, proceeding of hot Fevers. The Juice of the Root weakens the Poison of *Mercurius Sublimatus. H. M.*

XIX. *Canna Indica radice alba alexipharmaca. Cat. p.* 122. *Tab.* 149. *Fig.* 2. *Canna Indica angustifolia, pediculis longis ad imum folium, nodo singulari geniculatis. Pluken. Alm. p.* 79. Indian Arrow-Root.

This has a two or three Inches long, jointed Root, as big as ones Thumb; white, tapering, each *internodium* being half an Inch long, and at the joint having several two or three Inches long fibers to draw its Nourishment from the Earth. From this Root rise several Leaves, having three Inches long, broad, Foot-Stalks, inclosing one another with a white Ring at the Leaves setting on, they are four Inches long, and two broad, near the round Base, where broadest: thin, nervous, grassie, and of a yellowish green colour, in every thing like the *Canna Indica*. The Flowers, by their Buds, seem to agree in every thing with the foregoing, only are smaller.

This Plant was first brought from the Island *Dominica*, by Colonel *James Walker*, to *Barbados*, and there planted. From hence it was sent to *Jamaica*, being very much esteem'd for its Alexipharmack qualities. That Gentleman observed the Native *Indians* used the Root of this Plant with success, against the Poison of their Arrows, by only mashing and applying it to the poison'd Wounds.

The Root of this bruised and applied remedies the Poison of the *Mançanzel* and Wasps of *Guadaloupe*, even stopping a begun Gangreen. *Tertre. Rochf.*

I am inclinable to think this to be mentioned by *Harcourt* to grow in *Guiana*, where he says that the Juice of the Leaf called *Uppee*, cures the Wound of the Poisoned Arrows. *Harcourt ap. Purchas, lib. 6. cap. 6. p. 1276.* and by Sir *Walter Rawleigh*, where he tells us that there is a Root called *Tupara*, the Juice serving for ordinary Poison, quenching the heat of burning Fevers, healing inward Wounds, and Veins bleeding within the Body, Sir *Walter Raleigh*, *p. 59. ap. Hakl. p. 649.* I believe this also to be that Root spoke of by *Lopez de Gomara*, to be a Counter-poison to the *Mançanzel*, which he says grows in *Cartagena*, and was said to be the Herb wherewith *Alexander* heal'd *Ptolomy*, and which was discovered by a *Moor* in *Catalonia*, and was called *Scorçonera*, in which he might be easily mistaken, there being some resemblance between the Root of this Plant and of that.

Joh. Horton. ap. Hakl. p 3. p. 487. says that eight of their Men, with their General and Captain *Dudley*, going ashore at *Cape Verd*, were by the *Negros* there wounded with poisoned Arrows, amongst which the eight died, the General being cured by a Clove of Garlick drawing the Poison out of his Wound, he being taught it by a *Negro*.

XX. *Alsinefolia per terram sparsa, flore hexapetalo purpureo. Cat. p. 122.*

This from a small fibrous Root, sends out several Branches lying along the Ground, red, of about a Foot in length, having Leaves like those of Chickweed, set one against another at equal distances along the Stalk: the Flowers stand at the ends of the Branches are few hexapetalous, of a purple colour. After these follows a green Head, inclosing several roundish, flat, brown Seeds.

It grows every where in the *Savannas*, especially in dry places.

Chap. XVII.

Of Herbs whose Flowers are composed of several Flowers.

I. Sonchus Lævis Cord. hist. Cat. p. 122. Common smooth Sowthistle.
This is common every where through the whole Island.
It is cooling and adstringent, and good for hot and burning Stomachs. It increases Milk. It is proper for pains and gnawings of the Stomach. It is a remedy against the bitings of Scorpions. *Fuchs.*
It begets Milk in Nurses, in Pessaries it is good for Inflammations of those parts. *Lon.*
They are eat in *Italy*, especially the tender Roots by way of Sallet. *Math.*

II. *Sonchus asper laciniatus.* Park. Cat. p. 123.
I found this on the side of a Hill near Mr. *Batchelor*'s House, about four Miles from the Town of St. *Jago de la Vega.*
It is good against cold Pains of the Stomach, and Obstructions of the Liver, and Gall Bladder. *Adv.*

III. *Hieracium fruticosum, angustissimis graminis foliis, capitulis parvis.* Cat. p. 123. Tab. 149. Fig. 3.
This had a whitish oblong Root, with several fibers to draw its Nourishment, from whence rose a solid, straight, striated, green, small Stalk, about a Foot high, having Branches set opposite one to the other, going out of the *Alæ* of the Leaves. The Leaves were about two or three Inches long, and very hard, like those of Grass, without any Foot-Stalks, ending in points, by which this may be sufficiently distinguish'd from all I have hitherto seen of this Kind. The tops of the Twigs have small Heads of Flowers and after them a long pappous Seed, as others of this Kind.
I am not certain where I found this.

IV. *Hieracium minimum, longis integris & angustis foliis.* Cat. p. 123. Tab. 150. Fig. 1.
From a small, fibrous, oblong Root, springs one round, red Stalk, three or four Inches high, having here and there, without any order, little Branches set with many long, not indented, nor sinuated, narrow Leaves. Their edges are hairy, and their backsides spotted with blackish spots. At the tops of the Branches stand yellow Flowers, in every thing like those of the other *Hieraciums.*
It grows every where in the drier or sandy places of the Town *Savanna.*

V. *Dens leonis, folio subtus incano, flore purpureo.* Cat. p. 123. Tab. 150. Fig. 2.
This has several reddish, Inch long Roots, which united make an half Inch long white one, sending forth round the top of the Root, on the surface of the Ground, a great many Leaves three Inches long, and one broad, near the end where broadest. The Leaves have near the Root several deep Incisures or jags, and there they are narrow, as the others of this Kind. The upper side of the Leaf is of a dark green colour, and under it is very white or woolly, and in every thing for shape it is like the Leaf of our common

Dens

Dens Leonis. In the middle of these Leaves rise one or more Stalks, they are naked, pale green, covered with Wool or Down, a Foot and an half high, and the Head has on the outside a *Calyx,* made up of many green Leaves, some purple *Petala* standing round a whitish hairy Matter, after which follows pappous Seeds, standing round, ready to fly away when ripe, like the Seeds of the common *Dens Leonis.*

It grew near Colonel *Cope*'s Plantation at *Guanaboa.*

The Decoction is given to Women in Childbed. It dissipates Wind, provokes the *Catamenia,* is good against Convulsions, takes away Gripes, and is a remedy against all sorts of Cold, for it is hot and bitter. *Hernand.*

VI. *Conyza major inodora, helenii folio integro sicco & duro, cichorii flore albo è ramorum lateribus exeunte.* Cat. p. 123. Tab. 150. Fig. 3, 4.

This at first coming up has a great many Leaves, like those of the *Jacobæa folio integro,* five Inches long, and one and an half broad near the end where broadest, beginning very narrow, it continues so for two Inches of its length, and ends in a round point, 'tis hard, smooth, of a dark green colour, snipt or indented about the Edges. In a while after these Leaves rises a round, strong, green Stalk, four Foot high, from every joint, at a quarter of an Inches interval, goes one of these Leaves, inclosing the Stalk where it is join'd to it. It has Branches towards the top, standing round at every joint, divided into others, which are beset with lesser Leaves. *Ex alis foliorum* come the Flowers, without any Foot-Stalk, standing in several green Leaves, being a great many white, long, *Petala,* standing round like those of Cichory. After these follow pappous Seeds.

This is very anomalous, but I think it comes nearest to this place, although, if I rightly remember, 'tis not milky.

It grew on the other side of the *Rio Cobre,* near the Town of St. *Jago de la Vega,* in *Guanaboa,* near Colonel *Cope*'s House in his Plantation, and in the Thickets, near the *Moneque Savanna,* very plentifully.

The Stalks and Leaves of this Plant being hard, are made use of for Brooms to sweep and clean Houses withal.

VII. *Conyza inodora, helenii folio, integro, duro, angusto, oblongo, capitulis in lateribus ramorum conglomeratis.* Cat. p. 123. Tab. 148. Fig. 4.

This had a large oblong Root, with some lateral fibrils, from whence rose a single, round, striated, hollow Stalk, about two Foot high, having Leaves set on it alternatively, without any Foot-Stalks, their lower part whereby they are joined to the Stalk, having a membrane inclosing it. These Leaves are about five Inches long, and half an Inch broad near their top where broadest, from the Foot-Stalk, increasing to near the top, where they are broadest, ending round, being of a pale green colour, and corrugated on their surface. Towards the top come the Flowers in a Spike, standing without any Foot-Stalks, being inclosed with a *Perianthium,* made of some few dry membranes of a brown colour, sometimes one of these, and sometimes many being conglomerated together. After each of these follows small, brown cannulated Seeds, having much *Pappus* on it.

I found it about Mount *Diablo* very plentifully.

Any body who compares this Description and Figure with those of *Chrysanthemum Virginianum caule alato ramosius flore minore,* Pluk. Phyt. Tab. 139. Fig. 3. Alm. p. 100. will find them very different from one another, though Dr. *Plukenet* in his *Almagest.* p. 46. thinks they may be the same.

VIII. *Conyza*

VIII. *Conyza fruticosa, cisti odore, floribus pallide purpureis, summitatibus ramulorum insidentibus, capitulis & semine minoribus.* Cat. p. 123. Tab. 151. Fig. 2.

This Shrub rises to about six or seven Foot high, it has several Stems as big as ones Thumb, covered with a reddish brown Ropy, or membranaceous tough Bark, and the Branches go out opposite one to the other, or sometimes three together, they are thick set with Leaves, standing on a quarter of an Inch long Foot-Stalks; they are Inch long, half as broad near the round Base where broadest, the Nerves running from the Foot-Stalks end, as from a common Center, they are somewhat rough, viscid and smell like those of *Cistus.* The tops are branched out into several Foot-Stalks, sustaining several naked Heads like those of *Jacobæa* of a pale purple colour. After these follow many small, light brown, oblong, canulated pappous Seeds. There are some small varieties of this.

It grows by the way going to *Guanaboa* on the Red Hills, and on Mount *Diablo* on a small *Savanna* very plentifully.

IX. *Conyza fruticosa, cisti odore, floribus pallide purpureis summitatibus ramulorum insidentibus, capitulis & semine majoribus.* Cat. p. 124. Tab. 151. Fig. 3.

'Tis in every thing like the former, only seemed somewhat larger, the Heads were also much larger, being inclosed by several Leaves of a brown colour, surrounding *Squammatim* the Flowers and Seed.

X. *Conyza fruticosa, folio hastato, flore pallide purpureo.* Cat. p. 124. *Eupatorium Americanum, foliis urticæ mollibus & incanis.* Tournef. Inst. p. 456. *An Conyza Americana urticæ folio flore caruleo.* hort. Amst. p. 99?

This by a large woody Stalk rises to about seven Foot high, the Bark is of a whitish colour, the Branches are quadrangular, and set opposite one to another. The Leaves stand likewise on the ends of the Twigs on an Inch long Foot-Stalks, opposite one to the other, being almost triangular, an Inch and an half long, and more than an Inch broad at Base, a little hairy, having Ribs run through the Leaf to the several parts of it from the end of the Foot-Stalk. They are of a yellowish green colour, and very odoriferous. *Ex alis foliorum* towards the tops are small two Inches long Stalks, having here and there smaller Leaves of the same shape with the larger, supporting several naked Flowers, each whereof have an half Inch long Foot-Stalk, is naked, and composed of many pale purple Flowers with their several *Stamina,* standing very close by one another, and all inclosed by many green, long, scaly Leaves. After these follow several small, oblong, cannulated, or striated Seeds, of a light brown or gray colour, having some stiff pappous hairs on their ends.

It grows every where about the Town.

It is counted an admirable vulnerary, being only beaten and applied, having cur'd one who was Lanc'd through the Body at the taking of the Island.

There is a variety in this, the Leaves being sometimes more hairy, and smaller.

XI. *Conyza fruticosa flore pallide purpureo, capitulis e lateribus ramulorum spicatim exeuntibus.* Cat. p. 124. *An Corino affinis arbor Americana, Tremate Brasiliensibus.* Marcgr. Pluken. Alm. p. 121?

U u u This

This Shrub rises to four or five Foot high, having a round, whitish, woody Stalk, its Branches come out at top, and are reflected, or bow downwards. The Twigs are thick set with small Leaves, coming out alternatively on short, or no Foot-Stalks, they are somewhat curled, or uneven on their surface like Sage Leaves, an Inch and an half long, and half an Inch broad in the middle where broadest, ending in a point, and being whiter on the under side. *Ex alis foliorum* come the Flowers, which are of a light purple colour, like the others of this Kind, inclosed with some pale brown Leaves. After these follow small, oblong, white pappous Seeds.

It grows in the clear'd Woodlands at the Crescent Plantation, and in the *Caribe* Islands.

The bruised Leaves are good against Pains and Inflammations of the Eyes. The Leaves and pappous Seeds, because of their being Aromatick, are good in Baths to cleanse and scour. *Piso.*

XII. *Conyza major odorata, seu baccharis, floribus purpureis nudis.* Cat. p. 124. Tab. 152. Fig. 1.

This has a large, woody, short Root, having very many fibrils on every side. The Stalk rises as high as that of *Baccharis Monspeliensium.* Ger. Park. being round purple, solid, having Leaves standing without any order, on an Inch long Foot-Stalks, being three Inches long, one broad in the middle where broadest, rough, and notch'd about the edges. The Stalks are divided towards the top into several Branches, each being subdivided into several others, on the tops of which are round, purple Heads, of the bigness of those of the *Baccharis Monspelienfium,* consisting of an innumerable company of dry *Petala,* and white Down, almost like those of *Gnaphalium Americanum,* after which in small time follow many small, brownish, cannulated, pappous Seeds.

The whole Plant is very gratefully odoriferous.

It grew by the Sea-side in the Marish Grounds by Mr. *Delacree*'s in *Liguanee.*

XIII. *Conyza urtica folio.* Cat. p. 124. Tab. 152. Fig. 2.

This had several white, strong filaments for Roots, with lateral fibers, from which went up a square, reddish coloured, woody Stalk, a Foot and an half high, more or less. The Leaves, as well as Branches, stand opposite to one another, the first on three quarters of an Inch long Foot-Stalks, being about an Inch and an half long, and three quarters of an Inch broad in their middle where broadest, hairy, from their Foot-Stalks end increasing to the middle, and thence decreasing to a point, being very much serrated about the edges, and like the Leaves of Nettles. The Flowers and Seed come at top, the latter being cannulated, small, black and pappous, inclosed with small Leaves for their *Calyx,* set round them *squammatim,* as in others of this Kind.

It grows in *Jamaica* and the *Caribes.*

XIV. *Conyza folio hastato, seu triangulari, serrato, glabro.* Cat. p. 124. Tab. 153. Fig. 1, 2.

This Plant has a long, white Root, with several lateral fibrils, sending up a green, smooth foursquare Stalk, one Foot and an half high, bigger than a Swans-Quill. At an Inch and an half's distance the Branches come out set opposite one to the other, and the Leaves on the Branches in like manner, standing on Inch long Foot-Stalks, being almost triangular, they are an Inch broad at Base, and a little longer from the Foot-Stalks end to the opposite point, serrated pretty deep on the edges of a yellowish green colour,

being

being nervous, having Ribs running through them like those of Plantain. The Flowers come at top of the Branches, are whitish and naked like those of Groundsel, and after them follows a small, pappous, cannulated, brown Seed.

There is a variety of this, having the Leaves as it were eared at Base. It grew in Colonel *Nedham*'s Plantation in Sixteen Mile Walk.

XV. *Conyza minor procumbens fœtida, flore luteo, seminibus tomento obductis.* Cat. p. 124. Tab. 153. Fig. 3.

Several Strings, or brown, small fibrils, send forth on all Hands square Stalks, of about a Foot in length, at every two or three Inches of which are joints, and from these proceed the Branches, Twigs and Leaves. The Leaves are hairy all over, without any Foot-Stalks, an Inch long, and half as broad in the middle, where broadest, snipt, or indented about the edges, of a yellowish green colour, and unsavory smell. The Flowers stand many together on the tops of the Branches, and *ex alis foliorum*, and are of a yellow colour. After these follow a great many long, black Seeds, covered all over with a white Wool.

It grows in several places where Woods have been clear'd in most Plantations of the Island.

XVI. *Aster folio oblongo, integro, flore pallide cæruleo.* Cat. p. 124.

This rises about a Foot and an half high, with a reddish, round, smooth Stalk, having long pointed Leaves, broadest in the middle, a little whitish, smooth, without any notches on the edges. The Flowers are many, of a very light blue colour, the middle and *Barbula* being both of the same colour. I did not observe whether the Seeds were pappous or not, so to know not whether it be a true *Aster* or *Chrysanthemum asteris facie.*

It grows in moist watery places in the *Savannas* about the Town.

The Description of this shews it to be very different from the *Aster novæ Belgiæ latifolius umbellatus floribus dilute violaceis.* Herm. Hort. Leyd. This I take notice of, because Doctor *Plukenet* in his *Mantiss.* p. 29. thinks they may be the same.

XVII. *Aster canadensis annuus non descriptus.* Brunyer. Hort. Bles. Cat. p. 124. *Conyza acris annua alba hirsuta major.* Pluken. Alm. p. 117.

I found this in several places of this Island, as likewise plentifully in the *Caribes.*

XVIII. *Senecio major florum calyce purpureo.* Cat. p. 124.

This has several white fibrils, going out of every side of an oblong, reddish Root, from whence springs up a round, green, juicy Stalk, about one Foot and an half high. It has Leaves irregula ly plac'd, and thick set on every side without any Foot-Stalks, about four Inches long, and an Inch Broad near the farther end where broadest, beginning narrow, increasing to near the end, whence it straitens, ending in a point, being very much sinuated, or jagged on the edges like ordinary Groundsel. At the top come the Flowers, being many, naked, of a yellowish white colour. The calycular Leaves are purple, and each stands on sometimes a longer, and sometimes a shorter Foot-Stalk.

It grew in the clear'd Ground in Colonel *Nedham*'s Plantation in sixteen Mile Walk.

This seems to me to be very different from the *Senecio viscosus Æthiopicus flore purpureo.* Breyn. cent. As may be easily gathered from their Titles and Descriptions. Dr. *Plukenet* in his *Mantissa,* p. 170. thinks they may be the same.   XIX. *Senecio*

XIX. *Senecio minor, bellidis majoris folio. Cat. p.* 125. *Tab.* 152. *Fig.* 3.

This has several Inch and an half long, smooth, small, white Roots, no bigger than Thread. The Stalk is round, hoary, about five or six Inches high, having some few Leaves on them, without any order. The Leaves have no Foot-Stalks, but stick to the Stalk with a narrow beginning, and augment to their round end, being in all about an Inch long, rough, hoary, and of a whitish green colour, very often having two or three notches in them, and like in shape to the Leaves of the *Bellis major caule folioso, C. B. pin.* The top of the Stalk and small Branches coming out *ex alis foliorum*, support the Flowers, which are in every thing like those of Groundsel, being made up of many small, yellow Flowers, close set together, and encircled by many whitish, long, narrow Leaves.

It grew on the Banks of the *Rio Cobre*, under the Town, on the same side of the River.

The Figures and Descriptions of this Plant, and the *Senecio trixaginis specie ac mollitia cauliculis subrubicundis. Plin. Pluken. Alm. p.* 343. *Phyt. Tab.* 315. *Fig.* 1. shew them to be very different, though Dr. *Plukenet* in his *Mamissa*, p. 170. thinks they may be the same.

XXI. *Virga aurea major, sive herba Doria, folio sinuato hirsuto. Cat. p.* 125. *Tab.* 152. *Fig.* 4.

This has a very strong, Inch thick, striated, green Stalk, as high, or higher than a Man, having along the Stalk, several Leaves larger than those on its Branches, which are four Inches long, and one broad in the middle where broadest, rough, sinuated about the edges, and of a dark green colour. Towards the top of the Stalk are many Branches and Twigs, every one of which sustains a great many naked, yellow Flowers, like those of *Jacobaea*.

It grew on the Road to Mountain River, in Colonel *Cope*'s Plantation.

XXI. *Helichrysum caule alato, floribus spicatis. Cat. p.* 125. *Tab.* 152. *Fig.* 5.

This has several straight Stalks rising two Foot high from the same Root. The Stalks are round, though the two long, one tenth part of an Inch broad *Fimbria*, belonging to each Leaf, makes it look as if it were four square, being on every side of the Stalk. The Leaves are set at about an Inches distance from one another, on every side of the Stalk, having the two *Fimbriæ*, or ledges aforesaid under them on each side. They are three Inches long, and not over one quarter of an Inch broad, indented slightly about the edges, of a very dark green colour above, and woolly or white underneath, with one eminent Nerve running longways. The tops of the Stalks, and the small Branches near the top coming out *ex alis foliorum* are Spikes of Flowers standing sometimes singly, and sometimes three or four in a Tuft, on the very Stalk its self, without any *Petioli*, or Foot-Stalks, being naked, the outwardmost *Calicular* Leaves inclosing the Flowers, and the tops at first coming out being purple. After these come a great many small pappous Seeds, as in the others of this Kind.

This grows in the dry *Savannas*, near Mr. *Batchelors*, and over the *Rio Cobre* by the Angels.

XXII. *Chrysanthemum fruticosum maritimum, foliis glaucis oblongis, flore luteo. Cat. p.* 125.

This Shrub rose to about four Foot high, having under a whiteish, smooth Bark, a white Wood, being about the bigness of ones little Finger, divided into several Branches, towards the top set opposite to one another, whereon

whereon grow the Leaves by Tufts, or several together, one Tuft being set opposite to another, and made up of longer and shorter Leaves, which when longest are about an Inch in length, and one third part of an Inch broad near the end, where they are round and broadest, having at the end a short, scarce discernible prickle. They are at their beginning narrow, without any Foot-Stalks, all covered with a Down, making them look white all over, without any incisures on the edges. On the tops of the Branches stand the Flowers in large Heads, on the outside of which are many whitish small Leaves, inclosing the Flowers, which are many, close set together, of a yellow colour, having a Circle of the like colour'd *Petala*, or *Barbula* standing above them. After these follow many solid Seeds, like the others of this Kind.

I found it growing near St. *Christopher's* Cave, not far from the old Town of *Sevilla*, on the Rocks by the Sea-side.

XXIII. *Chrysanthemum Salviæ folio rugoso, scabro, oblongo. Cat. p. 125. Tab. 154. Fig. 1.*

The Stalks were jointed, woody, and had Leaves set on them, which were about two Inches long, and more than half as broad, being narrow both at beginning and point, equal on the edges, rough, and bullated, or like the Leaves of wild Sage. The Heads were as large as those of *Chrysanthemum Segetum*, but because the Specimen from which I described it was imperfect, I can say no more of it.

I found it in one of the *Caribes*.

XXIV. *Chrysanthemum trifoliatum scandens, flore luteo, semine longo, rostrato bidente. Cat. p. 125. Tab. 154. Fig. 2, 3. An Chrysanthemum Americanum, ciceris folio glabro, flore bellidis majoris. Herm. par. Bat. p. 124. An Bidens trifolia, Americana, leucanthemi flore. Tournefort. Inst. p. 462? Bellis major Americana frutescens trifoliata glabra. Bob. hist. Ox. part. 3. p. 30 ?*

This has a slender Stalk, four or five Foot high, needing the help of neighbouring Shrubs or Trees on which it leans, and among which it always grows, being divided into Branches, three or four Inches long, having Leaves standing on three-quarters of an Inch long common Foot-Stalks, being three always together; that Leaf in the middle, or opposite to the Foot-Stalk, is an Inch long, and half an Inch broad at Base, very deeply jagged or cut on the edges, smooth, of a grass green colour, and thin, the other two Leaves at Base are of the same shape, &c. only smaller. The Flowers are several, standing on an Inch and an half long Foot-Stalks, coming out *ex alis foliorum*, each of which has on the outside several green, long Leaves on the top of which are five or more yellow, pretty long *Petala*, within which are many Flowers crowded close together, as in the others of this Kind. After these follow several half an Inch long, rough Seeds, having two horns standing on their ends, in place of the *Pappus* in other Seeds.

It grows among the Trees and Shrubs on the *Red Hills* going to *Guanaboa*, on Mount *Diablo*, and other the inland woody parts of this Island.

The colour of the Flowers of this makes it seem to be different from the *Chrysanthemum Americanum, ciceris folio, &c.*

XXV. *Chrysanthemum cannabinum Americanum alatum, flore aphyllo, globoso, aurantio, baccharidis foliis. Breyn. Cat. p.126. Chrysanthemum Americanum caulalato, flore aphyllo, globoso, aurantio, foliis baccharidis. Commelin. hort. Amst. part. 1. p. 5. Bidens Indica, Hieracii folio, caule alato. Tournef. p. Inst. 462. Chrysanthemum conyzoides caule alato Curassavicum. Herm. par. Bat. p. 125.*

X x x *Chrysan-*

Chryfanthemum Curaffavicum alato caule, flore aurantiaco bullato. Bob. hift. Ox. part. 3. p. 25.

This has a large Root with many white strings strongly fixed in the Earth, from which rise several Stalks a Foot high, having a green Selvedge, or being alated. The Leaves grow along the Stalk, are rough, indented like those of *Cichory*, or *Jacea*, growing larger towards their top, and ending in a point about an Inch and an half long, and half an Inch broad near the end where broadest, of a dark green colour. The Flowers come many together in Heads like those of the others of this Kind, they are of a deep yellow, or Orange colour. After these follow broad Seeds, something like those of Parsnips, black in the middle, white about their edges, and having two prickles or horns on their ends.

It grows near the Bridge over *Black River* in St. *Dorothies* Parish.

XXVI. *Chryfanthemum Conyzoides nodiflorum, femine roftrato bidente. Cat. p. 126. Tab. 154. Fig. 4.*

This had a round brownish Stalk, about a Foot long, which was jointed at every two or three Inches, and at the joints had Leaves, set opposite to one another, standing on very short, if any Foot-Stalks. The Leaves were about an Inch and an half long, and half as broad in the middle where broadest, beginning narrow, increasing to the middle, and thence decreasing, and ending in a point, being hairy on their upper side, having many appearing Nerves, and being of a dark green colour. *Ex alis foliorum* come the Branches, set opposite to one another, and the Flowers, which are small Heads, without any Foot-Stalks, being one or more standing together, each being surrounded by a few dry, brown Membranes. After some time follows the Seed, which is small, gray oblong, having two horns, standing out as others of this Kind, which are *Bidentis*.

It grew in the inland part of this Island, and in *Barbados*.

XXVII. *Chryfanthemum paluftre, repens, minus, odoratum, folio fcabro trilobato. Cat. p. 126. Tab. 155. Fig. 1.*

The Stalk is jointed, creeping along the surface of the Earth, every joint sending into the Earth many hairy fibers and strings of a blackish brown colour, and Leaves opposite to one another. They are very like the Leaves of *Caryophyllata*, only very rough, having several points or notches, and being very aromatically sweet smelled. At, or towards the tops *ex alis foliorum* go out some long Foot-Stalks, having yellow Flowers exactly like the other *Chryfanthemums*.

It grows on the Banks of *Rio Cobre*, and in all the moist places of the Town, and other *Savannas*.

The Description and Figure of this, shew it to be different from the *Chryfanthemum hirfutum Virginianum auriculato dulcamare folio octopetalon, Pluken. Phyt. Tab. 242. Fig. 6.* though Dr. *Plukenet* doubts in his *Mantiffa, p. 48.* if it be not the same.

XXVII. *Chryfanthemum fylvaticum repens minus, Chamadryos folio, flore luteo nudo, femine roftrato. Cat. p.126. Tab. 155. Fig. 2. An Chryfanthemum Maderafpatanum menthe arvenfis folio & facie floribus ligemellis ad foliorum alas pediculis curtis. Pluken. Phyt. Tab. 118. Fig. 5. Alm. p. 100?*

This had several hairy fibrous Roots, which it struck into the Earth from the lowermost joint, it usually lay along on the surface of the Earth, having joints, and a Stalk about nine Inches long. The Leaves stand on half an Inch long Foot-Stalks, opposite to one another, being about three quarters of an Inch long, and as broad at round Base where broadest, from whence they

decreased

decreased to the point, being serrated about the edges, and of a dark green colour, with several appearing Ribs, going through the Leaf, some whereof take their Original from the end of the Foot-Stalk, as from a common Center. At the tops on three quarters of an Inch long *Petioli* stand the Heads, being made up of many Flowers set close together, naked, small, and yellowish, with many *Stamina* appearing on them. After these follow several oblong, short, brown, cannulated Seeds, having two or three prickles at their ends.

It grew in the inland Woods in several places of this Island.

XXVIII. *Chrysanthemum palustre minimum repens, apii folio.* Cat. p. 126. Tab. 155. Fig. 3. *Chrysanthemum Americanum, humile, ranunculi folio.* Plum. Tournef. Inst. p. 492.

This has small, stringy, dark brown Roots, sending out several Branches, spread round on the surface of the Ground, having Leaves cut in on the edges, resembling Smallage Leaves, of a pale green colour. The Flowers are made up of several small, yellow Flowers, standing close by one another in the manner of other *Chrysanthemums*, being surrounded by *Barbula*, of the same yellow colour. The Seeds are long, greenish brown, striated, and to each Flower follows one Seed, standing in the Heads just as the Flowers did.

It grows in the moister places of the Town *Savanna's*, and in the *Caribes*; and is in perfection some time after rainy weather.

Dr. *Plukenet* in his *Mantissa*, p. 47. doubts whether this be the *Chrysanthemum Chinense foliis plurifariam divisis, &c.* Phyt. Tab. 22. Fig. 3. whoever compares the Figures and Descriptions of these Plants, will soon see they differ.

XXIX. *Artemisia humilior flore majore albo.* Cat. p. 127. *Absynthium ersysmi folio, Achoavan Alpini quodammodo accedens ex insula Jamaicensi.* Pluk. Alm. p. 2. Wild Wormwood.

This from a small, whitish brown Inch long Root, having some few fibers, thereby to draw its Nourishment, rises one Foot high, several, striated, whitish, solid Stalks, supporting several Branches, coming from the inside of the Leaves, very deeply cut in on the edges, after the manner of Wormwood, from whence its name, or Mugwort, only whiter in colour than those of the last, the lower the Leaves the larger they are. The Flowers at top stand in Heads on the Branches ends, and are made up of a great many single white ones, making much larger Flowers than those of ordinary Mugwort.

It grows by the Town of St. *Jago de la Vega*, in the claiy and gravelly Grounds of the *Savannas*.

It flowers most part of the year.

It is made use of as a good vulnerary Herb, and accounted very effectual.

XXX *Scabiosa affinis anomala sylvatica, cnula folio, singulis flosculis albis in eodem capitulo perianthia habentibus, semine papposo.* Cat. p. 127. Tab. 156. Fig. 1, 2.

This has a round striated, rough, pretty large Stem, rising to three Foot high, having towards the bottom several Leaves set without any order, on half an Inch long Foot-Stalks. They are five Inches long, and two broad in the middle where broadest, from a narrow beginning increasing to the middle, and thence decreasing to the end, indented about the edges, being rough above, having the surface scabrous, or corrugated after the manner of Sage or Foxglove, and woolly underneath. Towards the top the Leaves are smaller, out of whose *Ala* come hoary inch long Foot-Stalks, supporting a
round

round Head, of many white tubulous, oblong Flowers, each Flower standing in a chaffy *Calyx*, or *Perianthium*, made up of several dry, brownish membranes, which afterwards contains three or four small, oblong, smooth, and shining gray Seeds, having a few pappous hairs on their upper ends.

It grew in the Woods on the Road to Colonel *Cope's* Plantation in *Guanaboa*, and in several other Woods of this Island very plentifully.

This is not the Plant called *Eupatorio affinis Americana, bulbosa floribus scariosis calyculis contestis. Plukenet. Alm.* p. 142. *Phyt. Tab.* 177. *Fig.* 4. as Doctor *Plukenet* thinks it may be in his *Mantissa,* p. 73.

XXXI. *Eryngium foliis angustis serratis fœtidum. Cat.* p. 127. *Tab.* 156. *Fig.* 3, 4. *An Eryngium fœtidum oblongis capitulis Americanum. Pluken. Alm.* p 137?

This Plant has six or seven round, smooth, whitish Roots, about ten Inches long, going straight down into the Earth, taking very firm rooting therein, which uniting in one towards the surface of the Earth, there sends forth Leaves spread on the Ground on every hand, to the number of five or six, eight Inches long, and one broad near the end where broadest, very deeply serrated, and having on its edge soft prickles. From the middle of the Leaves rise one or two Stalks about a Foot and an half high, being round, green, hollow, smooth, always divided into two, or observing a Dichotomy, and having at parting two deeply cut, prickly, short, Leaves. The top or Heads are like those of other *Eryngiums*, having several long, narrow Leaves under them, which are prickly; they are at first greenish, afterwards brown, and have several brown Seeds set round a small column. All parts of the Plant have a very penetrating strong, though not very unsavoury smell.

It grows at the *Crescent* Plantation near the Orange Walk on the Banks of the *Rio Cobre*, in moist, low, flat Grounds in several other places, by Colonel *Bourden's* House, in his Plantation, and in *Barbados*.

It is counted one of the greatest *Alexipharmaca's* of these parts. The Destill'd water of it is reckon'd a very great *Antepileptick*, and extremely to resist Hysterick Fits.

I question not this being the Plant mentioned by *Hernandez*, called *Cohayelli*, every thing agreeing to it. He tells us that,

It is hot in the fourth Degree, tasts like Skirrets, though a little sharp and smelling. The Root powdered, and taken to the quantity of three Drams in ten Ounces of water, strengthens the weak and cold Stomach, eases pains of the Belly and other parts from Colds, dissipates Wind, is good for Colick and Iliack Diseases, is Diuretick, and helps the *Catamenia*, cures Surfeits, incites to Venery, and is good against the Bites of Venemous Serpents. It has a better effect, if it be given out of a hot and strengthening Liquor, it dissipates preternatural Tumours, and humours in the Joints, and remedies all cold intemperatures. *Hernandez*.

It is called *Itubu* in *Surinam*, or *Fuga Serpentum*, because they come not where it grows. 'Tis *Alexipharmac* from its volatile Salt, and the smell of the Leaves cures Hysterick Fits. *Hern.*

The End of the first Volume.

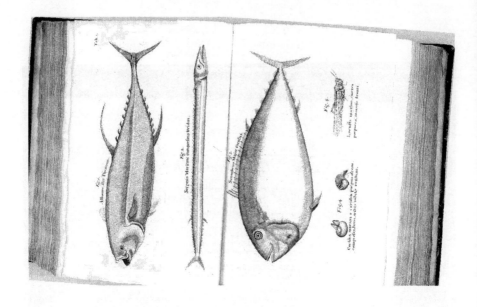

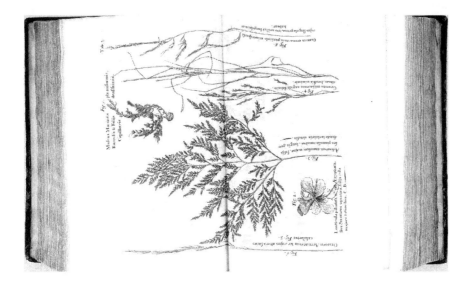

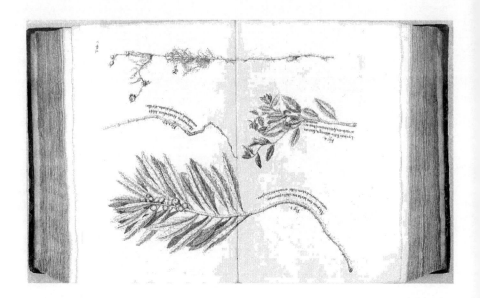

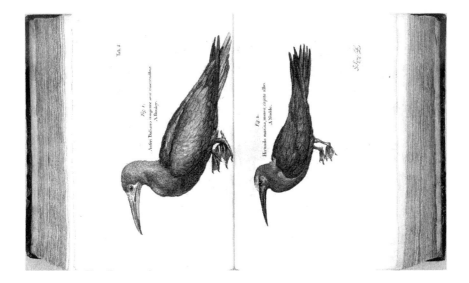

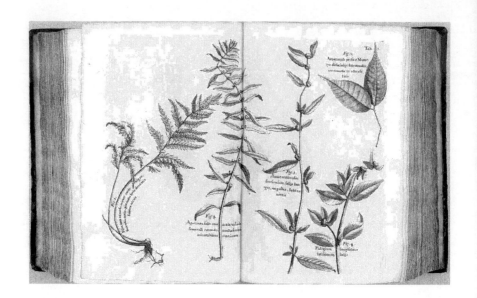

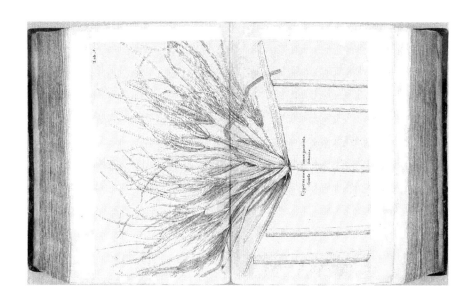

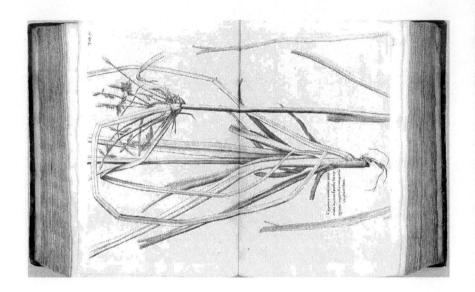

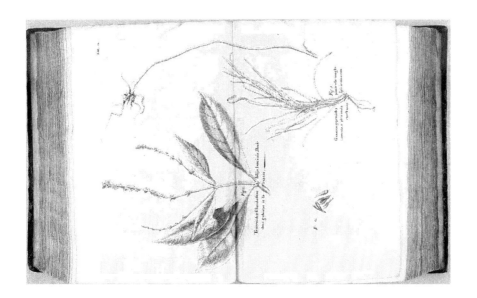

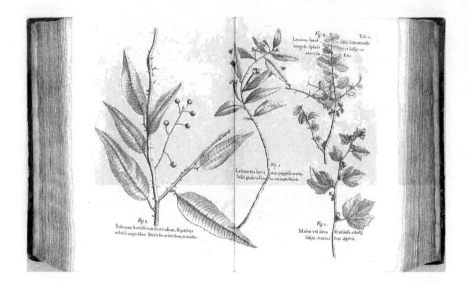

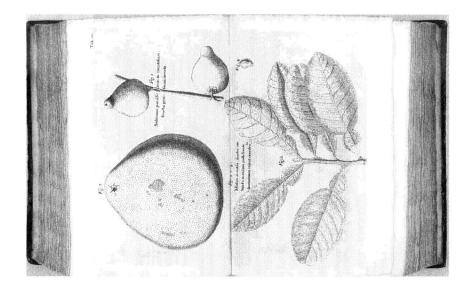

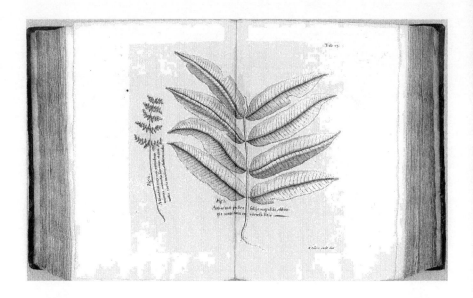

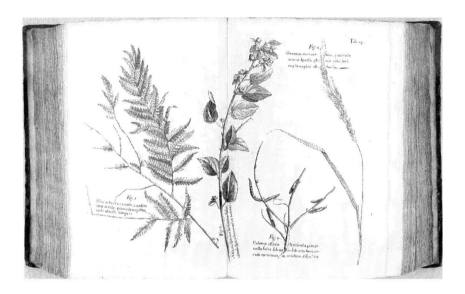

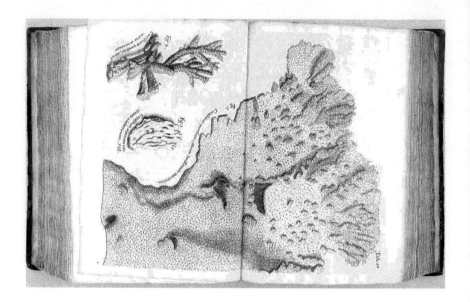

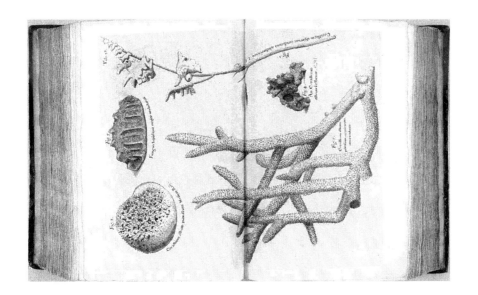

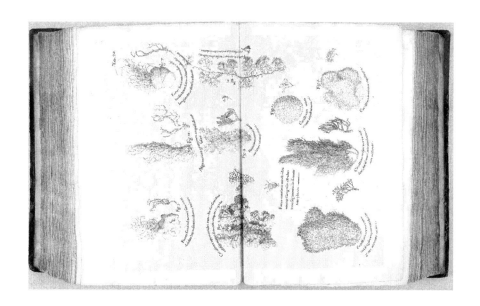

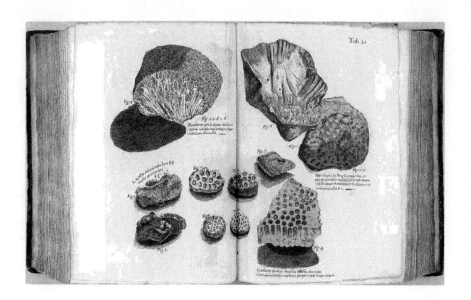

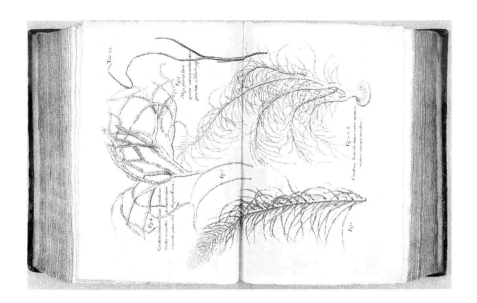

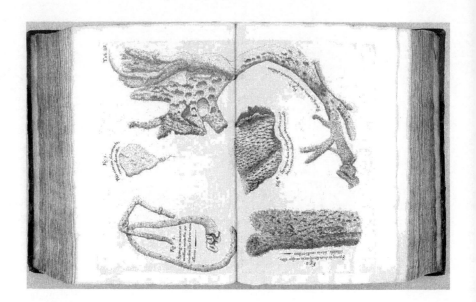

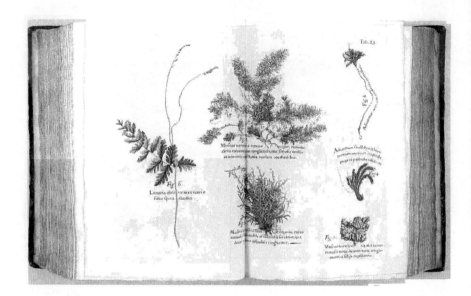

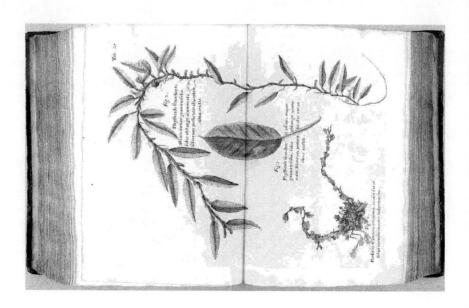

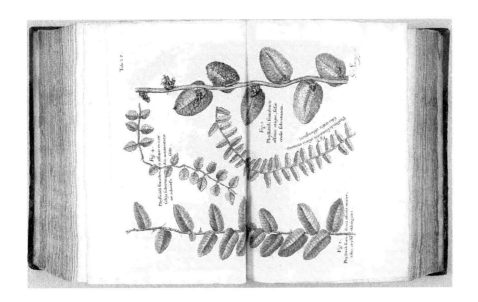

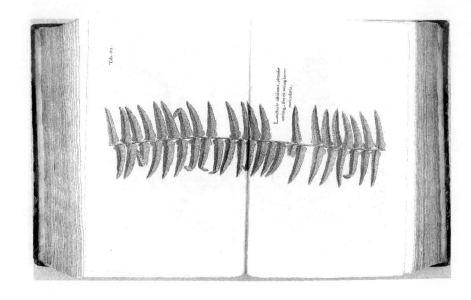

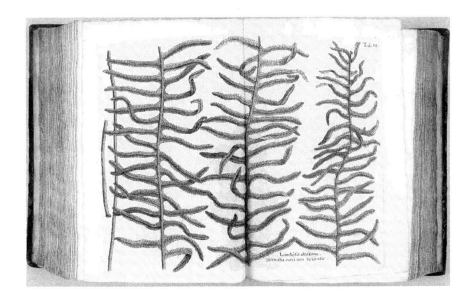

Lonchitis altissima, pinnulis raris non laciniatis

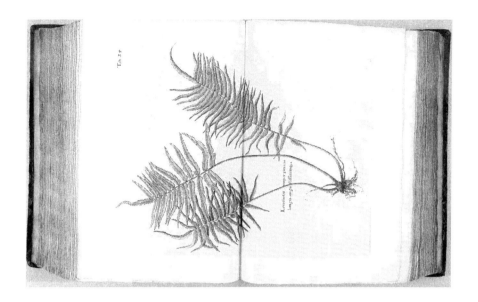

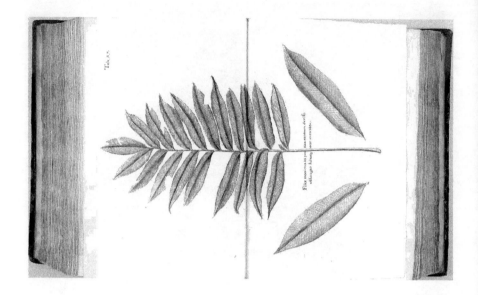

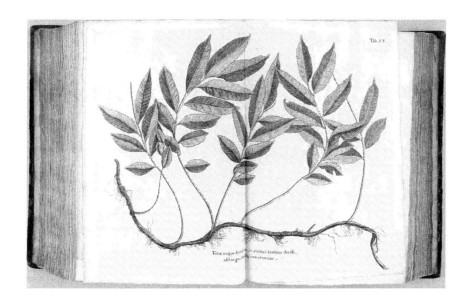

Tab. 5 V.

Filix major foliorum pinnis tantum divisis, oblongis, non crenatis.

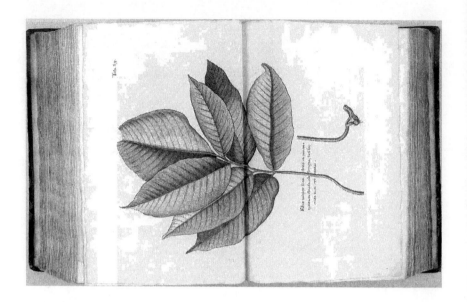

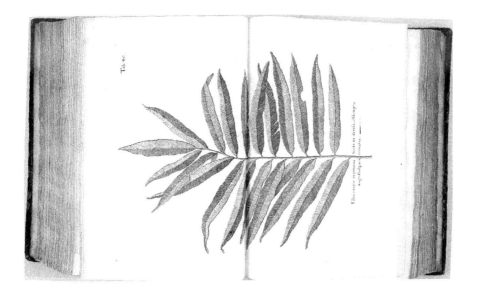

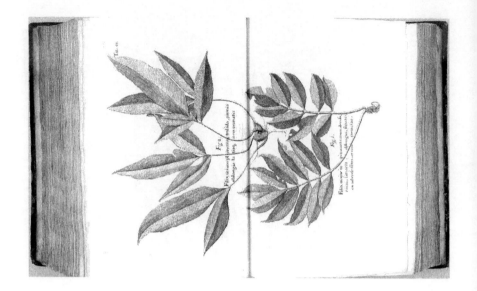

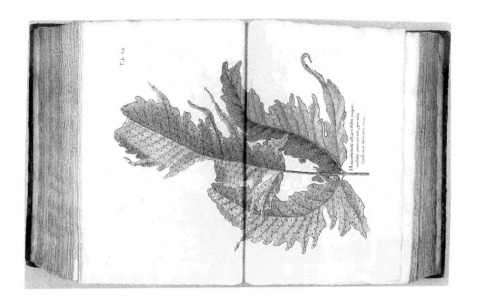

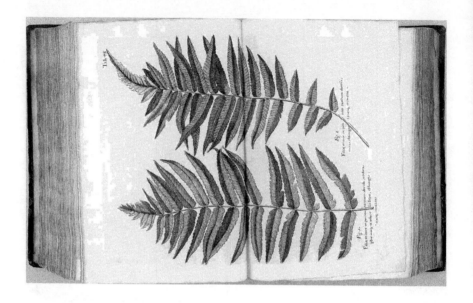

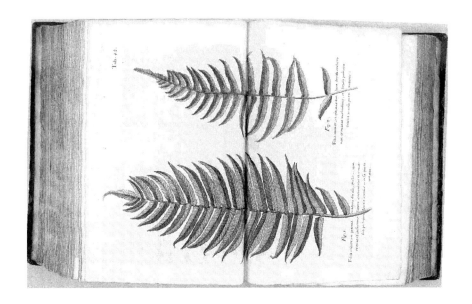

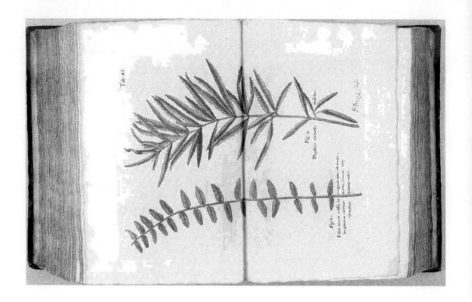

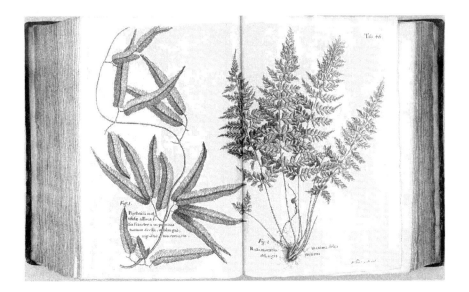

Tab. 46.

Fig.1. Phyllitidi aut bifidæ affinis f. lis foemina a se primum tantum divisa, oblonga, angusta, mucronata.

Fig.2. Ruta muraria maxima folio oblongis trenuis.

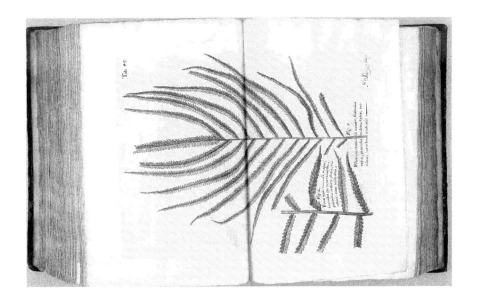

Fig. 1. Filix non ramosa, densissime varie, pinnulis longis capillaceis...

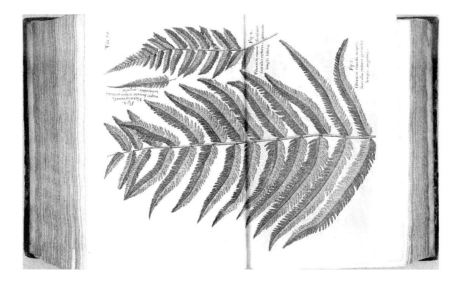

Tab. 12.

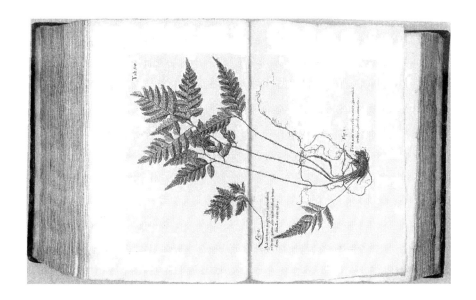

Tab. 10.

Filix dumosa, ramosa, aspera, ferulacea, major, pinnulis laciniatis, obtusis, per margines ferrulatis, mediorum Sinensium.

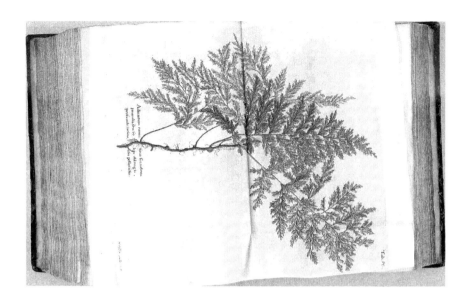

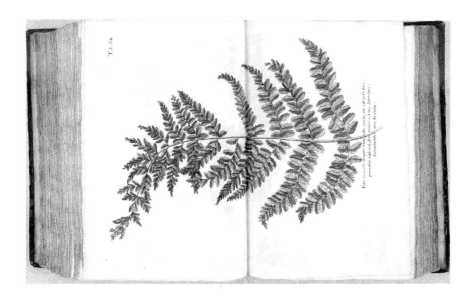

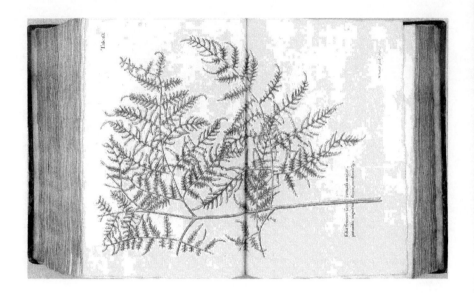

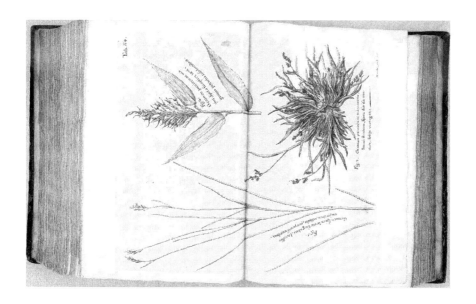

Tab. XV.

Fig. 2. Gramen paniculatum bicoccineum locustis villosis, Spica divisa, der alte onder, kortige vaaragare.

Fig. 1. Gramen Spicâ brevi, E Spicâ divisa, locustis angustioribus villosis, non penné ramosis.

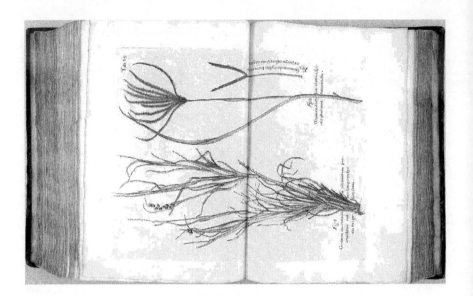

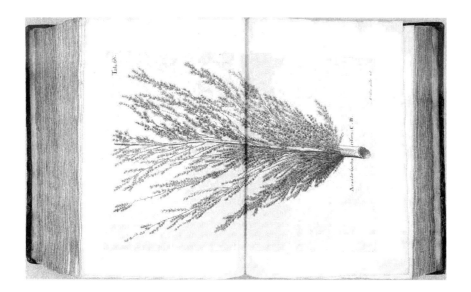

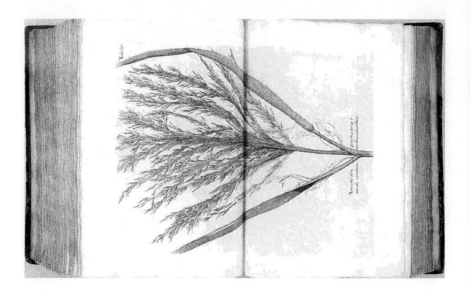

Tab. 67.

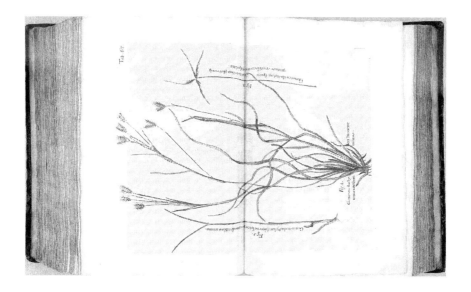

Fig. 3. Gramen daktyli, laterale fiesso, ramosssissum minus.
Fig. 4. Gramen dactylon, granum referens.
Fig. 5. Gramen ischaemum species, spiculis floribus parvis, ramosis umbellam effigies.

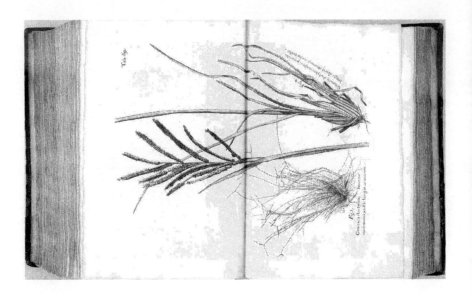

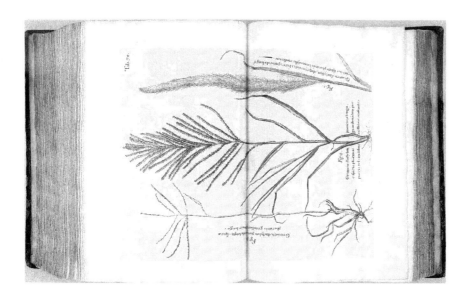

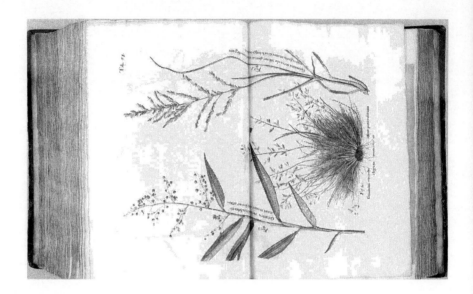

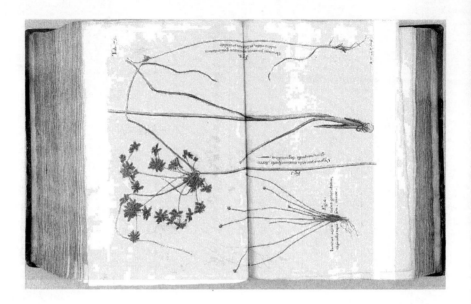

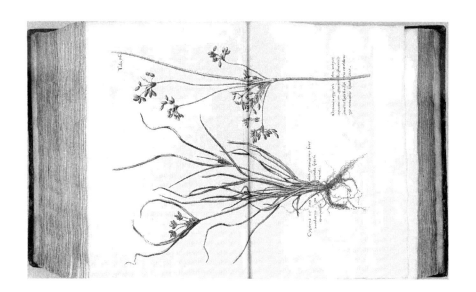

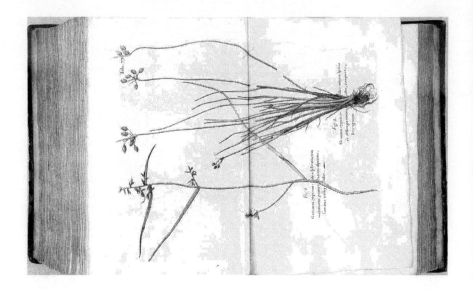

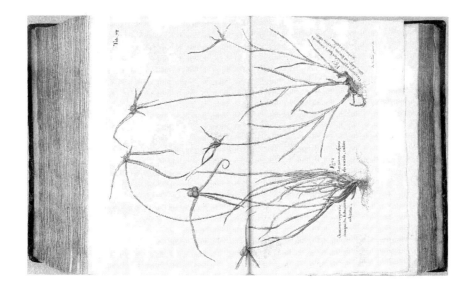

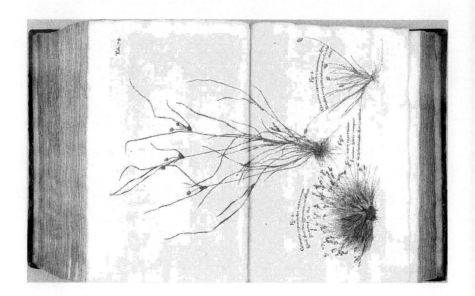

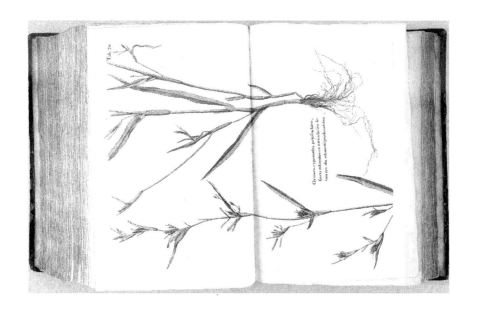

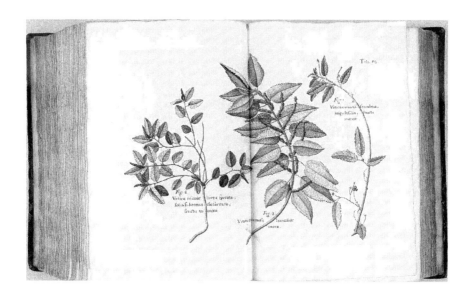

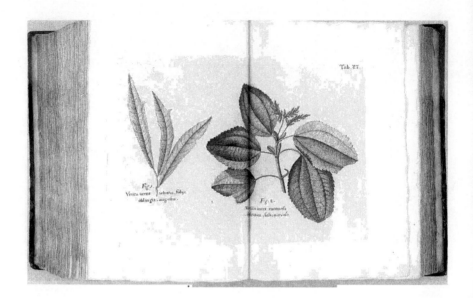

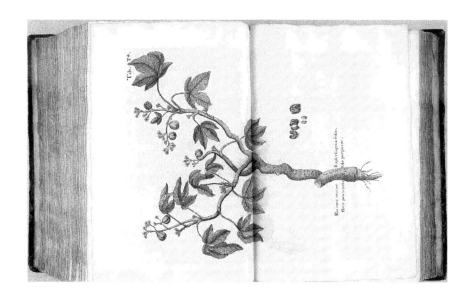

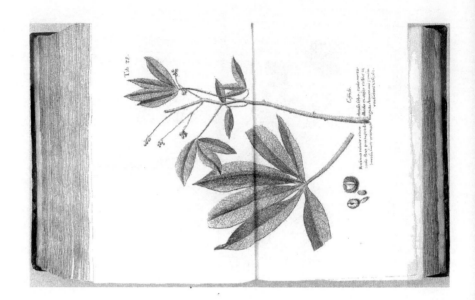

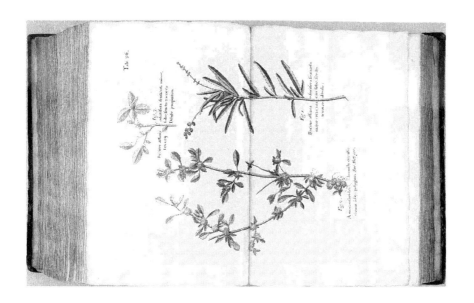

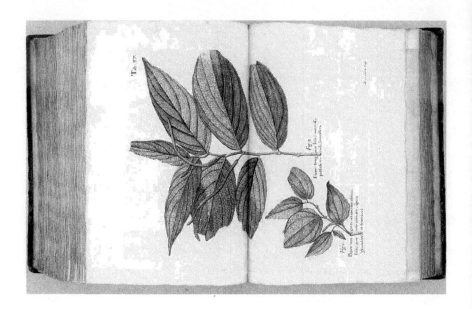

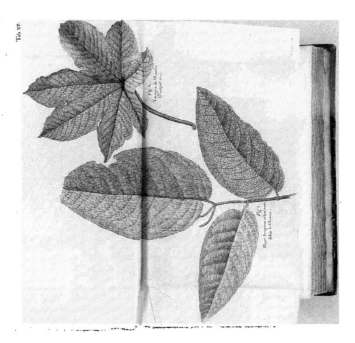

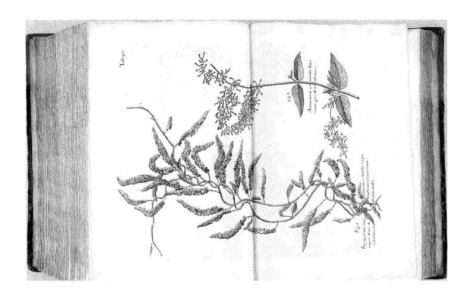

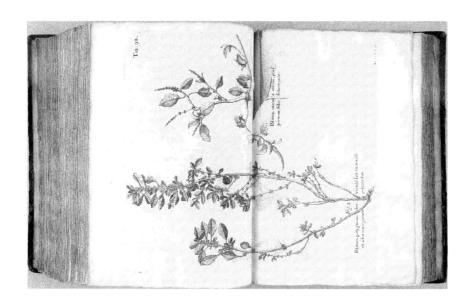

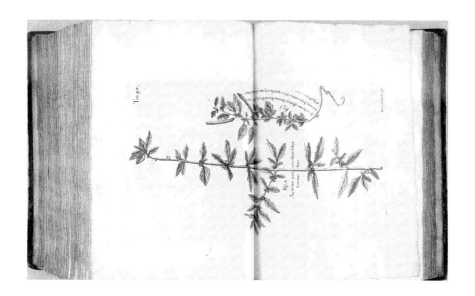

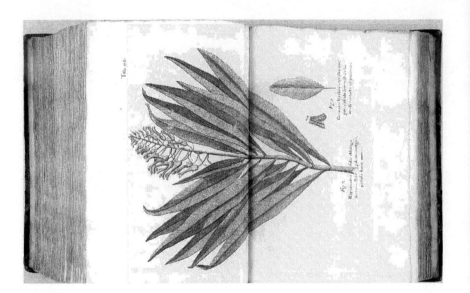

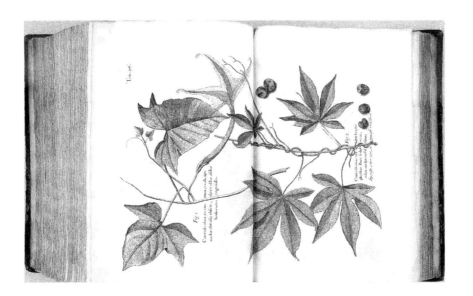

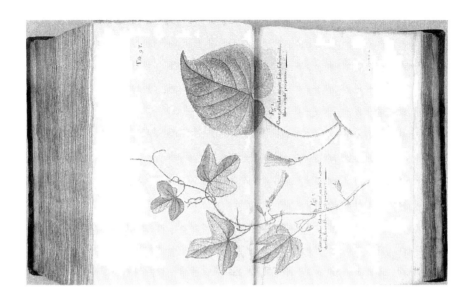

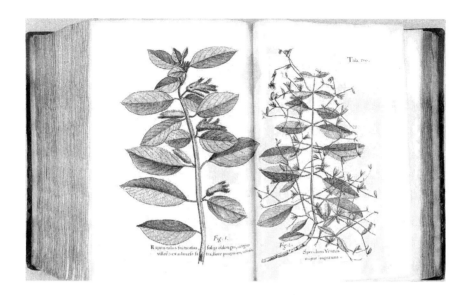

Tab. 100.

Fig. 1. Rapunculus fruticosus, foliis oblongis, utrinque villosis et adversis sitis, flore purpureo, silvestris.

Fig. 2. Speculum Veneris majus riparium.

Tab. 101.

Fig. 1. Rapunculus fruticosus latifolius, flore luteo favorolo, foliis ex adverso sitis.

Fig. 2. Rapunculus aquaticus folio oblongo, flore albo, radice longissima.

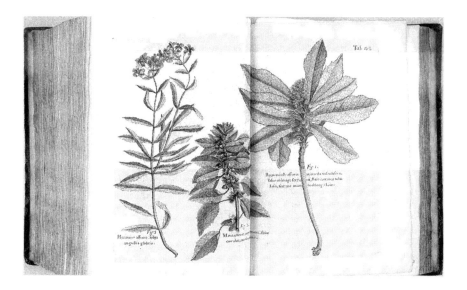

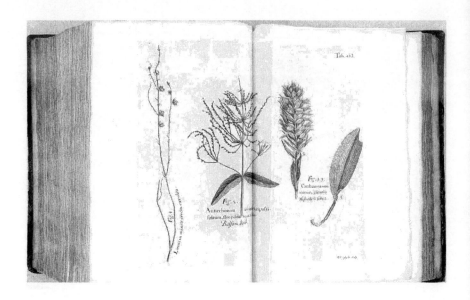

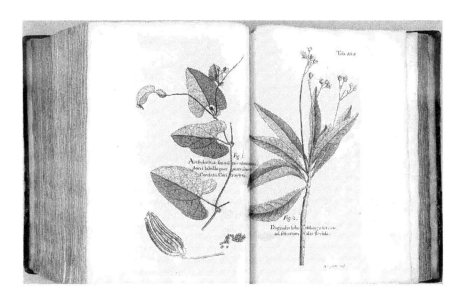

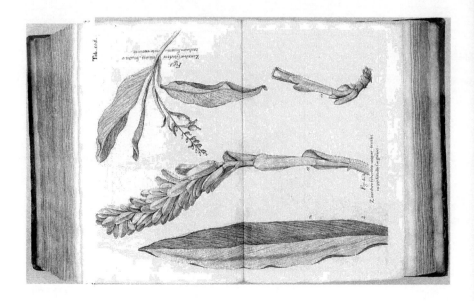

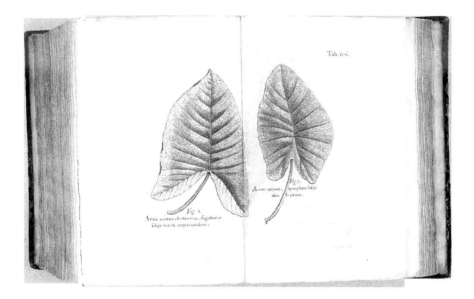

Tab. 106.

Fig. 1.
Arum minus, Nymphææ foliis, dictum Aegyptium.

Fig. 2.
Arum minus clodesianum, sagittariæ foliis viridi ingynvantibus.

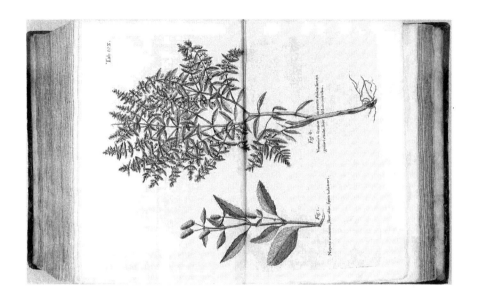

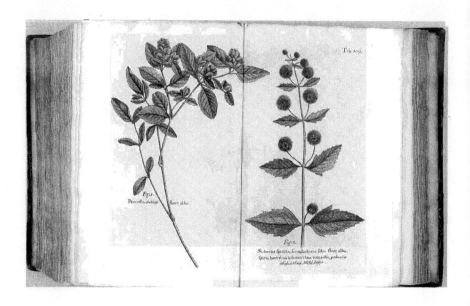

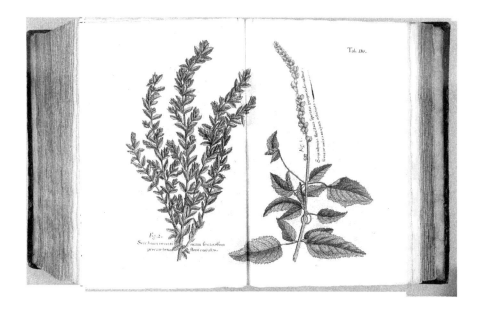

Tab. 110.

Fig. 2. Scorbium recenti... mum frutecofum procumbens... flore cœruleo.

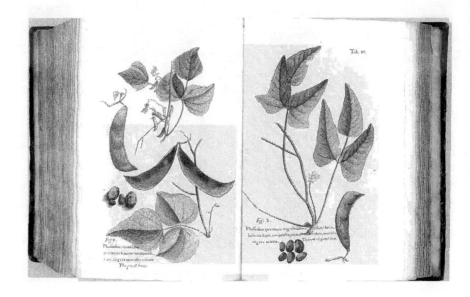

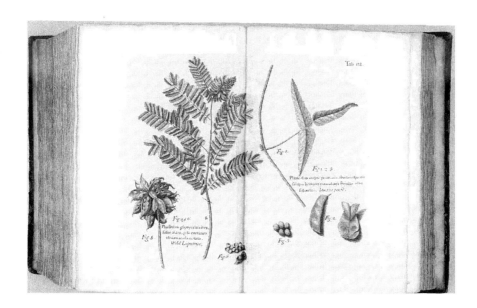

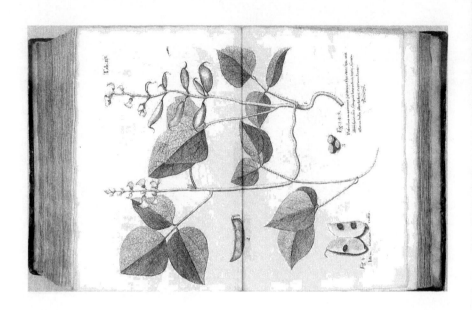

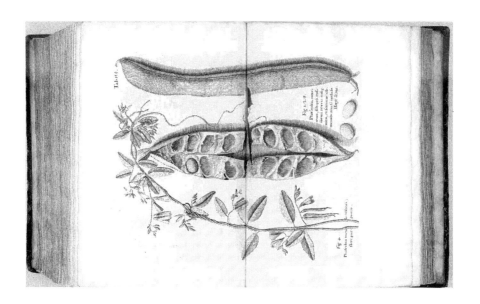

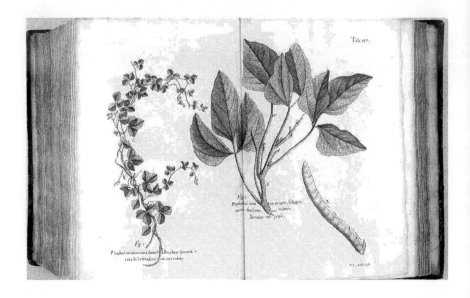

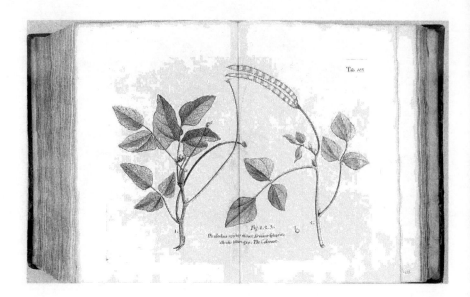

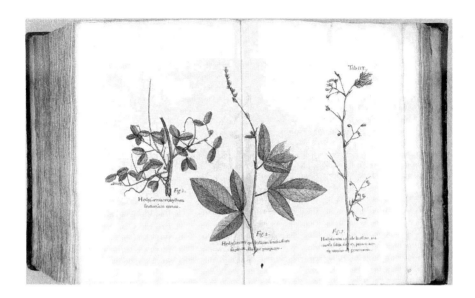

Tab. 119.

Fig. 1.
Lotus pentaphyllos siliqua quadrivalvi hirsuta,
minori non fœtens, floribus, foliis cisti-
niilli e pluribus in fasciculis inferioribus.

Fig. 2.
Anonis non spinosa minor, glabra,
procumbens, flore luteo.

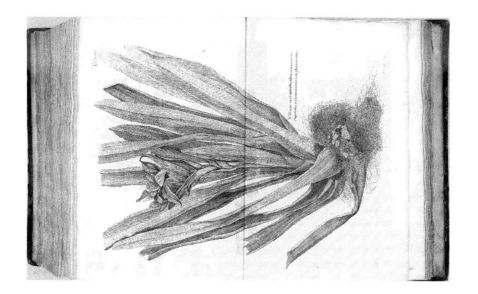

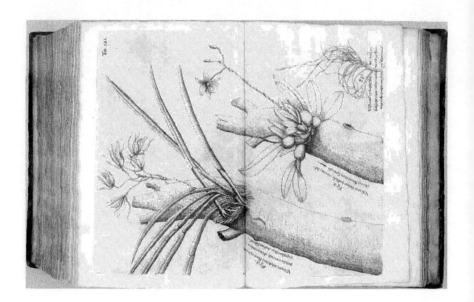

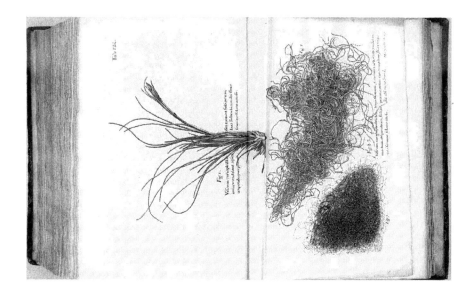

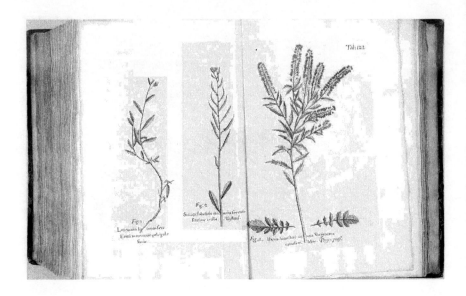

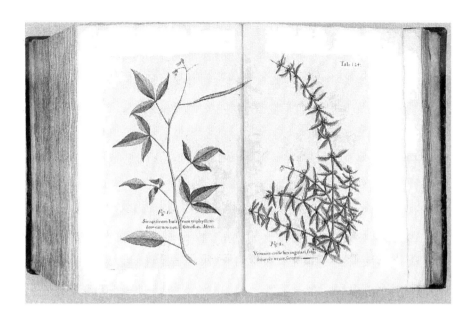

Tab. 124.

Fig. 1.
Sinapistrum Indicum triphyllum
flore carneo non spinosum. Here.

Fig. 2.
Veronica caule hexangulari, foliis
Asterios veris, Georgii.

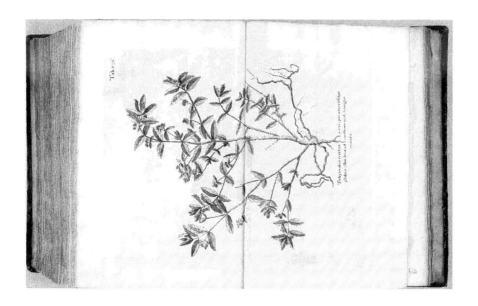

Tab. 26

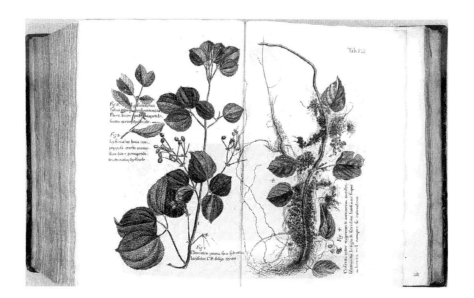

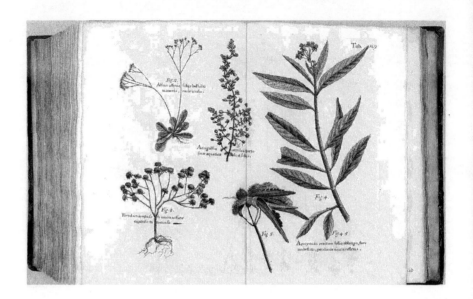

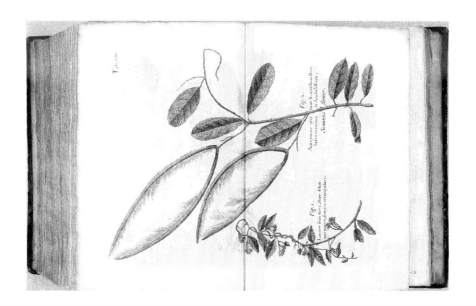

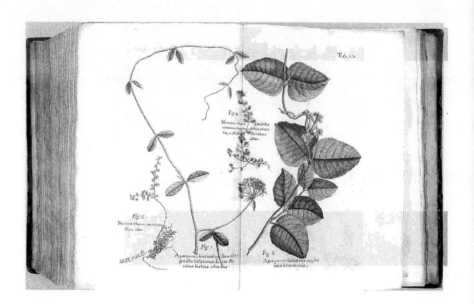

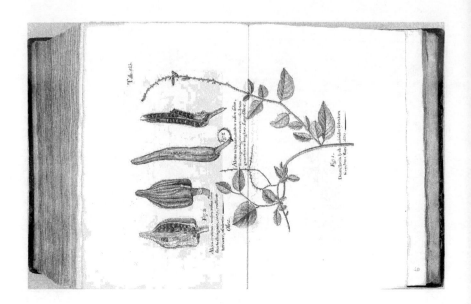

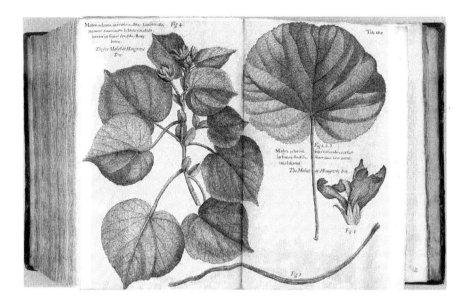

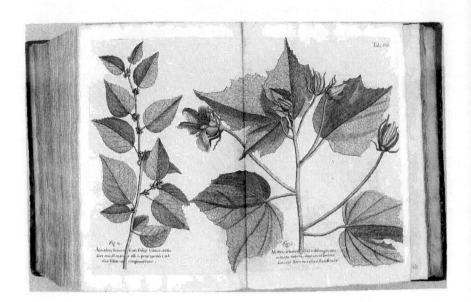

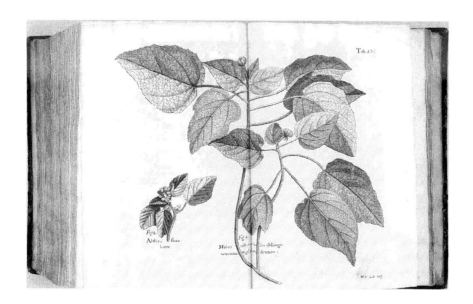

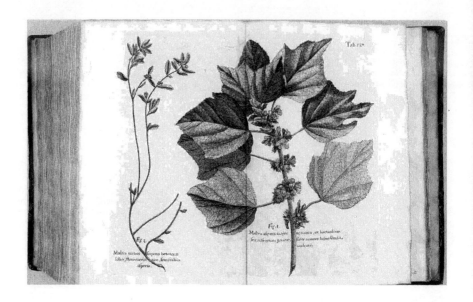

Tab. 137.

Fig. 1. Malva aspera major ignatica, ex hortensium seu reliquum genere, flore minore labio fendit, aculeato.

Fig. 2. Malva minor supina betonicæ folio, flosculo ex cæruleo semicubus aspera.

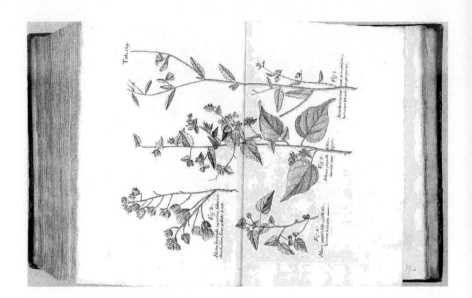

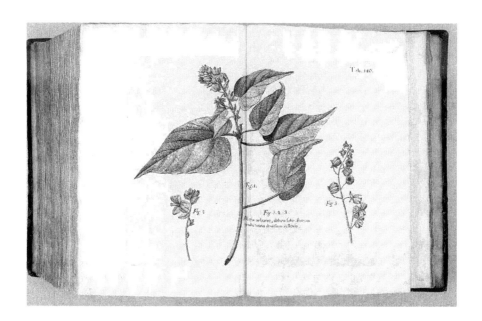

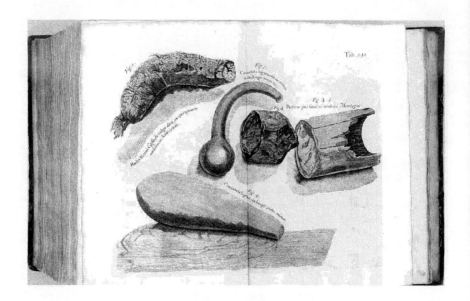

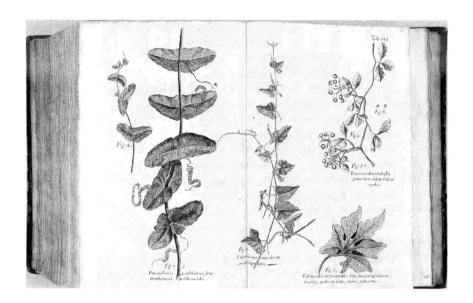

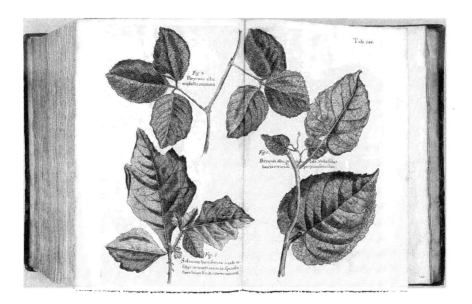

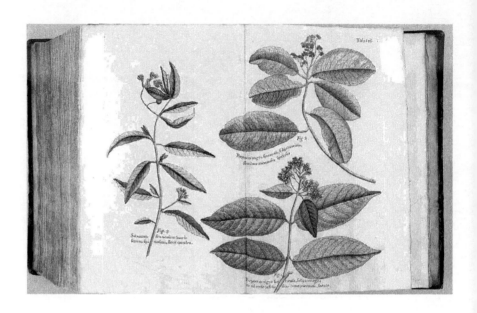

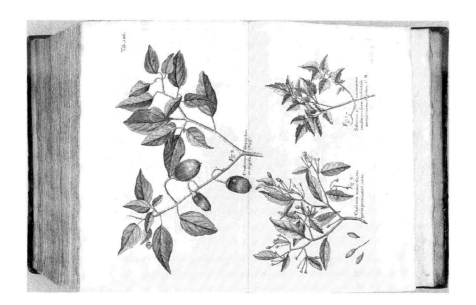

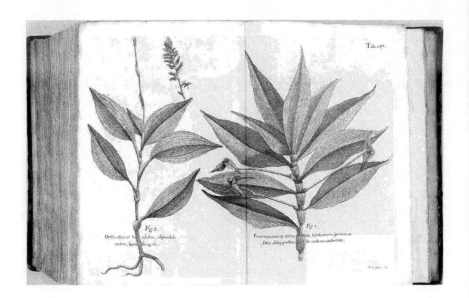

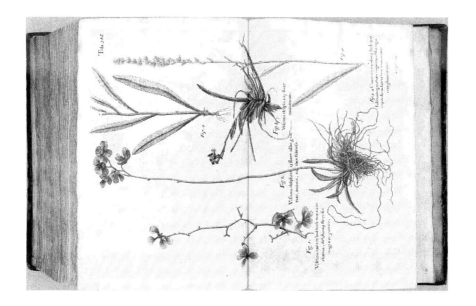

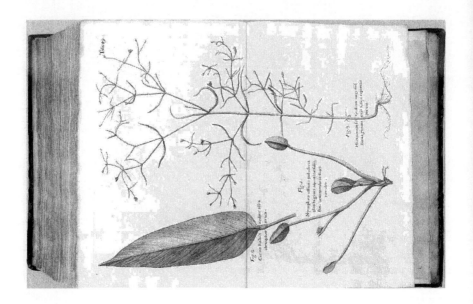

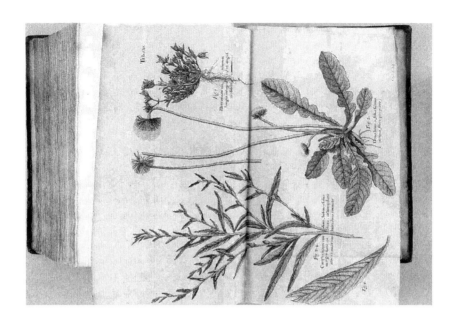

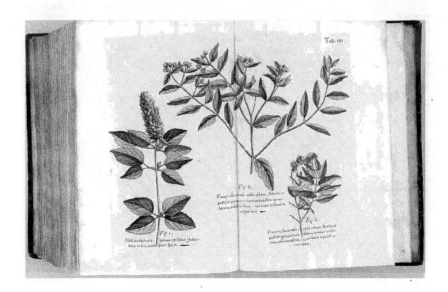

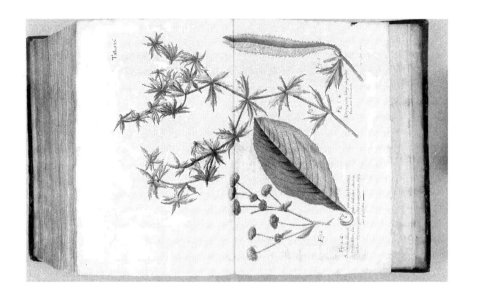

ERRORS of the Press.

Page 2. line 11. read *Guanahani*. p. 7. l. 23. for 100000, r. 1000000. p. 18. l. 31. r. *Gridiron or Plate of Iron*. p. 24. l. 28. r. *Rats or Dormice*. p. 29. 22. r. *or as most*. l. 23. r. *More of*. l. ult. r. *sometimes the* 1 *for Nights*, r. *Nights*. p. 30. l. 16. r. *all tre*. p. 37. ad Col. Sept. 22. r. *and great*. p. 38. ad Col. Octr. 26. r. *a very rash Sea-Breeze*. p. 30. Dec. 6. add *Dysenteries very common*. p. 43. l. 20. dele *after much Rain*. p. 46. l. 6. dele *as and as they can*. 24. l. 25. r. *its*. p. 47. l. 16. r. *their*. p. 48. l. 29. r. *after them, when*. Id. l. 41. *bandly or*. p. 49. l. 9. d. *if it can be*. p. 52. l. 26. r. *sometimes*. Ib. 72. after *Pigeonny* d. the Comma, p. 56. l. 13. and 15. *Curaffau*. Ib. l. 34. r. *or Creolians descended from them*. p. 60. l. 23. r. *some other Places or*. p. 61. l. 37. r. *(as well as the Scion or Juice from the best Canes)* p. 62. l. 14. *after you to the Quantity of*. l. 18. r. *Mane, tains*. l. 34. after *Pews* dele the Comma, p. 73. l. 37. r. *as sit*. p. 75. l. 17. r. *Finisera*. p. 76. l. 12. dele *it*. p. 83. l. 2. after *together* add a Comma, l. 6. dele *or*. l. 18. r. *their*. l. 29. r. *falling*, l. 24. after *Englisso* add a Colon: after *since* add *them*. l. 41. after *Water* add a Colon: p. 95. l. 2. r. *Particles which*: r. *out again*. p. 96. l. 36. r. 25 *Years*. p. 97. l. 29. r. *Distempers which*. l. 40. r. *Pustules which*. p. 106. l. 13. r. *Disease which*. p. 108 l. 31. for the same put me. l. 35. r. *Physitian*. l. 36 l. by *which the Operation of the Calentuck was stop'd*. p. 111. antepen. for of r. or. p. 132. l. 19. d. *first of*. l. 35. for by r. *for*. p. 134. l. 12. r. *his bed*. l. 40. r. *most who*. p. 123. pen. d. *flow*. p. 133. l. 22. after *in* add *an*. p. 134. l. 12. r. *Persons*. p. 135. l. 19. for her r. *there*. p. 136. l. 29. after *be* add *ho*. p. 140. l. 23. r. *her*. p. 144. l. 14. r. *Thought*. p. 146. l. 29. for *alternative* r. *alterative*. l. 2. *Bar, several times to*. p. 147. l. 21. *the other*. l. 27. r. *it in the Hammock*. p. 150. l. 2. for *not done* r. d. *omitted*. l. 4. r. *Chirurgeons*. l. 32. r. *used when they were*.

Page 17. line 1. for *West* read *East*. p. 47. l. *penult*. for *westwardly* read *eastwardly*. p. 57. l. 18. r. *Tournefort*. antepen. d. *the Leaves* and. p. 62. antepen. r. *Spongia*. ibid. r. *digitata*. p. 83. 37. d. *radio*. p. 93. l. 4. r. 53. p. 95. r. LXXVI, LXXVII, and so to the End of the Chapter. p. 116. l. 27. for 84 read 74. p. 130. l. 14. r. 8, *&*, 141. Fig. 1. p. 134. l. 19. r. *kind*. p. 149. l. 17. read *stupehed Benge*. p. 152. l. 9. for *it* r. *they*. p. 161. l. 7. for *Seds* read *Seed*. l. 12. r. *the Leaves of this*, and *for they* read *it*. p. 165. 30. add *alter or, nine*, and d. *nine the end of the same Line*. p. 167. l. 5. d. *Cajalp*. l. 43. r. XLII. p. 175. l. 27. r. *very much*. p. 179. l. 23. r. *which fee*. p. 184. l. ult. r. Fig. 2. p. 187. l. 18. for *are* read *is*. p. 198. l. 24. r. *Seed-Vessel*, p. 201. l. 43. r. 128. p. 212. l. 2. r. Tab. 131. Fig. 1. p. 230. l. 24. r. *jointed*. p. 252. l. 22. r. *rosaceus*. p. 259. l. 39. r. *irregulary*. p. 260. l. 35. r. XX.

Milton Keynes UK
Ingram Content Group UK Ltd.
UKHW020630260923
429382UK00005B/267